NUCLEASES

**COLD SPRING HARBOR
MONOGRAPH SERIES**

The Lactose Operon
The Bacteriophage Lambda
The Molecular Biology of Tumour Viruses
Ribosomes
RNA Phages
RNA Polymerase
The Operon
The Single-Stranded DNA Phages
Transfer RNA:
    Structure, Properties, and Recognition
    Biological Aspects
Molecular Biology of Tumor Viruses, Second Edition:
    DNA Tumor Viruses
    RNA Tumor Viruses
The Molecular Biology of the Yeast *Saccharomyces*:
    Life Cycle and Inheritance
    Metabolism and Gene Expression
Mitochondrial Genes
Lambda II
Nucleases
Gene Function in Prokaryotes
Microbial Development
The Nematode *Caenorhabditis elegans*
Oncogenes and the Molecular Origins of Cancer
Stress Proteins in Biology and Medicine
DNA Topology and Its Biological Implications
The Molecular and Cellular Biology of the Yeast *Saccharomyces:*
    Genome Dynamics, Protein Synthesis, and Energetics
    Gene Expression
Transcriptional Regulation
Reverse Transcriptase
The RNA World
Nucleases, Second Edition

NUCLEASES
SECOND EDITION

Edited by

Stuart M. Linn
University of California, Berkeley

R. Stephen Lloyd
University of Texas Medical Branch, Galveston

Richard J. Roberts
New England Biolabs, Beverly, Massachusetts

COLD SPRING HARBOR LABORATORY PRESS
1993

NUCLEASES, Second Edition

Monograph 25
© 1993 by Cold Spring Harbor Laboratory Press
All rights reserved
Printed in the United States of America
Book design by Emily Harste

Library of Congress Cataloging-in-Publication Data

Nucleases / edited by Stuart M. Linn, R. Stephen Lloyd, and Richard J.
  Roberts. -- 2nd ed.
     p. cm. -- (Cold Spring Harbor monograph series ; 25)
  Originally published: 1982.
  Includes bibliographical references and index.
  ISBN 0-87969-426-2
  1. Nucleases.   I. Linn, Stuart M.   II. Lloyd, R. Stephen.
  III. Roberts, Richard J.   IV. Series.
  QP609.N78N8   1993   93-38094
  574.19'25--dc20   CIP

Authorization to photocopy items for internal or personal use, or the internal or personal use of specific clients, is granted by Cold Spring Harbor Laboratory Press for libraries and other users registered with the Copyright Clearance Center (CCC) Transactional Reporting Service, provided that the base fee of $5.00 per article is paid directly to CCC, 222 Rosewood Dr., Danvers MA 01923. [0-87969-426-2/93 $5 + .00]. This consent does not extend to other kinds of copying, such as copying for general distribution, for advertising or promotional purposes, for creating new collective works, or for resale.

All Cold Spring Harbor Laboratory Press publications may be ordered directly from Cold Spring Harbor Laboratory Press, 10 Skyline Drive, Plainview, New York 11803. Phone: 1-800-843-4388 in Continental U.S. and Canada. All other locations: (516) 349-1930. FAX: (516) 349-1946.

Contents

Preface, vii

1 **Mechanistic Principles of Enzyme-catalyzed Cleavage of Phosphodiester Bonds, 1**
J.A. Gerlt

2 **Type II Restriction Endonucleases, 35**
R.J. Roberts and S.E. Halford

3 **The ATP-dependent Restriction Enzymes, 89**
T.A. Bickle

4 **Homing Endonucleases, 111**
J.E. Mueller, M. Bryk, N. Loizos, and M. Belfort

5 **The Nucleases of Genetic Recombination, 145**
S.C. West

6 **Fungal and Mitochondrial Nucleases, 171**
M.J. Fraser and R.L. Low

7 **Topoisomerases, 209**
T.-S. Hsieh

8 **Proofreading Exonucleases: Error Correction during DNA Replication, 235**
M.F. Goodman and L.B. Bloom

9 **Nucleases Involved in DNA Repair, 263**
R.S. Lloyd and S. Linn

| 10 | Artificial Nucleases, 317
D. Pei and P.G. Schultz |
|---|---|
| 11 | Ribonucleases H, 341
Z. Hostomsky, Z. Hostomska, and D.A. Matthews |
| 12 | RNA Maturation Nucleases, 377
M.P. Deutscher |
| 13 | Nucleases That Are RNA, 407
M.D. Been |

APPENDIX A
The Restriction Enzymes, 439
R.J. Roberts and D. Macelis

APPENDIX B
Some Well-characterized DNA-repair Nucleases/Glycosylases, 445
R.S. Lloyd

APPENDIX C
The Nucleases of *Escherichia coli*, 455
S. Linn and M.P. Deutscher

APPENDIX D
Compilation of Commercially Available Nucleases, 469
I. Schildkraut

Index, 485

Preface

In August 1981, the first meeting devoted to the subject of nucleases was held at Cold Spring Harbor Laboratory. A number of the speakers at that meeting had been cajoled into writing a chapter on their area of expertise. The idea was to produce a book that would provide a comprehensive account of this subject area. The result was the first edition of the book, *Nucleases*. A paperback edition, which contained a series of small addenda to some of the chapters, was produced in 1985. The seeds for this present edition were sown over the last few years as various individuals suggested updates and expansions of the book. The culmination was a second meeting on the subject of nucleases, which was held in February 1993 in Tamarron, Colorado, as a Keystone Symposium. Once again, potential authors were invited to the meeting and chapters were later written. The result is a complete re-write of the old book, although some of the authors are the same.

The nuclease field has witnessed a great deal of progress since the Cold Spring Harbor meeting in 1981, and this is reflected in new chapters describing synthetic nucleases, homing endonucleases, and ribozymes. Indeed, even the definition of a "nuclease" has elicited debate. Most dramatic has been the appearance of crystal structures for many nucleases, which have led to a deeper understanding of structure and function of various enzymes. Mechanistic studies have progressed apace and common themes are emerging as discussed in Chapter 1.

There are several deliberate omissions from the book. The single-strand-specific nucleases such as Bal31 were given comprehensive coverage in the earlier edition and little has changed. We had originally hoped that progress on the nucleases involved in mRNA splicing would have been sufficient to include a chapter on this topic. However, our understanding of the mechanistic aspects of this process is being hampered by the extremely complicated apparatus that accomplishes such splicing, and thus no clear examples of nucleases that might be involved in this process have yet been described.

Within the appendiceal material, we have chosen to highlight enzymes that have practical applications. Thus, the restriction enzymes and the less well-known DNA-repair enzymes are tabulated as are a range of other enzymes useful for DNA manipulations in vitro. We also present a list of all nucleases so far identified in *E. coli*, which gives an inkling of the complicated array of these enzymes that are necessary for the metabolism of even a simple organism.

The original idea for the 1982 book on nucleases was Jim Watson's, and, as always, his encouragement (sometimes quite direct) has been a great stimulus to produce the present book. Our objective is to provide a comprehensive coverage of the many facets of nucleases in a manner that will prove interesting to those already immersed in the field while serving as a useful overview to the newcomer interested in some particular class of nucleases. These enzymes have proved to be key tools for the molecular biologist, and we hope that this book will supplement the superficial view of these enzymes as merely cold reagents.

In addition to being the first comprehensive account of nucleases, the 1982 edition was the very first to be prepared "electronically" by Cold Spring Harbor Laboratory Press. Although the present book can claim no such "firsts," electronic communication and data retrieval have become an integral part of our lives. Communication between the editors was greatly enhanced by access to the Internet and to electronic resources such as Medline, Entrez, WAIS, and other information servers that now greatly enhance the productivity of the researcher. Among the nucleases, a comprehensive database of restriction enzymes, REBASE, is available electronically. Others are sure to follow. Perhaps the next edition of *Nucleases* will be disseminated totally electronically if we are to wait another 11 years!

We thank the Keystone Symposia as well as New England Biolabs and GIBCO Life Sciences for generously providing financial support for the meeting. We especially thank the authors for their contributions and for the good grace with which they accepted our editorial peregrinations. Finally, we acknowledge the help of the publications staff at Cold Spring Harbor Laboratory: John Inglis, Nancy Ford, and especially Dorothy Brown and Joan Ebert.

Stuart M. Linn
R. Stephen Lloyd
Richard J. Roberts

NUCLEASES

1
Mechanistic Principles of Enzyme-catalyzed Cleavage of Phosphodiester Bonds

John A. Gerlt
Department of Chemistry and Biochemistry
University of Maryland
College Park, Maryland 20742

    I. Introduction
   II. Hydrolytic (P-O) Cleavage of Phosphodiester Bonds: $S_N1(P)$ and $S_N2(P)$ Mechanisms
  III. Stereochemical Consequences of $S_N2(P)$ Mechanisms
      A. The Stereochemical Course of the Ribonuclease A Reaction: The Lack of Importance of Pseudorotation
      B. The Utility of Stereochemical Experiments to Detect Covalent Catalysis by Active Site Residues
      C. The Stereochemical Consequences of Nuclease-catalyzed Reactions
  IV. Relationships between $S_N2(P)$ Mechanisms and Active Site Structure
   V. The Active Sites of Nucleases
      A. Exonuclease Site of DNA Polymerase I
      B. Staphylococcal Nuclease
      C. Ribonuclease A
      D. Ribonuclease T_1
  VI. An Explanation for the Rate Accelerations in $S_N2(P)$ P-O Bond Cleavage Mechanisms
 VII. Eliminative (C-O) Cleavage of Phosphodiester Bonds: β-elimination Mechanisms
VIII. The β-elimination Reactions Catalyzed by Abasic Site DNA-repair "Endonucleases"
      A. The Active Sites of Abasic Site DNA-repair "Endonucleases"
          1. Endonuclease V
          2. Endonuclease III from *E. coli*
      B. Proposed Mechanisms for the Reactions Catalyzed by Abasic Site DNA-repair Endonucleases
  IX. An Explanation for the Rapid Rates of β-elimination C-O Bond Cleavage Reactions
   X. Summary

I. INTRODUCTION

Phosphodiester bonds, such as those found in both DNA and RNA, are extraordinarily resistant to hydrolysis. For example, the halftime for hy-

Nucleases, 2nd Edition
© 1993 Cold Spring Harbor Laboratory Press 0-87969-426-2/93 $5 + .00

drolysis of the simplest phosphodiester, dimethyl phosphate, in 1 M NaOH is approximately 15 years at 35°C (Chin et al. 1989). Although this stability is important for maintenance of genomic integrity, it poses a challenging problem to enzymes involved in the synthesis, repair, and degradation of both DNA and RNA. If reactions involving the phosphodiester backbones of nucleic acids are to occur rapidly in living organisms, the catalytic power of these enzymes must be formidable. In fact, hydrolysis of the phosphodiester bonds in DNA catalyzed by staphylococcal nuclease is accelerated by a factor of $\geq 10^{16}$ relative to the rate of the spontaneous reaction in the absence of enzyme (Serpersu et al. 1987). Although large, this rate acceleration is typical of protein enzymes that catalyze both hydrolysis and transesterification reactions of nucleic acids. The impressive catalytic power of these enzymes demands both mechanistic analysis and quantitative description. If the rate accelerations characteristic of enzyme-catalyzed phosphodiester bond hydrolysis could be understood quantitatively, the principles so elucidated could aid in the design of synthetic catalysts for (site-specific?) DNA and RNA degradation.

The extreme resistance of phosphodiester bonds to nucleophilic attack (as evidenced by their extremely slow hydrolysis in concentrated base) can be attributed, in part, to electrostatic repulsion between the phosphodiester anion and the approaching nucleophile, either neutral water or hydroxide anion (Westheimer 1987). This simple consideration is not, however, sufficient to explain the slow rates of hydrolysis of phosphodiester bonds in uncatalyzed reactions. The relative rates of hydrolysis of neutral trimethyl phosphate and monoanionic dimethyl phosphate by hydroxide ion differ by about 10^7 (Guthrie 1977), not $\geq 10^{16}$. Thus, if the rates of the enzyme-catalyzed reactions are to be understood, we must be able to explain not only the rate acceleration of 10^7 associated with neutralization of the negative charge of the substrate phosphodiester, but also the remaining and significant additional rate acceleration of as large as $\geq 10^9$ that appears to be a unique feature of enzyme active sites.

Until very recently, it was assumed that enzyme-catalyzed phosphodiester bond cleavage proceeded exclusively by P–O bond cleavage, i.e., hydrolysis reactions. However, the reactions catalyzed by a number of enzymes involved in the repair of damaged DNA are now known to proceed by C–O bond cleavage, i.e., β-elimination reactions involving aldehydic abasic sites (Bailly and Verly 1987; Mazumder et al. 1990; see Chapter 9, this volume). In retrospect, this diversity in mechanism is not surprising given the resistance of phosphodiester bonds to hydrolysis: A large number of enzymes unrelated to DNA and RNA synthesis, repair,

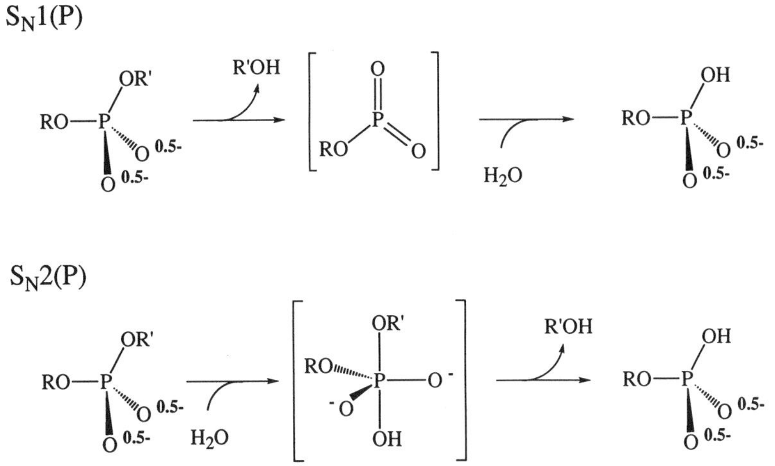

$S_N1(P)$

[chemical scheme showing $S_N1(P)$ mechanism with metaphosphate intermediate]

$S_N2(P)$

[chemical scheme showing $S_N2(P)$ mechanism with trigonal bipyramidal intermediate]

Figure 1 Comparison of $S_N1(P)$ (*top*) and $S_N2(P)$ (*bottom*) mechanisms for hydrolyses of phosphodiesters showing the structure of metaphosphate and trigonal bipyramidal intermediates, respectively.

and degradation, e.g., fumarase and enolase, are efficient catalysts for β-elimination reactions that are initiated by C–H bond cleavage adjacent to a carbonyl or carboxylic acid group. In these special cases when phosphodiester bonds can be broken by a β-elimination reaction involving C–O cleavage (damaged DNA containing aldehydic abasic sites), nature has apparently taken advantage of a mechanism that avoids, in part, the resistance of phosphodiesters to direct nucleophilic attack. In this chapter, the mechanisms and principles of both hydrolytic (P–O) and eliminative (C–O) cleavage of phosphodiester bonds are discussed.

II. HYDROLYTIC (P-O) CLEAVAGE OF PHOSPHODIESTER BONDS: $S_N1(P)$ AND $S_N2(P)$ MECHANISMS

Two quite distinct mechanisms are available for the hydrolyses of phosphate esters: a dissociative mechanism, termed $S_N1(P)$, and an associative mechanism, termed $S_N2(P)$. These are distinct because the $S_N1(P)$ mechanism involves the generation of an electron-deficient metaphosphate species as a reactive intermediate, whereas the $S_N2(P)$ mechanism involves the generation of an electron-rich pentacoordinate phosphorane as a reactive intermediate (Fig. 1).

The $S_N1(P)$ mechanism is not thought to be relevant to enzyme-catalyzed reactions of phosphodiester anions. This is because the highly electron-deficient species that would be generated would be a monoester of metaphosphoric acid; these species do not enjoy resonance stabilization and, as such, are less stable than metaphosphate anion (Fig. 1). For this reason, the $S_N1(P)$ mechanism is relevant only to hydrolysis reactions of phosphate monoesters and, in particular, phosphate monoester monoanions (monoacids) in which a proton is available to facilitate the departure of the alcohol ester leaving group. Considerable controversy has focused on whether the electron-deficient metaphosphate anion is a true intermediate in the hydrolysis reactions of phosphate monoesters. The current consensus is that it is not (Gerlt 1992). The reason for this is that metaphosphate anion has been judged to be so unstable that it reacts instantaneously (within the frequency of a bond vibration, 10^{-13} sec) with a nucleophile (water) to form the reaction product. If it is so unstable that it cannot exist, then it cannot be an intermediate (Jencks 1981). Thus, even though the transition state for the reaction is considered to be electron-deficient and to have metaphosphate character, the transfer of the monoester phosphoryl group from the leaving group to the acceptor nucleophile is concerted. The term "preassociative concerted" has been given to this mechanism by Jencks since if the phosphate monoester is not properly situated (preassociated) with the acceptor nucleophile, the reaction will not occur; if it is properly preassociated, the reaction will be concerted.

In contrast, the $S_N2(P)$ mechanism is considered to be relevant to reactions of phosphodiester anions. Attack of a nucleophile on a tetrahedral, tetracoordinate phosphodiester monoanion generates a trigonal bipyramidal, pentacoordinate phosphorane species (Fig. 1). Decomposition of this species with expulsion of the alcohol ester leaving group yields the reaction product; in the case of hydrolysis reactions, the product is a phosphomonoester. Whether the pentacoordinate species has sufficient lifetime to be an actual intermediate on the reaction coordinate or exists only as an unstable transition state has been controversial, and many biochemists have avoided the controversy by referring to this species as a "transition state/intermediate." However, as discussed in this chapter, the distinction between a transition state and intermediate is important since it has implications with respect both to the potential stereochemical course of the nucleophilic displacement reaction and, more importantly, to understanding the missing factor of $\geq 10^9$ in the observed rate accelerations of enzyme-catalyzed reactions of phosphodiester monoanions.

III. STEREOCHEMICAL CONSEQUENCES OF $S_N2(P)$ MECHANISMS

If the $S_N2(P)$ reaction described in the previous section is concerted, i.e., the trigonal bipyramidal, pentacoordinate species is a transition state, the reaction is quite analogous to the familiar S_N2 displacement reaction of tetrahedral carbon compounds. The nucleophile necessarily attacks the tetrahedral center (phosphorus or carbon) opposite to the leaving group substituent. If four different substituents are bonded to the tetrahedral atom, so that the tetrahedral center is chiral, the *relative* configuration of the center is inverted. In other words, if the nucleophile/new substituent and leaving group are considered equivalent, the reactant and product are enantiomers (mirror images). (If one of the substituents contains additional chiral centers, as is the case in nucleic acids, the reactant and product are diastereomers.) Thus, a concerted $S_N2(P)$ reaction necessarily occurs with inversion of configuration at phosphorus.

If the $S_N2(P)$ reaction described in the previous section is stepwise, the trigonal bipyramidal, pentacoordinate species is an intermediate. There is no analogy to this mechanism in carbon compounds, since it is impossible to expand the valency of carbon from four to five. With the formation of a stable intermediate, the stereochemical course of the reaction is potentially more complex. The complexity is introduced by the fact that if the pentacoordinate species has a sufficiently long lifetime, it may undergo a thermodynamically allowed ligand reorganization process termed "pseudorotation" that will allow the stepwise $S_N2(P)$ reaction to proceed with *retention* of relative configuration.

The trigonal bipyramidal, pentacoordinate intermediate that is formed by addition of a nucleophile to a tetracoordinate phosphodiester is assumed to have a structure in which the two apical P–O bonds are longer and weaker than the three equatorial P–O bonds (Westheimer 1968). As a result of a series of studies on the reactivity of five-membered ring-containing esters of phosphoric, phosphonic, and phosphinic acids, Westheimer first suggested that the attacking nucleophile and the leaving group must occupy the apical positions. If this intermediate were thermodynamically favored and proceeded to products by the microscopic reverse of its formation, with the exception that the leaving group rather than the nucleophile depart, the stereochemical course of the reaction would be inversion of relative configuration, just as in the concerted version of the $S_N2(P)$ reaction (Fig. 2, top).

Westheimer further suggested, however, that the trigonal pyramidal structure can rearrange, with the two apical ligands becoming equatorial ligands and two of the equatorial ligands becoming apical ligands. Such rearrangement would be favored if the potential apical substituents were more electron donating than the potential equatorial substituents. Thus, if this "pseudorotated" intermediate were to break down to products, the leaving group could be one of the three equatorial ligands in the initially formed pentacoordinate intermediate. In this case, the leaving group

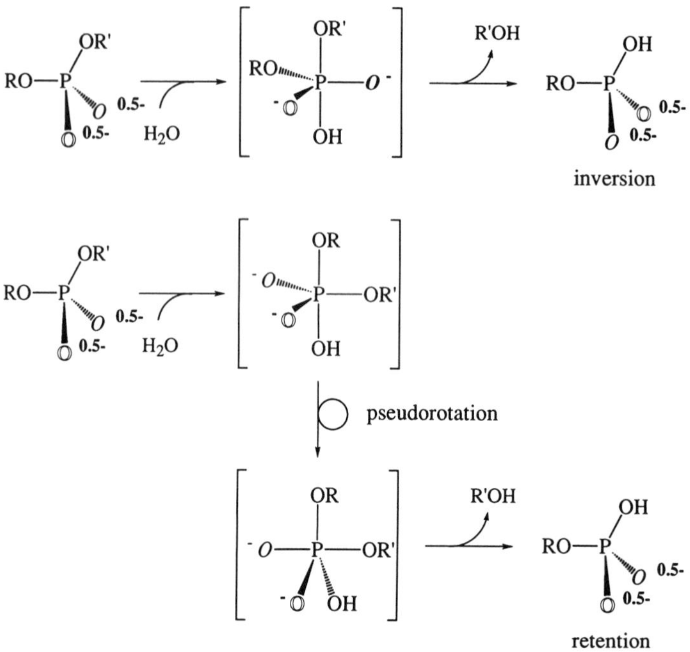

Figure 2 Analysis of the stereochemical outcomes expected for $S_N2(P)$ reactions without (*top*) and with (*bottom*) pseudorotation of a trigonal bipyramidal intermediate.

could, in effect, depart from a position adjacent to the incoming nucleophile. This would give the appearance of a "front side" displacement, whose stereochemical course would necessarily be retention of relative configuration (Fig. 2, bottom). Thus, the pseudorotation or ligand reorganization of the trigonal bipyramidal, pentacoordinate intermediate is stereochemically equivalent to an inversion of relative configuration of the reacting center. The overall displacement reaction can be simplistically thought of as two successive inversions of relative configuration, one associated with attack of the nucleophile and departure of the leaving group and the second with the pseudorotatory process.

Is there any reason to consider pseudorotatory processes in enzyme-catalyzed reactions? Now, in 1993, there is none. However, in 1970, shortly after Westheimer put forth his principles for the mechanisms of nonenzymatic displacement reactions of phosphate esters, which were based on experimental evidence for pseudorotation, and when Eckstein and his co-workers devised the first syntheses (Eckstein and Gindl 1968) and configurational analyses (Usher et al. 1970) of chiral nucleoside phosphorothioates, the mechanistic enzymology/bioorganic community had no idea as to whether pseudorotation would occur in an enzyme active site.

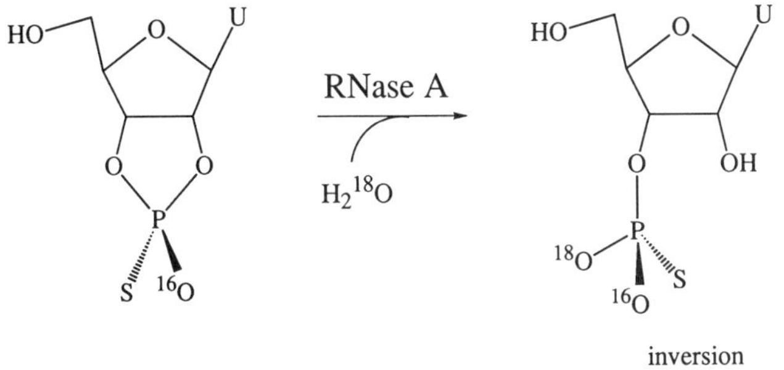

Figure 3 Stereochemistry of the reaction catalyzed by RNase A (Usher et al. 1970).

A. The Stereochemical Course of the Ribonuclease A Reaction: The Lack of Importance of Pseudorotation

In 1970, when Eckstein and co-workers studied the mechanism of the reaction catalyzed by ribonuclease A with chiral nucleoside phosphorothioates, the occurrence of 2′,3′-cyclic nucleotide intermediates in the hydrolysis of internucleotide bonds to form 3′-nucleotide products was already well established. It was also well established that active site functional groups participated in catalysis only as general acids and general bases, i.e., no kinetic or structural evidence was available suggesting that the mechanism involved covalent catalysis by active site residues. Thus, stereochemical characterization of the mechanism would provide unequivocal information regarding the importance of pseudorotation of pentacoordinate intermediates.

It is not important here to consider the details of the experimental procedures used by Eckstein and his co-workers in their studies (Usher et al. 1970); these have been reviewed many times by this and other authors. It is important, however, to consider the structures of the substrate and product in the stereochemical study. Ribonuclease A was used to catalyze the hydrolysis of the *endo* isomer of uridine 2′,3′-cyclic phosphorothioate in $H_2^{18}O$ to a diastereomer of uridine 3′-[$^{16}O,^{18}O$]phosphorothioate. Comparison of the configurations of the chiral phosphorous centers in the substrate and product reveals that the hydrolysis reaction proceeded with *inversion* of configuration at phosphorus (Fig. 3). This stereochemical course demonstrates that the $S_N2(P)$ reaction is not accompanied by pseudorotation of any trigonal bipyramidal, pentacoordinate intermediate, although the result does not rule out the formation of such an intermediate.

Whereas a single example certainly does not constitute sufficient evidence to discount the importance of pseudorotatory processes in enzyme-catalyzed reactions (they are important in nonenzymatic reactions, especially those of five-membered ring-containing cyclic esters [Hall and Inch 1980]), this and subsequent stereochemical studies of both protein and RNA enzymes that catalyze hydrolyses and/or transesterification reactions of phosphate esters without the involvement of covalent catalysis (formation of a nucleotidylated enzyme intermediate) have effectively discounted the importance of pseudorotation in enzyme-catalyzed reactions (Gerlt 1992). In retrospect, this is reasonable since pseudorotatory processes would necessarily be accompanied by significant movement of the substrate in the enzyme active site, and this would place considerable demands on the enzyme with respect to "gripping" the substrate throughout its conversion to product.

B. The Utility of Stereochemical Experiments to Detect Covalent Catalysis by Active Site Residues

Without the complication of pseudorotation, stereochemical studies of nuclease-catalyzed reactions have been used to establish whether phosphodiester hydrolysis reactions occur via covalent catalysis by active site functional groups. Since it can be expected that each $S_N2(P)$ reaction in an active site will occur by inversion of configuration, the stereochemical course that describes the conversion of substrate to product can be used to "count" the total number of displacement reactions that occur at phosphorus. Thus, an inversion of configuration indicates a single displacement reaction (in general, an odd number), and a retention of configuration indicates two successive displacement reactions (inversion followed by inversion is retention of configuration) (in general, an even number). Thus, a retention of configuration can be used as evidence for covalent catalysis by active site functional groups.

Although not a nuclease, bacterial (*Escherichia coli*) alkaline phosphatase is perhaps the best known example of a phosphohydrolase (phosphomonoesterase) that involves covalent catalysis (Coleman and Gettins 1983; Wyckoff et al. 1983). An active site serine (Ser-102) participates as a nucleophile to displace the ester leaving group, thereby leading to the formation of a phosphorylated serine intermediate. In a second step, water acts as a nucleophile to displace serine from the phosphorylated intermediate to generate the inorganic phosphate product and the active site serine. The phosphorylated serine intermediate has been observed by ^{31}P-labeled nuclear magnetic resonance (NMR) spectroscopy, as well as isolated and characterized by chemical degradation of the enzyme.

Alkaline phosphatase catalyzes a facile transesterification reaction in which 1,2- and 1,3-diols can compete effectively with water for the phosphorylated serine intermediate. Knowles and his co-workers established that the stereochemical course of a transesterification reaction catalyzed by alkaline phosphatase is retention of relative configuration (Jones et al. 1978). Again, the precise experimental details are not important, except to note that the phosphate monoester substrate was made chiral by isotopic substitution with the three stable isotopes of oxygen, ^{16}O, ^{17}O, and ^{18}O. However, this experiment does establish the validity of using a stereochemical approach to determine whether covalent catalysis must be considered in formulating a mechanism for an enzyme-catalyzed reaction.

C. The Stereochemical Consequences of Nuclease-catalyzed Reactions

To date (mid 1993), the stereochemical courses of at least 12 enzyme-catalyzed hydrolyses of phosphodiesters have been reported (Gerlt 1992). In contrast to the absolute stereochemical uniformity that has been documented in reactions catalyzed by kinases (all inversion of relative configuration), the reactions catalyzed by two nucleases, snake venom phosphodiesterase (Mehdi and Gerlt 1981a) and spleen exonuclease (Mehdi and Gerlt 1981b), are accompanied by retention of relative configuration, implying that the mechanisms for the hydrolysis of the internucleotidic bonds in nucleic acids need not be the same, at least in terms of covalent catalysis.

Although the structures of the active sites of these two enzymes are not yet available, Frey (1989) has suggested that the retention of configuration observed in these nuclease-catalyzed reactions may be the result of evolutionary pressure to conserve the number of binding sites that are necessary for biological function. If the function of an enzyme is to transfer a nucleotidyl group from a donor to an acceptor, rather than or in addition to from a donor to water (i.e., hydrolysis), and if the donor and acceptor groups have similar structures, the simplest enzyme active site would have a single binding site that would recognize both the donor and the acceptor. In other words, the enzymes that are observed to proceed with retention of configuration may function in vivo as nucleotidyl transferases rather than nucleotidyl hydrolases.

In the first half-reaction, the binding site would bind a donor nucleotide with the nucleotidyl group to be transferred, and an active site functional group would displace the donor nucleotide to yield a nucleotidylated enzyme intermediate. Following dissociation of the donor nucleotide from the binding site, the acceptor nucleotide would bind to the site previously occupied by the donor nucleotide. Once bound, the acceptor nucleotide would displace the active site functional group from the nucleotidylated intermediate to yield the acceptor nucleotide with the nucleotidyl group. Just as in the much simpler case of reactions catalyzed by alkaline phosphatase, water may be able to compete with the transfer reaction, thereby producing the observed hydrolytic activity.

IV. RELATIONSHIPS BETWEEN $S_N2(P)$ MECHANISMS AND ACTIVE SITE STRUCTURE

Stereochemical experiments such as those described in the previous sections are not sufficient to propose detailed mechanisms for nuclease-catalyzed hydrolysis of phosphodiester bonds. Instead, these experiments simply place restrictions on the possible mechanisms that can be formulated for the P–O bond-breaking reaction. For example, if the stereochemical course of the reaction is retention of configuration, then an active site functional group must participate in catalysis as the initial nucleophile that displaces the alcohol ester leaving group. However, the stereochemical course does not identify (1) the active site nucleophile, if any, that participates in the reaction, (2) the active site general base catalyst, if any, that may assist in the attack of the nucleophile on substrate, (3) the active site general acid catalyst, if any, that may assist in the departure of the leaving group, and (4) the active site electrophilic catalysts, if any, that may stabilize the pentacoordinate intermediate that occurs on the reaction coordinate. This information is necessary and, in principle, should be sufficient to propose a mechanism for the reaction that explains both the bond-breaking/making reactions and the impressive rate accelerations that characterize nuclease-catalyzed reactions.

Proper identification of the catalytic groups can only be made if high-resolution (X-ray crystallographic and/or NMR spectroscopic) structural information is available. Although this requirement might be viewed by some investigators as being unrealistic, e.g., consensus sequences for metal ion binding sites might be detected by DNA homology searches and "verified" by site-directed mutagenesis experiments, three-dimensional structural information about the precise orientation of functional groups to the substrate is required for intelligent formulation of a detailed chemical mechanism.

At present, X-ray structural information is available for at least eight nucleases: staphylococcal nuclease (Cotton et al. 1979; Loll and Lattman 1989), DNase I (Lahm and Suck 1991), the 3′→5′ exonuclease site of the Klenow fragment of DNA polymerase I (Beese and Steitz 1991), *Eco*RI restriction endonuclease (Kim et al. 1990; see Chapter 2, this volume), *Eco*RV restriction endonuclease (Winkler et al. 1993; see

Chapter 2, this volume), nuclease P_1 (Volbeda et al. 1991), RNase H (Yang et al. 1990), RNase A (Borah et al. 1985; Campbell and Petsko 1987), RNase T_1 (Arni et al. 1988), and barnase (Mauguen et al. 1982). Of these eight nucleases, the first five (the DNases and one RNase) are absolutely dependent on divalent metal ions for activity, and the X-ray structural data reveal the presence of divalent metal ion binding sites in the presumed active sites. The last three (ribonucleases) have no known metal ion requirement. Although this difference in active site structure might be suggestive of a fundamental difference in mechanisms, what is presumably important in all of these enzymes is that the active site contain functional groups, either those of the amino acid functional groups or those of the divalent metal ions, that can both increase the nucleophilicity of the nucleophile (water in the case of hydrolytic reactions) and stabilize the phosphorane intermediate that is likely to occur on the reaction coordinate.

The discussion that follows focuses on four nucleases that have been structurally characterized: two DNases, the $3' \rightarrow 5'$ exonuclease site of the Klenow fragment DNA polymerase I and staphylococcal nuclease, and two RNases, A and T_1. The reason for this choice is that these enzymes have been crystallized with inhibitors and/or products, so some information is available about the probable orientation of substrates with the functional groups in the active site.

V. THE ACTIVE SITES OF DNASES

A. Exonuclease Site of DNA Polymerase I

Steitz and co-workers have determined moderately high-resolution structures of wild-type and mutant versions of the $3' \rightarrow 5'$ exonuclease (proofreading) active site of the Klenow fragment of DNA polymerase I (Beese and Steitz 1991). This exonuclease active site is contained within a discrete domain that is separate from the domain that catalyzes DNA polymerization. Structures have been described for a complex of the unmutated Klenow fragment with thymidine $5'$ monophosphate (dTMP) and for a complex of the D424A mutant with a thymidine tetranucleotide. The former structure is that of a product complex, and the latter structure was determined with a catalytically inactive mutant in which binding of one of the two required divalent metal ions, but not secondary and tertiary structures of the protein, had been disrupted.

The structure of the product complex reveals the presence of two divalent metal ions (Mg^{++}, Mn^{++}, or Zn^{++}) that are separated by 3.9 Å (Fig. 4). The binding site for one of the divalent metals, designated site

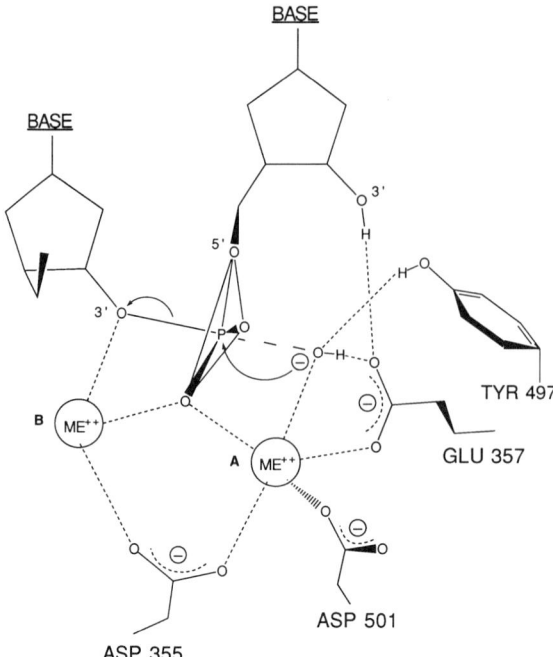

Figure 4 Exonuclease active site of DNA polymerase I. (Reprinted, with permission of Oxford University Press, from Beese and Steitz 1991.)

A, is provided by the carboxylate groups of three residues (Asp-355, Glu-357, and Asp-501), with one of the nonesterified phosphoryl oxygens of the bound dTMP providing an additional ligand. The fifth ligand in the distorted tetrahedral coordination geometry is provided by a water molecule. This water molecule, in turn, is hydrogen-bonded to the carboxylate group of Glu-357, as well as the hydroxyl group of Tyr-497. The second metal-binding site, designated site B, is provided by the carboxylate group of Asp-424 and two of the nonesterified phosphoryl oxygens of the bound dTMP. The remaining three ligands of the octahedral coordination geometry is provided by three water molecules.

On the basis of this structure, Beese and Steitz (1991) proposed that the pair of divalent metal ions may serve several catalytic roles in the mechanism of the hydrolysis reaction. The pK_a of a water molecule coordinated to one of the divalent metal ions, presumed to be that coordinated to the metal ion in site A, may be depressed, thereby generating a hydroxide ion in close proximity to the presumed binding site for the substrate internucleotidic phosphodiester bond. No obvious functional group candidate for a general base catalyst, which would assist in proton abstraction from water, is apparent from the structure of the product complex. In accord with this hypothesis, the exonuclease activity increases as the pH increases, as would be expected if the reaction were specific base-catalyzed, rather than general base-catalyzed. If an active site functional group were participating in catalysis as a general base, the rate of the reaction should become independent of pH above the pK_a of the conjugate acid of the general base catalyst.

The coordination of the anionic phosphodiester substrate to one or both of the divalent metal ions should reduce or even eliminate the electrostatic repulsion between the putative hydroxide ion nucleophile and the substrate that has often been cited as the major reason for the slow rates of nonenzymatic hydrolyses of phosphodiesters in basic solution.

The presence of two divalent metal ions also suggests that these may function to stabilize a pentacoordinate (phosphorane) *intermediate* in the active site. If such an intermediate were formed during the course of the reaction, the charge on the nonesterified phosphoryl oxygens would increase from 0.5 to 1, and this increase in charge in close proximity to the divalent metal ions would be expected to stabilize the dianionic phosphorane intermediate. Although the exonuclease active site is quite polar, the magnitude of the attractive electrostatic interactions between the dianionic phosphorane intermediate and either one or both of the divalent metal ions may be sufficient to allow efficient catalysis. Stabilization of a phosphorane intermediate in an enzyme active site is expected to allow the rates of the enzymatic reactions to greatly exceed those of the nonenzymatic reactions where stabilization of the dianionic phosphorane intermediate is not possible.

The exonuclease active site also appears to lack a general acid catalyst that could assist in the departure of the leaving group in the reaction, the 3'-oxygen atom of the adjacent nucleotide. Since the pK_a of the 3'-hydroxyl group of a 5'deoxynucleotide is expected to be about 16, this suggests that its rapid displacement must be catalyzed by an electrophile. Therefore, the presence of two divalent metal ions in the active site suggests that one of these (in site A) provides the hydroxide ion nucleophile and the second (in site B) coordinates with the 3'-oxygen atom of the leaving group, thereby allowing its rapid departure.

In summary, the divalent metal ions in the exonuclease active site of DNA polymerase I have been proposed to be *totally* responsible for rapid hydrolytic cleavage of the substrate phosphodiester bond.

B. Staphylococcal Nuclease

Cotton and his co-workers (Cotton et al. 1979) and, more recently, Lattman and his co-workers (Loll and Lattman 1989) have determined high-resolution structures of both wild-type and mutant versions of

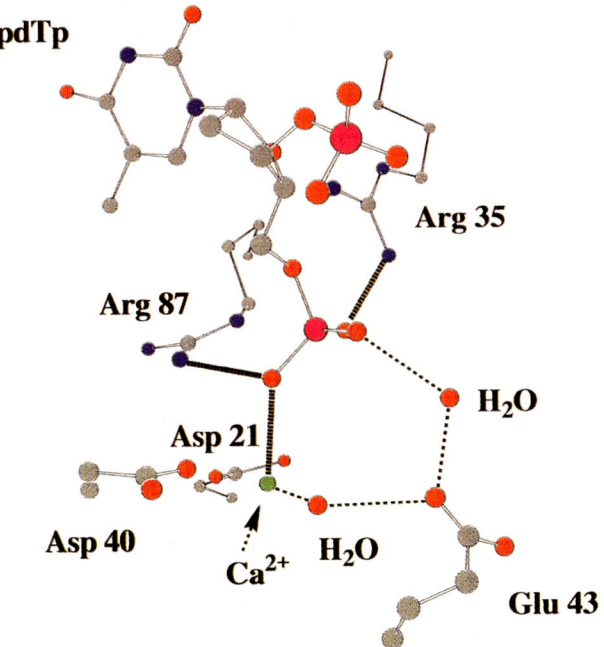

Figure 5 Active site of staphylococcal nuclease complexed with the inhibitor thymidine 3',5'-bisphosphate (Loll and Lattman 1989). Heavy broken lines indicate the hydrogen bonding and electrostatic interactions that neutralize the negative charge of the phosphate group of the inhibitor. Dashed lines indicate the interactions between Glu-43 and the active site water molecules.

staphylococcal nuclease (SNase) with both the catalytically essential Ca^{++} and the competitive inhibitor thymidine 3',5'-bisphosphate (pdTp) bound in the active site (Fig. 5).

In this structure, the binding site for the Ca^{++} is provided by the carboxylate groups of both Asp-21 and Asp-40, the carbonyl oxygen of Thr-41, three water molecules, and one of the nonesterified phosphoryl oxygens of the 5'-phosphate group of the bound pdTp, thereby producing a heptacoordinate, distorted octahedral coordination geometry for the metal ion. The remaining two nonesterified phosphoryl oxygens of this phosphate group are observed to be within hydrogen-bonding distance of the guanidinium groups of both Arg-35 and Arg-87. Since the carboxylate group of Glu-43 is hydrogen-bonded to two water molecules in the active site, one of those coordinated to the metal ion and a second that is also hydrogen-bonded to one of the nonesterified phosphoryl oxygens of the 5'-phosphate group, Cotton first proposed that the mechanism of phosphodiester bond cleavage involves nucleophilic activation of an active site water molecule by general base catalysis.

In the interpretation of the structure of the complex of pdTp bound to wild-type SNase, it was necessary to realize that pdTp is neither a substrate nor a product of the reaction catalyzed by SNase. The limit hydrolysis products from the degradation of DNA are 3′ nucleotides. Since the 5′-phosphate group of pdTp is observed to be coordinated to the essential Ca^{++}, it is not clear what step in the reaction the structure of this complex may mimic. Since a dianionic phosphorane could be an intermediate in the reaction and since the 5′-phosphate group is dianionic, the structure of the pdTp complex may be similar to that of the dianionic phosphorane bound in the active site. In addition, another complication is that an intermolecular contact between adjacent protein molecules in the crystals of SNase involves active site residues. The ε-ammonium groups of Lys-70 and Lys-71 from an adjacent molecule protrude into the active site, with the ionic contacts involving the putative general base in the reaction, Glu-43, as well as the 5′-phosphate group of pdTp.

With these caveats in mind, it is possible to propose that the anionic charge of the substrate internucleotidic phosphodiester bond is neutralized by coordination with the essential Ca^{++} and/or the guanidinium groups of Arg-35 and Arg-87. This charge neutralization should reduce the electrostatic repulsion between the putative hydroxide ion nucleophile and the anionic substrate, thereby increasing the rate of the enzymatic reaction relative to those of nonenzymatic reactions in basic solution.

Additionally, the essential Ca^{++} and/or the guanidinium groups of Arg-35 and Arg-87 may function to stabilize a pentacoordinate (phosphorane) *intermediate* in the active site. As discussed in the previous section, the charge on the nonesterified phosphoryl oxygens would increase from 0.5 to 1 when the phosphorane intermediate is formed, and this increase in charge, in close proximity to the Ca^{++} and both Arg-35 and Arg-87, would be expected to stabilize the dianionic phosphorane intermediate.

In contrast to the exonuclease active site of DNA polymerase I, the active site does appear to contain functional groups that may participate in catalysis either as general base catalysts to assist in the nucleophilic attack of water on the substrate phosphodiester or as general acid catalysts to assist in the departure of the 5′-hydroxyl group of the neighboring nucleotide from the phosphorane intermediate.

The hydrogen bonds between the carboxylate group of Glu-43 and water molecules that are in close proximity to the bound pdTp have led to the suggestion that this functional group participates in catalysis as a general base catalyst. In fact, studies employing site-directed substitutions of Glu-43 that have been performed in the author's laboratory support this proposal (Hibler et al. 1987), since the E43D and E43S substitutions are significantly reduced (>200-fold) in catalytic activity. However, the mechanistic interpretation of these reductions in catalytic activity is confused by the observation that alterations in the overall structure in the "conservative" E43D mutant protein could be easily detected by both NMR spectroscopy (Wilde et al. 1988) and X-ray crystallography (Loll and Lattman 1990). More recently, kinetic studies have led to the conclusion that general base catalysis is likely to be unimportant in the reactions catalyzed by wild-type SNase and several site-directed mutants. Instead, under the appropriate conditions, the k_{cat} for the SNase-catalyzed reaction increases in direct proportion to the hydroxide ion concentration (Hale et al. 1993). Whether the active site facilitates the formation of hydroxide ion in the active site by coordination of the nucleophilic water molecule to the essential Ca^{++} is unknown.

The guanidinium group of either Arg-35 or Arg-87 may also participate in the hydrolysis reaction as a general acid catalyst that facilitates the departure of the 5′ oxygen of the neighboring nucleotide. Since the pK_a values of Arg-35 and Arg-87 are likely to be ≥12, it is not necessarily intuitively obvious that these could function as *acidic* catalysts. However, since the pK_a of the departing alkoxide (~16) is certain to exceed those of the guanidinium groups, this does lead to the reasonable suggestion that one or both of these may function in catalysis as general acid residues. The roles of both of these residues in catalysis have been investigated by site-directed mutagenesis (both "conservative" lysine and non-conservative glycine substitutions), and the activities of these mutant proteins are reduced significantly from that of wild-type SNase ($\geq 10^4$). However, since the conservative lysine substitutions were observed to disrupt the structures of the mutant enzymes as revealed by NMR spectroscopy, mechanistic conclusions based on the activities of these mutants are difficult if not impossible (Pourmotabbed et al. 1990). Nevertheless, the presence of even weakly acidic proton donors in the active site of SNase (Arg-35 and Arg-87) does suggest that one or both of these may be involved in catalysis as general acid catalysts.

In summary, the structure of the active site of SNase reveals the presence of a divalent metal ion as well as amino acid functional groups that must be *totally* responsible for the rapid hydrolytic cleavage of the substrate phosphodiester bond.

C. Ribonuclease A

Three of the four RNases that have been structurally characterized do not require any divalent metal ions for activity, so their catalytic activity is necessarily associated solely with amino acid functional groups present in the active sites. RNase A obtained from bovine pancreas has been carefully scrutinized over the years with respect to protein folding pathways, catalytic mechanism, and structural analysis. The mechanism of RNA hydrolysis catalyzed by RNase A is known to proceed via two steps: (1) attack of the 2'-hydroxyl group on the adjacent internucleotidic phosphodiester bond to displace the 5'-hydroxyl group of the neighboring nucleotide and generate a 2',3'-cyclic nucleotide intermediate and (2) hydrolysis of the 2',3'-cyclic nucleotide intermediate to the final product, a 3' mononucleotide. The mechanisms of both of these half-reactions are thought to be similar, with the active site acid/base catalysts being similarly involved in both.

Wlodawer (Borah et al. 1985) and Petsko (Campbell and Petsko 1987) and their co-workers have independently determined the structures of RNase A complexed to a variety of inhibitors (Fig. 6). These structural analyses have revealed the presence of two histidine residues in the substrate-binding site, His-12 and His-119, and these are now widely believed to be involved in catalysis as acid/base catalysts. In the first half-reaction, His-12 acts as a general base catalyst to facilitate the attack of the 2'-hydroxyl group on the phosphodiester bond, and His-119 acts as a general acid catalyst to facilitate the displacement of the 5' oxygen of the adjacent nucleotide in the formation of the 2',3'-cyclic nucleotide intermediate. In the second half-reaction, His-119 acts as a general base catalyst to facilitate the attack of water on the 2',3'-cyclic nucleotide, and His-12 acts as a general acid catalyst to facilitate the displacement of the 2'-oxygen atom to form the 3'-mononucleotide product.

For many years, both half-reactions had been thought to be individually concerted processes. However, recent nonenzymatic model studies published by Breslow and his co-workers have led to the suggestion that both half-reactions may, in fact, proceed in a stepwise manner via the necessary formation of a stabilized monoanionic phosphorane intermediate. The kinetics of the nonenzymatic model reactions are consistent with the formation of a phosphorane intermediate (Anslyn and Breslow 1989).

The structure of the active site RNase A reveals that the ϵ-ammonium group of Lys-41 is in sufficiently close proximity to the active site that it could participate in catalysis as an electrophilic group that might stabilize a dianionic phosphorane intermediate by hydrogen bonding or electrostatic interactions. Although the interaction of the cationic functional group of Lys-41 with the anionic phosphodiester substrate might also decrease the electrostatic repulsion between the 2'-hydroxyl group and the phosphodiester in the first half-reaction and between the water and the phosphodiester in the second half-reaction, such a role for Lys-41 in catalysis has not been adequately considered.

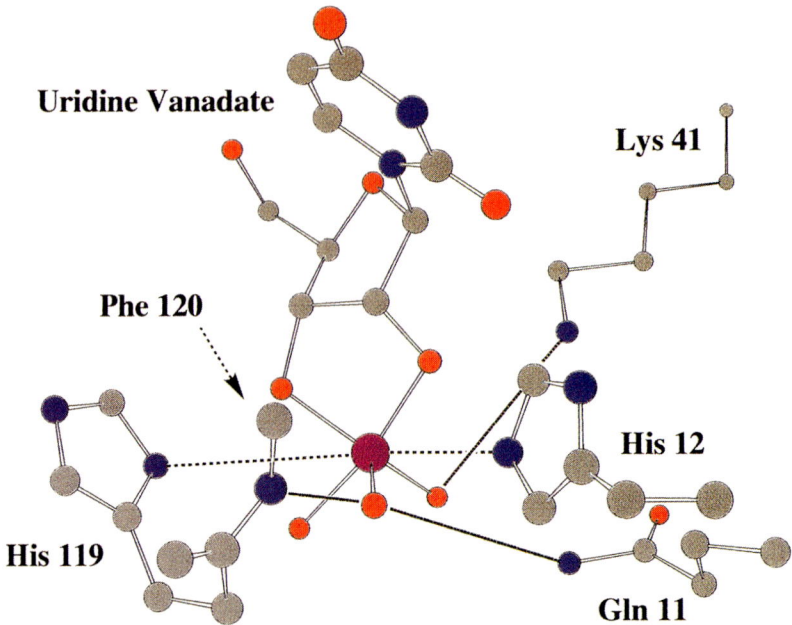

Figure 6 Active site of ribonuclease A complexed with the potential transition state analog uridine vanadate (Borah et al. 1985). Heavy broken lines indicate the hydrogen bonding and electrostatic interactions that neutralize the negative charge of the transition state analog. Dashed lines indicate the interactions between His-12 and His-119 and the transition state analog.

In addition, the structural studies reveal that both the carboxamide group of Gln-11 and the peptidic NH of Phe-120 are also close to the position of the bound substrate. Hydrogen bonding between one or both of these amide NH groups and an anionic phosphorane intermediate might also stabilize the intermediate, thereby contributing to the rapid rate of RNase-A-catalyzed reactions.

In summary, the structural studies reveal the presence of several amino acid functional groups that must be *totally* responsible for the rapid rate of RNase-A-catalyzed reactions. However, a quantitative explanation for the rate acceleration in terms of the contributions of each of these amino acid functional groups has been lacking.

D. Ribonuclease T$_1$

RNase T$_1$, a fungal enzyme, and barnase, a structurally homologous bacterial enzyme, also catalyze the hydrolysis of RNA via the formation of

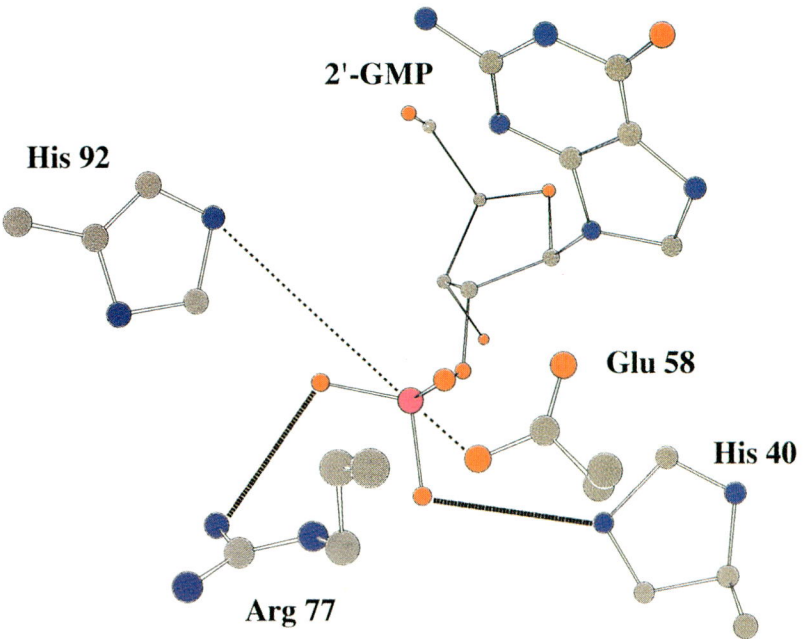

Figure 7 Active site of RNase T$_1$ complexed with the inhibitor 2'-GMP (Arni et al. 1988). Heavy broken lines indicate the hydrogen bonding and electrostatic interactions that neutralize the negative charge of the inhibitor. Dashed lines indicate the interactions between Glu-58 and His-92 and the transition state analog.

2',3'-cyclic nucleotide intermediates. However, their active sites are quite distinct from that of RNase A.

Saenger and his co-workers have determined high-resolution structures of RNase T$_1$ complexed with a number of inhibitors, including 2'-GMP (Fig. 7) (Arni et al. 1988). Thus, the structure of an inhibitor complex has been used to infer the identities and catalytic roles of essential functional groups in the active site.

The structural analyses, along with the results of site-directed mutagenesis investigations (Steyaert et al. 1990), suggest that Glu-58 serves as the general base catalyst and His-92 functions as the general acid catalyst in the formation of the 2',3'-cyclic nucleotide intermediate. The structural studies reveal the presence of two additional residues that are likely to be involved in catalysis, His-40 and Arg-77. Since both of these residues can be cationic, this suggests that they can participate in neutralization of the anionic charge of the substrate, thereby diminishing or eliminating electrostatic repulsion of the 2'-hydroxyl group that attacks the phosphodiester group. Furthermore, if in analogy to the proposal for the reaction catalyzed by RNase A, a phosphorane intermediate is involved in the formation of the 2',3'-cyclic nucleotide that is catalyzed by RNase T$_1$, His-40 and Arg-77 may interact electrostatically or via hydrogen bonding with the anionic phosphorane intermediate, thereby stabilizing it so that the rate of the reaction can be accelerated relative to that observed for nonenzymatic reactions under basic conditions.

Although the identities of the functional groups present in the active site of RNase T$_1$ differ from those present in the active site of RNase A, structural studies reveal the presence of several amino acid functional groups that must be *totally* responsible for the rapid rate of the catalyzed reactions. However, again, a quantitative explanation for the rate acceleration in terms of the contributions of each of these amino acid functional groups has been lacking.

VI. AN EXPLANATION FOR THE RATE ACCELERATIONS IN S$_N$2(P) P-O BOND CLEAVAGE MECHANISMS

As noted in the Introduction to this chapter, an understanding of the large rate accelerations that characterize nuclease-catalyzed reactions is important but has been lacking. Recently, the intellectual framework has been developed by which the mechanisms and rapid rates of a number of enzyme-catalyzed reactions can be understood (Gerlt and Gassman 1993). The approach is based on Marcus formalism (Cohen and Marcus 1968) for describing reaction coordinates that provides a simple mathematical relationship between the rate of a reaction ($\Delta G\ddagger$, the activation energy barrier for the reaction) and the stability/instability of transiently stable intermediates on the reaction pathway that are formed in the rate-limiting step of the reaction (ΔG^o, the thermodynamic barrier for formation of the intermediate). Marcus formalism was originally developed to describe the rates of electron transfer reactions. However, the same mathematical approach can be used to describe the rates of reactions involving the generation of enolate, tetrahedral, and pentacoordinate phosphorane intermediates.

In 1977, Guthrie provided estimates of the equilibrium constants for the addition of water and hydroxide ion to phosphodiester acids and their conjugate bases, phosphodiester monoanions, to form phosphoranes (Guthrie 1977). For example, the equilibrium constant for the addition of hydroxide ion to a monoanionic phosphodiester to form a dianionic phosphorane is 10^{-16} (ΔG^o = 22 kcal/mol). If one assumes that the dianionic phosphorane is a necessary intermediate on the reaction coordinate, the slow rate of the reaction is, in large part, due to the instability

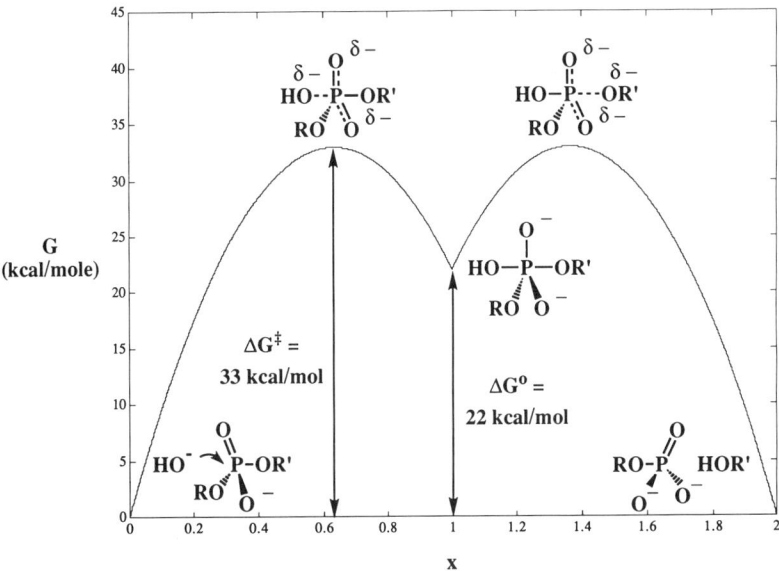

Figure 8 Reaction coordinate for the nonenzymatic hydroxide-mediated hydrolysis of a phosphodiester anion.

of the dianionic phosphorane intermediate (Fig. 8; $k = 2 \times 10^{-12}$ sec^{-1} for the hydrolysis of dimethyl phosphate in 1 N NaOH [$\Delta G\ddagger = 33$ kcal/mol]). However, $\Delta G\ddagger > \Delta G°$, i.e., the rate of the reaction is slower than that predicted on the basis of the thermodynamic barrier for the rate-determining formation of the dianionic phosphorane intermediate.

The difference $\Delta G\ddagger - \Delta G° = 11$ kcal/mol indicates that the addition of hydroxide ion to a monoanionic phosphodiester is intrinsically a slow process. The difference between $\Delta G\ddagger$ and $\Delta G°$ is not due solely to electrostatic repulsion between the hydroxide ion nucleophile and the monoanionic phosphodiester. This is also reflected in $\Delta G°$.

Marcus formalism assumes that the reaction coordinate for a reaction can be described adequately by the equation for an inverted parabola (Grunwald 1990), where $\Delta G°$ and $\Delta G\ddagger_{int}$ are independent energetic parameters and x describes the free energy, G, along the reaction coordinate, x ($x = 0$ for hydroxide ion and the substrate phosphodiester and $x = 1$ for the phosphorane intermediate):

$$G = -4\Delta G\ddagger_{int}(x - 0.5)^2 + \Delta G°(x - 0.5) \quad (1)$$

With this assumption, $\Delta G\ddagger$ can be partitioned between contributions from both $\Delta G°$ and $\Delta G\ddagger_{int}$. If an enzyme active site is to decrease $\Delta G\ddagger$, the active site *must* decrease either $\Delta G°$ or $\Delta G\ddagger_{int}$ from the values that together describe the nonenzymatic reaction. From Equation 1, an equation describing the magnitude of $\Delta G\ddagger$ in terms of both ΔG^o and $\Delta G\ddagger_{int}$ can be derived:

$$\Delta G\ddagger = \Delta G\ddagger_{int} (1 + \Delta G^o/4\Delta G\ddagger_{int})^2 \qquad (2)$$

From Equations 1 and 2, the activation energy barrier in excess of ΔG^o can be associated with $\Delta G\ddagger_{int}$. $\Delta G\ddagger_{int}$, the intrinsic activation energy barrier, is defined as $\Delta G\ddagger$ when $\Delta G^o = 0$. In this situation, if $\Delta G\ddagger_{int} > 0$, the reaction is intrinsically slow.

From Equation 2, if $\Delta G\ddagger = 33$ kcal/mol and $\Delta G^o = 22$ kcal/mol (the values for the nonenzymatic addition of hydroxide ion to a phosphodiester anion), $\Delta G\ddagger_{int} = 21$ kcal/mol. Thus, the rates of the nonenzymatic reactions are slow not only because the phosphorane intermediate that is formed is unstable, but also because the addition of a nucleophile to a phosphodiester to form an anionic phosphorane is intrinsically a very slow reaction.

Why is the addition of a nucleophile to a phosphodiester intrinsically slow? The most reasonable explanation is that as the nucleophile attacks the phosphorus atom, increased negative charge is localized on the nonesterified phosphoryl oxygen atoms, i.e., the charge increases from 0.5 per oxygen atom in a monoanionic phosphodiester to 1.0 per oxygen atom in a dianionic phosphorane. In aqueous solution, the developing negative charge on the nonesterified oxygen atoms is thought to require the reorientation of solvent molecules and their dipoles so that the charge can be optimally stabilized and the dianionic phosphorane intermediate can be formed (see, e.g., Bernasconi 1992). Reorientation of solvent molecules involves a decrease in the entropy of the solvent, and this contributes entropically to the activation energy barrier for formation of the phosphorane intermediate, i.e., it is responsible for the large value of $\Delta G\ddagger_{int}$ that characterizes the nonenzymatic reactions.

An analysis using Marcus formalism thus allows the $\Delta G\ddagger$s to be partitioned into two energetic components. Since the origins of both components can be understood (the instability of the phosphorane intermediate as measured by ΔG^o and the necessity for reorganization of solvent to stabilize the anionic charge as measured by $\Delta G\ddagger_{int}$), it is possible to provide more precisely an explanation for the large rate accelerations that characterize the enzymatic reactions.

The available physical organic chemistry on nucleophilic addition reactions to phosphodiesters indicates that the contribution of $\Delta G\ddagger_{int}$ to $\Delta G\ddagger$ can be significantly reduced or even eliminated by the association of electrophiles (protons and/or divalent metal ions) with *both* nonesterified phosphoryl oxygens of the substrate during the attack of the nucleophile. A similar conclusion is reached in mechanistically related reactions in which nucleophiles add to carbonyl groups to form a tetrahedral intermediate or a proton is abstracted from a C–H bond adjacent to a carbonyl group to form an enol(ate). Both of the latter reactions also involve localization of negative charge on the carbonyl oxygen as the nucleophile attacks or the proton is abstracted. The contribution of $\Delta G\ddagger_{int}$ to $\Delta G\ddagger$ can be significantly reduced or even eliminated by the association of an electrophile with the carbonyl group so that the developing negative charge can be stabilized without the requirement for entropically demanding reorganization of solvent molecules (Gerlt and Gassman 1993). Thus, it is not surprising that the active sites of nucleases contain electrophiles (protons donors and/or divalent metal ions) that can interact with both nonesterified phosphoryl oxygen atoms of the substrate phosphodiester.

A reduction in $\Delta G\ddagger_{int}$ alone is not, however, sufficient to explain the rates of the enzyme-catalyzed reactions. If we assume that $k_{cat} = 10^3$ sec^{-1} for a typical nuclease-catalyzed reaction ($\Delta G\ddagger$ = 14 kcal/mol), the value of ΔG^o that characterizes the formation of the (dianionic) phosphorane intermediate in nonenzymatic reactions (22 kcal/mol) must also be reduced. Significant reduction in the value of ΔG^o can be accomplished by the binding of the monoanionic phosphodiester substrate adjacent to one or more electrophiles. If the electrophile is a divalent metal ion, a reduction in ΔG^o can be accomplished by the increased affinity of the nonesterified phosphoryl oxygens for the metal ion as the charge on the oxygen increases (increased electrostatic interactions). If the electrophile is a general acid catalyst, the strength of the hydrogen bond between the nonesterified phosphoryl oxygen and the acidic catalyst may be significantly increased. In particular, unusually strong hydrogen bonds can be formed if a proton is shared by the conjugate bases of two acids that have nearly identical pK$_a$ values (Gerlt and Gassman 1993). Since the first and second pK$_a$ values of a neutral phosphorane are estimated to be 7.2 and 12.0 (Guthrie 1977), the electrophilic catalysts that are known to be present in the active sites of nucleases from X-ray crystallography (see above) may be able to stabilize significantly the phosphorane intermediate in the active site.

The same electrophiles that reduce $\Delta G\ddagger_{int}$ also reduce ΔG^o. Thus, the large rate accelerations that characterize nuclease-catalyzed reactions can be explained *totally* by active site geometries that position electrophiles in precise, fixed geometries relative to the substrate phosphodiesters. In retrospect, this explanation for the extraordinary rate accelerations of the enzymatic reactions relative to the nonenzymatic reactions is simple. However, until the enzymatic reactions were analyzed in terms of Marcus formalism (see Gerlt and Gassman 1993), the explanation was elusive.

VII. ELIMINATIVE (C-O) CLEAVAGE OF PHOSPHODIESTER BONDS: β-ELIMINATION MECHANISMS

In the past several years, a number of "endonucleases" that cleave phosphodiester bonds in damaged DNA molecules have been shown not to catalyze the hydrolytic cleavage of P–O bonds. These reactions are catalyzed by DNA-repair enzymes that accomplish DNA strand cleavage on the 3′ side of aldehydic abasic sites formed by hydrolysis of the N-glycosylic bond to the damaged DNA base. The damaged bases that are recognized and excised by these enzymes include cyclobutane pyrimidine dimers, oxidized pyrimidines, and oxidized purines (see Chapter 9, this volume).

The alternative pathway for cleavage of the phosphodiester bond 3′ to an aldehydic abasic site is C–O bond cleavage via a β-elimination reaction. For this reason, the strand cleavage is accomplished without the involvement of a water molecule. The presence of an aldehydic carbonyl group immediately adjacent to a C–H bond renders the C–H bond considerably more acidic than if the C–H bond were in an alkane. In fact, the reduction in pK_a by the carbonyl group is approximately 30 pK_a units (from ~50 in alkanes to ~20 in aldehydes). Thus, abstraction of the proton adjacent to the carbonyl group (a 2′ proton; henceforth referred to as the α-proton of the carbon acid) is facilitated. If elimination of the adjacent phosphodiester group (the 3′-phosphodiester bond; the β-substituent of the carbonyl group) then occurred, and electrons can be "pushed" to eliminate the β-substituent, the C–O bond would be cleaved, the DNA strand would be broken, and the aldehydic abasic site (the baseless deoxyribose residue) would be converted to an α,β-unsaturated aldehyde.

In principle, as described in the next sections, this alternative pathway for phosphodiester bond cleavage is followed by "endonucleases" in DNA repair. However, the chemical principles described in the previous paragraph are not necessarily sufficient to explain rapid rates of C–O bond cleavage.

VIII. THE β-ELIMINATION REACTIONS CATALYZED BY ABASIC SITE DNA-REPAIR "ENDONUCLEASES"

The DNA-repair enzymes that are known to catalyze 3′-strand cleavage at aldehydic abasic sites, endonuclease III from *E. coli* (Endo III), UV endonuclease V from bacteriophage T4 (Endo V), UV endonuclease from *Micrococcus luteus*, pyrimidine dimer endonuclease from *Saccharomyces cerevisiae*, and FPG protein from *E. coli* (see Chapter 9, this volume), catalyze at least two reactions: hydrolysis of a damaged base from a DNA duplex to produce an aldehydic abasic site (an *N*-glycosylase activity) and cleavage of the DNA strand on the 3′ side of the aldehydic abasic site (an "endonuclease" activity) to generate 5′-phosphate ends. In the case of the FPG protein and Endo V, a third activity is present in which the deoxyribose residue of the aldehydic abasic site is completely removed from the damaged DNA, thereby also generating 3′-phosphate ends.

Verly and his co-workers first provided experimental evidence that strand cleavage 3′ to aldehydic abasic sites by these enzymes did not proceed via a hydrolytic mechanism (Bailly and Verly 1987). In their studies of the strand cleavage reaction catalyzed by Endo III, Verly and co-workers examined the fate of tritium in the 2′-position of aldehydic abasic sites. Verly noted that tritium was released from the 2′-position of the labeled DNA, suggesting that a proton was abstracted from this position during the course of the reaction. The most reasonable explanation for the release of tritium from this position was hypothesized to be a β-elimination reaction, i.e., abstraction of the α-proton followed by elimination of the β-phosphodiester group that results in 3′-strand cleavage.

Release of tritium from the 2′-position of the isotopically labeled abasic site was also observed during the "endonuclease" reactions catalyzed by Endo V and the FPG protein. In addition, for the reactions catalyzed by these two enzymes, at least some "δ-elimination" of the 5′-phosphate group of the aldehydic abasic site was also observed.

The strand cleavage reactions catalyzed by Endo III, Endo V, and the FPG protein have also been investigated in the author's laboratory. However, these studies focused on establishing the chemical details of the reaction by rigorous mechanistic examination rather than circumstantial hypotheses based on only one aspect of the mechanism, i.e., apparent abstraction of the 2′ protons from aldehydic abasic sites. These studies have involved the use of ^{13}C-labeled aldehydic substrates and ^{13}C-labeled NMR spectroscopy to determine the structure of the reaction product derived from the labeled abasic site as well as aldehydic abasic sites labeled stereospecifically with ^{3}H in either the 2′-pro*R* or 2′-pro*S* positions (Mazumder et al. 1990).

In each of the enzyme-catalyzed reactions, the 2′-pro*S* proton of the abasic site was abstracted to produce a *trans*-α,β-unsaturated aldehyde. This information is sufficient to conclude that the substrate for the reaction *must* be an acyclic species, either the free aldehyde or an iminium ion derived from the aldehyde by reaction with an active site amino group. In addition, the stereochemical course of each β-elimination reaction was *syn*, i.e., the α-proton that is abstracted and the β-substituent that is eliminated lie on the same side of the σ-bond between the 2' and 3' carbons of the aldehydic abasic site. Interestingly, in two nonenzymatic model reactions, hydroxide ion- and amine-mediated cleavage reactions, the 2'-pro*R* proton is abstracted to produce a *trans*-α,β-unsaturated aldehyde (Mazumder et al. 1990). This information defines the stereochemical course of the reaction as *anti*; the α-proton that is abstracted and the β-substituent that is eliminated lie on opposite sides of the σ-bond between the 2' and 3' carbons of the aldehydic abasic site.

A. The Active Sites of Abasic Site DNA-repair "Endonucleases"

1. Endonuclease V

Since the mechanistic experiments described in the previous section were published, X-ray structures have been reported for Endo V by Morikawa and his co-workers (Morikawa et al. 1992; see Chapter 9, this volume) and for Endo III from *E. coli* by Tainer and his co-workers (Kuo et al. 1992; see Chapter 9, this volume).

The high-resolution structure reported for Endo V was determined in the absence of any bound substrate, product, or inhibitors. As such, the identification of the active site region is based on circumstantial evidence. The only surface of the "comma-shaped molecule" that is positively charged and is therefore likely to interact with polyanionic DNA is the internal curved surface. On this surface, ten cationic groups are present: Arg-3, Arg-22, Arg-26, Arg-32, Lys-35, Lys-39, Lys-86, Arg-117, Lys-121, and the amino-terminal α-amino group of Thr-2 (the amino-terminal methionine is posttranslationally cleaved). In the midst of this "sea of positive charges" is a lone carboxylate group, Glu-23. That this area does constitute the active site is supported by the results of recent site-directed mutagenesis experiments:

1. The E23Q mutation inactivates both the N-glycosylase reaction and the β-elimination reaction; however, the E23D mutation affects only the N-glycosylase activity (Doi et al. 1992).
2. The R3K, R3Q, R22K, R26K, R26Q, R117Q, and K121Q mutations all impair substrate binding and/or inactivate the catalytic activities (Dowd and Lloyd 1989a,b, 1990; Doi et al. 1992).
3. The T2S and T2V mutations both have N-glycosylase and β-elimination activities. In contrast, the T2P mutation and a mutation in which a glycine residue is inserted between Thr-2 and Arg-3 have negligible activities in both reactions (Schrock and Lloyd 1993).

On the basis of these results, Glu-23 could be an essential general base catalyst in the β-elimination reaction, and the amino-terminal α-amino group of Thr-2 may participate in both the N-glycosylase reaction and the β-elimination reaction by formation of an imine between the aldehyde functional group of the abasic site and the protein molecule (Schrock and Lloyd 1991).

2. Endonuclease III from E. coli

Endo III is an unusual DNA-repair endonuclease in that it contains a [4Fe-4S] center that has no obvious role in catalysis (the reactions catalyzed by Endo III are not redox reactions). The high-resolution structure of Endo III was determined in the presence of thymine glycol, an inhibitor for the N-glycosylase activity. The thymine glycol-binding site was located adjacent to a β-hairpin that has a sequence homologous to MutY, an enzyme that excises adenine in the correction of A-G, A-C, and A-8oxoG mismatched base pairs without β-elimination of the 3′-phosphate group from the aldehydic abasic site. Thus, this site was proposed to be the active site for the N-glycosylase and β-elimination reactions.

The thymine glycol-binding site is also within 3.5 Å of the side chains of Glu-112 and Lys-120. Just as in the case of Endo V, Glu-112 is surrounded by cationic groups (Arg-108 and Arg-119). Thus, even though Endo V and Endo III have no detectable sequence and structural homology, a carboxylate group (Glu-23 in Endo V and Glu-112 in Endo III) and a primary amino group (the amino-terminal α-amino group of Thr-2 in Endo V and the ε-amino group of Lys-120 in Endo III) have been implicated in catalysis.

B. Proposed Mechanisms for the Reactions Catalyzed by Abasic Site DNA-repair Endonucleases

Is it possible to propose mechanisms for both the N-glycosylase and β-elimination reactions catalyzed by Endo V and Endo III and, by analogy, the other enzymes that catalyze DNA strand cleavage by a β-elimination reaction? On the basis of the structural evidence now available, the mechanism proposed (Fig. 9) should be viewed as preliminary until additional experiments are performed.

The chemical precedents for the N-glycosylase reaction suggest that this reaction is likely to proceed via the intermediate formation of an N-acyl iminium ion intermediate (Cadet and Teoule 1974; Kelley et al. 1986). This intermediate would be formed in an acid-catalyzed reaction

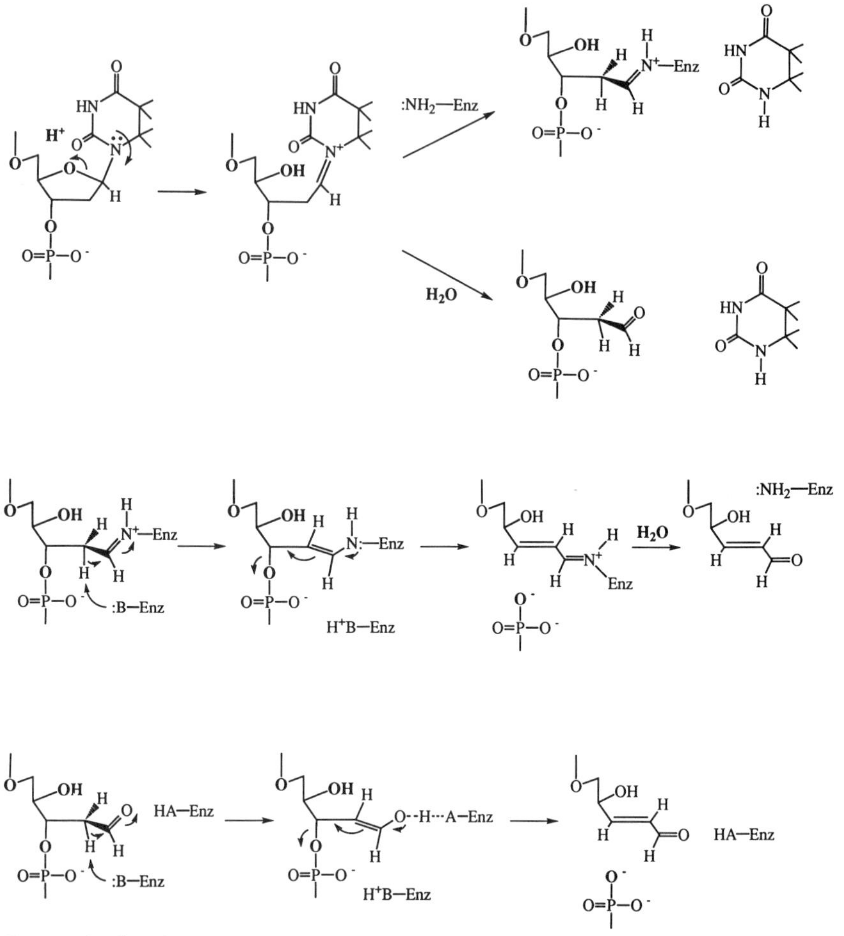

Figure 9 (*Top*) Proposed mechanism for N-glycosyl bond cleavage reactions catalyzed by Endo V, Endo III, and FPG protein. The reaction is shown for a ring-damaged pyrimidine that is a substrate for Endo III. (*Bottom*) Proposed mechanism for the β-elimination reaction catalyzed by Endo V, Endo III, and FPG protein.

that involves the opening of the furanoside ring of the deoxyribose moiety at the site of the damaged heterocyclic base. The acid catalyst in this reaction may be the active site carboxylate group (Glu-23 in Endo V and Glu-112 in Endo III); the pK_a values of these carboxylate groups are presently unknown.

Such an iminium ion intermediate has been implicated in the hydrolyses of the N-glycosylic bonds of various ring-saturated pyrimidine nucleosides. With these compounds, hydrolysis of the N-glycosylic bond is accompanied by epimerization at the 1′ carbon (interconversion of the α- and β-anomers of the furanose-ring-containing nucleoside) and ring expansion to form a mixture of the α- and β-anomers of pyranose-ring-containing nucleosides. That this mechanism can occur in the case of pyrimidines in which the 5,6-double bond has been saturated or purines in which the imidazole ring has been damaged (either by hydrolysis or by oxidation) is explained by the increased electron density of the nitrogen atom of the glycosylic bond. The damaged base substrates for the N-glycosylase activities catalyzed by Endo V, Endo III, and all of these other enzymes have increased electron density on the nitrogen atom of the N-glycosylic bond relative to that of undamaged purines and pyrimidines.

Given the presence of amino groups in the presumed active sites of both Endo V (amino-terminal α-amino group of Thr-2) and Endo III (Lys-120), it is reasonable to propose that the N-acyl iminium ion intermediate will react with the amino group and undergo a transimination reaction. In this reaction, the free aldehydic abasic site is never formed, since the product is the imine/iminium ion between the carbonyl group of the aldehydic abasic site and the amino group of the protein. In addition, such a reaction presumes that the pK_a of the amino group is lowered sufficiently that it can act as a nucleophile at neutral pH. The amino groups that have been implicated by the X-ray analyses are located near other cationic residues and provide an electrostatic explanation for the reduction in pK_a of the amino group that would be required for the transimination reaction.

An iminium ion has been long regarded by mechanistic enzymologists as a way in which abstraction of the α-proton (the 2′ proton of the abasic site) can be facilitated. Thus, on the basis of a large number of enzymological precedents (including acetoacetate decarboxylase, aldolase, and porphobilinogen synthase), it is reasonable to propose that an active site general base catalyst could abstract the α-proton from the iminium ion to generate a neutral enamine intermediate. In a subsequent vinylogous elimination reaction that is initiated by "pushing" electrons from the neutral nitrogen atom of the enamine, the phosphodiester leaving group can be eliminated from the 3′ carbon of the abasic site, thereby generating the iminium ion of the *trans*-α,β-unsaturated aldehyde generated by the elimination reaction. Hydrolysis of this iminium ion will produce the observed product and regenerate the free enzyme.

An analogous, but distinct, mechanism for these reactions would utilize a water molecule rather than an amino group as the nucleophile on the N-acyl iminium ion intermediate. Decomposition of the aminol adduct would generate a free aldehydic abasic site. Since, as previously noted, a carbonyl group is expected to reduce the pK_a of the 2'proton, an active site general base catalyst could participate in abstraction of the 2' proton. It has been noted elsewhere that the rapid abstraction of the 2' proton of the aldehyde would require the assistance of a general acid catalyst so that a neutral enol would be generated as an intermediate (Gerlt and Gassman 1992). Subsequent vinylogous elimination of the 3' phosphodiester initiated by pushing electrons from the oxygen of the enol OH groups could then occur, directly generating the *trans*-α,β-unsaturated aldehyde product.

IX. AN EXPLANATION FOR THE RAPID RATES OF β-ELIMINATION C-O BOND CLEAVAGE REACTIONS

Irrespective of whether the β-elimination reactions catalyzed by abasic site DNA-repair "endonucleases" occur via an iminium ion intermediate or a free aldehyde, it is important to point out that rapid abstraction of the 2'proton (α-proton of the aldehyde carbonyl or iminium group) requires further explanation. For example, if an active site carboxylate group is the general base catalyst that is participating in the β-elimination reactions, it is necessary to consider whether rapid proton abstraction is possible. As noted previously, the pK_a of the α-proton of an aldehyde is approximately 20; if a carboxylate group is the general base catalyst, its pK_a is likely approximately 5. A number of years ago, this large difference in pK_a values was suggested to require that such reactions must necessarily be concerted, since from thermodynamics, the basic catalyst cannot produce enough of the conjugate base of the carbon acid substrate to support the observed rates of the reactions (Thibblin and Jencks 1979). Similarly, although iminium groups have frequently been described by mechanistic enzymologists as "electron sinks," the exact quantitative meaning of this term has been elusive. In fact, the pK_a values of the α-protons of iminium ions are not much less than those of aldehydes, although the available data on this point are limited (Stivers and Washabaugh 1991).

The rates and mechanisms of β-elimination reactions can also be analyzed in terms of Marcus formalism (Gerlt and Gassman 1992, 1993). Just as the nonenzymatic reactions of phosphodiesters were previously noted in this chapter to be described by values of ΔG^o and $\Delta G^{\ddagger}_{int}$ that are too large to explain the observed rates of the enzyme-catalyzed reactions, the nonenzymatic abstraction of the α-protons of aldehydes are also described by values of ΔG^o and $\Delta G^{\ddagger}_{int}$ that are too large to explain the observed rates of enzyme-catalyzed β-elimination reactions (even though the pK_a values of the α-protons are reduced to about 20 by the carbonyl group). Thus, it is also necessary for enzymes to reduce the values of ΔG^o and/or $\Delta G\ddagger_{int}$ in the active sites of enzymes catalyzing β-elimination reactions.

As described above in the discussion of enzyme-catalyzed hydrolyses of phosphodiester bonds, $\Delta G\ddagger_{int}$ can be substantially reduced by the association of an electrophile (e.g., a proton) with the carbonyl group so that a neutral enol can be produced as the reaction intermediate. The interaction of this electrophile with the carbonyl group is also expected to reduce ΔG^o for these reactions, since an unusually strong hydrogen bond can be formed if a proton is shared by the conjugate bases of two acids that have nearly identical pK_a values. In the active sites of a number of enzymes that catalyze abstraction of the α-proton of a carbon acid, the pK_a values of the active site general acid catalyst and the enol OH group are matched, so a strong hydrogen bond can be formed and ΔG^o can be reduced (Gerlt and Gassman 1993).

The kinetic advantage of iminium ions as electron sinks in β-elimination reactions can be attributed to the observation that $\Delta G\ddagger_{int}$ for abstraction of the α-protons of iminium ions is much less than that for abstraction of the α-proton of an aldehyde (≤3 kcal/mol vs. 12 kcal/mol). Thus, whereas an iminium ion does not necessarily alter the thermodynamics for abstraction of the α-proton (i.e., reduce ΔG^o), it does substantially reduce $\Delta G\ddagger_{int}$ for the proton abstraction reaction. Thus, the involvement of the iminium ion does provide a kinetic advantage to the enzyme-catalyzed reaction, although the advantage appears to be less than that which would be available if a general acid catalyst were present to transfer a proton to the carbonyl group of the aldehydic form of the abasic site as a general acid catalyst abstracts the α-proton.

X. SUMMARY

Enzyme-catalyzed cleavage of phosphodiester bonds can occur via P–O or C–O bond cleavage reactions. The rates and mechanisms of both classes of reactions can be understood in terms of the structures of the active sites and the known principles of physical organic chemistry.

ACKNOWLEDGMENT

The author gratefully acknowledges the stimulating collaboration with the late Professor Paul Gassman, Department of Chemistry, University of Minnesota, in using Marcus formalism to understand the rates of enzyme-catalyzed reactions.

REFERENCES

Anslyn, E. and R. Breslow. 1989. On the mechanism of catalysis by ribonuclease: Cleavage and isomerization of the dinucleotide UpU catalyzed by imidazole buffers. *J. Am. Chem. Soc.* **111:** 4473-4482.

Arni, R., U. Heinemann, R. Tokuoka, and W. Saenger. 1988. Three-dimensional structure of the ribonuclease T_1*2'-GMP complex at 1.9 Å resolution. *J. Biol. Chem.* **263:** 15358-15368.

Bailly, V. and W. Verly. 1987. *Escherichia coli* endonuclease III is not an endonuclease but a β-elimination catalyst. *Biochem. J.* **242:** 565-572.

Beese, L.J. and T.A. Steitz. 1991. Structural basis for the 3'-5' exonuclease activity of *Escherichia coli* DNA polymerase I: A two metal ion mechanism. *EMBO J.* **10:** 25-33.

Bernasconi, C.F. 1992. The principle of non-perfect synchronization. *Adv. Phys. Org. Chem.* **27:** 119-238.

Borah, B., C.-W. Chen, W. Egan, M. Miller, A. Wlodawer, and J.S. Cohen. 1985. Nuclear magnetic resonance and neutron diffraction studies of the complex of ribonuclease A with uridine vanadate, a transition-state analogue. *Biochemistry* **24:** 2058-2067.

Cadet, J. and R. Teoule. 1974. Nucleic acid hydrolysis. I. Isomerization and anomerization of pyrimidinic deoxyribonucleosides in an acidic medium. *J. Am. Chem. Soc.* **96:** 6517-6519.

Campbell, R.L. and G.A. Petsko. 1987. Ribonuclease structure and catalysis: Crystal structure of sulfate-free native ribonuclease A at 1.5-Å resolution. *Biochemistry* **26:** 8579-8584.

Chin, J., F. Banaszczyk, V. Jubian, and X. Zou. 1989. Co(III) complex promoted hydrolysis of phosphate diesters: Comparison in reactivity of rigid cis-diaquotetraazacobalt(III) complexes. *J. Am. Chem. Soc.* **111:** 186-190.

Cohen, A.O. and R.A. Marcus. 1968. On the slope of free energy plots in chemical kinetics. *J. Phys. Chem.* **72:** 4249-4256.

Coleman, J.E. and P. Gettins. 1983. Alkaline phosphatase, solution structure and mechanism. *Adv. Enzymol. Relat. Areas Mol. Biol.* **55:** 381-452.

Cotton, F.A., E.E. Hazen, and M.J. Legg. 1979. Staphylococcal nuclease: Proposed mechanism of action based on structure of enzyme-thymidine 3',5'-bisphosphate-calcium ion complex at 1.5 Å resolution. *Proc. Nat. Acad. Sci.* **76:** 2551-2555.

Doi, T., A. Tecktenwald, Y. Karaki, M. Kikuchi, K. Morikawa, M. Ikehara, T. Inaoka, N. Hori, and E. Ohtsuka. 1992. Role of the basic amino acid cluster and Glu-23 in pyrimidine dimer glycosylase active of T4 endonuclease V. *Proc. Nat. Acad. Sci.* **89:** 9420-9424.

Dowd, R. and R.S. Lloyd. 1989a. Site-directed mutagenesis of the T4 endonuclease V gene: The role of arginine-3 in the target search. *Biochemistry* **28:** 8699-8705.

―――. 1989b. Biological consequences of a reduction in the non-target DNA scanning capacity of a DNA repair enzyme. *J. Mol. Biol.* **208:** 701-707.

―――. 1990. Biological significance of facilitated diffusion in protein-DNA interactions. Applications to T4 endonuclease V-initiated DNA repair. *J. Biol. Chem.* **265:** 3424-3431.

Eckstein, F. and H. Gindl. 1968. Uridin-2',3'-O,O-cyclothiophosphat. *Chem. Ber.* **101:** 1670-1673.

Frey, P.A. 1989. Chiral phosphorothioates: Stereochemical analysis of enzymatic substitution at phosphorus. *Adv. Enzymol. Relat. Areas Mol. Biol.* **62:** 119-201.

Gerlt, J.A. 1992. Phosphate ester hydrolysis. *Enzymes* **20:** 95-139.

Gerlt, J.A. and P.G. Gassman. 1992. Understanding enzyme-catalyzed proton abstraction from carbon acids: Details of stepwise mechanisms for β-elimination reactions. *J. Am. Chem. Soc.* **114:** 5928–5934.

———. 1993. An explanation for rapid enzyme-catalyzed proton abstraction from carbon acids: The importance of late transition states in concerted mechanisms. *J. Am. Chem. Soc.* **115:** (in press).

Grunwald, E. 1990. Reaction coordinates and structure-energy relationships. *Prog. Phys. Org. Chem.* **17:** 55–105.

Guthrie, J.P. 1977. Hydration and dehydration of phosphoric acid derivatives: Free energies of formation of the pentacoordinate intermediates for phosphate ester hydrolysis and of monomeric metaphosphate. *J. Am. Chem. Soc.* **99:** 3991–4001.

Hale, S.P., L.B. Poole, and J.A. Gerlt. 1993. Mechanism of the reaction catalyzed by staphylococcal nuclease: Identification of the rate-determining step. *Biochemistry* **32:** 7479–7487.

Hall, C.R. and T.D. Inch. 1980. Phosphorus stereochemistry. Mechanistic implications of the observed stereochemistry of bond forming and breaking processes at phosphorus in some 5- and 6-membered cyclic phosphorus esters. *Tetrahedron* **36:** 2059–2095.

Hibler, D W., N J. Stolowich, M.A. Reynolds, J.A. Gerlt, J.A. Wilde, and P.H. Bolton. 1987. Site-directed mutants of staphylococcal nuclease. Detection and localization by ^{1}H NMR spectroscopy of conformational changes accompanying substitutions for Glu 43. *Biochemistry* **26:** 6278–6286.

Jencks, W.P. 1981. How does a reaction choose its mechanism? *Chem. Soc. Rev.* **10:** 345–375.

Jones, S.R., L.A. Kindman, and J.R. Knowles. 1978. Stereochemistry of phosphoryl group transfer using a chiral [^{16}O,^{17}O,^{18}O] stereochemical course of alkaline phosphatase. *Nature* **275:** 564–565.

Kelley, J.A., J.S. Driscoll, J.J. McCormack, J.S. Roth, and V.E. Marquez. 1986. Furanose-pyranose isomerization of reduced pyrimidine and cyclic urea ribosides. *J. Med. Chem.* **29:** 2351–2358.

Kim, Y., J.C. Grable, R. Love, P.J. Greene, and J.M. Rosenberg. 1990. Refinement of *Eco*RI endonuclease crystal structure: A revised protein chain tracing. *Science* **249:** 1307–1309.

Kuo, C.-F., D.E. McRee, C.L. Fisher, S.F. O'Handley, R.P. Cunningham, and J.A. Tainer. 1992. Atomic structure of the DNA repair [4Fe-4S] enzyme endonuclease III. *Science* **258:** 434–440.

Lahm, A. and D. Suck. 1991. DNase I-induced DNA conformation. 2 Å structure of a DNase I-octamer complex. *J. Mol. Biol.* **221:** 645–667.

Loll, P. and E.E. Lattman. 1989. The crystal structure of the ternary complex of staphylococcal nuclease, Ca^{2+}, and the inhibitor pdTp, refined at 1.65 Å. *Proteins Struct. Funct. Genet.* **5:** 183–201.

———. 1990. Active site mutant Glu-43→Asp in staphylococcal nuclease displays non-local structural changes. *Biochemistry* **29:** 6866–6873.

Mauguen, Y., R.W. Hartley, D.J. Dodson, G.G. Dodson, G. Bricogne, C. Chothia, and A. Jack. 1982. Molecular structure of a new family of ribonucleases. *Nature* **297:** 162–164.

Mazumder, A., J.A. Gerlt, M.J. Absalon, J. Stubbe, J. Withka, and P.H. Bolton. 1990. UV endonuclease V from bacteriophage T4 catalyzes DNA strand cleavage at aldehydic abasic sites by a syn β-elimination mechanism. *Biochemistry* **29:** 1119–1126.

Mehdi, S. and J.A. Gerlt. 1981a. Stereochemical course of the hydrolysis of thymidine 5′-(4-nitrophenyl [^{17}O,^{18}O]phosphate) in H$_2$^{16}O catalyzed by the phosphodiesterase from snake venom. *J. Biol. Chem.* **256:** 12164–12166.

———. 1981b. Oxygen chiral phosphodiesters. 5. Stereochemical course of the hydrolysis of thymidine 3′-(4-nitrophenyl [$^{17}O,^{18}O$]phosphate) in $H_2^{16}O$ catalyzed by the exonuclease from bovine spleen. *J. Am. Chem. Soc.* **103:** 7018–7020.

Morikawa, K., O. Matsumoto, M. Tsujimoto, K. Katayangi, M. Ariyoshi, T. Doi, M. Ikehara, T. Inaoka, and E. Ohtsuka. 1992. X-ray structure of T4 endonuclease V: An excision repair enzyme specific for a pyrimidine dimer. *Science* **256:** 523–526.

Pourmotabbed, T., M. Dell'Acqua, J.A. Gerlt, S.M. Stanczyk, and P.H. Bolton. 1990. Kinetic and conformational effects of lysine substitutions for arginines 35 and 87 in the active site of staphylococcal nuclease. *Biochemistry* **29:** 3677–3683.

Schrock, R.D. and R.S. Lloyd. 1991. Reductive methylation of the amino terminus of endonuclease V eradicates catalytic activities. Evidence for an essential role of the amino terminus in the chemical mechanisms for catalysis. *J. Biol. Chem.* **266:** 17631–17639.

———. 1993. Site-direct mutagenesis of the NH2 terminus of T4 endonuclease V. The position of the αNH2 group affects catalytic activity. *J. Biol. Chem.* **258:** 880–886.

Serpersu, E.H., D. Shortle, and A.S. Mildvan. 1987. Kinetic and magnetic resonance studies of active-site mutants of staphylococcal nuclease: Factors contributing to catalysis. *Biochemistry* **26:** 1289–1300.

Steyaert, J., K. Hallenga, L. Wyns, and P. Stanssens. 1990. Histidine-40 of ribonuclease T_1 acts as base catalyst when the true catalytic base, glutamic acid-58, is replaced by alanine. *Biochemistry* **29:** 9064–9072.

Stivers, J.T. and M.W. Washabaugh. 1991. Normal acid behavior for C(α)-proton transfer from a thiazolium ion. *Bioorg. Chem.* **19:** 369–383.

Thibblin, A. and W.P. Jencks. 1979. Unstable carbanions. General acid catalysis of the cleavage of 1-phenylcyclopropanol and 1-phenyl-2-arylcyclopropanol anions. *J. Am. Chem. Soc.* **101:** 4963–4973.

Usher, D.A., D.I. Richardson, and F. Eckstein. 1970. Absolute stereochemistry of the second step of ribonuclease action. *Nature* **228:** 663–665.

Volbeda, A., A. Lahm, F. Sakiyama, and D. Suck. 1991. Crystal structure of *Penicillium citrinum* P1 nuclease at 2.8 Å resolution. *EMBO J.* **10:** 1607–1618.

Westheimer, F.H. 1968. Pseudorotation in the hydrolysis of phosphate esters. *Accts. Chem. Res.* **1:** 70–78.

———. 1987. Why nature chose phosphates. *Science* **235:** 1173–1178.

Wilde, J.A., P.H. Bolton, M. Dell'Acqua, D.W. Hibler, T. Pourmotabbed, and J.A. Gerlt. 1988. Identification of residues involved in a conformational change accompanying substitutions for glutamate 43 in staphylococcal nuclease. *Biochemistry* **27:** 4127–4132.

Winkler, F.K., D.W. Banner, C. Oefner, D. Tsernoglou, R.S. Brown, S.P. Heathman, R.K. Bryan, P.D. Martin, K. Petratos, and K.S. Wilson. 1993. The crystal structure of *Eco*RV endonuclease and of its complexes with cognate and non-cognate DNA fragments. *EMBO J.* **12:** 1781–1795.

Wyckoff, H.W., M. Handschumacher, H.M. Krishna Murthy, and J.M. Sowadski. 1983. The three dimensional structure of alkaline phosphatase from *E. coli*. *Adv. Enzymol. Relat. Areas Mol. Biol.* **55:** 453–480.

Yang, W., W.A. Hendrickson, R.J. Crouch, and Y. Satow. 1990. Structure of ribonuclease H phased at 2 Å resolution by MAD analysis of the selenomethionyl protein. *Science* **249:** 1398–1405.

2
Type II Restriction Endonucleases

Richard J. Roberts
New England Biolabs
Beverly, Massachusetts 01915

Stephen E. Halford
Department of Biochemistry, University of Bristol
Bristol, BS8 1TD, United Kingdom

I. **Introduction and History**
II. **Recognition Sequences and Cleavage Properties**
    A. Type IIs Enzymes
    B. Degenerate Recognition Sequences
    C. Unusual Type II Enzymes
    D. Determination of Cleavage Sites
    E. Effects of Methylation
    F. Single-stranded DNA Cleavage
III. **Genes and Their Organization**
    A. Cloning
    B. Genetic Location
    C. Sequences
IV. **DNA Binding**
    A. Enzymes That Bind Specifically to Their Recognition Sites
    B. Enzymes That Fail to Bind Specifically to Their Recognition Sites
    C. Transfer to Recognition Sites
V. **DNA Cleavage**
    A. Plasmid Substrates
    B. Oligonucleotide Substrates
    C. Specificity
VI. **Crystallography**
    A. Protein Structures
    B. DNA Structures
    C. DNA-Protein Interfaces
VII. **Phosphodiester Hydrolysis**
VIII. **DNA Recognition Functions**
    A. Altered Enzymes
    B. Altered Substrates
    C. Coupling Recognition to Catalysis
IX. **Evolution**
X. **Conclusions and Future Prospects**

I. INTRODUCTION AND HISTORY

The type II restriction endonucleases, commonly referred to as restriction enzymes, are among the most prolific and best known of all the nucleases. They are defined as double-strand nucleases that recognize specific DNA sequences and cleave at a defined point within or close to that sequence. They require Mg^{++} as a cofactor. They are the workhorses of the genetic engineer, and acronyms such as *Eco*RI and *Bam*HI have become essential vocabulary for the molecular biologist. Sadly, those same molecular biologists often know much less about the basic biology, biochemistry, and molecular biology of this fascinating class of endonucleases, which is the subject of this chapter. Modrich and Roberts (1982) have summarized progress in this field up to 1982. Recently, more specialized reviews of restriction-modification systems have focused on the genetics (Wilson and Murray 1991), biochemistry (Bennett and Halford 1989), biology (Heitman 1993), and structural aspects (Anderson 1993).

The first type II restriction endonuclease was called endonuclease R (Kelly and Smith 1970; Smith and Wilcox 1970) and is now known to have consisted of a mixture of *Hin*dII and *Hin*dIII. For this discovery, Hamilton Smith of Johns Hopkins University received the Nobel prize in medicine in 1977. A cosharer in that prize, Werner Arber, had previously postulated the existence of restriction endonucleases (Arber and Dussoix 1962; Dussoix and Arber 1962) as a means of explaining earlier genetic results (Luria and Human 1952; Bertani and Weigle 1953). Soon after Smith's discovery, a number of other restriction endonucleases were found, including endonuclease Z (now called *Hae*III) (Middleton et al. 1972), *Eco*RI (Yoshimori 1971; Hedgpeth et al. 1972), *Hpa*I, and *Hpa*II (Sharp et al. 1973). Beginning in 1973, a concerted effort was undertaken to assemble a collection of the known restriction endonucleases and to screen extensively for new ones. By 1976, almost 50 such endonucleases had been discovered (Roberts 1976), and several companies had emerged whose main product line was devoted to the type II restriction endonucleases. Their availability as off-the-shelf reagents aided the rapid development of recombinant DNA technology. This enabled many of the great discoveries of the 1970s, including the development of DNA sequencing technology and the enormous proliferation in the characterization of genes.

The original phenomenon of restriction-modification observed by Luria and explained by Arber is now known to have been mediated by the type I restriction systems discussed in more detail in Chapter 3 (this volume). The type II systems differ substantially. Most consist of two separate enzymatic activities. One is a restriction endonuclease that requires Mg^{++} as a cofactor and is able to cleave DNA at a specific recognition sequence; the second is a DNA methyltransferase, which is able to methylate that same sequence and render it insensitive to cleavage by the restriction endonuclease. In some systems, two methyltransferases are found (*Mbo*I, Ueno et al. 1993; *Dpn*I, *Dpn*II, de la Campa et al. 1987; *Alw*26I, *Eco*31I, *Esp*3I, Bitinaite et al. 1992). In a bacterium that contains both a restriction endonuclease and a methyltransferase, the host DNA is protected from the action of the restriction endonuclease by the methyltransferase. However, incoming DNA from a phage or a plasmid is unlikely to carry the appropriate methylation already and so it will be susceptible to cleavage by the restriction endonuclease. If there is one or more recognition sites on the incoming DNA, a competition will ensue between the methyltransferase and the restriction endonuclease because a single cleavage by a restriction endonuclease is likely to incapacitate the incoming DNA. The usual outcome will be the destruction of that DNA. Occasionally, an incoming DNA molecule will be completely methylated before the restriction endonuclease finds it and so some molecules will survive. The phage's capacity to grow in the cell is said to be restricted. Typically, the titer of a phage, as measured on a restricting host versus a nonrestricting host, may vary between one and five orders of magnitude depending on the number of target sites in the DNA and the relative activities of the restriction endonuclease and the methyltransferase in the cell.

An important aspect of restriction in vivo is that hemimethylated DNA, which is present in the host immediately after replication, should be resistant to double-strand cleavage by the restriction endonuclease. Often this is so, but in some systems, one strand may be nicked (Butkus et al. 1987). In these cases, DNA ligase can repair the nick easily and so the relative levels of restriction endonuclease, methyltransferase, and ligase all become important. The ability of ligase to repair nicked substrates can be critical when bacterial cells express high levels of restriction endonuclease and the cognate methyltransferase fails to provide complete protection (Heitman et al. 1989; M.D. Smith et al. 1992). Since hemimethylated DNA is the natural substrate, it might have been expected that it would be the preferred substrate for the methyltransferase. However, in most cases studied, there seems to be little difference between the rates of methylation of hemimethylated DNA versus unmethylated DNA by the type II systems. This is in contrast to the type I systems, where a hemimethylated substrate is greatly preferred over unmethylated DNA (Chapter 3).

Type II restriction endonucleases are widespread in nature. Over the years, many thousands of species of bacteria have been examined for the presence of these endonucleases, and they appear in all genera examined. Much large-scale screening has been carried out, and approximately one in four of all bacteria examined appear to have one or more type II restriction systems present within them. Often multiple systems are found, the current record being six in *Dactylococcopsis salina* (Laue et al. 1991). A complete listing of all known systems is available through REBASE, the restriction enzyme database (Roberts and Macelis 1993) (see Appendix A). Among the strains lacking detectable type II systems, it is likely that type I and/or type III systems are present, but it is much more difficult to assay these systems in vitro and so their presence would not have been detected.

II. RECOGNITION SEQUENCES AND CLEAVAGE PROPERTIES

Restriction enzymes typically recognize short sequences in DNA that vary from four to eight base pairs in length. The range of specificities is illustrated in Table 1, which lists the general patterns exhibited among the known recognition sequences. Some patterns are much more common than others. For instance, enzymes recognizing simple tetranucleotide and hexanucleotide palindromes are quite prevalent, and we are close to having examples of all possible specificities. For many classes, just a few examples are known, suggesting that future screening efforts will still be rewarded. Obviously, now that almost 2400 restriction enzymes are known representing 188 specificities (see Appendix A), it is becoming increasingly difficult to discover new specificities by random screening. For this reason, many laboratories have discontinued their screening programs and only a few systematic efforts remain. In recent years, the rate of discovery has dropped to about eight new specificities per year. This is accompanied by a steady discovery of isoschizomers (enzymes that recognize the same DNA sequence as a previously discovered enzyme). Interestingly, there are now many examples of two restriction enzymes recognizing the same DNA sequence but cleaving at different points within the sequence. These have been termed neoschizomers (Hamablet et al. 1989) and are listed in Table 2.

A. Type IIs Enzymes

Most of the enzymes now known recognize sequences with a dyad axis of symmetry, termed palindromes, and cleave within that sequence. However, there are a substantial number of enzymes that recognize asymmetric sequences and cleave at a short distance from that sequence. These are termed type IIs enzymes (Szybalski et al. 1991) and recognize sequences from the first class of patterns shown in Table 1. Usually, isoschizomers of type IIs enzymes exhibit cleavage properties identical with those of the prototype. However, recently, a type IIs enzyme, *Sts*I, has been found with the same recognition sequence as *Fok*I, but which cleaves differently (Kita et al. 1992). *Fok*I recognizes 5'-GGATG-3' and cleaves 9 nucleotides beyond this sequence on this strand and 13 nucleotides beyond the sequence on the complementary strand (Sugisaki and Kanazawa 1981). By convention, this is written GGATG (9/13). *Sts*I cleaves GGATG (10/14). It should be noted that the type IIs enzymes are usually listed with a specific point of cleavage away from the recognition sequence on both strands as above. In several cases now, some variability has been noted with respect to the exact site of cleavage. For instance, *Hph*I usually recognizes GGTGA (8/7) but has been shown to cleave GGTGA (9/8) in some cases (Kleid et al. 1976; Kang and Wu 1987; Cho and Kang 1990). This may reflect differences in the DNA sequence between the recognition site and the cleavage site and might be more general than has hitherto been reported. Interestingly, a mutant of *Fok*I has recently been isolated that also shows variable cleavage (Li and Chandrasegaran 1993). It cleaves the same DNA site as either GGATG (9/13) or (10/14).

B. Degenerate Recognition Sequences

One interesting aspect of the recognition sequences is that in many cases, the enzymes can accept several alternative nucleotides at a given position, whereas rigorously excluding others. Thus, *Acc*I (GT↓MKAC) can accept an A or a C (M) residue at position 3 but cannot cleave if G or T is present. At position 4, only G or T (K) can be present. However, the overall symmetric sequences either GTATAC or GTCGAC as well as the asymmetric sequence GTCTAC (complement GTAGAC) can all be cleaved. Thus, at both central positions, *Acc*I is recognizing two specific base pairs in a specific orientation. It is not known how this achieved. Similar examples can be found among the other patterns in Table 1.

C. Unusual Type II Enzymes

Among restriction enzymes currently classified as type II are several that have distinct and unusual properties, in some cases resembling the type I and type III enzymes (Chapter 3). One example is the enzyme *Bcg*I (Kong et al. 1993) that requires AdoMet as a cofactor and cleaves DNA symmetrically on both sides of its asymmetric recognition sequence (GCANNNNNNTCG). This results in the excision of a 34-bp fragment that is methylated and has the recognition sequence located centrally. Another example of this same kind of cleavage is probably demonstrated by *Sgr*20I (previously called *Sgr*II; recognition sequence CCWGG) (Orekhov et al. 1982). However, in this case, the exact site of cleavage is

Table 1 General Patterns Recognized by Restriction Endonucleases

General pattern	Specific example	Enzyme	Known	Possible
Asymmetric recognition sequences				
nnnn	CCGC	*Aci*I	2	256
nnnnn	ACGGC	*Bce*fI	16	1024
nnnnnn	ACCTGC	*Bsp*MI	21	4096
nnnnnnn	GCTCTTC	*Sap*I	1	16384
nnNnnn	CCTNAGC	*Bpu*10I	1	4096
nnNNNNNnnn	GCANNNNNTCG	*Bcg*I	1	4096
nnNNNNnnnYn	ACNNNNGTAYC	*Bae*I	1	4096
nnnRnn	CAARCA	*Tth*111II	2	1024
Ynnnnn	YGGCCG	*Gdi*II	1	1024
Symmetric recognition sequences				
Rnn'Y	RGCY	*Cvi*JI	1	4
nnn'n'	AATT	*Tsp*EI	14	16
nnNn'n'	CCNGG	*Scr*FI	6	16
nnNNn'n'	CCNNGG	*Sec*I	3	16
nnMKn'n'	GTMKAC	*Acc*I	1	16
nnSn'n'	CCSGG	*Cau*II	2	16
nnWn'n'	CCWGG	*Eco*RII	3	16
nMnn'Kn'	CMGCKG	*Nsp*BII	1	16
nDnn'Hn'	GDGCHC	*Sdu*I	1	16
nnWWn'n'	CCWWGG	*Sty*I	1	16
nnRYn'n'	ACRYGT	*Afl*III	4	16

nRnn′Yn′	GRCGYC	AcyI	2	16
Rnmm′n′Y	RAATTY	ApoI	5	16
RnnNn′n′Y	RGGNCCY	DraII	1	16
RnnWn′n′Y	RGGWCCY	PpuMI	1	16
nnYRn′n′	GGYRCC	HgiCI	2	16
nYnn′Rn′	CYCGRG	AvaI	1	16
Ynn′n′R	YACGTR	BsaAI	2	16
nnYNNNNRn′n′	CAYNNNNRTG	MslI	1	16
nWm′Wn′	GWGCWC	HgiAI	1	16
Wnm′n′W	WCCGGW	BetI	2	16
nnn′n′	AACGTT	AclI	55	64
nnNn′n′n′	CCTNAGG	SauI	3	64
nnNNn′n′n′	CACNNGTG	DraIII	3	64
nnNNNn′n′n′	GAANNNTTC	XmnI	3	64
nnnNNNNn′n′n′	CCANNNNTGG	PflMI	5	64
nnNNNNNNn′n′	CCNNNNNNGG	BsiYI	2	64
nnnNNNNNNn′n′n′	ACCNNNNNNGGT	HgiEII	3	64
nnnNNNNNNNNn′n′n′	CCANNNNNNNNTGG	XcmI	1	64
nRmn′n′Yn′	CRCCGGYG	SgrAI	1	64
nnnWn′n′n′	ACCWGGT	SexAI	2	64
nnnn′n′n′n′	ATTTAAAT	SwaI	8	256
nnnnNNNNNn′n′n′n′	GGCCNNNNNGGCC	SfiI	1	256

Within the column showing the general patterns the abbreviations are n = any nucleotide, n' = the complement of the nucleotide n. Thus, nnn′n′ signifies the set of 16 symmetric tetranucleotide palindromes AATT, ACGT, AGCT, ATAT, etc. Other abbreviations are the standard ones for designating degenerate nucleotides (*Eur. J. Biochem.* 150: 1–5 [1985]). R = A or G; Y = C or T; M = A or C; K = G or T; S = G or C; W = A or T; H = A or C or T; D = G or A or T; B = G or T or C; V = G or C or A; N = A or C or G or T.

Table 2 Type II Prototypes and Their Neoschizomers

*Bsp*120I *Apa*I G↓G G C C↓C	*Ava*I *Nli*387/7I C↓Y C G R↓G
*Ppu*10I *Bfr*BI *Eco*T22I A↓T G↓C A↓T	*Eco*HI *Cau*II ↓C C ↓S G G
*Dra*II *Pss*I R G↓G N C↓C Y	*Eco*RII *Aor*I ↓C C↓W G G
*Sel*I *Fnu*DII ↓C G↓C G	*Xma*I *Sma*I C↓C C↓G G G
*Lpn*I *Hae*II R G C↓G C↓Y	*Hin*P1I *Hha*I G↓C G↓C
*Asp*718I *Kpn*I G↓G T A C↓C	*Eco*56I *Nae*I G↓C C↓G G C
*Kas*I *Nar*I *Eco*78I *Bbe*I G↓G↓C↓G C↓C	*Cvi*AII *Nla*III C↓A T G↓
*Afa*22MI *Pvu*I C G A↓T↓C G	*Cvi*QI *Rsa*I G↓T↓A C
*Eco*ICRI *Sac*I G A G↓C T↓C	*Mlu*113I *Sac*II C C↓G C↓G G
*Sso*II *Scr*FI ↓C C↓N G G	*Xho*I *Sci*I C↓T C↓G A G
*Fok*I GGATG(9/13) *Sts*I GGATG(10/14)	*Sfa*NI GCATC(5/9) *Bsc*AI GCATC(4/6)

*Mly*I   *Ple*I
5' ↓NNNNNGAGTCNNNN↓ 3'
3' ↓NNNNNCTCAGNNNNN↓ 5'

The different cleavage sites for various neoschizomers are indicated by the arrows. Only one strand is shown when the recognition sequence is symmetric; cleavage is also symmetric. For the type IIs enzymes, *Fok*I, *Sts*I and *Sfa*NI, *Bsc*AI, only one strand is shown, but the cleavage sites on both strands are indicated using the usual nomenclature (see text). For *Mly*I and *Ple*I, both strands are shown since the two enzymes cleave on opposite sides of the recognition sequence as shown.

not known, but its fragmentation pattern has been published and compared with that of its prototype, *Eco*RII. *Sgr*20I fragments are systematically slightly smaller than those produced by *Eco*RII, consistent with the enzyme cleaving on both sides of its recognition sequence. Other enzymes with unusual properties include *Eco*57I (recognition sequence CTGAAG) and *Gsu*I (recognition sequence CTGGAG) (Petrusyte et al. 1987). Both sequences are asymmetric and cleavage occurs at the same distance (16/14) in each case. This is similar to the properties of the type III enzymes (Chapter 3). However, both enzymes have cofactor requirements and properties that distinguish them from typical type III enzymes. It has been suggested that these should be classified as type IV restriction enzymes (Janulaitis et al. 1992), although this has not yet been widely accepted.

D. Determination of Cleavage Sites

The methods used to characterize type II restriction enzyme recognition and cleavage sites have evolved considerably since the initial early and tedious methods were developed (for review, see Roberts 1976). Computer programs have been described (Gingeras et al. 1978; Tolstoshev and Blakesley 1982; Boyd et al. 1986) that will analyze fragment lengths produced by restriction enzymes on sequenced DNA. They can prove very useful in quickly identifying candidate sequences. The predictions can then be tested experimentally either by mapping experiments or, more usually, by using primed synthesis reactions and analyzing the products on a sequencing gel (Brown and Smith 1980). The basic technique uses a primer close to the site, which is extended with DNA polymerase on a suitable template. The product is cleaved with the restriction enzyme and the sample is divided into two. One sample is further treated with the Klenow DNA polymerase in the presence of all four deoxynucleoside triphosphates. If the initial cleavage was blunt, no further changes take place and the treated and untreated reactions will produce fragments of identical size. If the enzyme gives a staggered cleavage, then a 5′ extension will be repaired, resulting in a longer DNA fragment after Klenow treatment. Alternatively, if the enzyme left a 3′ extension, this will be trimmed back by the polymerase and the treated material will be shorter than the original fragment. The products of these two reactions are then run alongside a sequencing ladder prepared with the same primer. This allows the precise sites of cleavage on both strands to be determined. Comparison of cleavage products at several sites is important to avoid misidentifying degenerate sites and to permit unambiguous determination of the recognition sequence. Among the known restriction enzymes, there is considerable variation in the nature of the ends produced with examples of termini ranging from five-base 5'-nucleotide extensions through blunt ends to four-base 3'-nucleotide extensions (see Appendix A).

E. Effects of Methylation

Among known restriction systems, three types of methylations are used to provide protection against the cognate restriction enzyme: N6-methyladenine, N4-methylcytosine, and 5-methylcytosine. There is no way to predict which kind of methylation will be used or whether methylation of any given base in the recognition sequence will provide protection. This was dramatically illustrated by a pair of isoschizomers, *Mbo*I and *Sau*3AI. It was found that *Mbo*I could not cleave plasmid DNA obtained from normal strains of *Escherichia coli*, because of extensive methylation by the host methyltransferase, M·*Eco*Dam (Gelinas et al. 1977). Shortly after the discovery of *Mbo*I, an isoschizomer, *Sau*3AI, was discovered (Stobberingh et al. 1977) that could cleave DNA that had been modified by M·*Eco*Dam. Chromosomal DNA from *Moraxella bovis*, the source organism of *Mbo*I, could also be digested by *Sau*3AI, whereas chromosomal DNA from *Staphylococcus aureus*, the source of *Sau*3AI, could equally be digested by *Mbo*I (Brooks and Roberts 1982). Since it was known that M·*Eco*Dam produced N6-methyladenine (Marinus and Morris 1973), it was concluded that the *Mbo*I methyltransferase must be like *Eco*Dam and produce N6-methyladenine. It was speculated that the *Sau*3AI methyltransferase would produce 5-methylcytosine to protect against *Sau*3AI, and this has now been confirmed (Seeber et al. 1990). Thus, here were examples of two restriction enzymes that could be completely inhibited by methylation at one base within the recognition sequence but could cleave completely if any other base was methylated. The two enzymes can therefore be used to determine the methylation status of these two bases within their recognition sequence.

More useful though was the finding of the unusual properties of the isoschizomer pair *Hpa*II and *Msp*I (Waalwijk and Flavell 1978). Both enzymes recognize the sequence CCGG, but *Hpa*II is unable to cleave if the inner cytosine is methylated (Walder et al. 1983) and cleaves inefficiently if the outer cytosine is methylated (Butkus et al. 1987). *Msp*I can cleave if the inner cytosine is methylated but not if the outer cytosine is methylated (Walder et al. 1983). This has been of value because the cells of many higher organisms contain a methyltransferase that specifically modifies the CpG dinucleotide (for review, see Doerfler 1983), which is a part of the recognition site for *Hpa*II and *Msp*I. Thus, digestions by this pair of enzymes can be used to monitor the methylation status of the CG dinucleotide in and around certain genes. As a result, progress has been made in understanding the organization of eukaryotic genes and the importance of the methylation status in defining transcriptional activity. Many other enzymes are sensitive to methylation other than at the cognate site used by the companion methyltransferase, and a comprehensive database of current information is produced regularly (Nelson et al. 1993).

F. Single-stranded DNA Cleavage

Many restriction enzymes will cleave single-stranded DNA albeit with efficiencies that are very much less than cleavage of double-stranded DNA (Blakesley and Wells 1975; Blakesley et al. 1977). The precise mechanism of this cleavage remains unclear. Two compelling papers were published that offered strong support for two competing models. One model proposed that transient intramolecular duplex formation generates a site that can be recognized by the restriction enzyme (Blakesley et al. 1977). The competing model proposes that cleavage involves only the single strand of DNA (Yoo and Agarwal 1980). The "real" answer remains controversial. There has also been a report of cleavage of DNA-RNA hybrids by restriction enzymes (Molloy and Symons 1980). Unfortunately, the products of this reaction were not well characterized, and it is possible that the cleavage products seen resulted from cleavage of small amounts of double-stranded DNA present in the substrate used. No further reports of DNA-RNA hybrid cleavage have appeared.

III. GENES AND THEIR ORGANIZATION

A. Cloning

The first restriction system was cloned in 1978 (Mann et al. 1978), and since then, genes from more than 100 restriction systems have been cloned (Wilson 1991). Much of the success in cloning these systems has come from a clever selection technique proposed by Mann et al. (1978) and first used successfully by Szomolanyi et al. (1980). DNA is obtained from an organism known to contain a restriction system and a set of shotgun clones is made in a vector that contains at least one site for the restriction enzyme. The resulting clones are grown as a mixed population and plasmid DNA is prepared. This mixed population of cloned DNAs is then treated in vitro with the corresponding restriction endonuclease. If one of the clones encodes the appropriate methyltransferase and if that methyltransferase is expressed in *E. coli*, it will protect the plasmid DNA encoding it against the action of the restriction enzyme. However, all other DNAs will be cleaved by the restriction enzyme and destroyed to an extent determined by the number of recognition sites in the vector. Thus, following restriction enzyme cleavage of the mixed population and retransformation, any surviving clones should carry a functional methyltransferase gene. This, of course, is a prerequisite for the cloning of the restriction enzyme gene. At this point, nature has taken a hand. It turns out that as many as half of the clones carrying functional methyltransferase genes, embedded in large flanking sequences, coexpress the cognate restriction enzyme. In fact, we now know that in every system that has been examined in detail, the methyltransferase gene and the restriction enzyme gene are located very close to each other. In some instances, they actually overlap. Thus, this powerful selection technique for methyltransferase genes turns out to be a general trick for cloning restriction enzyme genes as first reported for *Bsu*RI (Kiss et al. 1985).

The method does not work immediately in every case, but usually this is because of problems associated with transcription and/or translation of foreign genes in *E. coli*. Thus, many genes from *Actinomycete* species are simply not expressed in *E. coli*, unless special attempts are made to introduce normal *E. coli* promoters and/or ribosome-binding sites. In some cases, this limitation has been overcome by cloning directly into an organism in which the genes can be expressed and using the methylation selection scheme in that organism. For instance, the *Sau*3AI system from *S. aureus* proved refractory to cloning in *E. coli* but was readily cloned into a homologous staphylococcal host lacking the *Sau*3AI restriction system (Seeber et al. 1990). In some cases, functional methyltransferase genes can be obtained fairly easily in *E. coli*, but the corresponding restriction enzyme genes appear to be missing. This can occur because they lack transcriptional and/or translational signals, but in some cases, problems exist because the two genes cannot be transferred into a naive *E. coli* host together. The host needs to be properly modified before the restriction enzyme gene is introduced. These kinds of problems can be overcome using a two-step procedure in which the methyltransferase is first introduced stably into *E. coli* and the flanking sequences that are presumed to contain the restriction enzyme gene are cloned separately in a compatible plasmid and then introduced into the protected *E. coli* host (*Dde*I, Howard et al. 1986; *Bam*HI, Brooks et al. 1989).

Other problems can ensue if the cloned methyltransferase gene fails to protect the host chromosome completely or is lethal to *E. coli*. For instance, in wild-type strains of *E. coli*, the *Mcr* (Raleigh and Wilson 1986) and *Mrr* (Heitman and Model 1987) systems specifically restrict methylated DNA, and so strains deficient in these systems are essential for these cloning experiments. So far, we know of only one clear example where the lethality of the methyltransferase has been shown to be the cause of a problem. This is the case of *Sau*3AI which recognizes GATC and forms 5-methylcytosine (Seeber et al. 1990). Undoubtedly, the problems here relate to the abnormal methylation of GATC sequences by the *Sau*3AI methyltransferase. In normal *E. coli*, M·*Eco*Dam forms N6-methyladenine in GATC sequences. There are a large number of these sequences close to the origin of replication of *E. coli*, and the abnormal methylation by M·*Sau*3AI may interfere with replication and induce lethality.

In principle, it should be possible to preprotect cells using a methyltransferase of low specificity to protect against a restriction enzyme of high specificity. For example, the *Sss*I methyltransferase modifies the sequence CG (Renbaum et al. 1990) and so in principle could be used to protect an *E. coli* host against all restriction enzymes that contain CG in their recognition sequence, provided they are inhibited by m5CG. So far, there are no published examples of the use of this approach to clone restriction endonuclease genes.

B. Genetic Location

Although some of the type II restriction systems such as *Eco*RI and *Eco*RII are known to be encoded on the plasmid (Yoshimori 1971), others are known to be encoded on the chromosome. Examples include *Dpn*I and *Dpn*II (Lacks et al. 1986) and *Sal*I (Chater and Wilde 1980), and there is no reason to suppose that one location would be preferable to another. For most systems, no detailed analysis has been carried out. Perhaps the most striking feature of the genetic organization of the restriction systems is that in every case where both the restriction enzyme and the methyltransferase gene have been characterized, they lie close to one another (Wilson 1991). Although this may reflect a bias of the cloning methods, it is not unexpected on evolutionary grounds (see below). In some systems, the two genes are separated by a single small open reading frame, which has been shown to be involved in the control of expression of the system (Tao et al. 1991; Ives et al. 1992).

The proximity of the two genes immediately suggests some kind of coordinate expression either at the transcriptional level or at the translational level, but only rarely has this been examined (see, e.g., Kiss et al. 1985). In part, this is because the genes have usually been studied in *E. coli*, rather than in their host organism, because of the ease of experimentation. It would be worthwhile to study more of these genes in their natural context to determine whether they are organized as an operon or if their expression is otherwise coordinated. Another explanation for their tight linkage could be that they are actually quite mobile and move readily from one organism to another. Since both genes are necessary to provide the restriction function, this would require that they were located close to one another, and furthermore, their expression would need to be tightly coordinated.

There are some clues to suggest that at least some of these systems are mobile. *Ngo*PII from *Neisseria gonorrhoeae* (Sullivan et al. 1987) and *Mth*TI from *Methanobacterium thermoformicicum* (Nolling and De Vos 1992) both recognize the sequence GG↓CC and cleave as indicated. The genes for both restriction enzymes and methyltransferases have been sequenced, and they are strikingly similar, suggesting a common evolutionary origin (Sullivan and Saunders 1989; Nolling and De Vos 1992). Another clue comes from a recent observation (G.G. Wilson, unpubl.) that in several strains of *Haemophilus*, the gene for valyl-tRNA-synthetase is found adjacent to the genes for several quite dissimilar restriction systems. This common location suggests that a homing mechanism may be operating to transfer them to similar chromosomal locations.

Interstrain transfer is well documented in *Diplococcus pneumoniae*. In one strain, the restriction enzyme, *Dpn*II, and two methyltransferases are present as a cassette (de la Campa et al. 1987). In a second strain, the same chromosomal location is occupied by an unusual type of restriction enzyme, *Dpn*I, that has a specificity complementary to that of *Dpn*II (Lacks and Greenberg 1977; Geier and Modrich 1979). *Dpn*II is an isoschizomer of *Mbo*I and recognizes the sequence GATC. It fails to cleave that sequence if the adenine is methylated. *Dpn*I, on the other hand, also recognizes GATC but will only cleave when the adenine is methylated. It is clear that switching takes place between the elements of the cassettes (Lacks et al. 1986). Interestingly, the isoschizomeric *Mbo*I system also contains two methylases (Ueno et al. 1993) and is similar in organization to that of *Dpn*II. However, no counterpart to *Dpn*I has yet been reported in the *Mbo*I system.

C. Sequences

Sequences are now available for more than 60 restriction enzyme genes and 100 methyltransferase genes. No significant similarity has been detected between the methyltransferase gene and the restriction enzyme gene of any cognate system, except for a small questionable similarity in the *Eco*RII system (Kossykh et al. 1993). Although this was at first surprising, it can be understood more readily for those enzymes that recognize symmetric sequences, since there the restriction enzyme usually acts as a dimer and the methyltransferase acts as a monomer. Thus, the modes of recognition are quite different, with the restriction endonuclease only needing to recognize half of the recognition sequence, whereas the methyltransferase must recognize the complete sequence. It is more surprising in the case of the type IIs enzymes where both methyltransferase and restriction enzyme act as monomers. In these cases, however, only two examples of completely sequenced systems exist (*Fok*I, Kita et al. 1989; *Hga*I, G.G. Wilson, unpubl.), and it may be that as more sequences are determined, similarities between the restriction enzyme and the methyltransferase will appear.

Among the methyltransferase genes, there are very clear examples of sequence similarities. Thus, the m5C-methyltransferases have a common architecture that has been well documented (Lauster et al. 1989; Posfai et al. 1989). The N6A-methyltransferases and the N4C-methyltransferases also show similarities, although the architecture is less well pronounced than for the m5C-methyltransferases (Klimasauskas et al. 1989; Lauster et al. 1989). For all DNA methyltransferases, the sequence similarities and identifying motifs are such that new genes can quickly be identified by comparative methods. However, N4C-methyltransferases and N6A-methyltransferases cannot yet be unambiguously distinguished without experimentation. In contrast, the restriction endonuclease genes show no discernible sequence similarities except for a few isolated cases of isoschizomers that are extremely similar and clearly have a common evolutionary origin, for example, *Eco*RI (Greene et al. 1981; Newman et al. 1981) and *Rsr*I (Stephenson et al. 1989).

Most of the restriction enzymes studied that recognize palindromic sequences have been shown to act as dimers or higher multimers. This is consistent with the idea that each monomer subunit interacts with one half of the recognition sequence so that the inherent symmetry in the dimer defines the symmetry of the recognition sequence. The structural studies of *Eco*RI and *Eco*RV support this view, although the simple model of each subunit interacting with just one half of the recognition sequence is not correct in detail (see below). In contrast, the type IIs enzymes that recognize asymmetric sequences interact with DNA as monomers and are suggested to have two domains, one responsible for DNA recognition and the other responsible for cleavage. Preliminary results support this model. Deletion of the carboxyl terminus of *Fok*I prevents cleavage but does not appear to affect binding (Li et al. 1992).

Restriction enzymes show a wide variation in molecular mass ranging from 18 kD for *Pvu*II (Athanasiadis et al. 1990; Tao and Blumenthal

1992) to 66 kD for *Bsu*RI (Kiss et al. 1985). No particular correlation exists between the length of the sequence recognized and the size of the monomer molecular mass, reinforcing the idea that most of these enzymes have evolved independently and use a variety of mechanisms to recognize DNA. As mentioned previously, most restriction enzymes interact with DNA as dimers that are preformed in solution in the absence of DNA. In some cases, higher oligomeric forms have been observed, but their relevance to the biochemical reaction is unknown.

IV. DNA BINDING

Before a nuclease can cleave DNA, it must bind to DNA. All type II restriction enzymes need a divalent metal ion, normally Mg^{++}, to cleave DNA. However, in the absence of divalent metal ions, many restriction enzymes have been shown to bind DNA to form stable DNA-protein complexes. Examples include *Bam*HI (Xu and Schildkraut 1991a), *Eco*RI (Terry et al. 1987), *Eco*RII (Gabbara and Bhagwat 1992), *Eco*RV (Taylor et al. 1991), *Rsr*I (Aiken et al. 1991a), and *Taq*I (Zebala et al. 1992a). For both *Eco*RI and *Eco*RV, the kinetics of DNA cleavage in the presence of Mg^{++} indicate ordered reactions; the enzyme binds first to DNA and then the enzyme-DNA complex binds Mg^{++} (Halford 1983; Taylor and Halford 1989). The DNA-protein complexes formed in the absence of Mg^{++} are therefore relevant to the reaction pathway.

The complexes have been analyzed by many of the standard techniques for DNA-protein interactions. The first studies on the association of a restriction enzyme with DNA used the filter-binding method (Halford and Johnson 1980; Jack et al. 1982), but this approach has generally been superseded by the gel-shift method, which is often more informative. For example, gel shifts on the binding of the *Eco*RI nuclease to a DNA with two *Eco*RI sites revealed two complexes, due to the binding of either one or two molecules of the protein (Terry et al. 1985), whereas the binding at either site alone retained the DNA on filters. DNA-binding sites have been characterized by several footprinting methods, including G- and A-methylation, phosphate ethylation, and UV photo-footprinting (Lu et al. 1983; Lesser et al. 1990; Becker et al. 1988). A special method, the preferential cleavage assay, has also been used to monitor restriction enzymes binding to their substrates (Jack et al. 1982; Terry et al. 1983). In this procedure, the nuclease is first mixed with DNA in the absence of Mg^{++}, and a solution that contains both $MgCl_2$ and a second DNA substrate is then added. The second substrate must be in large excess of both the enzyme and the first DNA so that any free enzyme will react with it rather than the first DNA. DNA cleavage is allowed to proceed for one or two turnovers, and the amount of the first DNA cut during this period is determined. The amount cut reveals the occupancy of the recognition site.

A. Enzymes That Bind Specifically to Their Recognition Sites

Binding of *Eco*RI to DNA either lacking *Eco*RI sites or containing one site was measured by filter binding (Halford and Johnson 1980; Terry et al. 1983). Even on DNA molecules 40,000 base pairs long, the DNA with the *Eco*RI site was bound more tightly than the DNA lacking sites, and, when the former DNA carried one molecule of *Eco*RI, it was located at the recognition site. To evaluate the effect of one-base-pair changes in the recognition sequence, DNA binding by *Eco*RI was studied with sets of short (14–17 bp) duplexes made from synthetic oligonucleotides (Lesser et al. 1990; Thielking et al. 1990). One duplex in each set contained the *Eco*RI recognition sequence, whereas the others covered all possible one-base-pair variations from this sequence. The equilibrium binding constants varied with each substitution but were typically 5000 times lower than the recognition sequence. Sequences that differed from the recognition site by two base pairs were bound even more weakly (Lesser et al. 1990).

The specificity of DNA binding seen with the *Eco*RI nuclease has also been seen with several other restriction enzymes, such as *Bam*HI and *Rsr*I (Aiken et al. 1991a; Xu and Schildkraut 1991a). In the absence of Mg^{++} ions, these enzymes again show a marked preference for binding to the recognition sequence over any other DNA sequence. *Rsr*I and *Eco*RI are homologous isoschizomers (see above), and their DNA-binding properties are very similar (Aiken et al. 1991a).

B. Enzymes That Fail to Bind Specifically to Their Recognition Sites

In gel-shift experiments on the binding of *Eco*RI to DNA, the number of complexes observed on the gel was equal to the number of *Eco*RI sites in the DNA (Terry et al. 1985). But similar experiments with *Eco*RV differed dramatically (Taylor et al. 1991). Irrespective of whether the DNA contained *Eco*RV sites, they revealed a series of complexes due to the binding of 1, 2, 3, ...n molecules of protein per molecule of DNA, where n is the maximum that can fit onto the DNA (Fig. 1). This suggests that in the absence of Mg^{++}, DNA binding by *Eco*RV is nonspecific and eventually covers the DNA from end to end. From the discrete ladder of complexes seen in Figure 1, the equilibrium constants for the binding of each successive molecule of the protein to the DNA were determined

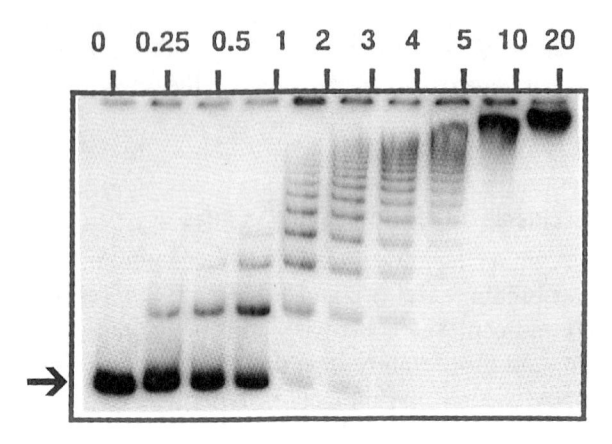

Figure 1 Gel-shift analysis of DNA binding by *Eco*RV. A 381-bp DNA fragment, with one *Eco*RV recognition site, was mixed with different amounts of the *Eco*RV restriction endonuclease, in a buffer lacking $MgCl_2$, and the DNA in each mix was then analyzed by electrophoresis through polyacrylamide. The enzyme concentrations (nM) are noted above each lane. The arrow marks the mobility of the free DNA. (Reprinted, with permission, from Taylor et al. 1991 [copyright American Chemical Society].)

(Taylor et al. 1991). The constants for the first, second, ...nth molecules of the *Eco*RV protein all had the same value, even when the DNA possessed an *Eco*RV site. Similar equilibrium constants were found using oligonucleotide duplexes that contained either the *Eco*RV recognition site or each possible one base-pair deviation (J. Alves and A. Pingoud, pers. comm.). The *Eco*RV restriction enzyme thus binds all DNA sequences, including its recognition site, with equal affinities.

This conclusion may appear to contradict the biological role of a restriction enzyme, which is to cleave DNA at a specific sequence. However, it is important to note that the above studies refer to reactions without Mg^{++}, a condition not encountered in the cell. Like *Eco*RV, *Taq*I (Zebala et al. 1992a) and *Cfr*9I (V. Siksnys, pers. comm.) also bind to DNA without preference for the recognition sites.

C. Transfer to Recognition Sites

Like many DNA-binding proteins, *Eco*RI finds its recognition sequence by first binding nonspecifically and then transferring to its specific site. This could occur by "sliding," which refers to linear diffusion from nonspecific sites to specific sites, or by "hopping," which refers to microscopic dissociations/reassociations within the same DNA molecule (von Hippel and Berg 1989). Hopping is unlikely for a restriction enzyme with one DNA-binding site but could be important for enzymes like *Nae*I (Conrad and Topal 1989; see below) or for enzymes that can oligomerize. *Eco*RI appears to use the sliding mechanism (Jack et al. 1982; Terry et al. 1985).

DNA outside the recognition site also plays a role in DNA cleavage reactions. Ehbrecht et al. (1985) found this by using mixtures of DNA fragments in which an *Eco*RI site was embedded in different lengths of nonspecific DNA as a substrate for DNA cleavage. Under some (but not all) experimental conditions, the rates of DNA cleavage increased with the length of the substrate DNA. Similar effects of flanking sequences have also been observed for *Bam*HI (Nardone et al. 1986). On a linear DNA, with two *Eco*RI sites close to one end, *Eco*RI cut the innermost site more readily than the outer site and yet, when the same DNA was circularized, both sites were cut with equal efficiency (Terry et al. 1985).

In vitro, linear diffusion along DNA explains the enhanced reaction rates (Terry et al. 1987) and the effects of increasing ionic strength (Jack et al. 1982) observed for *Eco*RI. However, extrapolating these results to the in vivo situation is complicated by the different ionic conditions found in the cytoplasm of a bacterial cell (Richey et al. 1987) and also the presence of cellular proteins that complex DNA in vivo and can block linear diffusion (Ehbrecht et al. 1985). For enzymes like *Eco*RV and *Taq*I that have no specificity for DNA binding in the absence of Mg^{++}, sliding has to be an important mechanism. In vitro experiments show that *Eco*RV at a nonspecific site can transfer to the specific site without dissociating from the DNA (Taylor et al. 1991).

V. DNA CLEAVAGE

When a DNA with several copies of the recognition site is used as substrate for a restriction enzyme, the reaction kinetics at one site often differ from those at other sites (*Eco*RI, Thomas and Davis 1975; *Pst*I, Armstrong and Bauer 1982; *Pae*7RI, Gingeras and Brooks 1983). It is therefore essential to use substrates with single sites for kinetic studies. Site variations are usually assigned to the differing flanking sequences surrounding the site. For both *Eco*RI and *Eco*RV, the length of DNA in contact with the protein is longer than the recognition sequence (Lu et al. 1983; Becker et al. 1988; Rosenberg 1991; Winkler et al. 1993). But only in a few cases have the effects of changing flanking sequences been examined systematically (Taylor and Halford 1992; Yang and Topal 1992).

*Eco*RII and *Nae*I typify restriction enzymes that must interact simultaneously with two copies of the recognition sequence before cleaving DNA (Krüger et al. 1988; Conrad and Topal 1989). Other examples include *Bsp*MI, *Hpa*II, *Nar*I, and *Sac*II (Oller et al. 1991). Both enzymes contain two distinct binding sites for their respective DNA sequences, one of which is an allosteric site that activates the other for DNA cleavage. Indeed, without this activation, the *Eco*RII-DNA complex is so devoid of activity that it is stable even in the presence of Mg^{++} (Gabbara and Bhagwat 1992). The two sequences can be supplied in *cis* if the sites are close and the DNA between the sites is then sequestered in a loop (Topal et al. 1991). Alternatively, they can be provided in *trans*, with the allosteric site being provided by an oligonucleotide that carries the recognition sequence (Conrad and Topal 1989; Pein et al. 1991). The activating function of the recognition sequence is distinct from its function as a substrate, as sites with certain flanking sequences can act as good activators but poor substrates or vice versa (Yang and Topal 1992). Indeed, the activating sequence does not even have to be a substrate for DNA cleavage (Pein et al. 1991).

A. Plasmid Substrates

Covalently closed circles of supercoiled DNA, typically 3000 to 6000 base pairs long, make convenient substrates for restriction enzymes. With a supercoiled substrate, the hydrolysis of a phosphodiester bond in one strand converts the DNA to its open circle form, whereas cleavage of both strands at the same site yields the linear form of the DNA. The three forms can be separated from one another by electrophoresis through agarose (Fig. 2). Hence, not only can one measure the reaction kinetics, by determining the amounts of each form of the DNA at various times during the reaction, but one can also observe the mode of DNA cleavage. If two consecutive single-strand breaks occur, then open circle DNA will be an intermediate in the reaction. But if cleavage occurs by a concerted double-strand break, no open circle form is produced (Fig. 2).

The amounts of the supercoiled, open circle, and linear forms of the DNA that are determined with this assay are the sums of both enzyme-bound and free DNA. The complete catalytic turnover of the endonuclease in steady-state reactions can be measured by this assay if the plasmid is in molar excess of the enzyme so that virtually all of the DNA is free rather than enzyme-bound. It can also be used to measure the cleavage of enzyme-bound DNA by studying single turnover reactions with the enzyme in excess of the substrate (see, e.g., Halford 1983; Terry et al. 1987; Zebala et al. 1992a). For a theoretical description of mechanistic pathways, see Bennett and Halford (1989).

With *Eco*RI, different studies on the kinetics of DNA cleavage appear to yield conflicting results, depending on which substrate was used

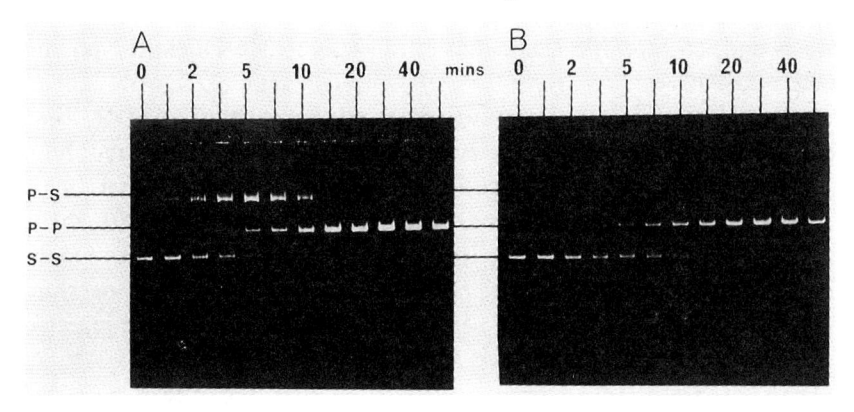

Figure 2 Plasmid substrates. The plasmid pMB9 was mixed with either the *Eco*RI endonuclease (*A*) or the *Sal*I endonuclease (*B*) in their optimal buffers for DNA cleavage. This plasmid has one recognition site for *Eco*RI and one for *Sal*I. Samples were withdrawn from the reactions at timed intervals (noted above the gel), mixed with EDTA to stop DNA cleavage, and then analyzed by electrophoresis through agarose. S-S indicates the mobility of the supercoiled form of pMB9, the covalently closed DNA intact in both strands; P-S indicates the mobility of the open circle form cleaved in one strand; P-P indicates the mobility of the linear form cut in both strands.

(Rubin and Modrich 1978). The lifetime of the open circle intermediates may be too short to detect (pBR322 or ColE1; Terry et al. 1987) or may be easily detectable (SV40 or pMB9; Halford 1983) (Fig. 2A). Changing the reaction conditions can also affect the accumulation of intermediates. With *Eco*RV, DNA cleavage under normal conditions gave no open circle DNA (Halford and Goodall 1988). However, at either low pH or low concentrations of $MgCl_2$, open circles were formed and dissociated from *Eco*RV during the reaction (Halford and Goodall 1988). At low pH, *Eco*RV has a reduced affinity for Mg^{++}, presumably due to the protonation of the aspartate residues that bind the metal ion (see below). Thus, the two sets of reaction conditions that give rise to open circle DNA are just the conditions where Mg^{++} may be bound to only one subunit of the dimeric enzyme. These same conditions also induced *Sal*I to generate an open circle intermediate when it had not done so under its optimal conditions (Fig. 2B) (Maxwell and Halford 1982). These observations are consistent with a general requirement for coupled reactions to have Mg^{++} bound to both subunits, and that failure to do so alters the mode of DNA cleavage.

The type of experiments described above have been carried out on many different restriction enzymes: Some use coupled reactions to cleave duplex DNA and others use uncoupled reactions (for review, see

Bennett and Halford 1989). Recent examples include a detailed kinetic analysis of DNA cleavage by *Bam*HI (Hensley et al. 1990) and an elegant series of both steady-state and single-turnover studies on *Taq*I (Zebala et al. 1992a). As expected for an enzyme from a thermophilic organism, *Taq*I has its optimal activity at high temperature but, at 60°C, its kinetic parameters for DNA cleavage are similar to those for other restriction enzymes at 37°C or less. However, its temperature dependence is surprisingly steep: Its K_m decreases 100-fold as the temperature is raised from 50°C to 60°C, and it appears to be "frozen" in an inactive conformation at low temperatures.

The kinetic constants from the steady-state reactions of restriction enzymes on plasmid or phage DNA vary from enzyme to enzyme and from substrate to substrate. But under their respective optimal conditions, these reactions nearly always produce k_{cat} values within an order of magnitude of 1 min^{-1} and K_m values within an order of magnitude of 1 nM (Bennett and Halford 1989). The K_m of a restriction enzyme for DNA cannot be very much higher than 1 nM in "sites," because these enzymes must be able to handle substrates present in one copy per cell, and one molecule in the volume of an *E. coli* cell is at a concentration of about 1 nM. This in turn limits k_{cat} to about 1 min^{-1} because the ratio of a k_{cat} of 1 min^{-1} and a K_m of 1 nM is 1.7×10^7 M^{-1}s^{-1}, which approaches the theoretical limit for k_{cat}/K_m in enzyme-catalyzed reactions (Jencks 1975).

B. Oligonucleotide Substrates

Instead of using plasmids as substrates, an alternative is to use duplexes made from synthetic oligonucleotides, typically 8–20 base pairs long. Reactions on oligonucleotides are usually monitored by sampling at timed intervals and then separating the substrate from the product(s) by either electrophoresis, homochromatography, or high-performance liquid chromatography (HPLC) (Aiken and Gumport 1991). But if the enzyme is assayed at a temperature where the substrate is double-stranded and the shorter products melt to single strands, the reaction can be monitored continuously in a UV spectrophotometer (Waters and Connolly 1992).

Oligonucleotide substrates pose fundamentally different problems for restriction enzymes compared to longer DNA molecules. An oligonucleotide challenges the enzyme with a unique DNA sequence, whereas a plasmid asks the nuclease to choose from a large number of alternative sequences. Consequently, the events that occur during reactions with oligonucleotides are different from those with plasmids. When the recognition site is part of a DNA macromolecule, the bulk DNA could act as a competitive inhibitor, but it can also help the enzyme find the recognition site, by facilitated transfer from nonspecific to specific sites. Thus, the rate constants for both the association and dissociation of the enzyme with its recognition site may differ between oligomeric and polymeric substrates. The rate-limiting step with a plasmid is often the dissociation of the final product cleaved in both strands (Terry et al. 1987), but this is not necessarily the case with an oligonucleotide (Alves et al. 1989a).

In nearly all cases, oligonucleotide substrates give k_{cat} values that are either similar or higher than those with large DNA molecules, and K_m values are often very much higher. This was first seen from the activity of the EcoRI enzyme on either SV40 DNA or an 8-bp duplex containing the EcoRI site (Greene et al. 1975). It has also been seen with other 8- or 10-bp substrates for EcoRI (Alves et al. 1984; Brennan et al. 1986; McLaughlin et al. 1987), although duplexes of more than 12 bp are cleaved by EcoRI with kinetics similar to those of DNA polymers (Lesser et al. 1990). However, a 12-bp duplex for EcoRV was cleaved with higher values of both k_{cat} and K_m compared to a plasmid substrate (Newman et al. 1990a). Part of these differences may be due to the oligonucleotide being too short to fill the DNA-binding site in the protein. Both EcoRI and EcoRV contact the sugar phosphate backbone of the DNA over a longer stretch than just the recognition site, and the interactions with the external phosphates can be essential for enzyme activity (Lu et al. 1983; Lesser et al. 1990). The addition of only one phosphate to an unphosphorylated oligonucleotide can alter the activity of a restriction enzyme (Van Cleve and Gumport 1992).

C. Specificity

One key property of a restriction endonuclease is its ability to discriminate its recognition sequence from all other DNA sequences. In vivo, a restriction enzyme would be lethal if it cleaved DNA readily at any sequence other than its recognition site. In a cell containing a restriction/modification (R/M) system, all copies of the recognition site on the chromosome are protected by the methyltransferase, but sequences that differ from the recognition site by one base pair remain unmethylated (Heitman and Model 1990a; Taylor et al. 1990; D.W. Smith et al. 1992). In addition, all of the applications of restriction enzymes in vitro depend on their ability to distinguish their recognition sequences from all other sequences. Yet, whenever it has been tested, restriction enzymes cleave DNA both at their recognition sites and at a limited number of additional sites. The additional sites generally differ from the recognition sequence by one base pair.

Under standard reaction conditions, the ratio of activities at cognate and noncognate sites is very large (Taylor and Halford 1989; Lesser et al. 1990; Thielking et al. 1990). But certain changes to the reaction conditions caused *Eco*RI to cleave DNA at many extra sites, in a process called "star activity" (Polisky et al. 1975). This "star activity" was originally assigned to DNA cleavage at the sequence AATT instead of its recognition site, GAATTC. It has since been shown that *Eco*RI* sites differ from the *Eco*RI site by just one base pair and that the difference can be anywhere within the 6-bp target sequence, although some *Eco*RI* sites are cleaved more readily than others (Rosenberg and Greene 1982). The changes to the reaction conditions include elevated pH and reduced ionic strength, the addition of water-miscible organic solvents such as glycerol or dimethylsulfoxide, and the replacement of Mg^{++} by Mn^{++} (Polisky et al. 1975; Hsu and Berg 1978). The same changes in reaction conditions also cause many other restriction enzymes to weaken the discrimination between their recognition sites and sites that are different by one base pair (Bennett and Halford 1989).

Sometimes, one noncognate site is cleaved more readily than another, and this allows the reaction to be measured at that particular site. For example, the cDNA from human immunodeficiency virus contains many sites that differ from the *Eco*RI site by one base pair, but one of these sites is cleaved by *Eco*RI more readily than any other: This site, GAATTA instead of GAATTC, occurs within a polypurine tract that may take up a non-B-DNA structure (Venditti and Wells 1991). In pAT153, the preferred noncognate site for *Eco*RV is GTTATC instead of GATATC and is flanked by alternating purines and pyrimidines, which confer flexibility to DNA structure (Taylor and Halford 1992). Under its standard reaction conditions, the activity of *Eco*RV at this site is about 10^6 times lower than that at the cognate site, but this ratio changes to 10^3 in the presence of 10% dimethylsulfoxide and to just 6 in the presence of Mn^{++} (Taylor and Halford 1989; Vermote and Halford 1992).

For *Eco*RI, systematic analyses of all nine possible single-base-pair changes within the recognition sequence have been carried out with substrates made from pairs of oligonucleotides (Thielking et al. 1990; Lesser et al. 1990). Each substitution caused a different reduction in k_{cat}/K_m for DNA cleavage, varying from 10^5-fold to 10^9-fold. For each of the nine substrates, the decline in DNA cleavage rate was larger than the decline in the equilibrium constant for DNA binding (Lesser et al. 1990). In similar experiments with *Eco*RV, the substitutions at each position in the recognition sequence had no significant effect on DNA binding, but they still reduced reaction rates by factors as large as those seen with *Eco*RI (J. Alves and A. Pingoud, pers. comm.).

Reactions of restriction enzymes at noncognate sites usually proceed via two successive single-strand breaks, even when the same enzyme under the same reaction conditions produces a double-strand break at the cognate site (Barany 1988; Taylor and Halford 1989; Thielking et al. 1990). At any site that differs from the EcoRI recognition sequence by one base pair, the EcoRI reaction in the unaltered half (G↓AA) is always much faster than that in the altered half, and often the only product detected is a DNA nick in this strand (Lesser et al. 1990; Venditti and Wells 1991). In sharp contrast, the majority of noncognate sites for TaqI are cut first in the altered half (Barany 1988).

DNA ligase seals single-strand breaks in duplex DNA more rapidly than it joins free ends from double-strand breaks (Lehman 1974). Ligase might thus be able to proofread restriction nucleases, by repairing single-strand breaks at noncognate sites before the endonuclease cuts the second strand, without repairing the double-strand breaks at cognate sites. When this scheme was modeled in vitro by adding *E. coli* DNA ligase to the EcoRV restriction enzyme, ligase made no difference to the rate at which EcoRV cleaved its cognate site, but it prevented any products from being formed at noncognate sites (Taylor and Halford 1989). This proofreading scheme has also been demonstrated in vivo by measuring the viability of temperature-sensitive ligase strains of *E. coli* carrying the EcoRV R/M system (Taylor et al. 1990). However, the complete picture for the interplay in vivo between R/M systems and all of the other cellular systems engaged in DNA metabolism is likely to be more complicated (Heitman 1993). Double-strand breaks in the chromosome, caused by a mutant EcoRI endonuclease, were repaired by DNA ligase, but they also induced the SOS response of *E. coli* (Heitman et al. 1989).

VI. CRYSTALLOGRAPHY

At present, crystal structures have been solved at high resolution for three restriction enzymes, EcoRI (Kim et al. 1990), EcoRV (Winkler et al. 1993), and BamHI (Strzelecka et al. 1990; A. Aggarwal, pers. comm.). Several others are in progress (PvuII, Anderson 1993; Athanasiadis and Kokkinidis 1991; MspI, D. Barford and I. Schildkraut, pers. comm.; FokI, SfiI, A. Aggarwal and I. Schildkraut, pers. comm.; DdeI, K. Grzeskowiak and I. Schildkraut, pers. comm.). One structure for EcoRI was solved as a DNA-protein complex containing the enzyme and a 12-bp duplex with the recognition site in the absence of Mg^{++} (Kim et al. 1990). Upon soaking with either Mg^{++} or Mn^{++}, the crystals survived DNA cleavage, and thus the structure of the postreactive enzyme-product complex could also be solved; structures of other DNA-protein complexes are in progress (J.M. Rosenberg, pers. comm.). For

EcoRV, three structures were solved: the free protein in the absence of DNA; the protein bound to a 10-bp duplex with the EcoRV site, and the protein bound to an unrelated DNA (Winkler et al. 1993). The structures of both the free protein and the specific DNA-protein complex for several DNA-binding proteins are now known (Freemont et al. 1991), but the EcoRV restriction enzyme is currently the only system for which both of these and a nonspecific DNA-protein complex are all available. The latter is crucial because the secret to specificity lies in the difference between cognate and noncognate complexes.

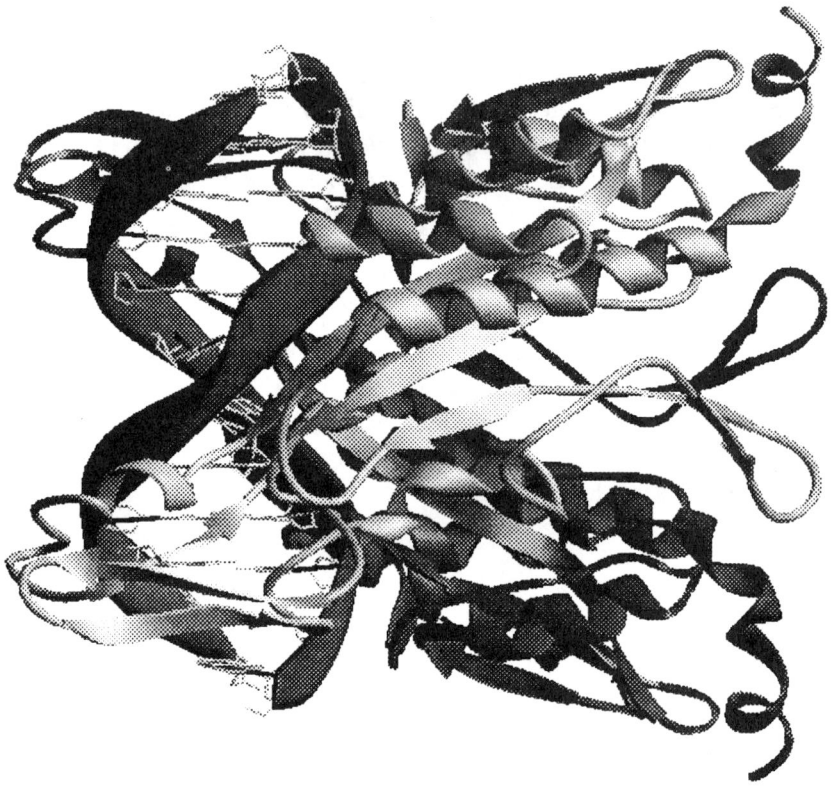

Figure 3 Crystal structure of the EcoRI-DNA complex. The peptide main chains of the two subunits in the EcoRI restriction enzyme are shown shaded light and dark; the twisted ribbons represent α helices; the arrows mark the β-sheets; the solid lines denote the loops that connect the helices and sheets. The DNA in this structure is the duplex form of CGCGAATTCGCG, which contains the EcoRI site (underlined). The phosphodiester backbone of the DNA is shown as two ribbons that are kinked at the position of the major deformation in the structure of this DNA. (Data from Y. Kim et al., in prep.; figure courtesy of J.M. Rosenberg.)

A. Protein Structures

The initial structure for *Eco*RI bound to DNA (McClarin et al. 1986) was incorrect and a revised model has since been presented (Kim et al. 1990 and in prep.). An excellent description of the revised model has appeared recently (Rosenberg 1991). Although the *Eco*RV restriction enzyme has no sequence homology with *Eco*RI (Bougueleret et al. 1984), it might have been expected that the structures of the two enzymes would be similar and that they might have the same motif for DNA recognition. Few similarities exist between *Eco*RI and *Eco*RV (Winkler et al. 1993). In contrast, the preliminary model for *Bam*HI shows that many sections are superimposable on the *Eco*RI structure, despite its lack of detectable sequence homology (A. Aggarwal, pers. comm.).

The current model for the complex of *Eco*RI with its DNA shows a dimeric protein with the DNA embedded into one side of the protein, spanning the dimer interface (Fig. 3). As expected (Kelly and Smith 1970), the complex has a single dyad axis relating the two subunits and the two halves of the palindromic DNA. At the dyad axis, the major groove of the DNA faces the protein. In each subunit, two arms extend from the main body of the protein to wrap around the DNA, but these remain within the major groove. No part of the protein enters the minor groove. Apart from these arms, the protein has a compact, comparatively normal structure for a globular protein. The core of each subunit is a five-stranded β-sheet lined on both sides with α helices. The components of the core are connected to each other by a succession of loops, together with some additional helices and sheets, that are located either at the "back" of the protein away from the DNA or toward the DNA on the DNA-protein interface (Fig. 3). Critical contacts to the DNA are made by a bundle of four α helices, two from each subunit, aligned almost perpendicular to the DNA with the amino terminus of each helix poking into the major groove. Contacts are also made by some of the loops noted above, positioned in and around the major groove (Kim et al. 1990 and in prep.).

In its complexes with either specific DNA (Fig. 4) or nonspecific DNA (Winkler et al. 1993), *Eco*RV consists of two L-shaped subunits that interact with each other over a small surface area, to create a U-shaped dimer with a deep cleft between the subunits. The fold of the peptide chain is completely different from that of *Eco*RI. In both of the *Eco*RV-DNA complexes, the DNA is located in the cleft, with its minor groove facing the base of the cleft, i.e., the opposite way round from *Eco*RI. The principal contacts to the DNA are made by two peptide loops per subunit, although other amino acids from the walls of the cleft contact the phosphodiester backbone. One loop, known as the R (for recog-

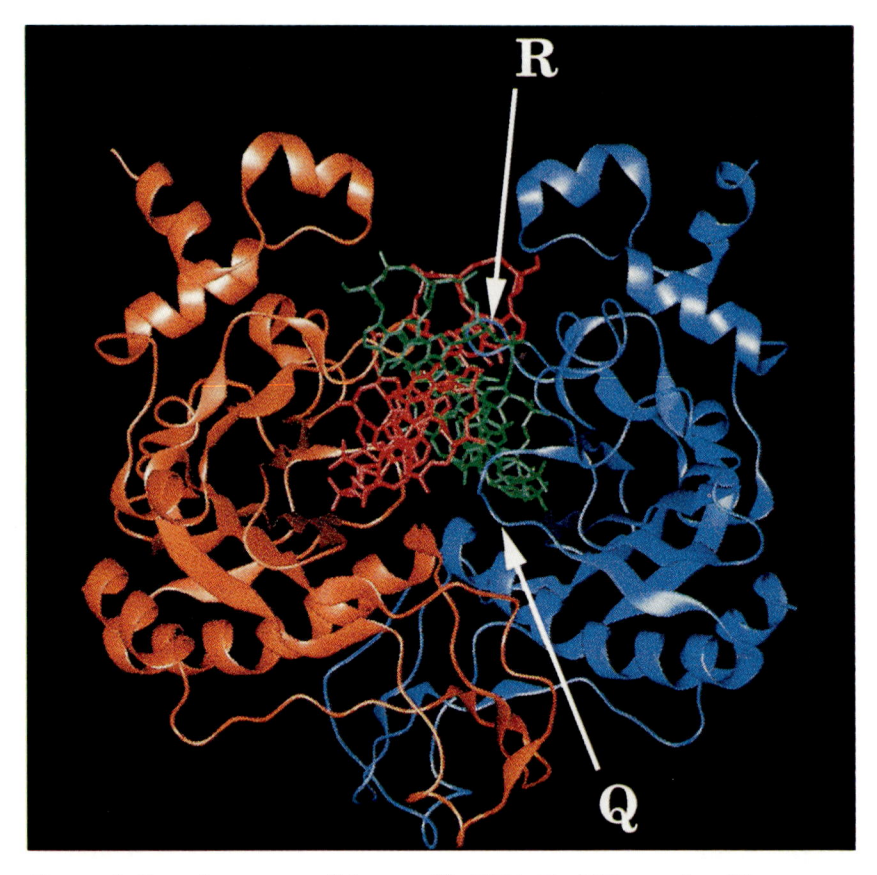

Figure 4 Crystal structure of the specific DNA-*Eco*RV complex. The structure shown is the *Eco*RV restriction enzyme bound to the duplex form of GGGATATCCC (*Eco*RV site underlined). The two strands of the DNA are colored red and green. The peptide main chains of the two subunits in the dimer are represented by blue and orange ribbons. The two peptide loops that are in direct contact with the DNA are marked R and Q. (Figure courtesy of I.B. Vipond, generated with data from F.K. Winkler.)

nition) loop, is located toward the top of the cleft above the DNA. In the complex at the recognition site (Fig. 4), the R-loop is positioned deep within the major groove, but, in the nonspecific complex, it is partly disordered and appears to lie distant from the DNA. The second loop at the base of the cleft contains several glutamines and is therefore known as the Q-loop (Winkler et al. 1993). The Q-loop approaches the minor groove of the DNA, contacting primarily phosphates rather than bases.

When *Eco*RV is bound to its cognate DNA (Fig. 4), the amino acids in the R-loop are in van der Waals' contact with the other subunit so that the protein surrounds the DNA (Winkler et al. 1993). DNA therefore cannot bind to *Eco*RV, at a site underneath the R-loops, by simply dropping the DNA into the cleft. This also applies to the structure of *Eco*RV in the absence of DNA: In the free protein, both the R-loops and Q-loops are disordered and virtually fill the space in the cleft that would otherwise be occupied by DNA. DNA binding by *Eco*RV must involve a series of conformational changes. The three separate structures for *Eco*RV reveal significant differences in protein conformation between the free enzyme in the absence of DNA, the specific complex, and the nonspecific complex (Winkler et al. 1993).

B. DNA Structures

When bound to *Eco*RI, the 12-bp duplex, CGCGAATTCGCG, is distorted from the regular B-structure for DNA (Kim et al. 1990). The distortion is primarily an untwisting at the center of the sequence, with concomitant unstacking of the two central base pairs, which occurs without bending the DNA to any marked extent. The untwisting widens the major groove, thus improving access to the bases. Crystals of the same 12-bp duplex, in the absence of *Eco*RI, yielded a structure that was close to that of orthodox B-DNA, with only minor variations in the helical parameters at each step (Dickerson and Drew 1981). The distortion of the DNA in the DNA-protein complex must therefore have been induced by its binding to *Eco*RI. The crystallography validates the earlier studies of Kim et al. (1984), who proposed that the binding of *Eco*RI to its recognition site in free solution unwinds the DNA by 25°. However, other solution studies have implied that the GAATTC sequence recognized by *Eco*RI may have an intrinsic deviation from B-DNA and that it is perhaps either kinked or bent before it binds the protein (Diekmann and McLaughlin 1988; Lane et al. 1991). Nevertheless, if deviations are present in the free DNA, they appear to be enhanced by *Eco*RI binding (Thomas et al. 1989).

When bound to *Eco*RV, the 10-bp duplex, GGGATATCCC, is radically distorted from B-DNA (Winkler et al. 1993). The most marked feature of the distortion is a sharp bend, directed toward the protein, in the axis of the DNA helix at the center of the recognition site (Fig. 5a). Solution studies have confirmed this bend (Stöver et al. 1993). Like *Eco*RI, the middle two base pairs in the *Eco*RV recognition site are unstacked, but, in this case, the roll is in the opposite direction. The bound substrate for *Eco*RV has a deep and narrow major groove and a correspondingly shallow minor groove.

*Eco*RV was also crystallized with an 8-bp duplex containing a nonspecific sequence, CGAGCTCG (Winkler et al. 1993). The cocrystals

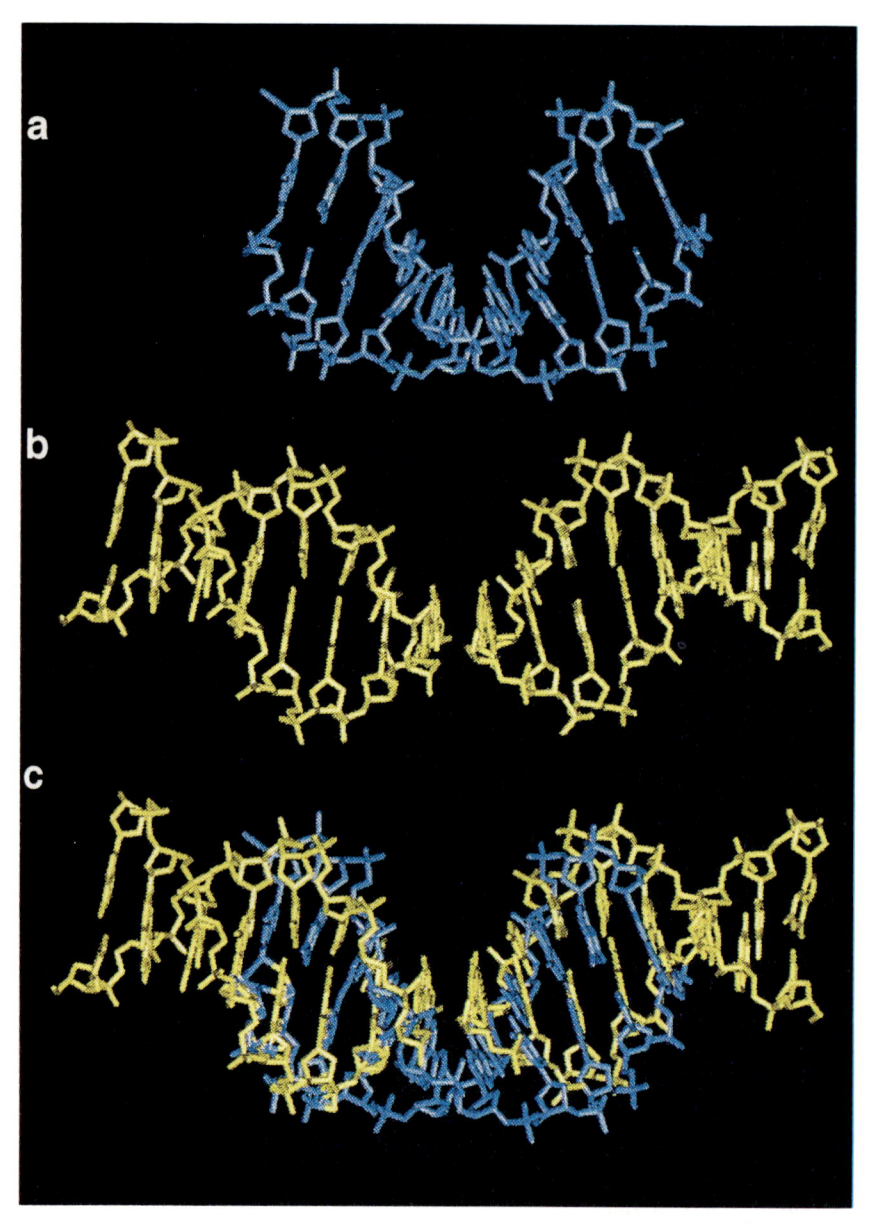

Figure 5 Specific (*a*) and nonspecific (*b*) DNAs bound to *Eco*RV, and an overlay of the two forms (*c*). The DNA in *a* is the duplex form of GGGATATCCC in the cocrystal with the *Eco*RV restriction enzyme. The DNA in *b* is the duplex form of CGAGCTCG in the cocrystal with *Eco*RV: The crystals contained two duplexes per protein dimer, stacked as shown. The overlay (*c*) was generated by superimposing the protein from one DNA-protein complex on the protein from the other complex. (Figure courtesy of I.B. Vipond, generated with data from F.K. Winkler.)

contained two duplexes per protein dimer, stacked end to end to give effectively a 16-bp DNA but lacking the phosphates between the eighth and ninth base pair (Fig. 5b). The *Eco*RV recognition site in the specific complex is thus replaced by TpCpG/CpGpA in the nonspecific complex. The DNAs in the specific and nonspecific complexes are located in the same overall position (Fig. 5c), but the structures of the bound DNAs differ. The 8-bp fragments of nonspecific DNA are close to ideal B-DNA, although the stack at the junction is not ideal: The helical axis in one fragment is displaced laterally from the other. The phosphates close to the ends of the specific DNA overlay the equivalent phosphates in the nonspecific DNA, but, toward the middle of the DNA, the phosphodiester backbone is placed differently from the nonspecific DNA (Fig. 5c). Yet the protein contacts almost all of the phosphates in the DNA within the cleft, in both specific and nonspecific complexes. *Eco*RV thus appears to have two distinct binding modes: one for a highly distorted DNA and another for a B-like structure, although a continuous DNA in the latter mode may have to be slightly kinked and untwisted (Winkler et al. 1993).

C. DNA-Protein Interfaces

Each base pair in duplex DNA possesses, on its edge facing the major groove, a unique array of three hydrogen-bonding functions (Seeman et al. 1976). The 5-methyl group of thymidine can also be used to distinguish DNA sequences. The *Eco*RI restriction enzyme uses 16 out of the possible 18 hydrogen-bonding functions in its 6-bp recognition sequence and makes van der Waals' contacts with all of the thymidine methyl groups (Fig. 6). No other DNA-binding protein characterized to date throws as many recognition functions at its DNA as *Eco*RI. The interactions include direct hydrogen bonds and also water-mediated bonds. The amino acids that recognize the bases in the DNA come from different parts of the polypeptide: Arg-145 and Arg-203 are at the ends of the two α helices that poke into the major groove, Met-137 and Ala-138 are on an extended arm that runs through the major groove, and Gln-115 is located next to the catalytic center. But almost all of the amino acids that interact with the bases also interact with other amino acids at the DNA-protein interface. For example, the guanine in the DNA makes a hydrogen bond via water to Arg-203, which in turn contacts Glu-144. But Glu-144 also contacts Asn-141 and Arg-145, two residues that make direct hydrogen bonds to the adenines (Fig. 6, but only a small fraction of the buttressing network is shown here) (R. Kim et al., in prep.).

The *Eco*RI methyltransferase protects the recognition site from the

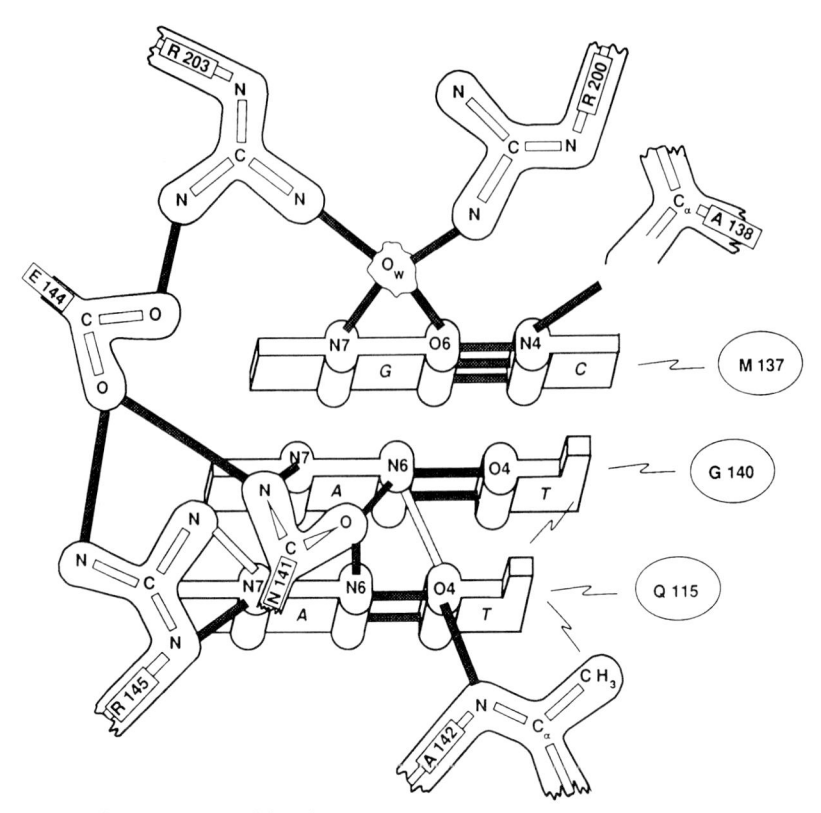

Figure 6 DNA recognition by *Eco*RI. The bases in one half of the *Eco*RI recognition site are shown as planks, with rounded protusions to denote the hydrogen-bonding moieties. The functional groups in the amino acids that interact with the bases are also shown. The solid lines represent hydrogen bonds, either between the amino acids and the bases or between two amino acids. Wavy lines illustrate the hydrophobic interactions between specified amino acids and the pyrimidine bases. (Data from Y. Kim et al., in prep.; figure courtesy of J.M. Rosenberg.)

endonuclease by N6 methylation of the second adenine (Dugaiczyk et al. 1974). Crystallography on the 12-bp DNA fragment used for the structure of the *Eco*RI-DNA complex, with N6-methyl adenine in place of the second adenine, showed that this protection cannot come from perturbing the structure of the DNA: The structure was unaltered by methylation (Frederick et al. 1988). However, the N6 amino group is recognized by Asn-141 and DNA methylation can be expected to disrupt not only the direct interactions, but also many other interactions within the buttressing network.

The recognition functions used by *Eco*RV are much simpler than those of *Eco*RI. The only amino acids in *Eco*RV that make sequence-specific hydrogen bonds to the DNA are Gly-182, Gly-184, Asn-185, and Thr-186 (Winkler et al. 1993). These are all in the R-loop that lies in the major groove of the DNA. The first three amino acids contact all three of the hydrogen-bonding functions in the GC base pair at the start of the EcoRV site, with either their carbonyl or amino groups from the peptide main chain. The last two amino acids use their side chains to contact all three groups in the adjacent AT base pair. Asn-185 uses its peptidyl amino group to bond to N7 in the guanine and its bifunctional side chain to interact with both N6 and N7 in the adenine. This adenine is the site of methylation by the EcoRV modification enzyme (McClelland and Nelson 1988), so the entire R-loop would be displaced from its proper position upon EcoRV methylation. However, there are no sequence-specific interactions that are visible in the crystal structure between EcoRV and the innermost TA base pair in each half of its recognition site. The two central base pairs in the EcoRV site are severely distorted when bound to the protein (Fig. 5), and the acquisition of this particular structure for any other DNA sequence may involve a substantial energy penalty (Winkler et al. 1993). This could be an example of indirect readout (Luisi and Sigler 1990).

VII. PHOSPHODIESTER HYDROLYSIS

The overall structure of EcoRI bound to its recognition sequence is radically different from that of EcoRV. Yet there is one striking similarity, which raises the possibility that these dissimilar enzymes use the same chemical mechanism to hydrolyze the phosphodiester bond. In both endonucleases, the bond that is cleaved in the reaction is surrounded by a proline, two acidic residues, and a lysine in the same relative positions: in EcoRI, Pro-90, Asp-91, Glu-111, and Lys-113; in EcoRV, Pro-73, Asp-74, Asp-90, and Lys-92 (Fig. 7) (Selent et al. 1992; Winkler 1992). The similarity with respect to the proline is probably coincidental. In EcoRI, Pro-90 has the favored *trans* configuration, whereas Pro-73 in EcoRV is the *cis* isomer. If EcoRV had the *trans* arrangement, the geometry of the connection between Asp-74 and the preceding section of polypeptide would be altered, and the section around Asn-70 makes several contacts to the phosphates in the DNA.

The other similarities in the two active sites are more significant. Glu-111 in EcoRI is essential for catalytic activity (King et al. 1989), and, in the postreactive DNA-protein complex after soaking the crystals with $MgCl_2$, the Mg^{++} ion lies between this residue and the phosphate of the cleaved bond (Rosenberg 1991). Likewise, Asp-74, Asp-90, and Lys-92 in EcoRV are all essential (Thielking et al. 1991; Selent et al. 1992). At the active site of EcoRV, the phosphate at the scissile bond is positioned

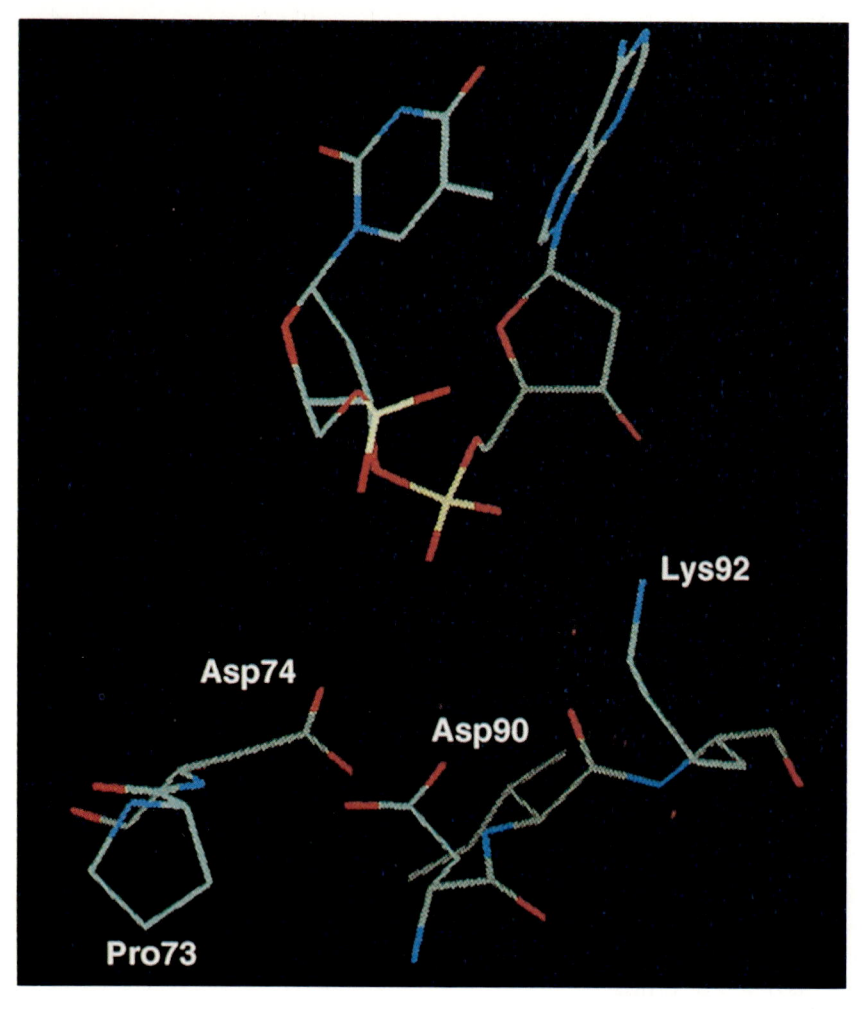

Figure 7 Active site of *Eco*RV. The diagram is taken from the structure of the *Eco*RV restriction enzyme bound to its recognition sequence in the absence of Mg++. It shows one strand of the DNA at the middle of the *Eco*RV recognition site, pTpA: *Eco*RV hydrolyzes the phosphodiester at the TpA step. The identified amino acids are residues in *Eco*RV that have equivalent residues in the active site of *Eco*RI. (Figure courtesy of I.B. Vipond, generated with data from F.K. Winkler.)

so that a Mg++ ion could be coordinated by this group and by both Asp-74 and Asp-90 (Winkler et al. 1993). This arrangement implies that the Mg++ ion polarizes the phosphorus and enhances its susceptibility to nucleophilic attack. The lysine can also interact with the phosphate at the scissile bond and perhaps stabilize the transition state (Fig. 7).

With both *Eco*RI and *Eco*RV, hydrolysis proceeds with stereochemical inversion at the phosphorus (Connolly et al. 1984; Grasby and Connolly 1992). This demands an odd number of chemical steps in the reaction and is most simply accounted for by the direct in-line displacement of the 3'-leaving group by water. But there is no obvious candidate for a base suitably positioned to activate the water in either *Eco*RI or *Eco*RV. It has been suggested that the phosphate next to the scissile bond could activate the water (Jeltsch et al. 1992; J.M. Rosenberg, pers. comm.), but if this phosphodiester had its normal pK_a of less than 2, it would virtually never accept a proton from water. An alternative is that the attack is by *aquo*-Mg^{++} in its deprotonated state, although it is far from certain that this would be deprotonated at neutral pH.

The precise mechanism of the chemical reaction catalyzed by *Eco*RI and *Eco*RV thus remains to be determined. The same applies to all other restriction enzymes. The sequence motif seen in *Eco*RI and *Eco*RV, PD...(D/E)-K, appears in several other restriction enzymes, but its significance has yet to be established (Anderson 1993). *Eco*RI has two copies of this motif, but only one is at the active site. An additional complication is that the active site of *Eco*RV contains two potential binding sites for divalent metal ions: one between Asp-74 and Asp-90, as noted above, and another between Asp-74 and Glu-45 (Winkler et al. 1993). Many nucleases contain two or more metal ions at their active sites (Steitz 1993) and *Eco*RV is perhaps another example. If so, then the chemical mechanisms of *Eco*RV and *Eco*RI would not be the same, because *Eco*RI lacks a counterpart to Glu-45.

VIII. DNA RECOGNITION FUNCTIONS

Two major approaches have been used to dissect DNA recognition and catalysis by restriction enzymes. One is to alter the protein by mutagenesis, replacing one amino acid with another. The other is to alter the substrate, replacing either a base with a base analog (see, e.g., Brennan et al. 1986) or a phosphate with a phosphorothioate (Connolly et al. 1984). For example, *Eco*RI is less active toward GAAUTC than GAATTC, which suggests a role for the methyl group on the inner thymidine (Brennan et al. 1986). In contrast, a mutant of *Eco*RI with Gln-115 replaced by alanine has the same activity at GAAUTC and GAATTC: This activity is equal to that of the wild-type enzyme on GAAUTC (Jeltsch et al. 1993). In the crystal structure of *Eco*RI (Rosenberg 1991), the methylene side chain of Gln-115 interacts hydrophobically with the methyl group on the inner thymidine (Fig. 6).

A. Altered Enzymes

In *Eco*RV, all of the amino acids in the R-loop have been mutated except for Gly-182 and Gly-184 (Thielking et al. 1991; Vermote et al. 1992). The effects of these mutations concur with the crystal structure. For example, substitution of Thr-186, in direct contact with the DNA, destroyed activity, whereas substitution of Thr-187, which is not in contact with the DNA, left wild-type levels of activity. The importance of the amino acids at the active site of *Eco*RV, noted in the preceding section, was also determined by site-directed mutagenesis (Selent et al. 1992). A fuller account of the relationship between the structure and the activity of *Eco*RV has been given elsewhere (Halford et al. 1993).

In the initial model for the *Eco*RI-DNA complex, Glu-144, Arg-145, and Arg-200 were believed to be the only amino acids responsible for DNA sequence recognition (McClarin et al. 1986). Consequently, many mutants were prepared at these positions, but they retained specificity for the *Eco*RI site (Alves et al. 1989b; Needels et al. 1989; Hager et al. 1990; Heitman and Model 1990a). The effects of these mutations have been reinterpreted on the current structure for *Eco*RI (Rosenberg 1991; Heitman 1992). For instance, Glu-144 is no longer placed to interact directly with the bases in the DNA; instead, it supports the amino acids involved in the direct interactions (see Fig. 6). The fact that mutations at "secondary" functions have just as much impact on activity as mutations at "primary" functions demonstrates that the buttressing network at the DNA-protein interface is indeed crucial.

Random mutagenesis, followed by genetic selection, has produced useful information about a number of restriction enzymes. In several studies (Yanofsky et al. 1987; King et al. 1989; Xu and Schildkraut 1991a), the DNA coding for the restriction enzyme was subjected to random mutagenesis and was then used to transform a strain that lacks the cognate modification methyltransferase. This procedure yields mutants that have either zero or very low DNA cleavage activity, since genes coding for active nucleases are lethal in the absence of the relevant methyltransferase. Interesting full-length mutants were identified by Western blotting with an antibody against wild-type *Eco*RI (Yanofsky et al. 1987). Gel-shift experiments have identified mutants of *Eco*RI (King et al. 1989) and *Bam*HI (Xu and Schildkraut 1991a), which retain the ability to bind DNA specifically.

In one study of this type on the *Eco*RI endonuclease, the transformation of a strain lacking the *Eco*RI modification enzyme with randomly mutagenized DNA yielded 240 colonies. Only one of these contained a protein that was still able to bind specifically to the *Eco*RI recognition site (King et al. 1989). This mutant had a single substitution, Glu-

111→Gly-111. Glu-111 lies right at the catalytic center (Kim et al. 1990), and its identification as a key residue in catalysis (King et al. 1989) is a testament to the power of random genetic methods.

A screen for mutants with perturbed specificity used randomly mutagenized DNA to transform a strain that carried the methyltransferase and also a reporter gene such as *lacZ* linked to an SOS-inducible promoter (Heitman and Model 1990b). The rationale of this strategy is that the enzyme cannot cleave its recognition site because these are fully methylated in vivo, but if a mutant were to have a higher activity at noncognate sites than the wild-type enzyme, the resultant DNA damage would activate the SOS response and thus induce the reporter gene. This procedure yielded several promiscuous mutants of *Eco*RI that were more active at noncognate sites than the wild-type enzyme (Heitman and Model 1990b) and whose properties can now be accounted for on the crystal structure (Heitman 1992).

B. Altered Substrates

Each base in duplex DNA has several hydrogen-bonding groups that can interact with a protein (Seeman et al. 1976). Therefore, base analogs can be used to remove just one of these groups. The appropriate analogs for 2-deoxyadenosine, for example, are shown in Figure 8. The complete series of analogs that delete one by one all of the accessible hydrogen-bonding groups on the base have also been synthesized for T, G, and C (Newman et al. 1990b; Waters and Connolly 1993). For several restriction endonucleases, oligonucleotide substrates have been made in which a base analog replaced one of the bases in their respective recognition sequences, and the cleavage properties were examined. Examples include *Eco*RI (Brennan et al. 1986; McLaughlin et al. 1987; Lesser et al. 1993), *Eco*RV (Newman et al. 1990a; Waters and Connolly 1993), *Rsr*I (Aiken et al. 1991b), and *Taq*I (Zebala et al. 1992b). For both *Eco*RI and *Eco*RV, the loss of almost any one of the functional groups in the DNA, which is seen in the crystal structure to interact with the protein, reduced DNA cleavage rates relative to the cognate oligonucleotide. However, the loss of functional groups in the DNA that are not directly contacted by the protein can also alter the activity of the enzyme. This illustrates one of the problems with interpreting these experiments. The disruption of one interaction may well have a cooperative effect throughout the DNA-protein interface. Simple explanations based on the anticipated disruption of a single recognition function are naive in the absence of detailed structural information on the new complex.

Two methods have been used to incorporate phosphorothioates into

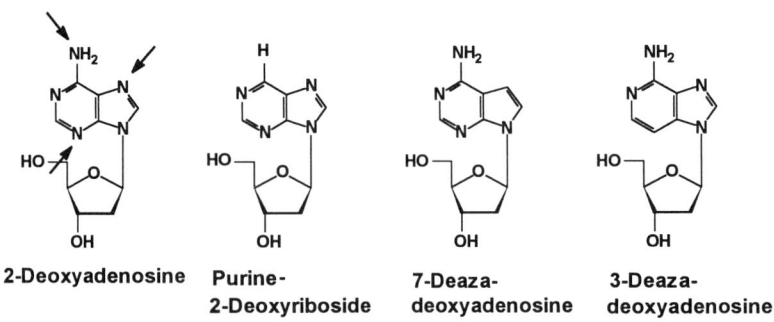

Figure 8 Adenine analogs. In duplex DNA, the base from 2-deoxyadenosine has two hydrogen-bonding functions that are accessible from the major groove (marked with arrows above the base) and one that is accessible from the minor groove (marked with an arrow below the base). Three analogs of 2-deoxyadenosine have been synthesized, each of which lacks one of these hydrogen-bonding moieties. (Data from Newman et al. 1990a; figure courtesy of B.A. Connolly.)

DNA substrates. One is by the enzymatic synthesis of DNA with three dNTPs and one dNTPαS (Potter and Eckstein 1984). This gives a duplex DNA where one strand contains phosphorothioates, stereoselectively in the Rp form, at all of the phosphodiester bonds preceding the nucleotide supplied in the αS form. A phosphorothioate at the scissile bond either severely diminishes DNA cleavage by the restriction enzyme or abolishes it entirely, and thus these substrates amplify the difference between cleaving the first and second strands of the DNA (Potter and Eckstein 1984). This difference was exploited to produce a novel and efficient method for site-directed mutagenesis (Taylor et al. 1985). However, phosphorothioates elsewhere in the recognition site or even in the flanking DNA can also reduce the activity of a restriction enzyme (Olsen et al. 1990; Lesser et al. 1992).

The second method for the incorporation of phosphorothioates is by the chemical synthesis of oligonucleotide substrates, and this allows the derivative to be placed at one specified site in the chain (Connolly et al. 1984). However, chemical synthesis will yield both the Rp and the Sp diastereoisomers of the phosphorothioate, unless chiral synthons are used (Lesser et al. 1992). Fortunately, the oligonucleotide containing an Rp phosphorothioate can generally be separated by HPLC from that with the Sp configuration (Grasby and Connolly 1992). Substrates of this type have yielded valuable information about the stereochemical path of phosphodiester hydrolysis by restriction enzymes (Connolly et al. 1984; Grasby and Connolly 1992) and about the stereochemistry of their interactions with the phosphates in the distorted DNA backbone (Lesser et al. 1992).

C. Coupling Recognition to Catalysis

Can we now account for the ability of restriction enzymes to distinguish their recognition sites from all other DNA sequences? For *Eco*RI, several processes appear to be involved in the recognition of the cognate site and the rejection of noncognate sites (Lesser et al. 1990; Heitman 1992). *Eco*RI binds more tightly to its recognition site than to other sites, but the reduction in binding affinity caused by changing one base pair in the DNA sequence is too small to explain the difference in reaction rates (Thielking et al. 1990). However, *Eco*RI interacts strongly with several phosphates in and around its recognition site, but these interactions are either altered or missing in its complexes with noncognate sites (Lesser et al. 1990). This implies that the conformation of the DNA (and/or the protein) in the specific complex differs from that in nonspecific complexes. Hence, at least part of the DNA cleavage specificity of *Eco*RI stems from indirect readout, in that only the cognate DNA appears to undergo the distortion to give the active DNA-protein complex (as shown in Fig. 3).

Isosteric base analogs, lacking a hydrogen-bonding function (Fig. 8), cause much smaller reductions to *Eco*RI activity than changing one base pair to a different base pair. The substrates with certain base analogs give the same pattern of phosphate interactions, as judged from ethylation experiments, as the canonical sequence, and in some cases, the effect of the analog can be accounted for by the loss of just one hydrogen bond between the protein and the DNA (Lesser et al. 1990, 1993). Conversely, the change in *Eco*RI activity caused by changing one base pair to another base pair is larger than can be accounted for by the loss of hydrogen-bonding and hydrophobic interactions. The discrimination against incorrect DNA sequences appears to be magnified by the resultant apposition of inappropriate functions. Instead of juxtaposing a hydrogen bond acceptor on the DNA with a donor in the protein, the incorrect complex may try to place two hydrogen bond donors against each other (Lesser et al. 1990). *Eco*RI thus generates its specificity by a subtle combination of both direct and indirect readout.

*Eco*RV shows no preference for binding to its recognition site over other sites (Taylor et al. 1991), so its DNA cleavage specificity cannot come from the same mechanism as *Eco*RI, although, in both cases, the distortion of the enzyme-bound DNA plays a key role. It may be impossible for an enzyme to acquire sufficient specificity from a DNA sequence as short as a restriction site without distorting the DNA (Winkler 1992). Although the free energy for the formation of the cognate complex with *Eco*RV is the same as that for the noncognate complex, the complexes differ from each other both mechanistically and structurally. When located at its recognition site on DNA, the *Eco*RV nuclease has a high affinity for Mg^{++} ions but, when located at a site different by one base pair, it has a very low affinity for Mg^{++} (Taylor and Halford 1989). This now appears to be the main factor determining the different rates of DNA cleavage (Vipond and Halford 1993). The intrinsic activity of *Eco*RV at its recognition site is probably similar to that at noncognate sites, but, since it has an absolute requirement for a divalent metal ion, the activity that is observed at each DNA sequence depends on the fractional saturation of each DNA-protein complex with Mg^{++}. This view is supported by the lack of discrimination by *Eco*RV in the presence of Mn^{++} (Vermote and Halford 1992). Both the cognate and noncognate complexes with *Eco*RV have high affinities for Mn^{++}, and under these conditions, the ratio of the DNA cleavage activities is 6, whereas it had been close to 10^6 with Mg^{++} as the cofactor.

The way in which the affinity of the *Eco*RV-DNA complex for Mg^{++} varies with the DNA sequence can be explained on the structures of the enzyme-DNA complexes (Winkler et al. 1993). In the specific complex with *Eco*RV bound to its recognition sequence, the distortion of the DNA leaves the phosphate at the scissile bond placed near two aspartates in the active site (see Fig. 7). The distortion therefore not only inserts the relevant phosphodiester bond into the active site, but it also creates part of the binding site for Mg^{++} between the phosphate and the aspartates (Winkler et al. 1993). Without this distortion, the analogous section of the B-like DNA in the nonspecific complex lies distant from the active site (see Fig. 5c). The energy required to deform the DNA must come from the direct interactions between the amino acids on the R-loop of *Eco*RV and the bases in the recognition sequence (Halford et al. 1993). The reason mutations in the R-loop reduce activity is not because they disrupt DNA binding, but rather because they disrupt Mg^{++} binding elsewhere in the protein (Vermote et al. 1992). The mutant proteins presumably fail to distort the DNA.

IX. EVOLUTION

The type II restriction-modification systems pose interesting problems for the evolutionist. On one hand, the methyltransferases, especially the m5C methyltransferases, show strong conservation of sequence, suggesting a common heritage (Lauster et al. 1989; Posfai et al. 1989). Their counterpart restriction enzymes have defied all attempts to detect similarity at the primary sequence level, except for a few isoschizomers that may have a recent common progenitor. However, in almost all cases known, the genes for the restriction enzymes and their methyltransferases are tightly linked, suggesting a common evolutionary thread. Until recently, this apparent paradox was brushed aside and it was thought that the individual elements in a system had evolved separately from a common ancestral methyltransferase gene, which had acquired a variety of different restriction enzyme genes (Heitman 1993). A possible resolution of the paradox has been suggested by K. Carlson (pers. comm.) on the basis of the differing evolutionary pressures that are imposed on the two genes.

In the presence of a functional restriction enzyme, the accurate and efficient functioning of the methyltransferase is tested at every round of replication. Under these conditions, it can easily be imagined that change is difficult. However, the restriction enzyme is under much less stringent constraints than the methyltransferase. Within the host cell, mutations that led to promiscuous cleavage would be strongly counter-selected, but mutations that decreased the catalytic efficiency of the endonuclease or resulted in its loss of activity would be essentially neutral in the host cell (except as below) until the system was challenged by phage infection or other invasion by foreign DNA. Assuming that this is an infrequent event, it would be possible for the endonuclease to accumulate mutations much more rapidly than the methyltransferase. Of course, within the original host, a mutation leading to a nonfunctional restriction enzyme would relieve the pressure on the methyltransferase temporarily. Thus, the two enzymes coevolve but under quite different selective pressures.

In the scheme outlined above, there is scope for methyltransferases to evolve new specificities, either independently as orphan enzymes or in conjunction with a temporarily incapacitated restriction enzyme. As the genes move from one member of the population to another or from one species to another, they may pick up different counterpart genes. For instance, if a common ancestor gave rise to the *Eco*RI and *Bam*HI restriction endonucleases, as might be possible given their similar structural features, then one might anticipate that they became associated with different methyltransferases, since *Bam*HI cleavage is prevented by N4 methylation on cytosine, whereas *Eco*RI cleavage is blocked by N6 methylation on adenine. Since both N4C and N6A methylation take place at an exocyclic amino group, and the chemistry is expected to be similar, it is within the bounds of possibility that these two classes of methyltransferases are themselves derived from a common ancestor. This would nicely explain the two conserved sequence motifs that are found in both families of enzymes (Klimasauskas et al. 1989; Lauster et al. 1989).

X. CONCLUSIONS AND FUTURE PROSPECTS

With almost 2400 type II restriction enzymes now cataloged, this group of nucleases is the largest of any known. Although only a few of these enzymes have been examined in detail, it is clear that the mechanisms employed to recognize DNA differ significantly from the well-studied recognition elements exemplified by zinc fingers and helix-turn-helix motifs found in other systems. Nevertheless, they show exquisite specificity of recognition, and it is likely that the mechanisms that they employ for DNA recognition have been tailored to accommodate the accompanying enzymatic activity that requires not only sequence-specific binding, but also release from the DNA after cleavage. In this respect, they differ significantly from their better-studied counterparts, such as transcription factors, where only binding to DNA is required for function. Still, it has been possible to isolate mutants of *Eco*RI (King et al. 1989), *Bam*HI (Xu and Schildkraut 1991a,b), and *Fok*I (Li et al. 1992) that retain the ability to bind to DNA and yet can no longer cleave. In several cases, such binding mutants can behave as sequence-specific-DNA-binding proteins with respectable dissociation constants. The superficial view that DNA interactions are mediated by a few known motifs has to be reconsidered. Further studies of the structure and function of the type II restriction enzymes will be essential in appreciating the full range of mechanisms by which DNA and proteins can interact. For example, Zebala et al. (1992b) have suggested that there should be at least two classes of mechanisms for the type II restriction enzymes.

One curious and interesting feature of certain type II restriction enzymes is their ability to recognize, with considerable specificity, degenerate sequences within the recognition site. This has been known for many years, but in none of these systems do we yet have any mechanistic or structural information to indicate how recognition of the degenerate position(s) is achieved. In most cases, the acceptable alternatives conform to the predictions of the Seeman-Rosenberg-Rich rules (Seeman et al. 1976), although several exceptions should be noted. Enzymes such as *Sdu*I (recognition sequence GDGCH↓C), *Acc*I (recognition sequence GT↓MKAC), and *Hgi*AI (recognition sequence GWGCW↓C) are obvious candidates for more detailed studies. Another interesting set of enzymes are those in which the two halves of the symmetric recognition sequence are separated by bases that are completely nonspecific. The series *Nco*I (recognition sequence: C↓CATGG), *Pfl*MI (recognition sequence: CCANNNN↓NTGG), *Bst*XI (recognition sequence: CCANNNNN↓NTGG), and *Xcm*I (recognition sequence: CCANNNNN↓NNNNTGG) each recognizes the same specific hexanucleotide sequence CCATGG, but the two halves are separated by 0, 5, 6, or 9 bp, respectively. One could imagine that recognition in this case is achieved by sequence motifs within the protein that are themselves separated by a stretch of protein sequence that determines the spacing. This situation would be analogous to the relationship between the type I enzymes *Eco*R124I and *Eco*R124/3I, which recognize similarly related sequences (Price et al. 1989; Chapter 3).

So far, only one type IIs system, *Fok*I, has been studied in any detail. The enzyme appears to contain two domains: one responsible for sequence recognition and the other for DNA cleavage. If this model is correct, it would seem an ideal candidate for efforts to change specificities. It would be useful to have more detailed mechanistic and structural information about other type IIs systems to see if they are built on common architectures or if, like *Eco*RI and *Eco*RV, each uses a quite different mechanism for DNA recognition and/or cleavage. Detailed comparisons of the neoschizomers *Fok*I and *Sts*I, which differ only in their position of cleavage, could also prove most informative. A few oddball systems, such as *Bcg*I and *Eco*57I, which do not fit cleanly within the present classification of restriction systems, may also prove interesting for biochemical study. Systems such as *Eco*RII and *Nae*I, which require interaction with two recognition sequences, can be expected to provide some new twists on DNA recognition and may lead to new ideas about the transfer of proteins along DNA.

One area that deserves further attention is the small controlling elements first discovered in the *Pvu*II system (Tao et al. 1991), which remains unexplained at the functional level. Undoubtedly, they will be important for the biology of the restriction systems, particularly if the systems are mobile. They may provide an experimental path into the biology and reveal new insights into the control of expression of these systems. This will undoubtedly be important in understanding the evolution of the systems, a subject that has been greatly neglected in a formal sense. Most defense systems have evolved to accommodate changes quickly, and evolutionary pressure on the restriction systems can be expected to follow this pattern. Much greater progress in this area is being made for the type I systems (Sharp et al. 1992; see Chapter 3).

A goal of many workers in this field has been to generate new specificities for restriction enzymes, particularly longer recognition sequences, by genetic manipulation of existing systems. This goal has proved elusive. It may be that systems such as *Eco*RI are simply not genetically malleable and that multiple mutations would be required before true change in specificity could be achieved. Other systems may prove more tractable, but a great deal more work will be required to obtain basic information about mechanisms of DNA-protein interactions in these systems before rational attempts to create new specificities are likely to be fruitful. A few years ago, excitement was high about the possibility for making artificial restriction enzymes using a variety of reagents such as catalytic antibodies or chemical reagents to cleave DNA in combination with oligonucleotide-directed reagents to achieve specificity. The demise of restriction enzymes seemed inevitable. However, despite significant success with these other reagents (see Chapter 10), new restriction enzymes continue to be discovered, their applications are increasing, and they are providing new and unexpected insights into DNA-protein interactions.

ACKNOWLEDGMENTS

We thank all of our colleagues in restriction enzymes for advice and discussions and for unpublished data. The work by S.E.H. was supported by the Science and Engineering Research Council and by a Royal Society Leverhulme Trust Research Fellowship. Work from the laboratory of R.J.R. was supported by the National Institutes of Health (HG-00303) and the National Science Foundation (DMB-8917650). Special thanks are due to Drs. D. Comb, I. Schildkraut, and their colleagues at New England Biolabs, who have been extremely generous in sharing materials and providing financial support to foster research in this field.

REFERENCES

Aiken, C.R. and R.I. Gumport. 1991. Base analogs in the study of restriction enzyme-DNA interactions. *Methods Enzymol.* **208:** 433–457.

Aiken, C.R., E.W. Fisher, and R.I. Gumport. 1991a. The specific binding, bending and unwinding of DNA by *Rsr*I endonuclease, an isoschizomer of *Eco*RI endonuclease. *J. Biol. Chem.* **266:** 19063–19069.

Aiken, C.R., L.W. McLaughlin, and R.I. Gumport. 1991b. The highly homologous isoschizomers *Rsr*I endonuclease and *Eco*RI endonuclease do not recognize their target sequences identically. *J. Biol. Chem.* **266:** 19070–19078.

Alves, J., C. Urbanke, A. Fliess, G. Maass, and A. Pingoud. 1989a. Fluorescence stopped-flow kinetics of the cleavage of synthetic oligodeoxynucleotides by the *Eco*RI restriction endonuclease. *Biochemistry* **28:** 7879–7888.

Alves, J., T. Rüter, R. Geiger, A. Fliess, G. Maass, and A. Pingoud. 1989b. Changing the hydrogen-bonding potential in the DNA binding site of *Eco*RI by site-directed mutagenesis drastically reduces the enzymatic activity, not, however, the preference of this restriction endonuclease for cleavage within the site -GAATTC-. *Biochemistry* **28:**

2678-2684.
Alves, J., A. Pingoud, A. Haupt, J. Langowski, F. Peters, G. Maass, and C. Wolff. 1984. The influence of sequences adjacent to the recognition site on the cleavage of oligodeoxynucleotides by the EcoRI endonuclease. *Eur. J. Biochem.* **140:** 83-92.
Anderson, J.E. 1993. Restriction endonucleases and modification methylases. *Curr. Opin. Struct. Biol.* **3:** 24-30.
Arber, W. and D. Dussoix. 1962. Host specificity of DNA producted by *Escherichia coli*: I. Host controlled modification of bacteriophage λ. *J. Mol. Biol.* **5:** 18-36.
Armstrong, K.A. and W.R. Bauer. 1982. Preferential site-dependent cleavage by restriction endonuclease *Pst*I. *Nucleic Acids Res.* **10:** 993-1007.
Athanasiadis, A. and M. Kokkinidis. 1991. Purification, crystallization and preliminary X-ray diffraction studies of the *Pvu*II endonuclease. *J. Mol. Biol.* **222:** 451-453.
Athanasiadis, A., M. Gregoriu, D. Thanos, M. Kokkinidis, and J. Papamatheakis. 1990. Complete nucleotide sequence of the *Pvu*II restriction enzyme gene from *Proteus vulgaris*. *Nucleic Acids Res.* **18:** 6434.
Barany, F. 1988. The *Taq*I "star" reaction: Strand preferences reveal hydrogen-bond donor and acceptor sites in canonical sequence recognition. *Gene* **65:** 149-165.
Becker, M.M., D. Lesser, M. Kurpiewski, A. Baranger, and L. Jen-Jacobson. 1988. "Ultraviolet footprinting" accurately maps sequence-specific contacts and DNA kinking in the EcoRI endonuclease-DNA complex. *Proc. Natl. Acad. Sci.* **85:** 6247-6251.
Bennett, S.P. and S.E. Halford. 1989. Recognition of DNA by type II restriction enzymes. *Curr. Top. Cell. Regul.* **30:** 57-104.
Bertani, G. and J.J. Weigle. 1953. Host controlled variation in bacterial viruses. *J. Bacteriol.* **65:** 113-121.
Bitinaite, J., Z. Maneliene, S. Menkevicius, S. Klimasauskas, V. Butkus, and A.A. Janulaitis. 1992. *Alw*26I, *Eco*31I and *Esp*3I-type IIs methyltransferases modifying cytosine and adenine in complementary strands of the target DNA. *Nucleic Acids Res.* **20:** 4981-4985.
Blakesley, R.W. and R.D. Wells. 1975. "Single-stranded" DNA from φX174 and M13 is cleaved by certain restriction endonucleases. *Nature* **257:** 421-422.
Blakesley, R.W., J.B. Dodgson, I.F. Nes, and R.D. Wells. 1977. Duplex regions in "single-stranded" φX174 DNA are cleaved by a restriction endonuclease from *Haemophilus aegyptius*. *J. Biol. Chem.* **252:** 7300-7306.
Bougueleret, L., M. Schwarzstein, A. Tsugita, and M. Zabeau. 1984. Characterization of the genes coding for the EcoRV restriction/modification system of *Escherichia coli*. *Nucleic Acids Res.* **12:** 3659-3676.
Boyd, A.C., I.G. Charles, J.W. Keyte, and W.J. Brammar. 1986. Isolation and computer-aided characterization of *Mme*I, a type II restriction endonuclease from *Methylophilus methylotrophus*. *Nucleic Acids Res.* **14:** 5255-5274.
Brennan, C.A., M.D. Van Cleve, and R.I. Gumport. 1986. The effects of base analogue substitutions on the cleavage by the EcoRI restriction endonuclease of octadeoxyribonucleotides containing modified EcoRI recognition sequences. *J. Biol. Chem.* **261:** 7270-7278.
Brooks, J.E. and R.J. Roberts. 1982. Modification profiles of bacterial genomes. *Nucleic Acids Res.* **10:** 913-934.
Brooks, J.E., J.S. Benner, D.F. Heiter, K.R. Silber, L.A. Sznyter, T. Jager-Quinton, L.S. Moran, B.E. Slatko, G.G. Wilson, and D.O. Nwankwo. 1989. Cloning the *Bam*HI restriction modification system. *Nucleic Acids Res.* **17:** 979-997.
Brown, N.L. and M. Smith. 1980. A general method for defining restriction enzyme cleavage and recognition sites. *Methods Enzymol.* **65:** 391-404.

Butkus, V., L. Petrauskiene, Z. Maneliene, S. Klimasauskas, V. Laucys, and A.A. Janulaitis. 1987. Cleavage of methylated CCCGGG sequences containing either N4-methylcytosine or 5-methylcytosine with MspI, HpaII, SmaI, XmaI and Cfr9I restriction endonucleases. *Nucleic Acids Res.* **15**: 7091-7102.

Chater, K.F. and L.C. Wilde. 1980. *Streptomyces albus* G mutants defective in the SalGI restriction-modification system. *J. Gen. Microbiol.* **116**: 323-334.

Cho, S.-H. and C. Kang. 1990. DNA sequence-dependent cleavage sites of restriction endonuclease HphI. *Mol. Cells* **1**: 81-86.

Connolly, B.A., F. Eckstein, and A. Pingoud. 1984. The stereochemical course of the restriction enzyme EcoRI—Catalysed reaction. *J. Biol. Chem.* **259**: 10760-10763.

Conrad, M. and M.D. Topal. 1989. DNA and spermidine provide a switch mechanism to regulate the activity of restriction enzyme NaeI. *Proc. Natl. Acad. Sci.* **86**: 9707-9711.

de la Campa, A.G., P. Kale, S.S. Springhorn, and S.A. Lacks. 1987. Proteins encoded by the DpnII restriction gene cassette: Two methylases and an endonuclease. *J. Mol. Biol.* **196**: 457-469.

Dickerson, R.E. and H.R. Drew. 1981. Structure of a B-dodecamer II: Influence of base sequence on helix structure. *J. Mol. Biol.* **149**: 761-786.

Diekmann, S. and L.W. McLaughlin. 1988. DNA curvature in native and modified EcoRI recognition sites and possible influence upon the endonuclease cleavage reaction. *J. Mol. Biol.* **202**: 823-834.

Doerfler, W. 1983. DNA methylation—A regulatory signal in eukaryotic gene expression. *Annu. Rev. Biochem.* **52**: 93-124.

Dugaiczyk, A., J. Hedgpeth, H.W. Boyer, and H.M. Goodman. 1974. Physical identity of the SV40 deoxyribonucleic acid sequence recognized by the EcoRI restriction endonuclease and modification methylase. *Biochemistry* **13**: 503-512.

Dussoix, D. and W. Arber. 1962. Host specificity of DNA produced by *Escherichia coli*: II. Control over acceptance of DNA from infecting phage λ. *J. Mol. Biol.* **5**: 37-49.

Ehrbrecht, H.-J., A. Pingoud, C. Urbanke, G. Maass, and C. Gualerzi. 1985. Linear diffusion of restriction endonucleases on DNA. *J. Biol. Chem.* **260**: 6160-6166.

Frederick, C.A., G.J. Quigley, G.A. van der Marel, J.H. van Boom, A.H. Wang, and A. Rich. 1988. Methylation of the EcoRI recognition site does not alter DNA conformation. *J. Biol. Chem.* **263**: 17872-17879.

Freemont, P.S, A.N. Lane, and M.R. Sanderson. 1991. Structural aspects of protein-DNA recognition. *Biochem. J.* **278**: 1-23.

Gabbara, S. and A.S. Bhagwat. 1992. Interaction of EcoRII endonuclease with DNA substrates containing single recognition sites. *J. Biol. Chem.* **267**: 18623-18630.

Geier, G.E. and P. Modrich. 1979. Recognition sequence of the dam methylase of *Escherichia coli* K12 and mode of cleavage of DpnI endonuclease. *J. Biol. Chem.* **254**: 1408-1413.

Gelinas, R.E., P.A. Myers, and R.J. Roberts. 1977. Two sequence-specific endonucleases from *Moraxella bovis*. *J. Mol. Biol.* **114**: 169-179.

Gingeras, T.R. and J.E. Brooks. 1983. Cloned restriction/modification system from *Pseudomonas aeruginosa*. *Proc. Natl. Acad. Sci.* **80**: 402-406.

Gingeras, T.R., J.P. Milazzo, and R.J. Roberts. 1978. A computer assisted method for the determination of restriction enzyme recognition sites. *Nucleic Acids Res.* **5**: 4105-4127.

Grasby, J.A. and B.A. Connolly. 1992. The stereochemical outcome of the hydrolysis reaction catalysed by the EcoRV restriction endonuclease. *Biochemistry* **31**: 7855-7861.

Greene, P.J., M. Gupta, H.W. Boyer, W.E. Brown, and J.M. Rosenberg. 1981. Sequence analysis of the DNA encoding the *Eco*RI endonuclease and methylase. *J. Biol. Chem.* **256:** 2143–2153.

Greene, P.J., M.S. Poonian, A.L. Nussbaum, L. Tobias, D.E. Garfin, H.W. Boyer, and H.M. Goodman. 1975. Restriction and modification of a self-complementary octanucleotide containing the *Eco*RI substrate. *J. Mol. Biol.* **99:** 237–261.

Hager, P.W., N.O.Reich, J.P. Day, T.G. Coche, H.W. Boyer, J.M. Rosenberg, and P.J. Greene. 1990. Probing the role of glutamic acid 144 in the *Eco*RI endonuclease using aspartic acid and glutamine replacements. *J. Biol. Chem.* **265:** 21520–21526.

Halford, S.E. 1983. How does *Eco*RI cleave its recognition site on DNA. *Trends Biochem. Sci.* **8:** 455–460.

Halford, S.E. and A.J. Goodall. 1988. Modes of DNA cleavage by the *Eco*RV restriction endonuclease. *Biochemistry* **27:** 1771–1777.

Halford, S.E. and N.P. Johnson. 1980. The *Eco*RI restriction endonuclease with bacteriophage λ DNA: Equilibrium binding studies. *Biochem. J.* **191:** 593–604.

Halford, S.E., J.D. Taylor, C.L.M. Vermote, and I.B. Vipond. 1993. Mechanism of action of restriction endonuclease *Eco*RV. *Nucleic Acids Mol. Biol.* **7:** 47–69.

Hamablet, L., G.C.C. Chen, A. Brown, and R.J. Roberts. 1989. *Lpn*I, from *Legionella pneumoniae*, is a neoschizomer of *Hae*II. *Nucleic Acids Res.* **17:** 6417.

Hedgpeth, J., H.M. Goodman, and H.W. Boyer. 1972. DNA nucleotide sequence restricted by the RI endonuclease. *Proc. Natl. Acad. Sci.* **69:** 3448–3452.

Heitman, J. 1992. How the *Eco*RI endonuclease recognizes and cleaves DNA. *Bioessays* **14:** 445–454.

———. 1993. On the origins, structures and functions of restriction-modification enzymes. *Genet. Eng.* **15:** 57–108.

Heitman, J. and P. Model. 1987. Site-specific methylases induce SOS DNA repair response in *Escherichia coli*. *J. Bacteriol.* **169:** 3243–3250.

———. 1990a. Substrate recognition by the *Eco*RI endonuclease. *Proteins Struct. Funct. Genet.* **7:** 185–197.

———. 1990b. Mutants of the *Eco*RI endonuclease with promiscuous substrate specificity implicate residues involved in substrate recognition. *EMBO J.* **9:** 3369–3378.

Heitman, J., N.D. Zinder, and P. Model. 1989. Repair of the *Escherichia coli* chromosome after *in vivo* scission by the *Eco*RI endonuclease. *Proc. Natl. Acad. Sci.* **86:** 2281–2285.

Hensley, P., G. Nardone, J.G. Chirikjian, and M.E. Wastney. 1990. The time-resolved kinetics of superhelical DNA cleavage by *Bam*HI restriction endonuclease. *J. Biol. Chem.* **265:** 15300–15307.

Howard, K.A., C. Card, J.S. Benner, H.L. Callahan, R. Maunus, K. Silber, G. Wilson, and J.E. Brooks. 1986. Cloning the *Dde*I restriction-modification system using a two-step method. *Nucleic Acids Res.* **14:** 7939–7951.

Hsu, M-T. and P. Berg. 1978. Altering the specificity of restriction endonuclease: Effect of replacing Mg^{++} with Mn^{++}. *Biochemistry* **17:** 131–138.

Ives, C.L., P.D. Nathan, and J.E. Brooks. 1992. Regulation of the *Bam*HI restriction-modification system by a small intergenic open reading frame, *bamHIC*, in both *Escherichia coli* and *Bacillus subtilis*. *J. Bacteriol.* **174:** 7194–7201.

Jack, W.E., B.J. Terry, and P. Modrich. 1982. Involvement of outside DNA sequences in the major kinetic path by which *Eco*RI endoculease locates and leaves its recognition sequence. *Proc. Natl. Acad. Sci.* **79:** 4010–4014.

Janulaitis, A., M. Petrusyte, Z. Maneliene, S. Klimasauskas, and V. Butkus. 1992. Purification and properties of the *Eco*57I restriction endonuclease and methylase—Prototypes of a new class (type IV). *Nucleic Acids Res.* **20:** 6043-6049.

Jeltsch, A., J. Alves, G. Maass, and A. Pingoud. 1992. On the catalytic mechanism of *Eco*RI and *Eco*RV: A detailed proposal based on biochemical results, structural data and molecular modelling. *FEBS Lett.* **304:** 4-8.

Jeltsch, A., J. Alves, T. Oelgeschläger, H. Wolfes, G. Maass, and A. Pingoud. 1993. Mutational analysis of the function of Gln115 in the *Eco*RI restriction endonuclease. *J. Mol. Biol.* **229:** 221-234.

Jencks, W.P. 1975. Binding energy, specificity, and enzyme catalysis; the Circe effect. *Adv. Enzymol. Relat. Areas Mol. Biol.* **43:** 219-410.

Kang, C. and C.-W. Wu. 1987. Studies on SP6 promoter using a new plasmid vector that allows gene insertion at the transcription initiation site. *Nucleic Acids Res.* **15:** 2279-2294.

Kelly, T.J. and H.O. Smith. 1970. A restriction enzyme from *Haemophilus influenzae*. II. Base sequence of the recognition site. *J. Mol. Biol.* **51:** 393-409.

Kim, R., P. Modrich, and S.-H. Kim. 1984. "Interactive" recognition in *Eco*RI restriction enzyme-DNA complex. *Nucleic Acids Res.* **12:** 7285-7292.

Kim, Y., J.C. Grable, R. Love, P. Greene, and J.M. Rosenberg. 1990. Refinement of *Eco*RI endonuclease crystal structure: A revised protein chain tracing. *Science* **249:** 1307-1309.

King, K., S.J. Benkovic, and P. Modrich. 1989. Glu-111 is required for activation of the DNA cleavage center of *Eco*RI endonuclease. *J. Biol. Chem.* **264:** 11807-11815.

Kiss, A., G. Posfai, C.C. Keller, P. Venetianer, and R.J. Roberts. 1985. Nucleotide sequence of the *Bsu*RI restriction-modification system. *Nucleic Acids Res.* **13:** 6403-6420.

Kita, K., H. Kotani, H. Sugisaki, and M. Takanami. 1989. The *Fok*I restriction-modification system. I. Organization and nucleotide sequences of the restriction and modification genes. *J. Biol. Chem.* **264:** 5751-5756.

Kita, K., H. Kotani, H. Ohta, H. Yanase, and N. Kato. 1992. *Sts*I, a new *Fok*I isoschizomer from *Streptococcus sanguis* 54, cleaves 5'GGATG(N)10/14 3'. *Nucleic Acids Res.* **20:** 618.

Kleid, D., Z. Humayun, A. Jeffrey, and M. Ptashne. 1976. Novel properties of a restriction endonuclease isolated from *Haemophilus parahaemolyticus*. *Proc. Natl. Acad. Sci.* **73:** 293-297.

Klimasauskas, S., A. Timinskas, S. Menkevicius, D. Butkiene, V. Butkus, and A.A. Janulaitis. 1989. Sequence motifs characteristic of DNA[cytosine-N4]methylases: Similarity to adenine and cytosine-C5 DNA-methylases. *Nucleic Acids Res.* **17:** 9823-9832.

Kong, H., R.D. Morgan, R.E. Maunus, and I. Schildkraut. 1993. A unique restriction endonuclease, *Bcg*I, from *Bacillus coagulans*. *Nucleic Acids Res.* **21:** 987-991.

Kossykh, V.G., A.V. Repyk, and S. Hattman. 1993. Sequence motifs common to the *Eco*RII restriction endonuclease and the proposed sequence specificity domain of three DNA-[cytosine-C5] methyltransferases. *Gene* **125:** 65-68.

Krüger, D.H., G.J. Barcak, M. Reuter, and H.O. Smith. 1988. *Eco*RII can be activated to cleave refractory DNA recognition sites. *Nucleic Acids Res.* **16:** 3997-4008.

Lacks, S. and B. Greenberg. 1977. Complementary specificity of restriction endonucleases of *Diplococcus pneumoniae* with respect to DNA methylation. *J. Mol. Biol.* **114:** 153-168.

Lacks, S.A., B.M. Mannarelli, S.S. Springhorn, and B. Greenberg. 1986. Genetic basis of the complementary *Dpn*I and *Dpn*II restriction systems of *S. pneumoniae*: An intercellular cassette mechanism. *Cell* **46:** 993-1000.

Lane, A.N., T.C. Jenkins, T. Brown, and S. Neidle. 1991. Interaction of berenil with the *Eco*RI dodecamer d(CGCGAATTCGCG)2 in solution studied by NMR. *Biochemistry* **28:** 1372–1385.

Laue, F., L.R. Evans, M. Jarsch, N.L. Brown, and C. Kessler. 1991. A complex family of class-II restriction endonucleases, *Dsa*I-VI, in *Dactylococcopsis salina*. *Gene* **97:** 87–95.

Lauster, R., T.A. Trautner, and M. Noyer-Weidner. 1989. Cytosine-specific type II DNA methyltransferases: A conserved enzyme core with variable target-recognizing domains. *J. Mol. Biol.* **206:** 305–312.

Lehman, I.R. 1974. DNA ligase: Structure, mechanism and function. *Science* **186:** 790–797.

Lesser, D.R., M.R. Kurpiewski, and L. Jen-Jacobson. 1990. The energetic basis of specificity in the *Eco*RI endonuclease-DNA interaction. *Science* **250:** 776–786.

Lesser, D.R., M.R. Kurpiewski, T. Waters, B.A. Connolly, and L. Jen-Jacobson. 1993. Facilitated distortion of the DNA site enhances *Eco*RI endonuclease-DNA recognition. *Proc. Natl. Acad. Sci.* **90:** 7548–7552.

Lesser, D.R., A. Grajkowski, M.R. Kurpiewski, M. Koziolkiewicz, W.J. Stec, and L. Jen-Jacobson. 1992. Stereoselective interaction with chiral phosphorothiates at the central DNA kink of the *Eco*RI endonuclease-GAATTC complex. *J. Biol.Chem.* **267:** 24810–24818.

Li, L. and S. Chandrasegaran. 1993. Alteration of the cleavage distance of *Fok*I restriction endonuclease by insertion mutagenesis. *Proc. Natl. Acad. Sci.* **90:** 2764–2768.

Li, L., L.P. Wu, and S. Chandrasegaran. 1992. Functional domains in *Fok*I restriction endonuclease. *Proc. Natl. Acad. Sci.* **89:** 4275–4279.

Lu, A.-L., W.E. Jack, and P. Modrich. 1983. DNA determinants important in sequence recognition by *Eco*RI endonuclease. *J. Biol. Chem.* **256:** 13200–13206.

Luisi, B.F. and P.B. Sigler. 1990. The stereochemistry and biochemistry of the *trp* repressor-operator complex. *Biochim. Biophys. Acta* **1048:** 113–126.

Luria, S.E. and M.L. Human. 1952. A nonhereditary, host-induced variation of bacterial viruses. *J. Bacteriol.* **64:** 557–569.

Mann, M.B., R.N. Rao, and H.O. Smith. 1978. Cloning of restriction and modification genes in *E. coli*: The *Hha*II system from *Haemophilus haemolyticus*. *Gene* **3:** 97–112.

Marinus, M.G. and N.R. Morris. 1973. Isolation of deoxyribonucleic acid methylase mutants of *Escherichia coli* K-12. *J. Bacteriol.* **114:** 1143–1150.

Maxwell, A. and S.E. Halford. 1982. The *Sal*GI restriction endonuclease: Mechanism of DNA cleavage. *Biochem. J.* **203:** 85–92.

McClarin, J.A., C.A. Frederick, B.-C. Wang, P. Greene, H.W. Boyer, J. Grable, and J.M. Rosenberg. 1986. Structure of the DNA-*Eco*RI restriction endonuclease recognition complex at 3Å resolution. *Science* **234:** 1526–1541.

McClelland, M. and M. Nelson. 1988. The effect of site-specific DNA methylation on restriction endonucleases and DNA modification methyltransferases—A review. *Gene* **74:** 291–304.

McLaughlin, L.W., F. Bensler, E. Graeser, N. Pile, and S. Scholtissek. 1987. Effects of functional group changes in the *Eco*RI recognition site on the cleavage reaction catalysed by the endonuclease. *Biochemistry* **26:** 7238–7245.

Middleton, J.H., M.H. Edgell, and C.A. Hutchison III. 1972. Specific fragments of φX174 deoxyribonucleic acid produced by a restriction enzyme from *Haemophilus aegyptius*, endonuclease Z. *J. Virol.* **10:** 42–50.

Modrich, P. and R.J. Roberts. 1982. Type-II restriction and modification enzymes. In *Nucleases* (eds. S.M. Linn and R.J. Roberts), pp. 109–154. Cold Spring Harbor

Laboratory, Cold Spring Harbor, New York.
Molloy, P.L. and R.H. Symons. 1980. Cleavage of DNA.RNA hybrids by type II restriction enzymes. *Nucleic Acids Res.* **8:** 2939-2946.
Nardone, G., J. George, and J.G. Chirikjian. 1986. Differences in the kinetic properties of *Bam*HI endonuclease and methylase with linear DNA substrates. *J. Biol. Chem.* **261:** 12128-12133.
Needels, M.C., S.R. Fried, R. Love, J.M. Rosenberg, H.W. Boyer, and P. Greene. 1989. Determinants of *Eco*RI endonuclease sequence discrimination. *Proc. Natl. Acad. Sci.* **86:** 3579-3583.
Nelson, M., E. Raschke, and M. McClelland. 1993. Effect of site-specific methylation on restriction endonucleases and DNA modification methyltransferases. *Nucleic Acids Res.* **21:** 3139-3154.
Newman, A.K., R.A. Rubin, S.H. Kim, and P. Modrich. 1981. DNA sequences of structural genes for *Eco*RI DNA restriction and modified enzymes. *J. Biol. Chem.* **256:** 2131-2139.
Newman, P.C., D.M. Williams, R. Cosstick, F. Seela, and B.A. Connolly. 1990a. Interaction of the *Eco*RV restriction endonuclease with the deoxyadenosine and thymidine bases in its recognition hexamer d(GATATC). *Biochemistry* **29:** 9902-9910.
Newman, P.C., V.U. Nwosu, D.M. Williams, R. Cosstick, F. Seela, and B.A. Connolly. 1990b. Incorporation of a complete set of deoxyadenosine and thymidine analogues suitable for the study of protein nucleic acid interactions into oligodeoxynucleotides. *Biochemistry* **29:** 9891-9901.
Nolling, J. and W.M. De Vos. 1992. Characterization of the archaeal, plasmid-encoded type II restriction-modification system *Mth*TI from *Methanobacterium thermoformicicum* THF: Homology to the bacterial *Ngo*PII system from *Neisseria gonorrhoeae*. *J. Bacteriol.* **174:** 5719-5726.
Oller, A.R., W. Vanden Broek, M. Conrad, and M.D. Topal. 1991. Ability of DNA and spermidine to affect the activity of restriction endonucleases from several bacterial species. *Biochemistry* **30:** 2543-2549.
Olsen, D.B., G. Kotzorek, and F. Eckstein. 1990. Investigation of the inhibitory role of phosphorothioate internucleotide linkages on the catalytic activity of the restriction endonuclease *Eco*RV. *Biochemistry* **29:** 9546-9551.
Orekhov, A.V., B.A. Rebentish, and V.G. Debabov. 1982. A new site-specific endonuclease from Streptomyces—*Sgr*II. *Dokl. Akad. Nauk SSSR* **263:** 217-220.
Pein, C.-D., M. Reuter, A. Meisel, D. Cech, and D.H. Krüger. 1991. Activation of restriction endonuclease *Eco*RII does not depend on the cleavage of stimulator DNA. *Nucleic Acids Res.* **19:** 5139-5142.
Petrusyte, M.P., J.B. Bitinaite, D.R. Kersulyte, C.J. Menkavicius, V.V. Butkus, and A.A. Janulaitis. 1987. New types of restriction endonucleases. *Dokl. Akad. Nauk SSSR* **295:** 1250-1253.
Polisky, B., P. Greene, D.E. Garfin, B.J. McCarthy, H.M. Goodman, and H.W. Boyer. 1975. Specificity of substrate recognition by the *Eco*RI restriction endonuclease. *Proc. Natl. Acad. Sci.* **72:** 3310-3314.
Posfai, J., A.S. Bhagwat, G. Posfai, and R.J. Roberts. 1989. Predictive motifs derived from cytosine methyltransferases. *Nucleic Acids Res.* **17:** 2421-2435.
Potter, B.V.L. and F. Eckstein. 1984. Cleavage of phosphorothioate-substituted DNA by restriction endonucleases. *J. Biol. Chem.* **259:** 14243-14248.
Price, C., J. Lingner, T.A. Bickle, K. Firman, and S.W. Glover. 1989. Basis for changes in DNA recognition by the *Eco*R124 and *Eco*R124/3 type I DNA restriction and modification enzymes. *J. Mol. Biol.* **205:** 115-125.

Raleigh, E.A. and G. Wilson. 1986. *Escherichia coli* restricts DNA containing 5-methylcytosine. *Proc. Natl. Acad. Sci.* **83**: 9070-9074.

Renbaum, P., D. Abrahamove, A. Fainsod, G.G. Wilson, S. Rottem, and A. Razin. 1990. Cloning, characterization, and expression in *Escherichia coli* of the gene coding for the CpG DNA methylase from *Spiroplasma sp.* strain MQ1 (M.*Sss*I). *Nucleic Acids Res.* **18**: 1145-1152.

Richey, B., D.S. Caley, M.C. Mossing, C. Kolka, C.F. Anderson, T.C. Farrar, and M.T. Record, Jr. 1987. Variability of the intracellular ionic environment of *Escherichia coli. J. Biol. Chem.* **262**: 7157-7164.

Roberts, R.J. 1976. Restriction endonucleases. *CRC Crit. Rev. Biochem.* **4**: 123-164.

Roberts, R.J. and D. Macelis. 1993. Restriction enzymes and methylases. *Nucleic Acids Res.* **21**: 3125-3137.

Rosenberg, J.M. 1991. Structure and function of restriction endonucleases. *Curr. Opin. Struct. Biol.* **1**: 104-113.

Rosenberg, J.M. and P. Greene. 1982. *Eco*RI* specificity and hydrogen bonding. *DNA* **1**: 117-124.

Rubin, R.A. and P. Modrich. 1978. Substrate dependence of the mechanism of *Eco*RI endonuclease. *Nucleic Acids Res.* **5**: 2991-2997.

Seeber, S., C. Kessler, and F. Gotz. 1990. Cloning, expression and characterization of the *Sau*3AI restriction and modification genes in *Staphylococcus carnosus* TM300. *Gene* **94**: 37-43.

Seeman, N.C., J.M. Rosenberg, and A. Rich. 1976. Sequence specific recognition of double-helical nucleic acids by proteins. *Proc. Natl. Acad. Sci.* **73**: 804-808.

Selent, U., T. Rüter, E. Köhler, M. Liedtke, V. Thielking, J. Alves, T. Oelgeschläger, H. Wolfes, F. Peters, and A. Pingoud. 1992. A site-directed mutagenesis study to identify amino acid residues involved in the catalytic function of the restriction endonuclease *Eco*RV. *Biochemistry* **31**: 4808-4815.

Sharp, P.A., B. Sugden, and J. Sambrook. 1973. Detection of two restriction endonuclease activities in *Haemophilus parainfluenzae* using analytical agarose-ethidium bromide electrophoresis. *Biochemistry* **12**: 3055-3063.

Sharp, P.M., J.E. Kelleher, A.S. Daniel, G.M. Cowan, and N.E. Murray. 1992. Roles of selection and recombination in the evolution of type I restriction-modification systems in enterobacteria. *Proc. Natl. Acad. Sci.* **89**: 9836-9840.

Smith, D.W., S.W. Crowder, and N.O. Reich. 1992. *In vivo* specificity of *Eco*RI methyltransferase. *Nucleic Acids Res.* **20**: 6091-6096.

Smith, H.O. and K.W. Wilcox. 1970. A restriction enzyme from *Hemophilus influenzae*. I. Purification and general properties. *J. Mol. Biol.* **51**: 379-391.

Smith, M.D., M. Longo, G.F. Gerard, and D.K. Chatterjee. 1992. Cloning and characterization of genes for the *Pvu*I restriction and modification system. *Nucleic Acids Res.* **20**: 5743-5747.

Steitz, T.A. 1993. DNA- and RNA-dependent DNA polymerases. *Curr. Opin. Struct. Biol.* **3**: 31-38.

Stephenson, F.H., B.T. Ballard, H.W. Boyer, J.M. Rosenberg, and P.J. Greene. 1989. Comparison of the nucleotide and amino acid sequences of the *Rsr*I and *Eco*RI restriction endonucleases. *Gene* **85**: 1-13.

Stobberingh, E.E., R. Schiphof, and J.S. Sussenbach. 1977. Occurrence of a class II restriction endonuclease in *Staphylococcus aureus. J. Bacteriol.* **131**: 645-649.

Stöver, T., E. Köhler, U. Fagin, W. Wende, H. Wolfes, and A. Pingoud. 1993. Determination of the DNA bend angle induced by the restriction endonuclease *Eco*RV in the presence of Mg^{++}. *J. Biol. Chem.* **268**: 8645-8650.

Strzelecka, T., L. Dorner, I. Schildkraut, and A. Aggarwal. 1990. Structural studies of the *Bam*HI restriction enzyme. *Biophys. J.* **57**: 68.

Sugisaki, H. and S. Kanazawa. 1981. New restriction endonucleases from *Flavobacterium okeanokoites* (*Fok*I) and *Micrococcus luteus* (*Mlu*I). *Gene* **16**: 73–78.

Sullivan, K.M. and J.R. Saunders. 1989. Nucleotide sequence and genetic organization of the *Ngo*PII restriction-modification system of *Neisseria gonorrhoeae*. *Mol. Gen. Genet.* **216**: 380–387.

Sullivan, K.M., H.J. Macdonald, and J.R. Saunders. 1987. Characterization of DNA restriction and modification activities in *Neisseria* species. *FEMS Microbiol. Lett.* **44**: 389–393.

Szomolanyi, I., A. Kiss, and P. Venetianer. 1980. Cloning the modification methylase gene of *Bacillus sphaericus* R in *Escherichia coli*. *Gene* **10**: 219–225.

Szybalski, W., S.C. Kim, N. Hasan, and A.J. Podhajska. 1991. Class-IIs restriction enzymes—A review. *Gene* **100**: 13–26.

Tao, T. and R.M. Blumenthal. 1992. Sequence and characterization of *pvuIIR*, the *Pvu*II endonuclease gene, and of *pvuIIC*, its regulatory gene. *J. Bacteriol.* **174**: 3395–3398.

Tao, T., J.C. Bourne, and R.M. Blumenthal. 1991. A family of regulatory genes associated with type II restriction-modification systems. *J. Bacteriol.* **173**: 1367–1375.

Taylor J.D. and S.E. Halford. 1989. Discrimination between DNA sequences by the *Eco*RV restriction endonuclease. *Biochemistry* **28**: 6198–6207.

———. 1992. The activity of the *Eco*RV restriction endonuclease is influenced by DNA sequences both inside and outside the DNA-protein complex. *Biochemistry* **31**: 90–97.

Taylor, J.D., I.G. Badcoe, A.R. Clarke, and S.E. Halford. 1991. *Eco*RV restriction endonuclease binds all DNA sequences with equal affinity. *Biochemistry* **30**: 8743–8753.

Taylor, J.D., A.J. Goodall, C.L. Vermote, and S.E. Halford. 1990. Fidelity of DNA recognition by the *Eco*RV restriction/modification system *in vivo*. *Biochemistry* **29**: 10727–10733.

Taylor, J.W., J. Ott, and F. Eckstein. 1985. The rapid generation of oligonucleotide-directed mutations at high frequency using phosphorothioate-modified DNA. *Nucleic Acids Res.* **13**: 8765–8785.

Terry, B.J., W.E. Jack, and P. Modrich. 1985. Facilitated diffusion during catalysis by *Eco*RI endonuclease. *J. Biol. Chem.* **260**: 13130–13137.

———. 1987. Mechanism of specific site location and DNA cleavage by *Eco*RI endonuclease. In *Gene amplification and analysis: Restriction endonucleases and methylases* (ed. J.G. Chirikjian), vol. 5, pp. 103–118. Elsevier, Amsterdam.

Terry, B.J., W.E. Jack, R.A. Rubin, and P. Modrich. 1983. Thermodynamic parameters governing interactions of *Eco*RI endonuclease with specific and non-specific DNA sequences. *J. Biol. Chem.* **258**: 9820–9825.

Thielking, V., J. Alves, A. Fliess, G. Maass, and A. Pingoud. 1990. Accuracy of the *Eco*RI restriction endonuclease: Binding and cleavage studies with oligodeoxynucleotide substrates containing degenerate recognition sequences. *Biochemistry* **30**: 4682–4691.

Thielking, V., U. Selent, E. Köhler, H. Wolfes, U. Pieper, R. Geiger, C. Urbanke, F.K. Winkler, and A. Pingoud. 1991. Site-directed mutagenesis studies with *Eco*RV restriction endonuclease to identify regions involved in recognition and catalysis. *Biochemistry* **30**: 6416–6422.

Thomas, G.A., W.L. Kubasek, W.L. Peticolas, P. Greene, J. Grable, and J.M. Rosenberg. 1989. Environmentally induced conformational changes in B-type DNA. *Biochemistry* **28**: 2001–2009.

Thomas, M. and R.W. Davis. 1975. Studies on the cleavage of bacteriophage λ DNA with EcoRI restriction endonuclease. *J. Mol. Biol.* **91:** 315-328.

Tolstoshev, C.M. and R.W. Blakesley. 1982. RSITE: A computer program to predict the recognition sequence of a restriction enzyme. *Nucleic Acids Res.* **10:** 1-17.

Topal, M.D., R.J. Thresher, M. Conrad, and J. Griffith. 1991. NaeI endonuclease binding to pBR322 DNA induces looping. *Biochemistry* **30:** 2006-2010.

Ueno, T., H. Ito, F. Kimizuka, H. Kotani, and K. Nakajima. 1993. Gene structure and expression of the MboI restriction-modification system. *Nucleic Acids Res.* **21:** 2309-2313.

Van Cleve, M.D. and R.I. Gumport. 1992. Influence of enzyme-substrate contacts located outside the EcoRI recognition site on cleavage of duplex oligodeoxyribonucleotide substrates by EcoRI endonuclease. *Biochemistry* **31:** 334-339.

Venditti, S. and R.D. Wells. 1991. A DNA conformational alteration induced by a neighboring oligopurine tract on GAATTA enables nicking by EcoRI. *J. Biol. Chem.* **266:** 16786-16790.

Vermote, C.L.M. and S.E. Halford. 1992. EcoRV restriction endonuclease: Communication between catalytic metal ions and DNA recognition. *Biochemistry* **31:** 6082-6089.

Vermote, C.L.M., I.B. Vipond, and S.E. Halford. 1992. EcoRV restriction endonuclease: Communication between DNA recognition and catalysis. *Biochemistry* **31:** 6089-6097.

Vipond, I.B. and S.E. Halford. 1993. Structure-function correlation for the EcoRV restriction enzyme: From non-specific binding to specific DNA cleavage. *Mol. Microbiol.* **9:** 225-231.

von Hippel, P.H. and O.G. Berg. 1989. Facilitated target location in biological systems. *J. Biol. Chem.* **264:** 675-678.

Waalwijk, C. and R.A. Flavell. 1978. MspI, an isoschizomer of HpaII which cleaves both unmethylated and methylated HpaII sites. *Nucleic Acids Res.* **5:** 3231-3236.

Walder, R.Y., C.J. Langtimm, R. Catterjee, and J.A. Walder. 1983. Cloning of the MspI modification enzyme. *J. Biol. Chem.* **258:** 1235-1241.

Waters, T.R. and B.A. Connolly. 1992. Continuous spectrophotometric assay for restriction endonucleases using synthetic oligodeoxynucleotides and based on the hyperchromic effect. *Anal. Biochem.* **204:** 204-209.

―――. 1993. Interactions of the EcoRV restriction endonuclease with the dG and dC bases in its recognition site. *Biochemistry* (in press).

Wilson, G.G. 1991. Organization of restriction-modification systems. *Nucleic Acids Res.* **19:** 2539-2566.

Wilson, G.G. and N.E. Murray. 1991. Restriction and modification systems. *Annu. Rev. Genet.* **25:** 585-627.

Winkler, F.K. 1992. Structure and function of restriction endonucleases. *Curr. Opin. Struct. Biol.* **2:** 93-99.

Winkler, F.K., D.W. Banner, C. Oefner, D. Tsernoglou, R.S. Brown, S.P. Heathman, R.K. Bryan, P.D. Martin, K. Petratos, and K.S. Wilson. 1993. The crystal structure of EcoRV endonuclease and of its complexes with cognate and noncognate DNA fragments. *EMBO J.* **12:** 1781-1795.

Xu, S.-Y. and I. Schildkraut. 1991a. Isolation of BamHI variants with reduced cleavage activities. *J. Biol. Chem.* **266:** 4425-4429.

―――. 1991b. Cofactor requirements of BamHI mutant endonuclease E77K and its suppressor mutants. *J. Bacteriol.* **173:** 5030-5035.

Yang, C.C. and M.D. Topal. 1992. Nonidentical DNA-binding sites of endonuclease NaeI recognize different families of sequences flanking the recognition site. *Biochemistry* **31:** 9657-9664.

Yanofsky, S.D., R. Love, J.A. McClarin, J.M. Rosenberg, H.W. Boyer, and P.J. Greene. 1987. Clustering of null mutants in the *Eco*RI endonuclease. *Proteins Struct. Funct. Genet.* **2**: 273–282.

Yoo, O.J. and K.L. Agarwal. 1980. Cleavage of single strand oligonucleotides and bacteriophage φX174 DNA by *Msp*I endonuclease. *J. Biol. Chem.* **255**: 10559–10562.

Yoshimori, R.N. 1971. "A genetic and biochemical analysis of the restriction and modification of DNA by resistance transfer factors." Ph.D. thesis, University of California, San Francisco.

Zebala, J.A., J. Choi, and F. Barany. 1992a. Characterization of steady-state, single-turnover and binding kinetics of the *Taq*I restriction endonuclease. *J. Biol. Chem.* **267**: 8097–8105.

Zebala, J.A., J. Choi, G. Trainor, and F. Barany. 1992b. DNA recognition of base analogues and chemically modified substrates by the *Taq*I restriction endonuclease. *J. Biol. Chem.* **267**: 8106–8116.

3
The ATP-dependent Restriction Enzymes

Thomas A. Bickle
Microbiology Department
Biozentrum, University of Basel
CH-4056 Basel, Switzerland

I. Introduction
II. Type I Restriction Enzymes
   A. Occurrence
   B. Structural Genes
   C. Family Relationships among Type I Restriction Enzymes
      1. DNA and Protein Sequence Homologies within Families
      2. DNA and Protein Sequence Homologies between Families
   D. Evolution of DNA Sequence Specificity in Type I Enzymes
      1. By Homologous Recombination
      2. By Unequal Crossing Over
      3. By Transposition
   E. Enzyme Structure and Mechanisms
      1. Enzyme Structures
      2. Enzyme Mechanisms
III. Type III Restriction Enzymes
   A. Occurrence
   B. Genetics
   C. Enzyme Mechanism
   D. DNA Cleavage
IV. Concluding Remarks

I. INTRODUCTION

Restriction and modification of phages by bacteria were first described more than 40 years ago by Luria and Human (1952) for T-even phages and by Bertani and Weigle (1953) for the phages P2 and λ. The basic observation was that phages grown on one bacterial strain would grow poorly when tested on certain other related bacterial strains. Upon closer investigation, the poor growth was found to be an all or nothing affair. Most infected cells produced no phage at all; in the terminology that was adopted, the phages were *restricted*, although a few cells produced a normal yield of phage particles. The few phages that did result from the infection were *modified* so that they grew normally on the new host when tested in a second cycle of infection. Often, they were now restricted by the original host. The phenomenon of modification is Lamarckian rather than Mendelian in character since it is an adaptive response of the virus to the host that is lost when the virus is passaged through other cells.

Some 10 years after the discovery of restriction and modification, the first of a series of papers from Arber's laboratory appeared that offered a molecular explanation for the effect (Arber and Dussoix 1962; Dussoix and Arber 1962). Both restriction and modification were properties of the DNA of the infecting phage. Restriction was due to the action of a nuclease that recognized specific sequences in incoming DNA as a signal to cleave it; modification was due to the action of another enzyme on the same sequences that made them resistant to restriction. The modification enzyme was suspected earlier to be a sequence-specific DNA methylase (Arber 1965), although this was not proven until much later (Kuhnlein and Arber 1972; Smith et al. 1972). Although early studies of restriction and modification were done with phages, it was subsequently shown that restriction and modification are active on foreign DNA irrespective of how it enters the cell: in a phage head, by conjugation, or by transformation (for reviews of this early work, see Arber 1968; Arber and Linn 1969; Boyer 1971).

By the beginning of the 1970s, several different restriction enzymes had been purified (Linn and Arber 1968; Meselson and Yuan 1968; Roulland-Dussoix and Boyer 1969; Greene et al. 1974), and it was clear that biochemically, they could be divided into two classes on the basis of their cofactor requirements. One group of enzymes required Mg^{++}, ATP, and S-adenosylmethionine (AdoMet) to cleave DNA, and the second group required only Mg^{++}. Boyer (1971) proposed that the first class be called type I and the second class, type II. The type II enzymes are dealt with in Chapter 2; the ATP-dependent enzymes are the subject of this chapter. It was soon realized that the ATP-dependent restriction enzymes included two fundamentally different groups, which are now called types I and III (Kauc and Piekarowicz 1978). For other recent reviews of these enzymes, see Bickle (1987), Wilson and Murray (1991), and Bickle and Krüger (1993).

II. TYPE I RESTRICTION ENZYMES

A. Occurrence

All of the known type I restriction systems are found in the Enterobacteriaceae (for a possible exception, see Gromkova and Goodgal 1976). The systematic searches that have been made for type II restriction enzymes would not have detected type I enzymes because type I DNA cleavage products do not give discrete bands after gel electrophoresis. However, there is good reason to suspect that the distribution of type I systems is taxonomically limited: They were detected in enteric bacteria by restriction of phages; if they were widespread, one would have expected them to have been discovered by now in other bacterial groups with well-defined genetics, such as Bacilli or Pseudomonads.

B. Structural Genes

Complementation analysis first showed that three genes were involved in type I restriction and modification. These genes are called *hsdS*, *hsdM*, and *hsdR* (where *hsd* derives from *h*ost *s*pecificity for *D*NA). The *hsdS* gene is necessary for both restriction and modification and is responsible for recognition of the DNA sequence specific for the system. The *hsdM* gene is necessary for modification, and the *hsdR* gene, together with the other two genes, is required for restriction (Boyer and Roulland-Dussoix 1969; Glover and Colson 1969; Hubacek and Glover 1970). The genes for many type I systems from *Escherichia coli* and various *Salmonella* species map close to *serB* at 98.5 minutes on the *E. coli* genetic map (Boyer 1964; Arber and Wauters-Willems 1970; Glover 1970; Bullas and Colson 1975; Bullas et al. 1980). More recently, the structural genes for many type I systems from *E. coli*, *Salmonella*, and *Citrobacter freundii* have been cloned and sequenced with results that entirely corroborate the results of the genetic analysis (Sain and Murray 1980; Gough and Murray 1983; Loenen et al. 1987; Cowan et al. 1989; Kannan et al. 1989; Price et al. 1989).

The *hsd* genes are tightly clustered and organized into two transcriptional units, with *hsdM* and *hsdS* transcribed from a promoter in front of the *hsdM* gene; *hsdR* has its own promoter (Sain and Murray 1980; Suri and Bickle 1985; Price et al. 1989). The relative order of the transcriptional units is, however, different among families of enzymes (see next section).

C. Family Relationships among Type I Restriction Enzymes

Fortuitously, all of the early studies on type I restriction were done with systems from *E. coli* and *Salmonella* that showed genetic complementation. These same systems were later shown, not surprisingly, to be physically similar to each other by techniques such as DNA hybridization or immunological methods. More recent studies have complicated this picture: Systems have been found that do not complement the original set and that have very little structural homology with them (Murray et al.

Table 1 The Families of Type I Restriction Enzymes

Enzyme	Recognition sequence[a]	References
Type IA family		
EcoBI	TGA(N$_8$)TGCT	Lautenberger et al. (1978); Ravetch et al. (1978); Sommer and Schaller (1979)
EcoKI	AAC(N$_6$)GTGC	Kan et al. (1979)
EcoDI	TTA(N$_7$)GTCY	Nagaraja et al. (1985c)
StySBI	GAG(N$_6$)RTAYG	Nagaraja et al. (1985b)
StySPI	AAC(N$_6$)GTRC	Nagaraja et al. (1985b)
StySQI	AAC(N$_6$)RTAYG	Nagaraja et al. (1985a)
StySJI[b]	GAG(N$_6$)GTRC	Gann et al. (1987)
Type IB family		
EcoAI	GAG(N$_7$)GTCA	Kröger and Hobom (1984); Suri et al. (1984b)
EcoEI	GAG(N$_7$)ATGC	Cowan et al. (1989)
CfrAI	GCA(N$_8$)GTGG	Kannan et al. (1989)
Type IC family		
EcoDXXI	TCA(N$_7$)RTTC	Piekarowicz and Goguen (1986); Gubler et al. (1992)
StyR124I	GAA(N$_6$)RTCG	Price et al. (1987b)
StyR124/3I	GAA(N$_7$)RTCG	Price et al. (1987b)
EcoRD2I[b]	GAA(N$_6$)RTTC	Gubler et al. (1992)
EcoRD3I[b]	GAA(N$_7$)RTTC	Gubler et al. (1992)
EcoDR2I[b]	TCA(N$_6$)RTCG	Gubler et al. (1992)
EcoDR3I[b]	TCA(N$_7$)RTCG	Gubler et al. (1992)
EcoDXXI::Tn5[c]	TCA(N$_8$)TGA	J. Meister (unpubl.)
EcoprrI	CCA(N$_7$)RTGC	C. Tyndall (unpubl.)

[a]N, any nucleotide; R, either purine; Y, either pyrimidine. The subscripted number within the brackets indicates the number of residues in the nonspecific spacer.
[b]These R/M systems are artificial hybrids generated in the laboratory.
[c]This enzyme has not yet received a formal name. It has a Tn5 transposable element in the middle of the *hsdS* gene.

1982; Price et al. 1987a). The new findings currently define three families of type I systems, known as types IA, B, and C. Within families, genetic complementation occurs and the structural genes have much homology (to be discussed in more detail below). Between families, there is neither genetic complementation nor similarity of the structural genes. All of the families are quite clearly type I: Two of the families are allelic (or pseudoallelic) and the third is plasmid-encoded. They have a similar genetic organization; the enzymes have similar subunit structures, cofactor requirements, and reaction mechanisms; and they all recognize

Type IA

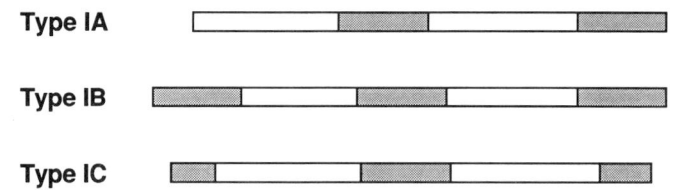

Type IB

Type IC

Figure 1 Structure of *hsdS* genes. (*Stippled bars*) Regions of the genes that are highly homologous among different family members. (*Open bars*) Regions where no homology is found (with the exceptions discussed in the text).

split, asymmetric DNA recognition sequences. The three families are described in Table 1.

1. DNA and Protein Sequence Homologies within Families

Within a given family, the DNA and therefore amino sequences of the *hsdM* and *hsdR* genes are quite highly conserved (Murray et al. 1982; Daniel et al. 1988; Gubler et al. 1992; Sharp et al. 1992). The most careful study has been done by Sharp et al. (1992), who compared the DNA and deduced amino acid sequences of *hsdM* genes and found that the proteins have on average 95% amino acid sequence identity. The situation is quite different for the *hsdS* genes and proteins. They contain two extensive regions of nonhomology, one at the 5′ end of the genes and the other toward the 3′ end, flanked by homologous regions (Fig. 1) (Gough and Murray 1983; Cowan et al. 1989; Gubler et al. 1992). As shown below, each of the nonhomologous, hypervariable regions encodes a protein domain that recognizes one half of the recognition sequence. The conserved regions are thought to provide protein-protein interactions with the HsdM and HsdR proteins.

2. DNA and Protein Sequence Homologies between Families

As stated earlier, there is no extensive sequence homology between members of different families. The weak homologies that are found, however, are sufficient to indicate that all families had a common ancestor in the distant past. For example, Sharp et al. (1992) found that the *hsdM* genes of different families have about 26–33% identical amino acids, depending on the particular pairwise comparison that was made. Although a few of these conserved residues are found in all DNA adenine methylases (Lauster et al. 1987; Guschlbauer 1988; Narva et al. 1988; Smith et al. 1990), it was suggested that the degree of homology was high enough to exclude an independent origin of the different families.

An important exception to the rule that different families do not resemble each other concerns those regions of the *hsdS* genes that *within* a family are hypervariable. The type IA enzyme *Sty*SBI and the type IB enzymes *Eco*AI and *Eco*EI all have the sequence GAG as the 5' moiety of their recognition sequences (Table 1). Cowan et al. (1989) compared the amino acid sequences of these three enzymes and found that there was considerable homology between the families (50% identity) in the region of the amino-terminal hypervariable region, supporting the idea, discussed in the next section, that this region of the protein recognizes the 5' moiety of the recognition sequence.

D. Evolution of DNA Specificity in Type I Enzymes

1. By Homologous Recombination

Type I restriction systems are the only specific DNA-binding proteins that have been shown to change their specificity spontaneously by purely natural means. This was first seen in the type IA family where transduction of the *Sty*SPI *hsd* genes into cells expressing *Sty*SBI led to the appearance of a new specificity, which was called *Sty*SQI (Bullas et al. 1976). It was later shown by both DNA heteroduplex analysis and DNA sequencing that the *Sty*SQI system arose by recombination within the central conserved region of the parental *hsdS* genes (Fuller-Pace et al. 1984; Fuller-Pace and Murray 1986). The recognition sequence of *Sty*SQI is shown in Table 1 (Nagaraja et al. 1985a,b). The sequence is a hybrid between that of the two parents, with a 5'-trinucleotide moiety identical to that recognized by *Sty*SPI and a 3' moiety identical to that of *Sty*SBI. These findings immediately led to the idea that *hsdS* genes encode two DNA-binding domains: an amino-terminal domain that recognizes the 5' moiety of the recognition sequences and a carboxy-terminal domain that recognizes the 3' moiety. The recombination event that produced the *Sty*SQI system would have reassorted these domains. This idea was strengthened when the reciprocal recombinant between the *Sty*SBI and *Sty*SPI *hsdS* genes was made (called *sty*SJI; Table 1) and shown to have the specificity predicted by the model (Gann et al. 1987).

2. By Unequal Crossing Over

Another kind of spontaneous change in sequence specificity is seen with the type IC systems, *Sty*R124I and *Sty*R124/3I. Cells expressing either one of these systems will, at low frequency, switch to expressing the other system with concomitant loss of expression of the first system (Glover et al. 1983). The DNA sequences recognized by *Sty*R124I and *Sty*R124/3I are shown in Table 1 (Price et al. 1987b). The only difference between them is in the length of the nonspecific spacer separating the two specific moieties, which is six base pairs long for *Sty*R124I and seven base pairs long for *Sty*R124/3I. The extra base pair in the *Sty*R124/3I spacer moves the two specific moieties 3.4 Å farther apart and rotates them by 36°, not a trivial difference in terms of enzyme recognition. The explanation both for the structure of the recognition sequences and for the spontaneous switch in specificity became clear after the *hsdS* genes of the two systems were sequenced (Price et al. 1989). The genes are identical except for a small region within the central conserved region of the genes, where *Sty*R124I has an inframe 12-bp sequence repeated twice, whereas *Sty*124/3I has the same sequence repeated three times. Unequal crossing over between these repeated sequences must be the mechanism of interconversion between the two forms of the gene. The extra four amino acids in the *Sty*R124/3I sequence, if in α-helical conformation, would span approximately 6 Å, leading to the hypothesis that this region of the protein forms a spacer separating the two DNA recognizing domains. Thus, the length of the spacer would determine the number of nonspecific base pairs separating the two specific moieties of the recognition sequence.

This hypothesis predicts that it is the length of the spacer rather than its exact amino acid sequence that is important for function, and, in fact, several amino acid changes could be made in the repeated sequences of *Sty*R124I without measurable effects on either activity or specificity (Gubler and Bickle 1991). Mutants were constructed with changes in the length of the repeated region, and these had drastic effects on both activity and specificity. Variants with either one or four copies of the repeat, for example, were virtually inactive in restriction (10^5 to 10^6 times less active than the wild type). However, they were still efficient modification enzymes: The mutant with a single copy of the repeat methylated the *Sty*R124I recognition sequence, whereas the mutant with four repeats methylated both the *Sty*R124I and the *Sty*R124/3I sequences but would not methylate a putative recognition sequence with eight base pairs in the nonspecific spacer (Gubler and Bickle 1991). These results explain why variants with one or four copies of the repeat, which would arise naturally by the same mechanism that switches between *Sty*R124I and *Sty*R124/3I, have never been recovered; they would score as restriction-negative. The fact that the mutations have severe effects on restriction without affecting modification is probably because they affect the structure of the central conserved regions of the *hsdS* gene products. This region of the protein most likely has a dual function: It acts as a spacer between the DNA-binding domains of the protein and it also provides the surface for interaction with the *hsdR* gene product.

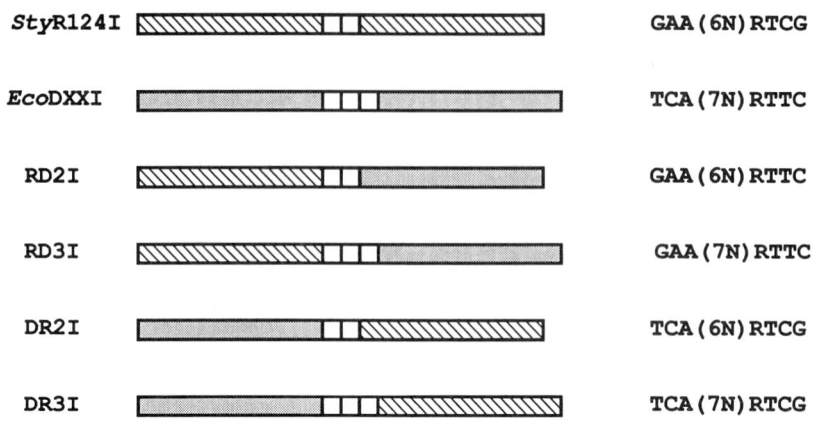

StyR124I	[diagram]	GAA(6N)RTCG
EcoDXXI	[diagram]	TCA(7N)RTTC
RD2I	[diagram]	GAA(6N)RTTC
RD3I	[diagram]	GAA(7N)RTTC
DR2I	[diagram]	TCA(6N)RTCG
DR3I	[diagram]	TCA(7N)RTCG

Figure 2 Hybrids between the *Eco*DXXI and *Eco*R124 *hsdS* genes. (*Open boxes*) Number of copies of the repeated sequence discussed in the text. The sequences recognized by the wild-type and hybrid proteins are shown to the right of the structures.

The idea that *hsdS* genes contain two DNA-binding domains separated by a spacer region whose length determines the number of base pairs separating the specific moieties of the recognition sequences was tested using the type IC enzyme, *Eco*DXXI (Piekarowicz et al. 1985; Skrzypek and Piekarowicz 1989). This enzyme recognizes a DNA sequence totally different from that recognized by *Sty*R124I or *Sty*R124/3I (Table 1). Its *hsdS* gene is homologous to the *hsdS* gene of *Sty*124/3I in the central region and it contains three copies of the 12-bp repeated sequence (J. Meister et al., unpubl.). All possible hybrids between the two halves of *Eco*DXXI and *Sty*R124I with either two or three copies of the 12-bp repeated sequence were constructed (Fig. 2). All of the constructs gave active restriction and modification systems that recognized DNA sequences entirely consistent with the postulate given above (Fig. 2).

3. By Transposition

A third way in which type IC restriction systems can modify their specificity has been discovered recently (E. Skrzypek and A. Piekarowicz, unpubl.). A Tn5 derivative of the *Eco*DXXI *hsd* region was isolated that appeared to have an altered DNA sequence specificity. Upon further investigation, it was found that the Tn5 had inserted into the *hsdS* gene, just 3' to the central conserved region. The *hsdS* gene product produced by the mutant had approximately half the molecular weight of the wild-type protein. This was confirmed by isolating the protein from an SDS-polyacrylamide gel and determining its amino-terminal amino acid sequence. The DNA sequence recognized by the mutant protein was determined to be TCA(N$_8$)TGA (J. Meister et al., unpubl.).

This sequence resembles type II restriction enzyme recognition sequences in that it is twofold rotationally symmetric. It consists of two copies of the trimeric moiety of the *Eco*DXXI recognition sequence in inverted repeat configuration. The conclusion is inescapable that the mutant enzyme must assemble two copies of the truncated *hsdS* gene product, each of which recognizes one half of the recognition sequence. The protein totally lacks the carboxy-terminal conserved region that is thought to be important for protein-protein interactions in other type I systems. However, for *Eco*DXXI, a short stretch of amino acid sequence at the carboxyl terminus of the *hsdS* gene product is repeated in the center of the protein, and it is perhaps this factor that allows the assembly of a functional enzyme.

E. Enzyme Structures and Mechanisms

1. Enzyme Structures

For all three classes of type I systems, both monofunctional modification methylases and multifunctional restriction/modification (R/M) enzymes have been isolated (Linn and Arber 1968; Meselson and Yuan 1968; Lautenberger and Linn 1972; Suri et al. 1984a,b; Price et al. 1987a; Gubler and Bickle 1991). The modification enzymes contain the products of the *hsdM* and *hsdS* genes in a stoichiometry that for the type IA enzyme *Eco*BI has been reported to be 1:1 and for the type IC enzyme *Sty*R124I has been reported to be 2:1 (Lautenberger and Linn 1972; Taylor et al. 1992). However, it should be noted that both kinds of type I enzymes have been reported to undergo aggregation and disproportionation upon storage (Eskin and Linn 1972; Lautenberger and Linn 1972). The multifunctional R/M enzymes contain the products of the *hsdM*, *hsdR*, and *hsdS* genes with a most likely stoichiometry of 2:2:1. For the type IB enzyme *Eco*AI, the association between the *hsdR* product and the modification enzyme seems to be particularly weak, such that an R/M enzyme cannot be purified. Instead, a modification enzyme and an *hsdR* subunit can be purified, and when mixed, reconstitute restriction activity (Suri and Bickle 1985).

It is clear from the genetic studies that the HsdR protein is necessary for restriction, the HsdM protein is necessary for modification, and the HsdS protein is required for both reactions because it mediates the sequence-specific DNA interactions. More recent studies corroborate this picture. For example, the *hsdS* subunit of *Sty*R124I has recently been isolated and shown to be a sequence-specific DNA-binding protein with no

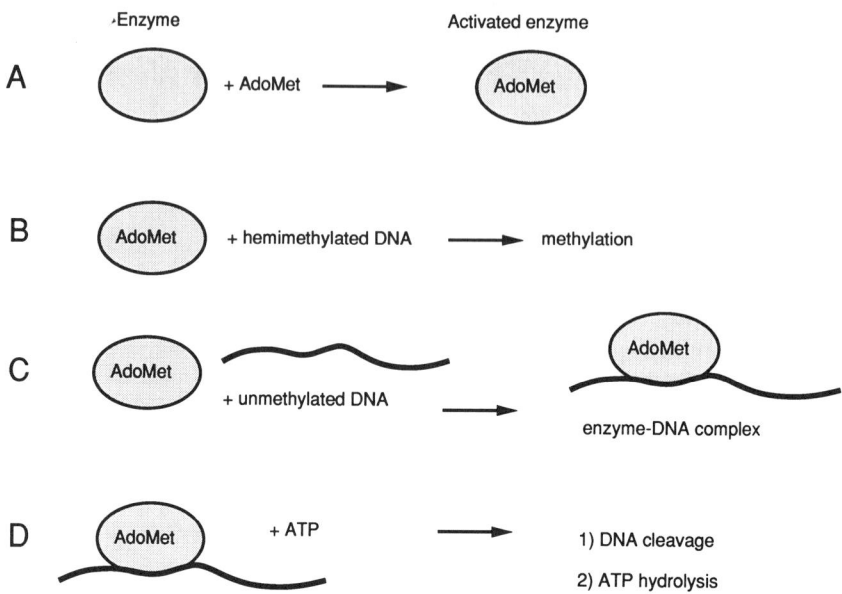

Figure 3 Reaction mechanism of *Eco*KI. (*A*) The enzyme first binds and is allosterically activated by AdoMet. (*B*) Hemimethylated substrate DNA is methylated in the other strand. (*C*) The activated enzyme forms stable complexes with unmethylated recognition sites. (*D*) Addition of ATP to the enzyme-DNA complex leads to DNA cleavage at random positions far from the recognition sites, followed by the transformation of the enzyme into a DNA-dependent ATPase.

discernible enzymatic activities (Kusiak et al. 1992). Again, amino acid sequence motifs typical of AdoMet- and ATP-binding proteins are found in the *hsdM* and *hsdR* subunits, respectively (Loenen et al. 1987).

2. Enzyme Mechanisms

Most investigations of enzymatic mechanisms have been with the type IA enzymes, *Eco*KI and *Eco*BI (for review, see Bickle 1982). The main features of the reaction mechanism of the R/M enzymes are summarized in Figure 3, which was derived mainly from data with *Eco*KI. The enzyme in the absence of AdoMet has no detectable affinity for DNA but is allosterically activated by AdoMet to a form that binds with high affinity to both modified and nonmodified recognition sites (Bickle et al. 1978). Free AdoMet can then be removed (e.g., by gel filtration), and the reactions leading to restriction can still go to completion. In the absence of ATP, these complexes with DNA are relatively stable, with a half-life of 6 minutes for a modified site and 22 minutes for a nonmodified site (Yuan et al. 1975).

The result of addition of ATP to the enzyme-AdoMet-DNA complexes depends on the methylation state of the recognition site. If the site is methylated on both strands, ATP reduces the affinity of the enzyme for the DNA such that the enzyme is released. If the site has a methyl group in one strand of the DNA only (like the product of DNA replication), ATP stimulates the methylation of the other strand and the DNA is not cleaved (Vovis et al. 1974; Burckhardt et al. 1981; Suri et al. 1984a). Even in the absence of ATP, hemimethylated sites are methylated relatively efficiently. These reactions show overall first-order reaction kinetics with rate constants of 3×10^{-3} sec^{-1} in the presence of ATP and 4×10^{-4} sec^{-1} in its absence. For comparison, methylation of nonmethylated sites has a rate constant of 6×10^{-5} sec^{-1}, and this reaction is inhibited about twofold by ATP (Suri et al. 1984a).

Addition of ATP to complexes with nonmethylated DNA led to DNA cleavage followed by massive ATP hydrolysis. This reaction is complex. The earliest discernible event is a massive conformational change in the enzyme accompanied by an increased ability of the enzyme to bind DNA to nitrocellulose filters (Meselson et al. 1972; Bickle et al. 1978). The enzyme then proceeds to cleave the DNA. This last aspect of the reaction mechanism is still mysterious and has generated considerable controversy because although the enzyme remains bound to the recognition site throughout the reaction, the actual cut sites are random and far from the recognition sites, rarely less than 400 and often as far as 7000 base pairs away (Horiuchi and Zinder 1972; Adler and Nathans 1973; Bickle et al. 1978; Rosamond et al. 1979; Yuan et al. 1980a). Three different models have been proposed for the mechanism whereby the enzyme finds its cleavage sites. Two of them, both based mainly on electron microscopy, have the enzyme "tracking" along the DNA, forming ever larger loops until the cleavage site is reached. The models differ in that for one of them, the enzyme can only cleave to one side of the asymmetric recognition site (Rosamond et al. 1979) and for the other, cleavage can occur on either side (Yuan et al. 1980a). The third model requires at least two recognition sites in the DNA. An enzyme molecule would bind to each site, and the molecules would move along the DNA until they bumped into each other, at which point, the DNA would be cut (Studier and Bandyopadhyay 1988). This last model claims that DNA molecules with single recognition sites are a special case and cut by a different mechanism. The mechanism of type I cleavage is perhaps worth reinvestigating with more modern technology. Whatever the exact mechanism, DNA cleavage is a two-step process; the DNA is first nicked in one strand and later cut in the second, most probably by another molecule of enzyme. The DNA ends produced by type I enzymes are poorly characterized. All that is known of them is that they cannot be labeled by polynucleotide kinase, even following phosphatase treatment (Eskin and Linn 1972; Murray et al. 1973), and that they have long 3' protrusions (Endlich and Linn 1985).

The reactions catalyzed by the type IB and IC enzymes are very similar to the type IA mechanism described above. The main differences concern the methylation of unmodified DNA. Type IA enzymes preferentially methylate hemimethylated substrates, the reaction with nonmethylated DNA is slow, and the reaction with both substrates is inhibited by ATP (Suri and Bickle 1985). Type 1C enzymes show the same substrate preferences as type IA enzymes; however, for these enzymes, the reaction is stimulated by ATP (Price et al. 1987a). The type IB enzyme *Eco*AI shows a completely different pattern: It methylates hemimethylated and nonmethylated substrates equally well, but the reaction is completely dependent on ATP (Suri and Bickle 1985).

III. TYPE III RESTRICTION ENZYMES

A. Occurrence

The type III enzymes constitute the smallest class of restriction systems with only four known members. The *Eco*PI and *Eco*P15I genes are carried by the *E. coli* phage P1 and the related P15 defective prophage, *Hin*fIII is found in *Haemophilus influenzae* strain Rf, and *Sty*LTI is chromosomally coded in many *Salmonella* species (Arber and Dussoix 1962; Arber and Wauters-Willems 1970; Colson et al. 1970; Piekarowicz and Kalinowska 1974; Bullas et al. 1980). All of these enzymes share a common reaction mechanism and enzyme structure and in this sense form a natural group. In addition, the three enzymes from the Enterobacteriaceae show a great deal of amino acid sequence homology (the *Hin*fIII genes have not been cloned and sequenced).

B. Genetics

Type III enzymes are coded by two contiguous structural genes called *mod* and *res* (Iida et al. 1983). The *mod* gene product recognizes the DNA sequences specific for the system and is, in addition, a modification methylase. The *res* gene product alone has no enzymatic activity, but as a complex with the *mod* gene product, it is necessary for restriction. There is some controversy concerning the transcriptional organization of these genes. Transposon insertions into the *mod* gene are completely polar for the expression of *res*, indicating that both genes are transcribed from a single promoter located in front of the *mod* gene (Iida et al. 1983). On the other hand, in vitro transcription (Iida et al. 1983) and in vivo studies with *lacZ* operon fusions (Sharrocks and Hornby 1991) indicate that the situation is more complex. Sharrocks and Hornby (1991) in particular find evidence for at least four promoters transcribing these two genes.

The *mod* genes of *Eco*PI, *Eco*P15I, and *Sty*LTI, as well as the *res* genes of *Eco*P1 and *Sty*LTI, have been sequenced (Hümbelin et al. 1988; Dartois et al. 1993). DNA heteroduplex analysis has shown that the *res* gene of *Eco*P15I is highly homologous to that of *Eco*PI (Iida et al. 1983)). The *mod* genes have a structure reminiscent of that of the *hsdS* genes of type I systems, in that they are a mosaic of conserved and non-conserved regions. The central third of the genes is totally dissimilar for the three genes, whereas the 5′ and 3′ thirds are conserved. It is thought that the conserved regions encode protein domains that interact with the *res* gene product and that the variable regions encode sequence-specific DNA-binding domains. Mutations that lead to loss of modification without affecting restriction (Rosner 1973) map in this region (Hümbelin et al. 1988). The phenotype of these mutants was later shown to be due to a failure to bind AdoMet (Rao et al. 1989a). The *Eco*PI and *Sty*LTI *res* gene product sequences show surprisingly little homology apart from a stretch of 50 amino acids toward the center, where the two proteins are virtually identical. This conserved region is a good candidate for forming the interface between the *res* and *mod* subunits (Dartois et al. 1993).

C. Enzyme Mechanism

Type III restriction enzymes characteristically recognize DNA sequences that lack symmetry and cleave the DNA 25–26 base pairs downstream from the sequence. The sequences are AGACC for *Eco*PI, CAGCAG for *Eco*P15I, CGAAT for *Hin*fIII, and CAGAG for *Sty*LTI (Bächi et al. 1979; Hadi et al. 1979; Piekarowicz et al. 1981; De Backer and Colson 1991). Modified DNA has the unusual property of containing a methylated base (an N^6meA) in one strand of the DNA only, and thus half of the sites following DNA replication completely lack modification. The peculiar cleavage mechanism of these enzymes (see below) explains why these unmodified sites do not lead to suicidal restriction.

Most of the details of the type III reaction mechanism have been worked out with *Eco*PI and *Eco*P15I. Insofar as they have been investigated, *Hin*fIII and *Sty*LTI have proved to have very similar mechanisms. The type III enzymes show an absolute requirement for ATP, but, unlike type I enzymes, they are not ATPases. In addition, DNA hydrolysis is stimulated by AdoMet, leading to a complicated situation where DNA cleavage and modification methylation are competing reactions (Haberman 1974; Risser et al. 1974; Reiser and Yuan 1977; Kauc and Piekarowicz 1978). Both ATP and AdoMet are allosteric effectors for DNA cleavage (Yuan and Reiser 1978), and nonhydrolyzable ATP analogs can substitute for ATP in the cleavage reaction, albeit poorly (Yuan et al. 1980b).

No allosteric effects are apparent in the methylation reaction, which follows a random mechanism in which either DNA or AdoMet can bind first to the enzyme. An unusual feature of the kinetics of methylation is that the enzyme is inhibited by concentrations of AdoMet that are not much higher than the K_m values for AdoMet (Rao et al. 1989b). This suggests nonproductive binding of AdoMet to the methylated DNA-enzyme complex, and interestingly, a mutant of the EcoPI methylase (S240A) is apparently more active than the wild type under standard assay conditions because it can no longer be inhibited by substrate AdoMet. A mutation of the same serine codon to proline leads to loss of AdoMet binding (Rao et al. 1989a).

D. DNA Cleavage

The asymmetric recognition sequences of type III restriction systems bear a methyl group in only one strand of the DNA when they are fully modified. For example, EcoP15I recognizes CAGCAG, and the second adenine is methylated in modified DNA. No methyl groups are found in the other strand of the DNA, and indeed, there are no adenines in this strand to be methylated (Hadi et al. 1979; Meisel et al. 1991). This experimental finding was difficult to reconcile with the viability of strains expressing type III restriction systems. DNA replication generates sites that are completely unmethylated and which ought to be restricted unless they were protected by an unknown mechanism. One mechanism that has been suggested was that the modification methylase would be built into the DNA replication complex and the newly replicated sites would be modified before they ever left the complex (Hadi et al. 1979).

The explanation for the viability of strains expressing type III restriction systems came from investigations stimulated by a fortuitous observation: Phage T3 is restricted by EcoP15I, whereas its close relative, T7, is not. The genome of T7 has been sequenced and it contains 36 EcoP15I sites (Dunn and Studier 1983). The most striking factor about these sites is that they all have the same orientation: The CAGCAG sequence is in one strand of the DNA and its CTGCTG complement is in the other strand (Schroeder et al. 1986). This led to the hypothesis that it is the peculiar orientation of the EcoP15I sites that makes T7 DNA refractory to

*Eco*P15I cleavage. This hypothesis has two testable corollaries: (1) *Eco*P15I restriction should require two recognition sites and (2) these two recognition sites should be in inverse orientation. The hypothesis was tested and vindicated using DNA of phage M13 constructs with different numbers and orientations of *Eco*P15I sites. Single or multiple sites with any orientation are substrates for modification methylation; only nonmodified sites in inverse orientation are substrates for restriction (Meisel et al. 1992). The true recognition site for type III restriction is therefore in some ways similar to some type II recognition sequences, consisting as it does of a twofold rotationally symmetrical sequence interrupted at the center of symmetry by a spacer of nonspecific sequence (in this case, of variable length). All of the nonmodified sites in freshly replicated DNA necessarily have the same orientation (with all of the modified sites in the other orientation), explaining the survival of strains expressing type III systems.

The experimental work described above was done with *Eco*P15I. However, there is good reason to believe that the phenomenon will be general for all type III systems. First, they all methylate a single strand of their recognition sequence and thus require some mechanism for protecting freshly replicated DNA from restriction. Second, much previously published work with *Eco*PI is readily interpretable in terms of the mechanism (very few facts are available concerning *Hin*fIII or *Sty*LTI). One striking example was the finding that both in vitro and in vivo heteroduplex phage λ DNA with one nonmodified strand and one *Eco*PI modified strand was totally resistant to *Eco*PI restriction (Arber et al. 1963; Hattman et al. 1978).

IV. CONCLUDING REMARKS

This chapter was concerned mainly with the genetics and biochemistry of the ATP-dependent type I and type III restriction and modification enzymes. No effort has been made to place these enzymes in their biological context. In particular, the major role of restriction enzymes is to protect the cell from assault by foreign DNA, the most obvious sources of which are conjugative plasmids and phage genomes. Phages and conjugative plasmids have responded to the threat of restriction by evolving many mechanistically different strategies for avoiding restriction, and these mechanisms have in their turn forced the evolution of the restriction and modification systems (Krüger and Bickle 1983; Bickle and Krüger 1993).

Most bacteria turn out have at least one DNA restriction system, and the best studied strain, *E. coli* K12, has at least four. Given that the vast majority of individual bacterial cells are unlikely to experience phage infection or conjugation, it is perhaps surprising that the cell should expend so much metabolic energy to produce enzymes that it will never use. It has often been suggested that restriction enzymes may play some other role in the cellular economy, in genetic recombination, for example (Price and Bickle 1986), and yet fully viable strains can be found that have no restriction systems (Daniel et al. 1988). I believe that the prevalence of DNA restriction systems is a sign of genetic selection operating at the population level: A population whose individual members can prevent phage propagation (even if the infected individual is killed in the process) is fitter than one that cannot. Paradoxically, selection for function takes place in those cells in which the function is not used.

ACKNOWLEDGMENTS

I thank Charles Colson and Noreen Murray for providing unpublished information and insights, and Rich Roberts for making his database of the restriction literature available to the scientific community. Work from my laboratory has been supported by the Swiss National Science Foundation.

REFERENCES

Adler, S.P. and D. Nathans. 1973. Studies of SV40 DNA. V. Conversion of circular to linear SV40 DNA by restriction endonuclease from *Escherichia coli* B. *Biochim. Biophys. Acta* **199:** 177-188.

Arber, W. 1965. Host-controlled modification of bacteriophage. *Annu. Rev. Microbiol.* **19:** 365-378.

―――. 1968. Host-controlled restriction and modification of bacteriophage. *Symp. Soc. Gen. Microbiol.* **18:** 295-314.

Arber, W. and D. Dussoix. 1962. Host specificity of DNA produced by *Escherichia coli*: I. Host controlled modification of bacteriophage λ. *J. Mol. Biol.* **5:** 18-36.

Arber, W. and S. Linn. 1969. DNA modification and restriction. *Annu. Rev. Biochem.* **38:** 467-500.

Arber, W. and D. Wauters-Willems. 1970. Host specificity of DNA produced by *Escherichia coli*. XII. The two restriction and modification systems of strain 15T⁻. *Mol. Gen. Genet.* **108:** 203-217.

Arber, W., S. Hattman, and D. Dussoix. 1963. On the host-controlled modification of bacteriophage λ. *Virology* **21:** 30-35.

Bächi, B., J. Reiser, and V. Pirrotta. 1979. Methylation and cleavage sequences of the *Eco*P1 restriction-modification enzyme. *J. Mol. Biol.* **128:** 143-163.

Bertani, G. and J.J. Weigle. 1953. Host controlled variation in bacterial viruses. *J. Bacteriol.* **65:** 113-121.

Bickle, T.A. 1982. The ATP-dependent restriction endonucleases. In *Nucleases* (ed. S.M. Linn and R.J. Roberts), pp. 85-108. Cold Spring Harbor Laboratory, Cold Spring Har-

bor, New York.
——. 1987. DNA restriction and modification systems. In Escherichia coli *and* Salmonella typhimurium: *Cellular and molecular biology* (ed. J.L. Ingraham et al.), pp. 692–696. American Society for Microbiology, Washington, D.C.
Bickle, T.A. and D.H. Krüger. 1993. The biology of DNA restriction. *Microbiol. Rev.* **57**: 434–450.
Bickle, T.A., C. Brack, and R. Yuan. 1978. ATP-induced conformational changes in the restriction endonuclease from *Escherichia coli* K 12. *Proc. Natl. Acad. Sci.* **75**: 3099–3103.
Boyer, H. 1964. Genetic control of restriction and modification in *Escherichia coli. J. Bacteriol.* **88**: 1652–1660.
——. 1971. DNA restriction and modification mechanisms in bacteria. *Annu. Rev. Microbiol.* **25**: 153–176.
Boyer, H.W. and D. Roulland-Dussoix. 1969. A complementation analysis of the restriction and modification of DNA in *Escherichia coli. J. Mol. Biol.* **41**: 459–472.
Bullas, L.R. and C. Colson. 1975. DNA restriction and modification systems in *Salmonella*. III. SP, a *Salmonella potsdam* system allelic to the SB system in *Salmonella typhimurium. Mol. Gen. Genet.* **139**: 177–188.
Bullas, L.R., C. Colson, and B. Neufeld. 1980. Deoxyribonucleic acid restriction and modification systems in *Salmonella*: Chromosomally located systems of different serotypes. *J. Bacteriol.* **141**: 275–292.
Bullas, L.R., C. Colson, and A. van Pel. 1976. DNA restriction and modification systems in *Salmonella*. SQ, a new system derived by recombination between the SB system of *Salmonella typhimurium* and the SP system of *Salmonella potsdam. J. Gen. Microbiol.* **95**: 166–172.
Burckhardt, J., J. Weisemann, D.L. Hamilton, and R. Yuan. 1981. Complexes formed between the restriction endonuclease *Eco*K and heteroduplex DNA. *J. Mol. Biol.* **153**: 425–440.
Colson, C., A.M. Colson, and A. van Pel. 1970. Chromosomal location of host specificity in *Salmonella typhimurium. J. Gen. Microbiol.* **60**: 265–271.
Cowan, G.M., A.A.F. Gann, and N.E. Murray. 1989. Conservation of complex DNA recognition domains between families of restriction enzymes. *Cell* **56**: 103–109.
Daniel, A. S., F. V. Fuller-Pace, D. M. Legge, and N. E. Murray. 1988. Distribution and diversity of *hsd* genes in *Escherichia coli* and other enteric bacteria. *J. Bacteriol.* **170**: 1775–1782.
Dartois, V., O. De Backer, and C. Colson. 1993. Sequence of the *Salmonella typhimurium Sty*LTI restriction-modification genes. Homologies with *Eco*P1 and *Eco*P15I type III R-M systems and presence of helicase domains. *Gene* **127**: 105–110.
De Backer, O. and C. Colson. 1991. Identification of the recognition sequence for the M.*Sty*LTI methyltransferase of *Salmonella typhimurium* LT7: An asymmetric site typical of type-III enzymes. *Gene* **97**: 103–107.
Dunn, J.J. and F.W. Studier. 1983. The complete nucleotide sequence of bacteriophage T7 DNA, and the locations of T7 genetic elements. *J. Mol. Biol.* **166**: 477–535.
Dussoix, D. and W. Arber. 1962. Host specificity of DNA produced by *Escherichia coli*. II. Control over acceptance of DNA from infecting phage lambda. *J. Mol. Biol.* **5**: 37–49.
Endlich, B. and S. Linn. 1985. The DNA restriction endonuclease of *Escherichia coli* B. II. Further studies of the structure of DNA intermediates and products. *J. Biol. Chem.* **260**: 5729–5738.
Eskin, B. and S. Linn. 1972. The deoxyribonucleic acid modification and restriction en-

zymes of *Escherichia coli* B. II. Purification, subunit structure, and catalytic properties of the restriction endonuclease. *J. Biol. Chem.* **247**: 6183-6191.

Fuller-Pace, F.V. and N.E. Murray. 1986. Two DNA recognition domains of the specificity polypeptides of a family of type I restriction enzymes. *Proc. Natl. Acad. Sci.* **83**: 9368-9372.

Fuller-Pace, F.V., L.R. Bullas, H. Delius, and N.E. Murray. 1984. Genetic recombination can generate altered restriction specificity. *Proc. Natl. Acad. Sci.* **81**: 6095-6099.

Gann, A.A.F., A.J.B. Campbell, J.F. Collins, A.F.W. Coulson, and N.E. Murray. 1987. Reassortment of DNA recognition domains and the evolution of new specificities. *Mol. Microbiol.* **1**: 13-22.

Glover, S.W. 1970. Functional analysis of host-specificity mutants in *Escherichia coli. Genet. Res.* **15**: 237-250.

Glover, S.W. and C. Colson. 1969. Genetics of host-controlled restriction and modification in *Escherichia coli. Genet. Res.* **13**: 227-240.

Glover, S.W., K. Firman, G. Watson, C. Price, and S. Donaldson. 1983. The alternate expression of two restriction and modification systems. *Mol. Gen. Genet.* **190**: 65-69.

Gough, J.A. and N.E. Murray. 1983. Sequence diversity among related genes for recognition of specific targets in DNA molecules. *J. Mol. Biol.* **166**: 1-19.

Greene, P.J., M.C. Betlach, H.W. Boyer, and H.M. Goodman. 1974. The *Eco*RI restriction endonuclease. *Methods Mol. Biol.* **7**: 87-105.

Gromkova, R. and S.H. Goodgal. 1976. Biological properties of a *Haemophilus influenzae* restriction enzyme, HindI. *J. Bacteriol.* **127**: 848-854.

Gubler, M. and T.A. Bickle. 1991. Increased protein flexibility leads to promiscuous protein-DNA interactions in type IC restriction-modification systems. *EMBO J.* **10**: 951-957.

Gubler, M., D. Braguglia, J. Meyer, A. Piekarowicz, and T.A. Bickle. 1992. Recombination of constant and variable modules alters DNA sequence recognition by type IC restriction-modification enzymes. *EMBO J.* **11**: 233-240.

Guschlbauer, W. 1988. The DNA and S-adenosylmethionine-binding regions of *Eco*Dam and related methyltransferases. *Gene* **74**: 211-214.

Haberman, A. 1974. The bacteriophage P1 restriction endonuclease. *J. Mol. Biol.* **89**: 545-563.

Hadi, S.M., B. Bächi, J.C.W. Shepherd, R. Yuan, K. Ineichen, and T.A. Bickle. 1979. DNA recognition and cleavage by the *Eco*P15I restriction endonuclease. *J. Mol. Biol.* **134**: 655-666.

Hattman, S., J.E. Brooks, and M. Masurekar. 1978. Sequence specificity of the P1 modification methylase (M.*Eco*P1) and the DNA methylase (M.*Eco*dam) controlled by the *Escherichia coli dam* gene. *J. Mol. Biol.* **126**: 367-380.

Horiuchi, K. and N. Zinder. 1972. Cleavage of bacteriophage f1 DNA by the restriction enzyme of *Escherichia coli. Proc. Natl. Acad. Sci.* **69**: 3220-3224.

Hubacek, J. and S.W. Glover. 1970. Complementation analysis of temperature-sensitive host specificity mutations in *Escherichia coli. J. Mol. Biol.* **50**: 111-127.

Hümbelin, M., B. Suri, D.N. Rao, D.P. Hornby, H. Eberle, T. Pripfl, S. Kenel, and T.A. Bickle. 1988. Type III DNA restriction and modification systems *Eco*P1 and *Eco*P15I. Nucleotide sequence of the *Eco*P1 operon, the *Eco*P15I *mod* gene and some *Eco*P1 *mod* mutants. *J. Mol. Biol.* **200**: 23-29.

Iida, S., J. Meyer, B. Bächi, M. Stålhammar-Carlemalm, S. Schrickel, T.A. Bickle, and W. Arber. 1983. DNA restriction-modification genes of phage P1 and plasmid p15B. *J. Mol. Biol.* **165**: 1-18.

Kan, N.C., J.A. Lautenberger, M.H. Edgell, and C.A. Hutchison III. 1979. The nucleotide sequence recognized by the *Escherichia coli* K12 restriction and modification enzymes. *J. Mol. Biol.* **130:** 191–209.

Kannan, P., G.M. Cowan, A.S. Daniel, A.A.F. Gann, and N.E. Murray. 1989. Conservation of organization in the specificity polypeptides of two families of type I restriction enzymes. *J. Mol. Biol.* **209:** 335–344.

Kauc, L. and A. Piekarowicz. 1978. Purification and properties of a new restriction endonuclease from *Haemophilus influenzae* Rf. *Eur. J. Biochem.* **92:** 417–426.

Kröger, M. and G. Hobom. 1984. The nucleotide sequence recognized by the *Escherichia coli* A restriction and modification enzyme. *Nucleic Acids Res.* **12:** 887–899.

Krüger, D.H. and T.A. Bickle. 1983. Bacteriophage survival: Multiple mechanisms for avoiding the DNA restriction systems of their hosts. *Microbiol. Rev.* **47:** 345–360.

Kuhnlein, U. and W. Arber. 1972. Host specificity of DNA produced by *Escherichia coli*. XV. The role of nucleotide methylation in in vitro B-specific modification. *J. Mol. Biol.* **63:** 9–19.

Kusiak, M., C. Price, D. Rice, and D.P. Hornby. 1992. The HsdS polypeptide of the type IC restriction enzyme *Eco*R124 is a sequence-specific DNA-binding protein. *Mol. Microbiol.* **6:** 3251–3256.

Lauster, R., A. Kriebardis, and W. Guschlbauer. 1987. The GATATC-modification enzyme *Eco*RV is closely related to the GATC-recognizing methyltransferases *Dpn*II and *dam* from *E. coli* and phage T4. *FEBS Lett.* **220:** 167–176.

Lautenberger, J.A. and S. Linn. 1972. The deoxyribonucleic acid modification and restriction enzymes of *Escherichia coli* B. I. Purification, subunit structure, and catalytic properties of the modification methylase. *J. Biol. Chem.* **247:** 6176–6182.

Lautenberger, J.A., N.C. Kan, D. Lackey, S. Linn, M.H. Edgell, and C.A. Hutchison III. 1978. Recognition site of *Escherichia coli* B restriction enzyme on φXsB1 and simian virus 40 DNAs: An interrupted sequence. *Proc. Natl. Acad. Sci.* **75:** 2271–2275.

Linn, S. and W. Arber. 1968. Host specificity of DNA produced by *Escherichia coli*, X. In vitro restriction of phage fd replicative form. *Proc. Natl. Acad. Sci.* **59:** 1300–1306.

Loenen, W.A.M., A.S. Daniel, H.D. Braymer, and N.E. Murray. 1987. Organization and sequence of the *hsd* genes of *Escherichia coli* K12. *J. Mol. Biol.* **198:** 159–170.

Luria, S.E. and M.L. Human. 1952. A nonhereditary, host-induced variation of bacterial viruses. *J. Bacteriol.* **64:** 557–569.

Meisel, A., D.H. Krüger, and T.A. Bickle. 1991. M.*Eco*P15I methylates the second adenine in its recognition sequence. *Nucleic Acids Res.* **19:** 3997.

Meisel, A., T.A. Bickle, D.H. Krüger, and C. Schroeder. 1992. Type III restriction enzymes need two inversely oriented recognition sites for DNA cleavage. *Nature* **355:** 467–469.

Meselson, M. and R. Yuan. 1968. DNA restriction enzyme from *E. coli*. *Nature* **217:** 1110–1114.

Meselson, M., R. Yuan, and J. Heywood. 1972. Restriction and modification of DNA. *Annu. Rev. Biochem.* **41:** 447–466.

Murray, N.E., J.A. Gough, B. Suri, and T.A. Bickle. 1982. Structural homologies among type I restriction-modification systems. *EMBO J.* **1:** 535–539.

Murray, N.E., P.L. Batten, and K. Murray. 1973. Restriction of bacteriophage λ by *Escherichia coli* K. *J. Mol. Biol.* **81:** 395–407.

Nagaraja, V., J.C.W. Shepherd, and T.A. Bickle. 1985a. A hybrid recognition sequence in a recombinant restriction enzyme and the evolution of DNA sequence specificity. *Nature* **316:** 371–372.

Nagaraja, V., J.C.W. Shepherd, T. Pripfl, and T.A. Bickle. 1985b. Two type I restriction enzymes from *Salmonella* species. Purification and DNA recognition sequences. *J.*

Mol. Biol. **182:** 579-587.

Nagaraja, V., M. Stieger, C. Nager, S.M. Hadi, and T.A. Bickle. 1985c. The nucleotide sequence recognized by the *Escherichia coli* D type I restriction and modification enzyme. *Nucleic Acids Res.* **13:** 389-399.

Narva, K.E., J.L. Van Etten, B.E. Slatko, and J.S. Benner. 1988. The amino acid sequence of the eukaryotic DNA [N6-adenine]methyltransferase, M.*Cvi*BIII, has regions of similarity with the prokaryotic isoschizomer M.*Taq*I and other DNA [N6-adenine] methyltransferases. *Gene* **74:** 253-259.

Piekarowicz, A. and J.D. Goguen. 1986. The DNA sequence recognized by the *Eco*DXXI restriction endonuclease. *Eur. J. Biochem.* **154:** 295-298.

Piekarowicz, A. and J. Kalinowska. 1974. Host specificity of DNA in *Haemophilus influenzae*: Similarity between host-specificity types of *Haemophilus influenzae* Re and Rf. *J. Gen. Microbiol.* **81:** 405-411.

Piekarowicz, A., J.D. Goguen, and E. Skrzypek. 1985. The *Eco*DXXI restriction and modification system of *Escherichia coli* ET7. Purification, subunit structure and properties of the restriction endonuclease. *Eur. J. Biochem.* **152:** 387-393.

Piekarowicz, A., T.A. Bickle, J.C.W. Shepherd, and K. Ineichen. 1981. The DNA sequence recognised by the *Hinf*III restriction endonuclease. *J. Mol. Biol.* **146:** 167-172.

Price, C. and T.A. Bickle. 1986. A possible role for DNA restriction in bacterial evolution. *Microbiol. Sci.* **3:** 296-299.

Price, C., T. Pripfl, and T.A. Bickle. 1987a. *Eco*R124 and *Eco*R124/3: The first members of a new family of type I restriction and modification systems. *Eur. J. Biochem.* **167:** 111-115.

Price, C., J.C.W. Shepherd, and T.A. Bickle. 1987b. DNA recognition by a new family of type I restriction enzymes: A unique relationship between two different DNA specificities. *EMBO J.* **6:** 1493-1497.

Price, C., J. Lingner, T.A. Bickle, K. Firman, and S.W. Glover. 1989. Basis for changes in DNA recognition by the *Eco*R124 and *Eco*R124/3 Type I DNA restriction and modification enzymes. *J. Mol. Biol.* **205:** 115-125.

Rao, D.N., H. Eberle, and T.A. Bickle. 1989a. Characterization of mutations of the bacteriophage P1 *mod* gene encoding the recognition subunit of the *Eco*P1 restriction and modification system. *J. Bacteriol.* **171:** 2347-2352.

Rao, D.N., M.G.P. Page, and T.A. Bickle. 1989b. Cloning, over-expression and the catalytic properties of the *Eco*P15I modification methylase from *Escherichia coli*. *J. Mol. Biol.* **209:** 599-606.

Ravetch, J.V., K. Horiuchi, and N.D. Zinder. 1978. Nucleotide sequence of the recognition site for the restriction-modification enzyme of *Escherichia coli* B. *Proc. Natl. Acad. Sci.* **75:** 2266-2270.

Reiser, J. and R. Yuan. 1977. Purification and properties of the P15 specific restriction endonuclease from *Escherichia coli*. *J. Biol. Chem.* **252:** 451-456.

Risser, R., N. Hopkins, R.W. Davis, H. Delius, and C. Mulder. 1974. Action of *Escherichia coli* P1 restriction endonuclease on simian virus 40 DNA. *J. Mol. Biol.* **89:** 517-544.

Rosamond, J., B. Endlich, and S. Linn. 1979. Electron microscopic studies of the mechanism of action of the restriction enzyme of *Escherichia coli* B. *J. Mol. Biol.* **129:** 619-635.

Rosner, J.L. 1973. Modification-deficient mutants of bacteriophage P1. I. Restriction by P1 cryptic lysogens. *Virology* **52:** 213-222.

Roulland-Dussoix, D. and H.W. Boyer. 1969. The *Escherichia coli* B restriction endonuclease. *Biochim. Biophys. Acta* **195:** 219-229.

Sain, B. and N.E. Murray. 1980. The *hsd* (host specificity) genes of *E. coli* K12. *Mol. Gen. Genet.* **180:** 35–46.

Schroeder, C., H. Jurkschat, A. Meisel, J.G. Reich, and D. Krüger. 1986. Unusual occurrence of *Eco*P1 and *Eco*P15I recognition sites and counterselection of type II methylation and restriction sequences in bacteriophage T7 DNA. *Gene* **45:** 77–86.

Sharp, P.M., J.E. Kelleher, A.S. Daniel, G.M. Cowan, and N.E. Murray. 1992. Roles of selection and recombination in the evolution of type I restriction-modification systems in enterobacteria. *Proc. Natl. Acad. Sci.* **89:** 9836–9840.

Sharrocks, A.D. and D.P. Hornby. 1991. Transcriptional analysis of the restriction and modification genes of bacteriophage P1. *Mol. Microbiol.* **5:** 685–694.

Skrzypek, E. and A. Piekarowicz. 1989. The *Eco*DXXI restriction and modification system: Cloning the genes and homology to type I restriction and modification systems. *Plasmid* **21:** 195–204.

Smith, H.O., T.M. Annau, and S. Chandrasegaran. 1990. Finding sequence motifs in groups of functionally related proteins. *Proc. Natl. Acad. Sci.* **87:** 826–830.

Smith, J.D., W. Arber, and U. Kuhnlein. 1972. Host specificity of DNA in *Escherichia coli*. XIV. The role of nucleotide methylation in *in vivo* B-specific modification. *J. Mol. Biol.* **63:** 1–8.

Sommer, R. and H. Schaller. 1979. Nucleotide sequence of the recognition site of the B-specific restriction modification system in *Escherichia coli*. *Mol. Gen. Genet.* **168:** 331–335.

Studier, F.W. and P.K. Bandyopadhyay. 1988. Model for how type I restriction enzymes select cleavage sites in DNA. *Proc. Natl. Acad. Sci.* **85:** 4677–4681.

Suri, B. and T.A. Bickle. 1985. *Eco*A: The first member of a new family of type I restriction modification systems—Gene organization and enzymatic activities. *J. Mol. Biol.* **186:** 77–85.

Suri, B., V. Nagaraja, and T.A. Bickle. 1984a. Bacterial DNA modification. *Curr. Top. Microbiol. Immunol.* **108:** 1–9.

Suri, B., J.C.W. Shepherd, and T.A. Bickle. 1984b. The *Eco*A restriction and modification system of *Escherichia coli* 15T$^-$: Enzyme structure and DNA recognition sequence. *EMBO J.* **3:** 575–579.

Taylor, I., J. Patel, K. Firman, and G. Kneale. 1992. Purification and biochemical characterisation of the *Eco*R124 type I modification methylase. *Nucleic Acids Res.* **20:** 179–186.

Vovis, G.F., K. Horiuchi, and N.D. Zinder. 1974. Kinetics of methylation of DNA by a restriction endonuclease from *Escherichia coli* B. *Proc. Natl. Acad. Sci.* **71:** 3810–3813.

Wilson, G.G. and N.E. Murray. 1991. Restriction and modification systems. *Annu. Rev. Genet.* **25:** 585–627.

Yuan, R. and J. Reiser. 1978. Steps in the reaction mechanism of the *Escherichia coli* plasmid P15-specific restriction endonuclease. *J. Mol. Biol.* **122:** 433–445.

Yuan, R., D.L. Hamilton, and J. Burckhardt. 1980a. DNA translocation by the restriction enzyme from *E. coli* K. *Cell* **20:** 237–244.

Yuan, R., T.A. Bickle, W. Ebbers, and C. Brack. 1975. Multiple steps in DNA recognition by restriction endonuclease from *E. coli* K. *Nature* **256:** 556–560.

Yuan, R., D.L. Hamilton, S.M. Hadi, and T.A. Bickle. 1980b. Role of ATP in the cleavage mechanism of the *Eco*P15I restriction endonuclease. *J. Mol. Biol.* **144:** 501–519.

4
Homing Endonucleases

John E. Mueller, Mary Bryk,[1] Nick Loizos,[1] and Marlene Belfort
Molecular Genetics Program
Wadsworth Center for Laboratories and Research
New York State Department of Health
David Axelrod Institute
Albany, New York 12201-2002

I. Introduction
II. Historic Review
III. General Characteristics of Homing Endonucleases
   A. Endonuclease-mediated Homing
   B. Endonuclease ORF Location
   C. Endonuclease Expression
   D. Sequence Motifs
      1. The LAGLI-DADG Motif
      2. The GIY-YIG Motif
      3. The Zinc Finger Motif
      4. Unclassified
   E. Target Recognition and Cleavage Specificity
      1. General Properties
      2. Genetic Studies
      3. Physical Studies
IV. Evolutionary Considerations
   A. Evidence for Mobility of Endonuclease Genes
   B. Invasion and Its Aftermath
   C. Endonuclease Properties That Potentiate Invasiveness
   D. Down-regulation of Endonuclease Expression
   E. Cross-species Transfer
V. Concluding Remarks

I. INTRODUCTION

Homing endonucleases are a group of enzymes whose catalytic activity results in self-propagation. The sequences that code for these endonucleases usually interrupt genes by localizing as open reading frames in introns or as inframe spacers in protein-coding sequences. The target of a

---

[1]Also affiliated with: Department of Microbiology, Immunology and Molecular Genetics, Albany Medical College, Albany, New York 12208.

Table 1 Characteristics of Intron and/or Homing Endonucleases

Kingdom	Organism	Endo[a]	Genome[b]	Gene[c]	Site[d]	Intron
Eukaryotes	S. cerevisiae	I-SceI	Mit	L-rRNA	GI	ω
		I-SceII	Mit	COXI	GI	aI4
		I-SceIII	Mit	COXI	GI	aI3
		I-SceIV	Mit	COXI	GI	aI5
		PI-SceI	Nuc	VMA1/TFP1	PI	–
	P. polycephalum	I-PpoI	Nuc	L-rRNA	GI	LSU-3
	Chlamydomonas	I-CsmI	Mit	COBI	GI	cobI1
	ssp.[m]	I-CreI	Chl	L-rRNA	GI	LSU
		I-CeuI	Chl	L-rRNA	GI	LSU-5
		I-ChuI	Chl	L-rRNA	GI	LSU-1
Eubacteria	phage T4	I-TevI	Phg	td	GI	td
		I-TevII	Phg	sunY	GI	sunY
	phage RB3	I-TevIII	Phg	nrdB	GI	nrdB
	phage SP01	I-HmuI[n]	Phg	pol	GI	pol
	phage SP82	I-HmuII[n]	Phg	pol	GI	pol
	M. tuberculosis	PI-MtuI	Chr	recA	PI	–
Archaea	D. mobilis	I-DmoI	Chr	L-rRNA	AI	LSU
	T. litoralis	PI-TliI	Chr	pol	PI	–
		PI-TliII[o]	Chr	pol	PI	–
	Pyrococcus GB-D	PI-PspI[o]	Chr	pol	PI	–

[a] Endonuclease name.

[b] (Mit) Mitochondrial; (Nuc) nuclear; (Chl) chloroplast; (Phg) bacteriophage; (Chr) chromosomal.

[c] (L-rRNA) Large ribosomal RNA; (COXI) cytochrome oxidase; (VMA1/TFP1) vacuolar ATPase; (COBI) cytochrome b; (td) thymidylate synthase; (sunY) anaerobic nucleotide reductase; (nrdB) small subunit ribonucleotide reductase; (pol) DNA polymerase.

[d] (GI) Group I intron; (PI) protein intein; (AI) archaeal intron.

[e] Position in group I intron (see Fig. 3).

[f] Size estimated from ORF sequences. For ORFs in-frame with the upstream exon, the endonuclease amino terminus corresponds to the 5′ splice site. Where two sizes are given, polypeptides of different lengths have endonuclease activity.

[g] (F) Freestanding; (I) in-frame with upstream exon; (C) continuous with both exons.

[h] (Common) Codons in ORF used frequently in other genes in the organism; (rare) codons in ORF used infrequently in other genes. n.a. indicates not available.

[i] (LAG) LAGLI-DADG; (GIY) GIY-YIG; (Zn) zinc finger; n.o. indicates none observed, and n.a. indicates sequence not available.

[j] Proximal indicates within five nucleotides of insertion site, and distal indicates more than ten nucleotides from insertion site. n.a. indicates cleavage site not available. Cleavage sites are illustrated in Fig. 7.

Table 1 (Continued.)

ORF position[e]	Size (aa)[f]	Frame[g]	Codon usage[h]	Motif[i]	Cleavage[j]	Mobile?[k]	Refs.[l]
P8	235	F	common	LAG	proximal	yes	1–6
P8	316	I	rare	LAG	proximal	yes	1,3,7–11
P1	334	I	rare	LAG	proximal	n.d.	1,3,7,9,12
P2	306	I	rare	LAG	proximal	yes	1,3,13,14
–	454	C	common	LAG	proximal	yes	15–17
P1	138/160	F	common	n.o.	proximal	yes	18,19
P2	237	I	rare	LAG	n.a.	yes	20
P6a	163	F	rare	LAG	proximal	yes	21–23
P6a	218	F	common	LAG	proximal	yes	24–26
P9.2	218	F	rare	LAG	proximal	n.d.	27
P6a	245	F	common	GIY	distal	yes	28–33
P9.1	258	F	common	n.o.	distal	yes	28,30,32–34
P6a	269	F	common	Zn	distal	n.d.	28,30,35
P8	176	F	n.a.	n.o.	proximal[n]	n.d.	36,37
P8	187	F	n.a.	n.o.	distal[n]	n.d.	36,37
–	440	C	n.a.	LAG	n.a.	n.d.	38–40
–	188/193	F	n.a.	LAG	proximal	n.d.	41–43
–	390	C	n.a.	LAG	proximal	n.d.	44,45
–	538	C	n.a.	LAG	proximal	n.d.	44–46
–	n.a.	C	n.a.	n.a.	n.a.	n.d.	46

[k]n.d. indicates mobility not demonstrated.

[l]Refs. (1) Michel and Westhof 1990; (2) Dujon 1980; (3) Hensgens et al. 1983; (4) Colleaux et al. 1988; (5) Jacquier and Dujon 1985; (6) Macreadie et al. 1985; (7) Bonitz et al. 1980; (8) Hanson et al. 1982; (9) Waring et al. 1982; (10) Wenzlau et al. 1989; (11) Delahodde et al. 1989; (12) Perea et al. 1993; (13) Moran et al. 1992; (14) Seraphin et al. 1992; (15) Hirata et al. 1990; (16) Kane et al. 1990; (17) Bremer et al. 1992; (18) Muscarella and Vogt 1989; (19) Muscarella et al. 1990; (20) Colleaux et al. 1990; (21) Rochaix et al. 1985; (22) Durrenberger and Rochaix 1991; (23) Thompson et al. 1992; (24) Turmel et al. 1991; (25) Marshall and Lemieux 1991; (26) Gauthier et al. 1991; (27) Cote et al. 1993; (28) Shub et al. 1988; (29) Chu et al. 1984; (30) Gott et al. 1988; (31) Michel and Dujon 1986; (32) Bell-Pedersen et al. 1990; (33) Quirk et al. 1989a; (34) Tomaschewski and Ruger 1987; (35) Eddy and Gold 1991; (36) Goodrich-Blair 1993; (37) H. Goodrich-Blair and D.A. Shub, pers. comm.; (38) Davis et al. 1991; (39) Davis et al. 1992; (40) E.O. Davis, pers. comm.; (41) Kjems and Garrett 1985; (42) Dalgaard et al. 1993; (43) Dalgaard and Garrett 1992; (44) Perler et al. 1992; (45) Hodges et al. 1992; (46) F.B. Perler, pers. comm.

[m]*Chlamydomonas* species: (*Csm*) *C. smithii*; (*Cre*) *C. reinhardtii*; (*Ceu*) *C. eugametos*; (*Chu*) *C. humicola*.

[n]These endonucleases make a single-strand break.

[o]These endonucleases are isoschizomers and inserted at the same position in their respective *pol* genes.

114   J.E. Mueller et al.

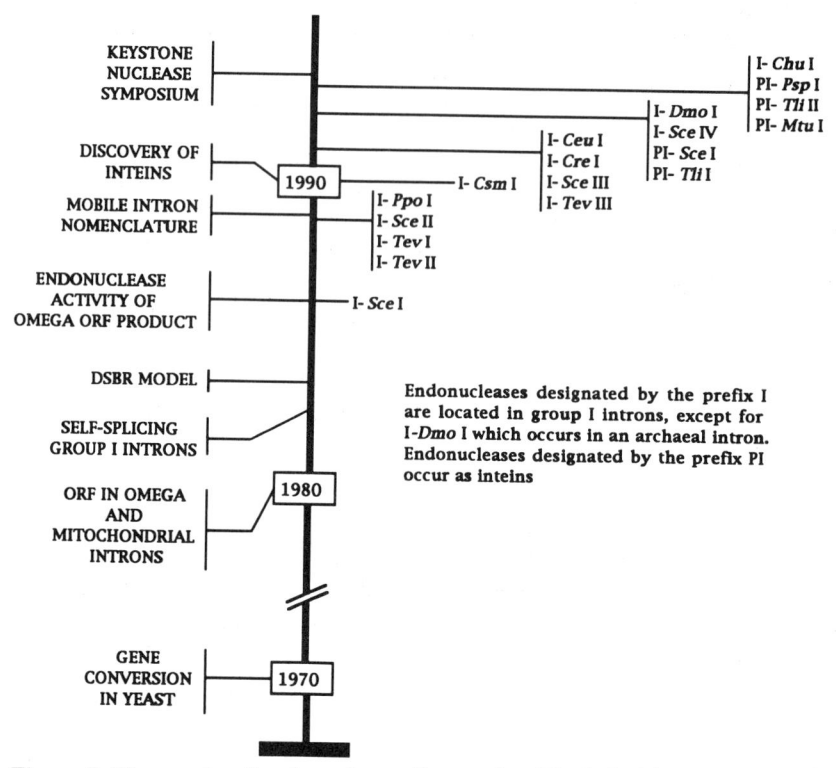

```
KEYSTONE
NUCLEASE                                                      I- Chu I
SYMPOSIUM                                                     PI- Psp I
                                                              PI- Tli II
                                               I- Dmo I       PI- Mtu I
                                               I- Sce IV
                                  I- Ceu I     PI- Sce I
DISCOVERY OF                      I- Cre I     PI- Tli I
INTEINS           ┌──────┐        I- Sce III
                  │ 1990 │── I- Csm I           I- Tev III
MOBILE INTRON     └──────┘  I- Ppo I
NOMENCLATURE                I- Sce II
                            I- Tev I
                            I- Tev II
ENDONUCLEASE
ACTIVITY OF              I- Sce I
OMEGA ORF PRODUCT

DSBR MODEL              Endonucleases designated by the prefix I
                        are located in group I introns, except for
SELF-SPLICING           I-Dmo I which occurs in an archaeal intron.
GROUP I INTRONS         Endonucleases designated by the prefix PI
                        occur as inteins
ORF IN OMEGA     ┌──────┐
AND              │ 1980 │
MITOCHONDRIAL    └──────┘
INTRONS


GENE             ┌──────┐
CONVERSION       │ 1970 │
IN YEAST         └──────┘
```

Figure 1 Time scale of endonuclease discoveries. The left side of the time scale lists hallmark events, and the right side lists the homing endonucleases in order of their discovery. (DSBR MODEL) Double-strand break-repair model. All events noted and individual enzymes are discussed and referenced in the text. The endonucleases are designated by the prefix I for *i*ntron-encoded. The intein endonucleases are designated by the prefix PI for *p*rotein *i*nsert (Lambowitz and Belfort 1993; F.B. Perler et al., in prep.). The prefix is followed by a three-letter *ge*nus-*sp*ecies abbreviation of the organism in which it is found and a Roman numeral suffix indicating the chronological order of discovery (Dujon et al. 1989). For details of the enzymes and the organisms from which they originate, see Table 1.

homing endonuclease is its cognate intronless or spacerless allele. The endonuclease initiates a DNA mobility or "homing" event by making a double-strand cut in its target. Repair of the cleaved allele results in the conversion of this gene to the interrupted endonuclease-encoding form. Homing endonucleases are widespread, found in all three kingdoms and in a range of genetic environments, which include mitochondrial, chloroplast, nuclear, and bacteriophage genomes (Table 1). Although the discovery of these endonucleases is recent, genetic consequences attributable to their presence have been observed for some time (see Fig. 1).

We review here the history and general properties of homing endonucleases, point out both similarities and differences among the individual enzymes, and address the evolutionary implications of endonuclease gene mobility.

II. HISTORIC REVIEW

In 1970, the unidirectional transfer of a *Saccharomyces cerevisiae* genetic marker, termed omega (ω), from ω⁺ to ω⁻ yeast strains, was reported (Coen et al. 1970), sparking a great deal of interest regarding the role of nonreciprocal recombination in yeast mitochondrial genetics. With the discovery of restriction enzymes and the introduction of advanced molecular techniques, the ω locus was mapped to an intron in the mitochondrial large ribosomal RNA gene (L-rRNA gene) (Dujon and Michel 1976; Bos et al. 1978; Heyting and Menke 1979). DNA sequencing of the gene revealed an open reading frame (ORF) within the intron (Dujon 1980). Later it was shown that both expression of this intron ORF and a double-strand break in the ω⁻ allele were required for the non-reciprocal transfer of the ω determinant (Jacquier and Dujon 1985; Macreadie et al. 1985; Zinn and Butow 1985). In 1986, the enzymatic activity of the intron-encoded protein was demonstrated (Colleaux et al. 1986). This activity, termed I-*Sce*I (for nomenclature, see legend to Fig. 1), resulted in the endonucleolytic cleavage of the intronless allele and the subsequent transfer of the endonuclease-encoding intron. The coconversion of genetic markers flanking the ω determinant exhibited a polarity similar to that observed during mating-type switching (Dujon and Jacquier 1983), a process that is presumed to take place via a double-strand break-repair mechanism (Orr-Weaver et al. 1981; Szostak et al. 1983). A model of this pathway of homologous recombination that appears to account for intron homing is illustrated in Figure 2 (for review, see Belfort 1990).

This endonuclease-promoted intron mobility phenomenon was viewed as a peculiarity of the ω system until the discovery of four other mobile introns in phylogenetically diverse systems. These include the aI4α intron in the mitochondrial *COXI* gene of *S. cerevisiae* (Delahodde et al. 1989; Wenzlau et al. 1989), the third intron in the nuclear L-rRNA gene of *Physarum polycephalum* (Muscarella and Vogt 1989), and the introns of the *td* and *sunY* genes of bacteriophage T4 (Quirk et al. 1989a). These studies, which stemmed from observations of the polymorphic distribution of introns in the different systems, demonstrated endonuclease-mediated double-strand breaks and unidirectional intron

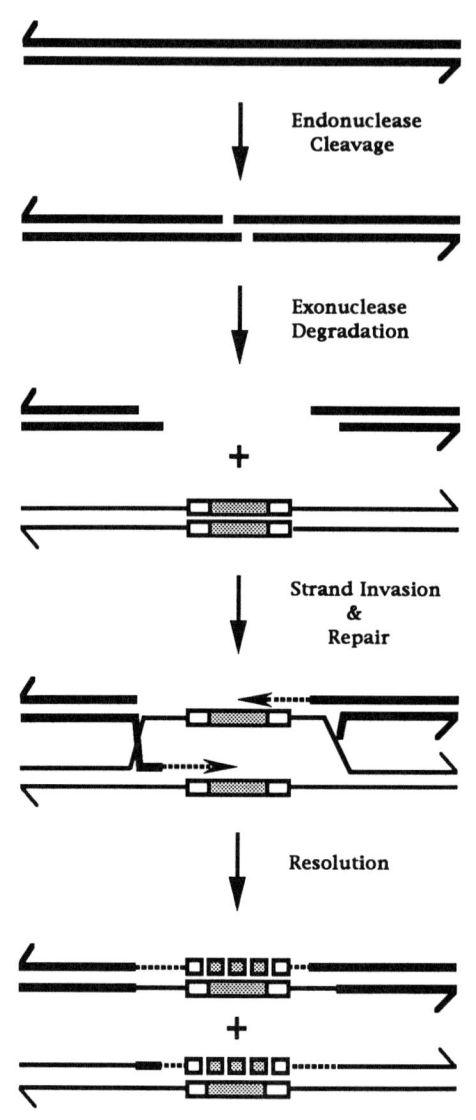

Figure 2 Model of the double-strand break-repair pathway for transfer of endonuclease-coding sequences. Following cleavage by the homing endonuclease, the recipient allele (*heavy lines*) is subjected to exonucleolytic degradation and aligned with homologous sequences of an intron-containing donor (*thin lines* with open and stippled boxes). Once aligned, a 3' tail of the recipient invades the donor, which acts as the template for DNA-repair synthesis (*dashed lines* plus arrows). The Holliday junctions formed during this process are resolved, resulting in two intron-containing duplexes. (*Half-arrows*) 3' end of the DNA strands; (*open boxes*) intron sequences; (*stippled boxes*) endonuclease coding sequences.

transfer in genetic crosses (Pedersen-Lane and Belfort 1987; Delahodde et al. 1989; Muscarella and Vogt 1989; Quirk et al. 1989a,b; Wenzlau et al. 1989; Bell-Pedersen et al. 1990; Muscarella et al. 1990). The endonucleases I-*Sce*II, I-*Ppo*I, I-*Tev*I, and I-*Tev*II correspond to ORF products of the above-named *S. cerevisiae*, *P. polycephalum*, and T-even phage introns, respectively.

These discoveries were followed rapidly by reports of mobility-type intron-encoded endonucleases in several *Chlamydomonas* organellar group I introns (I-*Csm*I, Colleaux et al. 1990; I-*Cre*I, Durrenberger and Rochaix 1991; I-*Ceu*I, Gauthier et al. 1991; I-*Chu*I, Cote et al. 1993), in another T-even phage group I intron (I-*Tev*III, Eddy and Gold 1991), and in other *S. cerevisiae* mitochondrial group I introns (I-*Sce*III, Sargueil et al. 1991; I-*Sce*IV, Moran et al. 1992; Seraphin et al. 1992). The discovery of a homing endonuclease (I-*Dmo*I) in the 23S rRNA of the archaeon *Desulfurococcus mobilis*, in a typical archaeal intron (Kjems and Garrett 1985) rather than a group I intron, has further expanded the taxonomic range of these endonucleases and has fueled speculation regarding the evolution of these enzymes given their occurrence in different intron families (Dalgaard et al. 1993; Doolittle 1993). Group I intron-encoded nucleases have also been discovered in two *Bacillus subtilis* phages (I-*Hmu*I, I-*Hmu*II; Table 1); however, these have been linked to neither double-strand nuclease activity nor intron mobility (Goodrich-Blair et al. 1990; Goodrich-Blair 1993; Goodrich-Blair and Shub 1993).

A remarkable class of endonuclease genes that reside directly within coding sequences have been reported recently. These "inteins," which are synthesized as fusions with flanking polypeptides termed the "exteins," self-excise from the extein sequences and then act as mobility-type endonucleases (for review, see Hendrix 1991; Grivell 1992; Shub and Goodrich-Blair 1992; Lambowitz and Belfort 1993; Wallace 1993). The intein/extein nomenclature is used to replace previous terms such as protein spacer, protein intron, protein insert, and intervening protein sequence (F.B. Perler et al., in prep.).

As the first reported example, the translation product of the *VMA1*/ *TFP1* gene in *S. cerevisiae* is a 119-kD preprotein containing an ATPase interrupted by an intein (Hirata et al. 1990; Kane et al. 1990). The intein is posttranslationally excised, and the exteins are fused to form the functional 69-kD ATPase (Kane et al. 1990). The 50-kD intein is a site-specific homing endonuclease termed VDE (*VMA1*-derived endonuclease) or PI-*Sce*I (for nomenclature see legend to Fig. 1), whose target is an uninterrupted *VMA1* gene (Bremer et al. 1992; Gimble and Thorner 1992). Cleavage of the inteinless *VMA1* allele results in its inheriting the homing endonuclease-coding sequences (Gimble and Thorner 1992). Inteins have also been identified in the DNA polymerases of the archaea *Thermococcus litoralis* and *Pyrococcus* species GB-D (PI-*Tli*I, PI-*Tli*II, and PI-*Psp*I, respectively) (Perler et al. 1992; F.B. Perler, pers. comm.), in the RecA protein of the eubacterium *Mycobacterium tuberculosis* (PI-*Mtu*I) (Davis et al. 1991, 1992; E.O. Davis, pers. comm.), and in the vacuolar ATPase subunit of *Candida tropicalis* (ATPase-Int) (Gu et al. 1993). Although there has not yet been a report of endonucleolytic activity in the *C. tropicalis* system, a number of common features with the mobile inteins suggest that similar homing activity will be demonstrated (Gu et al. 1993). These inteins extend the range of known enzymatic functions among homing endonucleases to include peptide hydrolase-ligase activities and support the hypothesis that homing endonuclease genes are mobile genetic elements (Lambowitz 1989; Perlman and Butow 1989; Belfort 1990; Bell-Pedersen et al. 1990).

III. GENERAL CHARACTERISTICS OF HOMING ENDONUCLEASES

In 1989, the homing endonucleases were first recognized as a distinct class of enzymes, and a definition of terms was introduced to describe these enzymes and intron mobility (Dujon et al. 1989). "Intron mobility" or "homing" is the event in which a sequence of DNA representing the intron of a gene is inserted into an intronless copy of that gene. The "intron insertion site" is the precise location in the intronless allele where the intron is integrated. The "cleavage site" is the exact position of cleavage by the homing endonuclease. The "homing site" includes the intron insertion site, sequences required for endonuclease recognition and binding, as well as cleavage. For a description of the nomenclature of the homing endonucleases, see legend to Figure 1.

A. Endonuclease-mediated Homing

Homing endonucleases generate double-strand breaks thereby promoting recombination events that result in gene conversion. A double-strand break-repair model, invoked as the basis of nonreciprocal recombination events in yeast (Orr-Weaver et al. 1981; Szostak et al. 1983), has been adapted to the unidirectional transfer of mobile introns to their intron-minus cognates (Fig. 2) (for review, see Belfort 1990). According to the model, repair of the break results in the production of two intron-containing alleles. Homology between the donor and recipient alleles is an essential component of this process. Unlike more typical transposases that participate in DNA transactions subsequent to cleavage (for review, see Berg and Howe 1989), the bacteriophage T4 intron endonuclease I-*Tev*I is not required for mobility beyond the cleavage event (Clyman and Belfort 1992). However, a post-cleavage function for homing endonucleases has not been eliminated in eukaryotic systems, where I-*Sce*I has been shown to bind to its downstream homing-site sequences after cleavage, leaving open the possibility of an additional role in recombination (Plessis et al. 1992; Perrin et al. 1993).

B. Endonuclease ORF Location

Group I introns require specific RNA secondary and tertiary structures to catalyze the transesterifications necessary for splicing. The interactions that constitute the catalytic core of the ribozyme include pairings P3, P4, P5, P6, P7, P8, and P9.0, whereas P1 and P10 pairings bring the exon sequences into the active site (Cech 1988, 1990; Michel and Westhof 1990; Saldanha et al. 1993). By looping out of peripheral elements of the intron core, the group I homing endonuclease ORFs do not perturb the folding of the catalytic ribozyme (Table 1; Fig. 3). However, the coding sequences of several homing endonucleases extend beyond these peripheral loops and into the catalytic core of the intron (Fig. 3). Notably, these sequences are consistent with both formation of the catalytic core and function of the endonuclease.

Positioning of the homing endonuclease ORFs within archaeal introns is likely to be less critical, since apart from a bulge-helix-bulge motif near the intron boundaries, the intron sequences play no known role in splicing (for review, see Lambowitz and Belfort 1993). In contrast, the endonuclease genes found within protein-coding sequences are tolerated by virtue of protein splicing. Optimal peptide contexts for intein localization may exist (see below), but the small number of examples precludes their precise identification. Additionally, endonucleases related by sequence to the homing endonucleases have been found in freestanding form between genes (see below), but none of these have been shown to be capable of homing.

C. Endonuclease Expression

The modes of expression of homing endonuclease genes are highly variable. Group I intron ORFs can be either freestanding or continuous with the upstream exon (Table 1; Fig. 4A,B). Freestanding endonuclease ORFs are expressed from either precursor RNA (e.g., I-*Sce*I; Zhu et al. 1987, 1989) or independent transcripts (e.g., I-*Tev*I, I-*Tev*II) (Fig. 4A). The latter freestanding ORFs are expressed from late promoters within

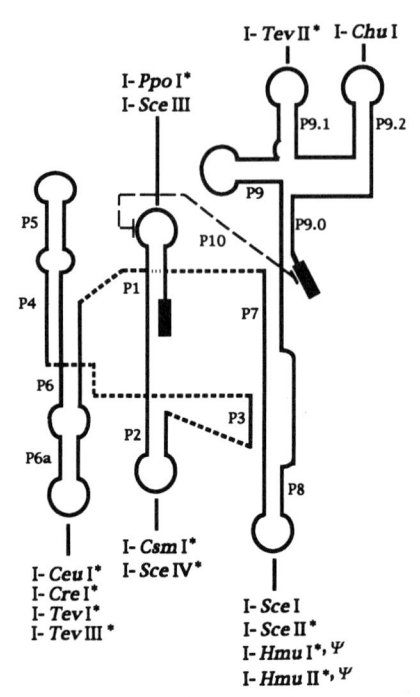

Figure 3 Location of endonuclease ORFs in group I introns. The location of individual homing endonuclease ORFs is shown within a depiction of the secondary-tertiary structure folding model for group I introns (Michel and Westhof 1990; Saldanha et al. 1993). (*Dotted and solid lines*) Continuous sequences; (*dashed thin lines*) long-range pairings; (*closed boxes*) exon sequences. Asterisks indicate ORFs that extend into intron stems, and Greek Psis indicate endonucleases that do not create double-strand breaks or home.

the introns (Gott et al. 1988). In contrast, those intron endonuclease ORFs that are inframe with the upstream exons are believed to be translated as fused precursors that are subsequently processed to generate functional endonucleases (Fig. 4B) (for review, see Perlman and Butow 1989). Furthermore, the intein endonuclease ORFs are inframe with both extein ORFs and are expressed as continuous precursors that are processed by protein splicing (Table 1; Fig. 4C) (for review, see Shub and Goodrich-Blair 1992). Interestingly, the same sequence motif (LAGLI-DADG, see below) occurs in all of the inteins, as well as in many of the group I intron endonucleases that are fused to the upstream exons (Table 1). These observations suggest that proteolytic processing of the intron-encoded precursors may occur by a mechanism similar to that of the inteins (Shub and Goodrich-Blair 1992; Lambowitz and Belfort 1993; Wallace 1993).

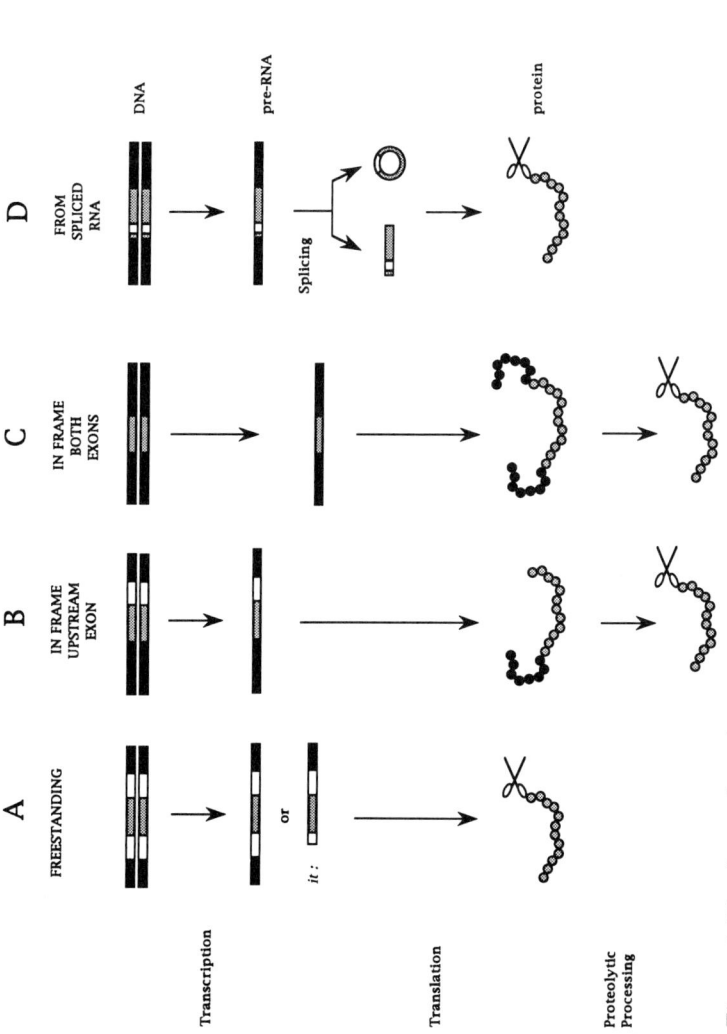

Figure 4 Homing endonuclease expression. Modes of expression of homing endonuclease ORFs are described in text. (*Closed boxes*) Exon sequences; (*open boxes*) intron sequences; (*stippled boxes*) homing endonuclease sequences; (*it*) independent transcript; (beads) protein sequence; (beads with scissors) active homing endonuclease.

Different modes of expression have also been reported for endonucleases located within introns in nontranslated rDNAs or rRNA genes. For example, I-*Ppo*I, encoded by the intron in the nuclear L-rRNA gene of *P. polycephalum*, is translated from one of two uniquely processed transcripts containing initiation codons in either the upstream exon or the intron (Ruoff et al. 1992). This results in the production of two different-sized I-*Ppo*I proteins that are both able to cleave an intronless L-rRNA gene and initiate intron homing (Muscarella et al. 1990). Expression of I-*Dmo*I, from the *D. mobilis* L-rRNA intron, is unusual in that its ORF is translated into active endonuclease from excised intron RNA but not from precursor RNA (Fig. 4D) (Dalgaard et al. 1993). Translation products from both the linear and circular intron species are active site-specific nucleases despite differences at their carboxyl termini. The adap-

```
A.     LAGLI-DADG MOTIF

                    P1                      P2
                 LAGLIDADG              LAGLIDADG
I-SceI       35 - AGIGLILGDAYI -  90 - LAYWFMDDGGKW -  86
I-SceII      86 - WLAGLIDGDGYF -  94 - WFVGFFDADGTI - 112
I-SceIII     30 - YLAGLIEGDGSI - 120 - WLAILTDADGNF - 160
I-SceIV     101 - IMTGILLTDGWI -  98 - SLAHMIMCDGSF -  83
PI-SceI     208 - YLLGLWIGDGLS -  96 - FLAGLIDSDGYV - 126
I-CsmI       42 - IAVGLLLSDAHA -  91 - ALAYWIAGDGCW -  80
I-CreI*      11 - YLAGFVDGDGSI - 140
I-ChuI        9 - LIFGSLLGDGNL -  99 - ALAYFYIDDGAL -  86
I CeuI*      57 - FLAGFLEGEASL - 149
I-DmoI       12 - YLLGLIIGDGGL -  84 - FIKGLYVAEGDK -  73
PI-TliI     146 - ELVGLIVGQGNW -  86 - FLRGLFSADGTV - 134
PI-TliII    281 - KLLGYYVSEGYA -  83 - FLEAYFTGDGDI - 150
PI-MtuI     113 - RLLGYLIGDGRD -  88 - LLFGLFESDGWV - 215
°HO Endo    214 - WMLGLWLGDGTT -  99 - FLAGLIDSDGYV - 250
°Endo.SceI  203 - YLSGLIEGDGYI - 105 - WLAGFTAADGSF - 145
ATPase-Int  200 - YLLGTWAGIGNV - 123 - LIAGLVDAAGNV - 124
Consensus     # - xloGllogdgxo -   # - olaglidadGxo -   #
                  i     ea                ffos

              INTEIN BOUNDARY RESIDUES

         PI-SceI       .. |C - 450 - VHN |C..
         PI-TliI       .. |S - 386 - VHN |T..
         PI-TliII      .. |S - 534 - AHN |S..
         PI-MtuI       .. |C - 436 - VHN |C..
         ATPase-Int    .. |C - 467 - VHN |C..


B.     GIY-YIG MOTIF

I-TevI       1 - KSGIYQIKNTLNNK.VYVGSA..KDFEKR - 218
SegA         8 - YNYTYVITNLVNNK.IYYGTHSTDDLNDG - 185
Consensus    # - ksGIYxxinkxnxkxvYIGsaxxxdlvkR -   #


C.     ZINC FINGER MOTIF

I-TevIII   210 - CDFC - 12    - HNDRC - 38
           241 - CPYC - 13    - HGDNC -  6
Consensus    # - CxxC - 12-13 - HxxxC -  #
```

Figure 5 (*See facing page for legend.*)

tation of endonuclease expression to the organism in these different biological systems is discussed further in Section IV.

D. Sequence Motifs

Comparative amino acid sequence analysis of the homing endonucleases has revealed that the individual enzymes fall into families, based on the presence of different sequence motifs.

1. The LAGLI-DADG Motif

This motif contains a region of about 100 amino acids flanked by dodecamer repeats, P1 and P2. The LAGLI-DADG consensus within the dodecamers are present in the majority of homing endonucleases including representatives from all three kingdoms (Table 1; Figs. 5 and 6). This motif was first observed in yeast mitochondrial intron maturases involved in RNA processing (Lazowska et al. 1980; Jacq et al. 1982; Michel et al. 1982; Waring et al. 1982; Hensgens et al. 1983) and later in

Figure 5 Amino acid sequence motifs of the homing enzymes and related endonucleases. Amino acids are represented by the one-letter code. (Bold-faced letters) Conserved amino acids; (#) number of amino acids flanking motif; (dots) gap in alignment; (x) any amino acid. (*A*) LAGLI-DADG. Consensus derived from sequences in figure. (Uppercase) Invariant; (lowercase) partially conserved (≥50% for single assignment or >60% of either amino acid for alternative residue assignment); (o) greater than 65% hydrophobic residues; (P1 and P2) dodecamer repeats, boxed; (*) only contains P1. Circled D in PI-*Tli*I P1 repeat represents residue important for endonucleolytic activity, see text. (°) Nonhoming endonucleases: HO endo is *S. cerevisiae* HO endonuclease (Russell et al. 1986); Endo.*Sce*I is *S. cerevisiae* mitochondrial endonuclease (Nakagawa et al. 1991); and ATPase-Int is *C. tropicalis* vacuolar ATPase subunit A intein (Gu et al. 1993). Intein Boundary Residues: Conserved amino acids at boundaries (*vertical lines*) are shown with conventions as above. (*B*) GIY-YIG. Consensus adapted from Tian et al. (1991). GIY and YIG elements are boxed. (SegA) Bacteriophage T4 intergenic endonuclease (Sharma et al. 1992). (*C*) Zinc finger. Motif is boxed (Klug and Rhodes 1987). Accession numbers: I-*Sce*I M11280, P03882, V00684; I-*Sce*II J01481; I-*Sce*III J01481, P03877, V00694; I-*Sce*IV for sequence (Hensgens et al. 1983); PI-*Sce*I J05409, P17255; I-*Csm*I S12023; I-*Cre*I P05725 X01977; I-*Chu*I for sequence (Cote et al. 1993); I-*Ceu*I Z17234; I-*Tev*I M12742, P13299; I-*Tev*II Y00122; I-*Tev*III X59078; I-*Dmo*I P21505, X03263; PI-*Tli*I and PI-*Tli*II M74198; PI-*Mtu*I P26345; HO endonuclease M14678, P09932; Endo.*Sce*I J05574, M55275; SegA M69268.

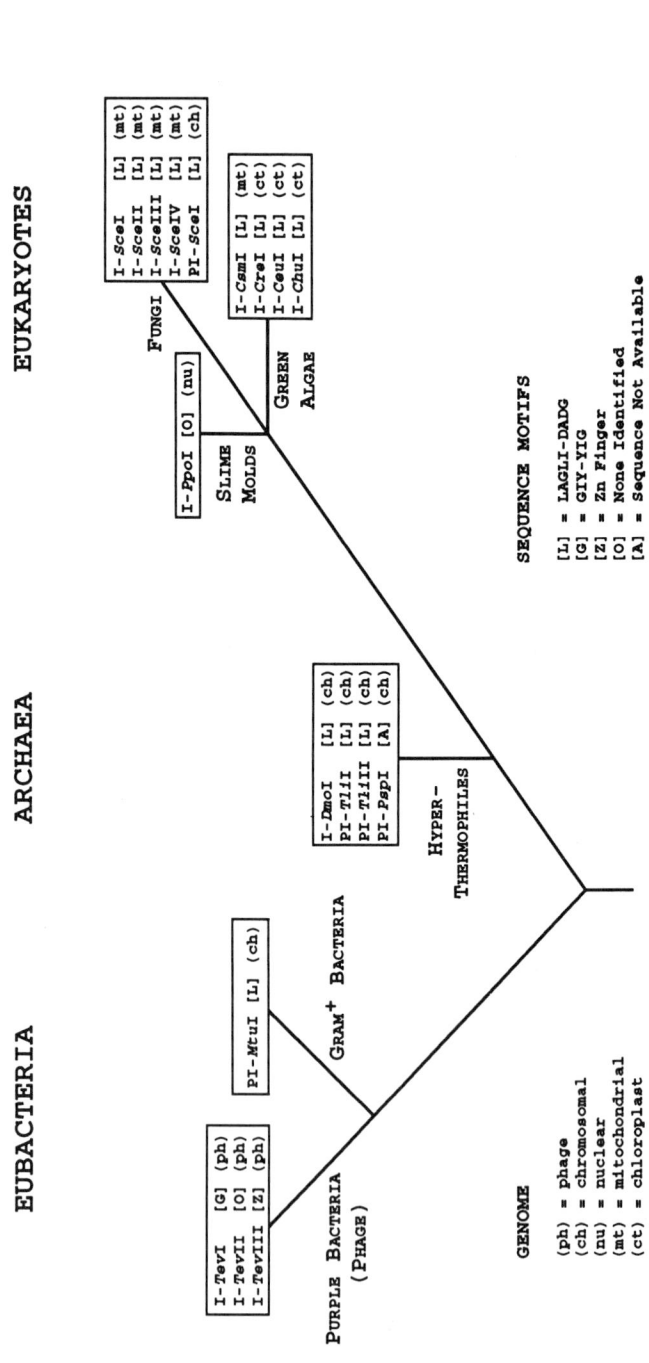

Figure 6 Distribution of homing-type endonuclease genes on the universal phylogenetic tree. (Adapted from Woese et al. 1990.)

numerous *Podospora anserina* and *Saccharomyces douglassi* mitochondrial intron ORFs (Cummings et al. 1989a,b, 1990; Tian et al. 1991). Furthermore, the motif has been identified in *S. cerevisiae* HO endonuclease (Russell et al. 1986) and in a mitochondrial nonintron-encoded nuclease Endo.*Sce*I (Nakagawa et al. 1991; Chapter 6, this volume). Thus, the LAGLI-DADG motif occurs in proteins associated with a number of functions, including double-strand DNA cleavage, RNA maturation, and protein processing.

DNA cleavage activity has been linked to a conserved acidic residue in the P1 repeat of PI-*Tli*I from the archaeon *T. litoralis* (Fig. 5A, circled D) (Hodges et al. 1992). This is the first direct evidence for a LAGLI-DADG-associated endonucleolytic activity. Interestingly, acidic amino acids have been implicated in cleavage by type II restriction enzymes *Eco*RI and *Eco*RV (Thielking et al. 1991; for review, see Heitman 1993; Chapter 2). Although a role for the LAGLI-DADG motif in RNA maturase activity has not yet been demonstrated, its presence in these proteins is suggestive of a similar function. Interestingly, a cobI4 maturase-defective variant is suppressed by a single-amino-acid substitution in the homing endonuclease I-*Sce*II. Considering that these proteins are members of the same motif family and share 60% sequence identity, this observation supports a link between endonuclease and maturase function, the latter of which is likely to have evolved to assist in the excision of splicing-defective introns (Lambowitz 1989; Goguel et al. 1992).

Although the LAGLI-DADG motif occurs in both inteins and several intron endonucleases that are translated as fusion proteins with the upstream exon and believed to be proteolytically processed to form active endonucleases (Colleaux et al. 1990; Wernette et al. 1990; Moran et al. 1992), it is still unclear whether the LAGLI-DADG motif is associated with protease activity. Indeed, the previously mentioned LAGLI-DADG mutation in PI-*Tli*I did not affect protein-splicing activity (Hodges et al. 1992). However, conservation of terminal amino acids in the eukaryotic, eubacterial, and archaeal inteins (Fig. 5A; conserved histidine [H] and asparagine [N] residues at the carboxyl end) as well as nucleophilic residues, cysteine (C), serine (S), or threonine (T), found within or proximal to the intein elements, suggests a common proteolytic mechanism (Fig. 5A). Indeed, recent evidence (Cooper et al. 1993) suggests that self-excision of the *VMA1/TFP1* intein involves optimal intein conformation, the conserved N residue (to initiate protein splicing), and the nucleophilic amino acids at the intein-extein junctions. Other mutational and biochemical studies have been initiated to address the protein-splicing mechanism (Davis et al. 1992; Hirata and Anraku 1992; Hodges et al. 1992; Gu et al. 1993).

2. The GIY-YIG Motif

The amino acid motif GIY-(10 or 11 amino acids)-YIG occurs in I-*Tev*I, the *td* intron-encoded homing endonuclease from phage T4 (Table 1; Fig. 5B) (Michel and Dujon 1986). This motif has been identified in mitochondrial group I intron ORFs of *Neurospora crassa* (Burger and Werner 1985) and in the optional group I introns of *Podospora* (Michel and Cummings 1985; Cummings et al. 1989b) and *S. douglassii* (Tian et al. 1991). In addition, nonintronic bacteriophage T4 proteins, including the site-specific endonuclease SegA, contain degenerate forms of this motif (Fig. 5B) (Sharma et al. 1992). Proteins that have the GIY-YIG motif are divided into two subgroups, based on the spacing (10 or 11 amino acids) between the GIY and YIG elements and the presence of flanking amino acid repeats (Cummings et al. 1989b). A consensus derived from GIY-YIG-containing proteins indicates a conserved arginine (R) residue and several partially conserved lysine (K) residues near the GIY and YIG elements (Fig. 5B). Interestingly, the amino acids Y, K, and R have been implicated in site-specific endonucleolytic catalysis in other systems, suggesting a possible mechanistic link (for review, see Heitman 1992; Vogel and Das 1992). Although there is no direct evidence yet that the GIY-YIG motif is associated with endonuclease activity, its occurrence in I-*Tev*I and other optional intron ORFs, as well as in SegA endonuclease, is compelling.

3. The Zinc Finger Motif

Two zinc finger domains are present at the carboxyl end of I-*Tev*III, the endonuclease encoded within the *nrdB* intron of the T-even phage RB3 (Eddy and Gold 1991). The zinc finger motif (Table 1; Fig. 5C) is generally associated with DNA recognition and binding (for review, see Klug and Rhodes 1987). Additionally, recent data indicate that zinc atoms may be involved in catalysis by nuclease P1 (Volbeda et al. 1991), although these are not associated with a zinc finger motif. It remains to be determined whether the zinc finger domains of I-*Tev*III play a role in target recognition and/or nucleolytic catalysis.

4. Unclassified

The phage intron endonucleases I-*Hmu*I and I-*Hmu*II, which nick only one DNA strand (Table 1), apparently have none of the above consensus sequences. Additionally, two typical homing endonucleases that make double-strand breaks, the *P. polycephalum* nuclear enzyme, I-*Ppo*I, and the T4 *sunY* endonuclease, I-*Tev*II, do not appear to contain any of the

Eukaryotes

```
                                      ⬇
I-SceI      5'-AGTTACGCTAGGGAT AA|CAGGGTAATATAG-3'
            3'-TCAATGCGATCCCTA| TTGTCCCATTATATC-5'

I-SceII     5'-TTCTGATTCTTTGGT CACCC|TGAAGTATAT-3'
            3'-AAGACTAAGAAACCA G|TGGGACTTCATATA-5'

I-SceIII    5'-CTTGGAGGTTTTGGT AAC|TATTTATTACCA-3'
            3'-GAACCTCCAAAACCA| TTGATAAATAATGGT-5'

I-SceIV     5'-AATTTTCTCATGATT A|GCTCTAATCCATGG-3'
            3'-TTAAAAGAGTAC|TAA TCGAGATTAGGTACC-5'

PI-SceI     5'-ATCTATGTCGGG|TGC| GGAGAAAGAGGTAAT-5'
            3'-TAGATACAGCC|CACG CCTCTTTCTCCATTA-3'

I-PpoI      5'-TAACTATGACTCTCT TAA|GGTAGCCAAAAG-3'
            3'-ATTGATACTGAGA|GA ATTCCATCGGTTTTC-5'

I-CsmI*     5'-GTACTAGCATGGGGT CAAATGTCTTTCTGG-3'
            3'-CATGATGGTACCCCA GTTTACAGAAAGACC-5'

I-CreI      5'-CTGGGTTCAAAACGT C|GTGA|GACAGTTTGG-3'
            3'-GACCCAAGTTTTGCA G|CACTCTGTCAAACC-5'

I-CeuI      5'-TCTAACTATAACGGT CCTAA|GGTAGCGAGG-3'
            3'-AGATTGATATTGCCA G|GATTCCATCGCTCC-5'

I-ChuI      5'-GAAGGTTTGGCACCT C|G|ATGTCGGCTCATC-3'
            3'-CTTCCAAACCGTG|GA GCTACAGCCGAGTAG-5'
```

Archaea

```
                                      ⬇
I-DmoI      5'-ATGCCTTGCCGGGTA A|GTTCCGGCGCGCAT-3'
            3'-TACGGAACGGCC|CAT TCAAGGCCGCGCGTA-5'

PI-TliI     5'-GTTCTTTATGCGGAC AC|TGACGGCTTTTAT-3'
            3'-CAAGAAATACGCC|TG TGACTGCCGAAAATA-5'

PI-TliII*   5'-AAATTGCTTGCAAAC AGCTATTACGGCTAT-3'
            3'-TTTAACGAACGTTTG TCGATAATGCCGATA-5'

PI-PspI*    5'-AAAATCCTGGCAAAC AGCTATTATGGGTAT-3'
            3'-TTTTAGGACCGTTTG TCGATAATACCCATA-5'
```

Eubacteria (Phage)

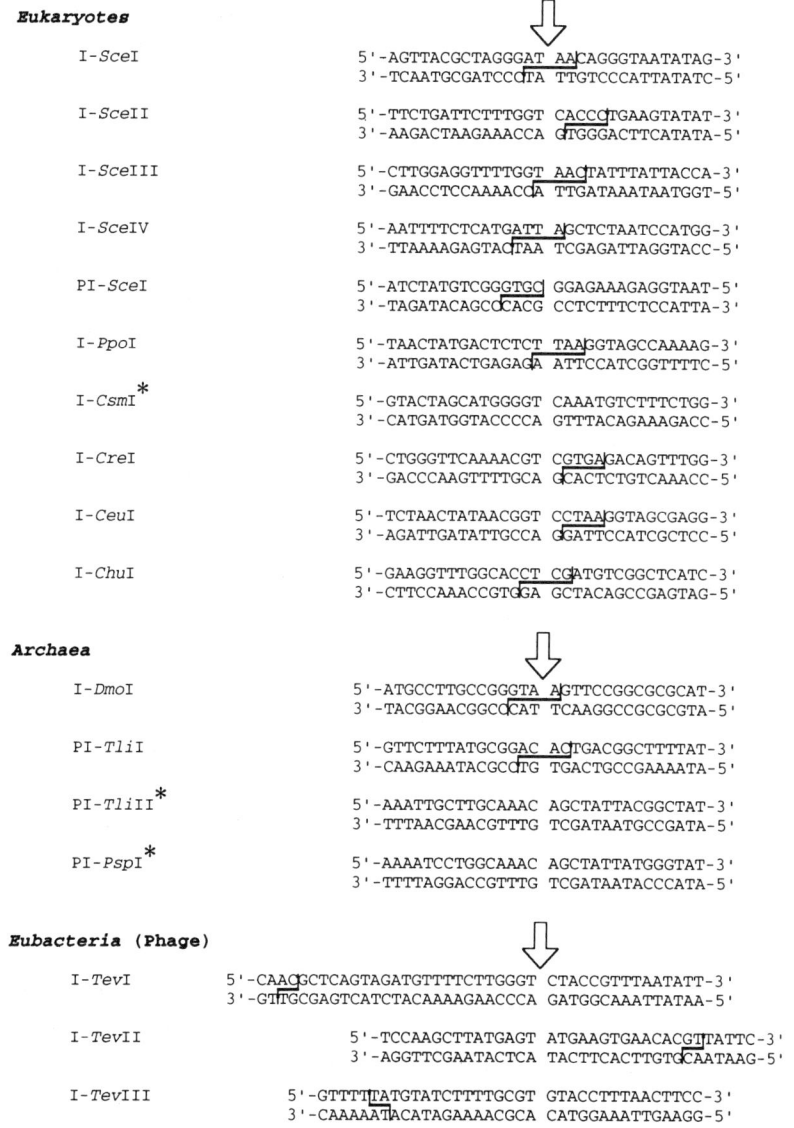

```
                                      ⬇
I-TevI      5'-CAAC|GCTCAGTAGATGTTTTCTTGGGT CTACCGTTTAATATT-3'
            3'-GTT|GCGAGTCATCTACAAAAGAACCCA GATGGCAAATTATAA-5'

I-TevII     5'-TCCAAGCTTATGAGT ATGAAGTGAACACGT|TATTC-3'
            3'-AGGTTCGAATACTCA TACTTCACTTGTG|CAATAAG-5'

I-TevIII    5'-GTTTT|TATGTATCTTTTGCGT GTACCTTTAACTTCC-3'
            3'-CAAAAT|ACATAGAAAACGCA CATGGAAATTGAAGG-5'
```

Figure 7 Target sites of mobility-type endonucleases. Arrows and gap indicate intron insertion sites. (*Staggered lines*) Cleavage sites; (*) cleavage sites not available. For complete descriptions of endonucleases and references, see Table 1.

characteristic motifs. It is unclear whether the consensus sequences in these proteins are so degenerate as to have defied detection or whether these latter homing endonucleases represent members of different enzyme families. Regardless, most of the homing endonucleases *are* members of protein families (Table 1), an observation discussed further in considering the phylogenetic diversity of these enzymes (Fig. 6).

E. Target Recognition and Cleavage Specificity

1. General Properties

The target sites of the homing endonucleases are long, ranging from 15 bp for I-*Ppo*I (Lowery et al. 1992) to approximately 40 bp for I-*Tev*I (Chu et al. 1991; Bryk et al. 1993). The homing endonucleases exhibit a broad range of sequence specificities and cutting frequencies. Highly selective endonucleases, such as I-*Sce*I, PI-*Sce*I, and I-*Ppo*I, cleave DNA at a relatively low frequency, once in approximately 10^7 bp (Thierry et al. 1991; Bremer et al. 1992; Lowery et al. 1992), whereas somewhat more tolerant endonucleases, such as I-*Sce*II, cleave at a higher frequency, once in 10^5–10^6 bp (Wernette et al. 1992). The permissive endonucleases, such as I-*Tev*I and I-*Tev*II, cleave DNA at a relatively high frequency, once in 10^4–10^5 bp (Lambowitz and Belfort 1993). Notably, the specificity of these endonucleases is not related to the length of their respective target sites.

The cleavage sites of homing endonucleases are generally asymmetric in sequence (Fig. 7), although the I-*Ppo*I (Lowery et al. 1992), I-*Cre*I (Thompson et al. 1992), and I-*Ceu*I (Marshall and Lemieux 1992) homing sites exhibit limited symmetry near the intron insertion site. Strikingly, the eukaryotic and archaeal endonucleases, most of which contain the LAGLI-DADG motif, cleave their substrates in close proximity to the intron insertion site by making staggered cuts that generate four-base 3′ extensions. In contrast, the eubacterial (phage) non-LAGLI-DADG intron-encoded endonucleases cut their targets at a distance from their respective intron insertion sites (Table 1; Fig. 7). Cleavage by I-*Tev*I and I-*Tev*II results in two-base 3′ overhangs, whereas cutting by I-*Tev*III generates two-base 5′ extensions (Fig. 7). Both genetic and biophysical approaches are being taken to study the recognition and cleavage reactions of several of these homing endonucleases.

2. Genetic Studies

The interaction of homing endonucleases with their substrates has been studied primarily via mutational analyses of the homing site. These studies provide a framework for examining the broad range of sequence specificities among the endonucleases. Such analyses indicate that the infrequent cutters, I-*Sce*I (Colleaux et al. 1988; Monteilhet et al. 1990), I-*Ceu*I (Marshall and Lemieux 1992), and I-*Ppo*I (Lowery et al. 1992), are most sensitive to sequence alterations within their homing sites. Fifty percent of single-base changes introduced into the I-*Sce*I or I-*Ceu*I homing sites completely block endonuclease activity (Colleaux et al. 1988; Marshall and Lemieux 1992). These highly site-specific endonucleases have an immediate value as tools for genome mapping.

Enzymes that exhibit intermediate sequence specificity are more permissive regarding sequence changes within their homing sites (Sargueil et al. 1991; Moran et al. 1992; Seraphin et al. 1992; Wernette et al. 1992; Durrenberger and Rochaix 1993). Indeed, for I-*Sce*II, most single-base changes are tolerated to varying extents. However, a small number of missense mutations in the I-*Sce*II homing site are capable of abolishing endonuclease sensitivity (Sargueil et al. 1991; Wernette et al. 1992).

Mutational studies with the I-*Tev*I homing site have provided the basis for understanding more promiscuous protein:DNA interactions. Although I-*Tev*I homing site recognition and cleavage are precise, I-*Tev*I endonuclease is extremely tolerant of base substitutions (Quirk et al. 1989a,b; Bell-Pedersen et al. 1991; Chu et al. 1991; Bryk et al. 1993). Indeed, no single base within a 48-bp region encompassing the homing site is essential for enzyme cleavage, although preferred sequence domains have been identified (Bryk et al. 1993). Furthermore, *td* homing site variants containing methylated *dam* sites were cleaved with wild-type efficiency, indicating that I-*Tev*I is tolerant of these major groove modifications. These results, combined with physical analyses described below, suggest that sequence-tolerant homing site recognition by I-*Tev*I involves atypical recognition of the substrate outside of the major groove of the DNA.

3. Physical Studies

Physical studies designed to determine protein:DNA contacts directly have been performed for I-*Sce*I (Perrin et al. 1993), I-*Ppo*I (Ellison and Vogt 1993; Ellison et al. 1993), and I-*Tev*I and I-*Tev*II (Bryk et al. 1993; N. Loizos and M. Belfort, unpubl.). Such analyses suggest that I-*Sce*I initially binds sequences downstream from the intron insertion site and then contacts upstream sequences, through interactions in the major groove of the target DNA. These contacts bring the catalytic site of the endonuclease into close proximity of the DNA cleavage site (Fig. 8A) (Perrin et al. 1993). The resulting cleavage event by I-*Sce*I involves a two-step kinetic reaction with product release as the rate-limiting determinant. It is believed that the endonuclease remains bound to the downstream cleavage product, accounting for its relatively low turnover

A

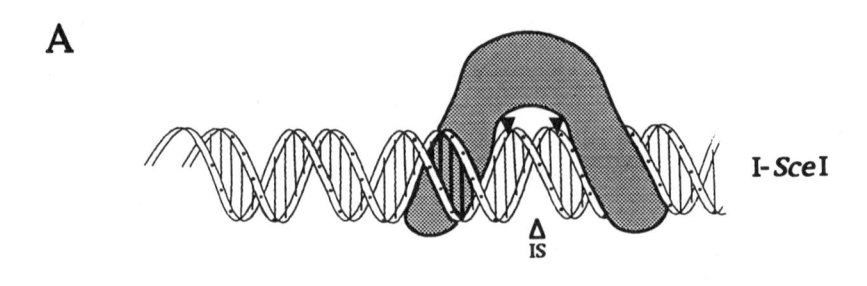

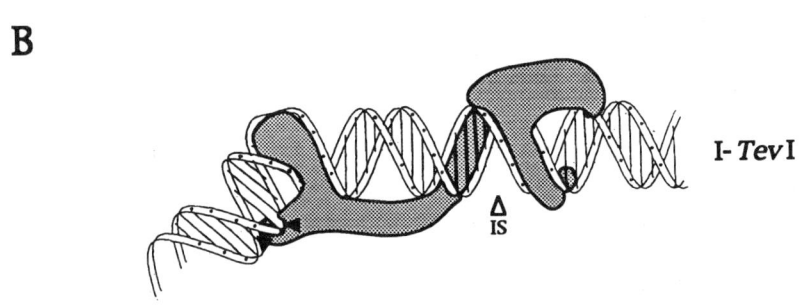

I-*Sce*I

B

I-*Tev*I

Figure 8 Endonuclease:homing site models. The working models, based on genetic and physical studies, were derived from data presented at the 1993 *Keystone Symposia on Molecular and Cellular Biology—Nuclease: Structure, Function and Biological Roles*. (*A*) I-*Sce*I (adapted from model presented by Perrin et al. 1993), depicting predominantly major groove interactions. (*B*) I-*Tev*I (adapted from model presented by M. Belfort), depicting predominantly minor groove interactions. The two regions contacted by the enzyme were inferred from genetic and physical studies (Bryk et al. 1993). These representations (*A* and *B*) are not meant to imply stoichiometries, which remain to be determined. (*Open arrowheads*) Intron insertion sites; (*closed arrowheads*) cleavage sites.

rate (Monteilhet et al. 1990; Perrin et al. 1993). This delayed release of I-*Sce*I has led to speculation that the endonuclease may be involved in the ensuing recombination events (Plessis et al. 1992; Perrin et al. 1993).

Deletion and footprinting analyses have revealed that I-*Ppo*I recognizes a 13–15-bp sequence flanking the insertion/cleavage sites (Ellison et al. 1993) but that the enzyme also makes contacts outside this region (Ellison and Vogt 1993). These distant sequence-tolerant contacts may play a role in stabilizing I-*Ppo*I binding. The studies indicate that I-*Ppo*I, like I-*Sce*I, binds its homing site primarily across the major groove, although minor groove interaction is also evident. The extended nature of the interaction domain and contacts in the minor groove of the DNA substrate are features reminiscent of the phage enzymes (see below).

For the expansive *td* intron homing site, DNase I and hydroxyl radical footprinting have identified the region flanking the *td* intron insertion site as the primary I-*Tev*I-binding domain, with a secondary region of contact, which approaches the cleavage site and is likely to direct catalysis (Bell-Pedersen et al. 1991; Bryk et al. 1993). Interestingly, these data, together with methylation and ethylation interference studies, indicate that I-*Tev*I binds DNA across the less discriminative minor groove in close proximity to the phosphate backbone in both of these contacted regions. Interference analysis of I-*Tev*II with its homing site also reveals contacts across the minor groove (N. Loizos and M. Belfort, unpubl.). Combined with the above-mentioned mutational analysis of the I-*Tev*I homing site, these data suggest that determinants other than single nucleotides are the basis for interactions between I-*Tev*I and its homing site. Such recognition may involve redundant base pair contacts and/or target structure (Bryk et al. 1993). Recently, I-*Tev*I has also been shown to induce a bend within its homing site, at the region of contact near the cleavage site, suggesting that distortion of the helix may allow the juxtaposition of the active site of I-*Tev*I to sequences at the cleavage site (Fig. 8B) (J.E. Mueller, unpubl.).

As discussed earlier, homing endonucleases I-*Tev*I and I-*Tev*II are rather permissive regarding target selection, whereas I-*Sce*I and I-*Ppo*I exhibit a greater degree of sequence selectivity. These variable specificities can be explained, in part, by the fact that sequence discrimination is more precise in the major groove (Seeman et al. 1976). Predictably, the more promiscuous enzymes, I-*Tev*I and I-*Tev*II, bind their DNA substrates across the less discriminative minor groove, whereas the more selective endonucleases, I-*Sce*I and I-*Ppo*I, make extensive major groove interactions. The different recognition properties of the enzymes would impact the ability of endonuclease genes to act as mobile elements, as discussed below.

IV. EVOLUTIONARY CONSIDERATIONS

A. Evidence for Mobility of Endonuclease Genes

There is compelling circumstantial evidence supporting the hypothesis that the phylogenetically widespread homing endonuclease ORFs (see Fig. 6) are mobile DNA elements and progenitors of the mobile introns. First, the striking sequence and structural conservation displayed by group I introns is contrasted by the haphazard positioning and/or variable sequence characteristics of the endonuclease ORFs within introns (see Fig. 3). For example, introns of *P. polycephalum* and *T. thermophila* share 70% sequence identity and exact positioning within the L-rRNA gene, yet only the former contains an endonuclease ORF (Muscarella and

Vogt 1993). In addition, the introns in the *td* and *sunY* genes of bacteriophage T4 exhibit sequence conservation (~60%), yet their ORFs, encoding I-*Tev*I and I-*Tev*II, respectively, are nonhomologous and located in different peripheral stem-loop structures (Table 1; Fig. 3) (Shub et al. 1988). A more striking example exists in two closely related *Neurospora* species; both have the highly conserved ND1 intron (97% identity), but these introns contain completely dissimilar ORFs located at different positions (Mota and Collins 1988). Although the disparity in conservation between the intron and ORF sequences could reflect a strict requirement for specific nucleic acid sequence in ribozymes that is relaxed in the ORFs, the variable location of the ORFs supports the theory that introns were invaded by endonuclease genes in independent integration events.

Second, rare codon usage is a property of many of the endonuclease ORFs (Table 1) (Bonitz et al. 1980; Colleaux et al. 1990; Durrenberger and Rochaix 1991; Cote et al. 1993). It has been proposed that a recently acquired gene may contain codons that are atypical of its newly occupied genome, whereas a more established gene will have adapted to having codons most frequently used by the organism. Although there may be selective pressure on an organism to maintain rare codons and consequently control the expression of potentially deleterious genes, rare codon usage by the homing endonucleases may indicate the recent lateral transfer of these genes (Lambowitz and Belfort 1993).

Third, the wide distribution of mobile endonucleases lends credence to the "mobile ORF" hypothesis (Fig. 6). Mobile endonuclease genes are found not only in group I introns, but also in archaeal introns (Dalgaard et al. 1993). Although there appears to be no mechanistic similarity between these two types of introns at the level of splicing, both encode LAGLI-DADG-containing mobility endonucleases, suggesting strongly that the endonuclease genes have invaded these disparate introns, which provide them with a phenotypically neutral site.

Finally, the positioning of intron-related endonuclease genes in non-intronic locations is suggestive of their invasive properties. Such enzyme-coding sequences occur intergenically at sites that might also be considered to be phenotypically neutral. For example, there exists a GIY-YIG-containing endonuclease gene, similar to that in the *td* intron, in an intergenic region of T4 (SegA, Sharma et al. 1992). Similarly, the LAGLI-DADG endonuclease genes exist not only in organellar group I introns, but also in freestanding form between genes, in both the yeast nucleus (HO endonuclease, Russell et al. 1986) and mitochondrion (Endo.*Sce*I, Nakagawa et al. 1991). Even more remarkably, this family of LAGLI-DADG endonucleases is found as inteins in archaea, eukaryotes, and bacteria (for review, see Shub and Goodrich-Blair 1992; E.O. Davis, pers. comm.). The finding of related endonucleolytic/proteolytic elements in diverse settings in all three kingdoms suggests further that these ORFs represent "selfish DNAs" that, by evolving protein hydrolase-ligase functions, have mastered the ability to invade genomes without disrupting essential gene function.

B. Invasion and Its Aftermath

The endonuclease genes may self-propagate via their products, which promote recombination by virtue of creating double-strand breaks. Since gene conversion at that site could involve insertion of any foreign DNA sequence, how might one rationalize specific integration of the endonuclease gene? The first such event might be a matter of pure chance; however, endonuclease genes would be at an advantage over other foreign DNA inserts because of the ability to self-perpetuate through the double-strand cleavage activity of their products. Endonuclease ORFs would therefore tend to persist in the population rather than be lost.

Those homing endonucleases that have colonized coding sequences seem to have carried the evolutionary process one step further by ensuring both endonuclease expression and splicing of the host protein. Inteins have developed mechanisms that allow recognition of the ends of the endonuclease, protein hydrolysis at these sites, and ligation of the flanking exteins. Once these mechanisms became established, selection for precise protein processing and endonuclease activity is likely to have helped maintain critical components of the insert (Fig. 5A) (Davis et al. 1992; Hodges et al. 1992; Cooper et al. 1993).

The ability of the intein ORFs to home to uninterrupted coding sequences is readily understood in terms of their DNA cleavage activity and the double-strand break-repair model (see Fig. 2). In contrast, where endonucleases have invaded introns, maintenance of gene function is more straightforward because of RNA splicing, but the homing property is more difficult to rationalize than for the inteins. Intron endonuclease ORFs require one of two circumstances for intron homing to be established, since homologous intronless (rather than ORF-less) alleles are required as recipients. Either the endonuclease must develop specificity for the intronless allele, if such exists, or a rare transposition event is required to translocate the intron to a new site that is recognized by the endonuclease. Once localized to the new site, the intron would be in a homing situation, with the availability of homologous intronless alleles containing the endonuclease target sequence (Belfort 1990; Bell-Pedersen et al. 1990).

C. Endonuclease Properties That Potentiate Invasiveness

Several features of the endonuclease:DNA interaction may promote endonuclease ORF or intron invasion at novel sites. First, many of the homing endonucleases studied thus far tolerate substantial sequence degeneracies within their homing sites. As discussed above, single-base substitutions at any position in the I-*Tev*I homing site are permitted (Bryk et al. 1993), whereas I-*Sce*II can recognize homing sites with greater than 25% degeneracy (Moran et al. 1992). Even homing endonucleases that are sequence-sensitive allow a limited number of base substitutions within their targets (e.g., I-*Sce*I, Thierry et al. 1991; I-*Ceu*I, Marshall and Lemieux 1992; I-*Ppo*I, Lowery et al. 1992). As another example, PI-*Sce*I cleaves the yeast genome only once but also cleaves a similar sequence in the bacteriophage λ genome in which 24 out of 40 bp are identical (Bremer et al. 1992; Gimble and Thorner 1992). Such recognition properties may facilitate rare cleavage at an ectopic site, as would be necessary for horizontal transfer events. Second, it has been shown that I-*Tev*I recognizes determinants outside of the major groove, which is often occluded by base modifications in native DNAs (Bryk et al. 1993). These unusual features would allow a broader range of substrates to be available to the endonuclease and may promote the rare event necessary to effect intron transposition. Third, the endonuclease target sequences are extensive (15–40 bp) and thus provide a framework of homology that would favor gene conversion at the new site.

D. Down-regulation of Endonuclease Expression

Like other mobile elements, endonuclease genes must be regulated to prevent their rampant spread and thus maintain host viability. For several endonucleases, regulatory control occurs at the level of RNA processing (see Fig. 4). As discussed above, the archaeal intron endonuclease, I-*Dmo*I, is only expressed from spliced intron RNA (Dalgaard et al. 1993). Furthermore, the expression of I-*Sce*I and subsequent mobility of the ω intron require 3′ processing of the I-*Sce*I ORF transcript (Zhu et al. 1987, 1989). These requirements for processing must impose temporal constraints on endonuclease expression. Such delays are likely to affect the frequency of intron homing. I-*Sce*II, which is inframe with the upstream exon and translated as a fused precursor, offers yet another example of the regulation of ORF expression via RNA splicing. Interestingly, in this case, RNA splicing has a negative effect on I-*Sce*II expression by eliminating the pre-mRNA from which the fused precursor protein is translated.

There is also evidence for regulation of intron endonuclease synthesis at the transcriptional and translational levels, after T4 phage infection.

Here, translation of the endonuclease from the pre-RNA is inhibited by an RNA structure that sequesters the ribosome-binding site. Therefore, translation awaits synthesis of an endonuclease-specific transcript from a late phage promoter that is regulated by a T4-specific transcription factor, gp55 (Fig. 4A) (Gott et al. 1988). Adaptation of endonuclease expression to the organism is reflected in a delay of endonuclease synthesis until after phage DNA replication, when, presumably, there is an increased availability of DNA substrates.

Little is known about the regulation of intein expression. Protein splicing itself may control the endonucleolytic activity of the intein. PI-*Tli*I variants that cannot splice produce only low levels of endonuclease activity (Hodges et al. 1992), suggesting that the flanking exteins interfere with enzyme function. Remarkably, PI-*Sce*I homing is observed only during meiosis, although functional endonuclease can be purified from both meiotic and mitotic cells (Gimble and Thorner 1992). No data are yet available to explain this cell-cycle-specific control, although substrate availability is one possibility.

E. Cross-species Transfer

The amino acid sequence motifs shared by the homing endonucleases suggest a common ancestry within the different families (Table 1; Fig. 6). The wide distribution of the LAGLI-DADG motif in endonucleases from group I and archaeal introns, inteins as well as freestanding genes, raises the possibility of the horizontal transfer of these sequences from one genome to another. Any of a number of processes might be responsible for the exchange of genetic material between organisms, even across kingdom lines (Heinemann 1991; Lambowitz and Belfort 1993). Horizontal transmission has also been invoked to explain the common occurrence of intron ORFs encoding GIY-YIG proteins in the filamentous fungi, *Neurospora* and *Podospora*, and bacteriophage T4 (Michel and Dujon 1986). Not only are homing endonucleases able to recognize and cleave sequences in unrelated genomes (I-*Sce*I, Thierry et al. 1991; I-*Ppo*I, Lowery et al. 1992; Muscarella and Vogt 1993; I-*Tev*I, Bell-Pedersen 1991), but the intron with flanking sequences from the L-rRNA gene of *P. polycephalum* was shown to home to L-rRNA sequences in *S. cerevisiae* as well (Muscarella and Vogt 1993). Although this is an artificially created situation, these experiments begin to address the feasibility of cross-species intron transfer. Nevertheless, phylogenetic studies of endonuclease families will be required to negate the possibility that these enzymes are ancient, having predated the divergence of the three kingdoms from a common ancestor (Doolittle 1993).

V. CONCLUDING REMARKS

Homing endonucleases are a remarkable class of enzymes that impart mobile properties on the DNA sequences that encode them. It was this dynamic homing property of the endonuclease element that led to the discovery of the enzymes. They are phylogenetically diverse, being found in all three kingdoms, in a variety of genetic environments, in nuclei and organelles of lower eukaryotes, and in archaeal and phage genomes. Homing endonuclease genes with common sequence motifs exist in group I introns, in archaeal introns, and as insertion elements in protein-coding sequences. The distribution and common properties of these elements suggest that they are invasive units, capable of extraordinary measures to adapt to their host genomes. Some of the enzymes in group I introns appear to have evolved maturase function to aid in splicing, whereas those expressed as protein fusions have acquired the uncanny ability to effect protein splicing.

Successful parasites must strike a balance between invasiveness, to potentiate their spread, and restraint, to preserve viability of their host. On the one hand, the recognition properties of the enzymes provide the basis for maximizing the invasive potential of the element, through their unusual target specificities and lengthy recognition sequences. On the other hand, there are many ways in which the expression of the different endonuclease elements is down-regulated, to safeguard the host genome from mass destruction. This equilibrium ensures cleavage and integration into a wide variety of substrates while controlling the frequency of such events.

Despite this coherent picture, questions remain. Why are homing endonuclease ORFs not found in the self-splicing group II introns, which exist in several of the same genomes as the endonuclease-coding group I introns but encode reverse-transcriptase-like proteins? The apparent inability of these introns to accommodate the homing endonuclease genes suggests that there are constraints imposed on endonuclease elements by intron structure and/or splicing pathway. In addition, it is puzzling that no intergenic endonucleases of the same families as the homing enzymes have been shown to facilitate mobility. Furthermore, how do these enzymes acquire and perform the multiple functions of DNA cleavage, RNA processing, and/or protein splicing? Answers to this latter question will be based on mechanistic studies and are therefore likely to be the first to emerge.

ACKNOWLEDGMENTS

We thank Drs. Elaine Davis, Francine Perler, Volker Vogt, and Eldora Ellison for sharing unpublished results, Tim Coetzee and Jacob Dalgaard for helpful discussions; Jonathan Clyman, Deborah Court, Victoria Derbyshire, Heidi Goodrich-Blair, Joyce Huang, Monica Parker, Francine Perler, and David Shub for critical comments on the manuscript; and Maryellen Carl for preparing the manuscript. Our laboratory is supported by grants GM-39422 and GM-44844 (to M.B.) and GM-15454 (to J.E.M.) from the National Institutes of Health.

REFERENCES

Belfort, M. 1990. Phage T4 introns: Self-splicing and mobility. *Annu. Rev. Genet.* **24:** 363-385.

Bell-Pedersen, D. 1991. "Intron mobility in bacteriophage T4." Ph.D. thesis, State University of New York, Albany.

Bell-Pedersen, D., S.M. Quirk, M. Bryk, and M. Belfort. 1991. I-*Tev*I, the endonuclease encoded by the mobile *td* intron, recognizes binding and cleavage domains on its DNA target. *Proc. Natl. Acad. Sci.* **88:** 7719-7723.

Bell-Pedersen, D., S. Quirk, J. Clyman, and M. Belfort. 1990. Intron mobility in phage T4 is dependent upon a distinctive class of endonucleases and independent of DNA sequences encoding the intron core: Mechanistic and evolutionary implications. *Nucleic Acids Res.* **18:** 3763-3770.

Berg, D.E. and M.M. Howe. 1989. *Mobile DNA*. American Society for Microbiology, Washington, D.C.

Bonitz, S.G., G. Coruzzi, B.E. Thalenfeld, A. Tzagoloff, and G. Macino. 1980. Assembly of the mitochondrial membrane system. Structure and nucleotide sequence of the gene coding for subunit 1 of yeast cytochrome oxidase. *J. Biol. Chem.* **255:** 11927-11941.

Bos, J.L., C. Heyting, P. Borst, A.C. Arnberg, and E.F.J. van Bruggen. 1978. An insert in the single gene for the large ribosomal RNA in yeast mitochondrial DNA. *Nature* **275:** 336-338.

Bremer, M.C.D., F.S. Gimble, J. Thorner, and C.L. Smith. 1992. VDE endonuclease cleaves *Saccharomyces cerevisiae* genomic DNA at a single site: Physical mapping of the *VMA1* gene. *Nucleic Acids Res.* **20:** 5484.

Bryk, M., S.M. Quirk, J.E. Mueller, N. Loizos, C. Lawrence, and M. Belfort. 1993. The *td* intron endonuclease makes extensive sequence tolerant contacts across the minor groove of its DNA target. *EMBO J.* **12:** 2141-2149.

Burger, G. and S. Werner. 1985. The mitochondrial *URF1* gene in *Neurospora crassa* has an intron that contains a novel type of URF. *J. Mol. Biol.* **186:** 231-242.

Cech, T.R. 1988. Conserved sequences and structures of group I introns: Building an active site for RNA catalysis—A review. *Gene* **73:** 259-271.

―――. 1990. Self-splicing of group I introns. *Annu. Rev. Biochem.* **59:** 543-568.

Chu, F.K., G.F. Maley, F. Maley, and M. Belfort. 1984. Intervening sequence in the thymidylate synthase gene of bacteriophage T4. *Proc. Natl. Acad. Sci.* **81:** 3049-3053.

Chu, F.K., F. Maley, A.-M. Wang, J. Pedersen-Lane, and G. Maley. 1991. Purification and substrate specificity of a T4 phage intron-encoded endonuclease. *Nucleic Acids Res.* **19:** 6863-6869.

Clyman, J. and M. Belfort. 1992. *Trans* and *cis* requirements for intron mobility in a prokaryotic system. *Genes Dev.* **6:** 1269-1279.

Coen, D., J. Deutch, P. Netter, E. Petrochillo, and P.P. Slonimski. 1970. Mitochondrial genetics. I. Methodology and phenomenology. *Symp. Soc. Exp. Biol.* **24:** 444-496.

Colleaux, L., L. D'Auriol, F. Galibert, and B. Dujon. 1988. Recognition and cleavage site of the intron-encoded *omega* transposase. *Proc. Natl. Acad. Sci.* **85:** 6022–6026.

Colleaux, L., M.-R. Michel-Wolwertz, R.F. Matagne, and B. Dujon. 1990. The apocytochrome *b* gene of *Chlamydomonas smithii* contains a mobile intron related to both *Saccharomyces* and *Neurospora* introns. *Mol. Gen. Genet.* **223:** 288–296.

Colleaux, L., L. D'Auriol, M. Betermier, G. Cottarel, A. Jacquier, F. Galibert, and B. Dujon. 1986. Universal code equivalent of a yeast mitochondrial intron reading frame is expressed into *E. coli* as a specific double strand endonuclease. *Cell* **44:** 521–533.

Cooper, A.A., Y-J. Chen, M.A. Lindorfer, and T.H. Stevens. 1993. Protein splicing of the yeast *TFP1* intervening protein sequence: A model for self-excision. *EMBO J.* **12:** 2575–2583.

Cote, V., J.-P. Mercier, C. Lemieux, and M. Turmel. 1993. The single group-I intron in the chloroplast *rrn*L gene of *Chlamydomonas humicola* encodes a site-specific DNA endonuclease. *Gene* **129:** 69–79.

Cummings, D.J., F. Michel, and K.L. McNally. 1989a. DNA sequence analysis of the 24.5 kilobase pair cytochrome oxidase subunit I mitochondrial gene from *Podospora anserina*: A gene with sixteen introns. *Curr. Genet.* **16:** 381–406.

Cummings, D.J., F. Michel, and K.L. McNally. 1989b. DNA sequence analysis of the apocytochrome *b* gene of *Podospora anserina*: A new family of intronic open reading frame. *Curr. Genet.* **16:** 407–418.

Cummings, D.J., F. Michel, J.M. Domenico, and K.L. McNally. 1990. DNA sequence analysis of the mitochondrial *ND4L-ND5* gene complex from *Podospora anserina*. Duplication of the *ND4L* gene within its intron. *J. Mol. Biol.* **212:** 269–286.

Dalgaard, J.Z. and R.A. Garrett. 1992. Protein-encoding introns from the 23S rRNA-encoding gene from stable circles in the hyperthermophilic archaeon *Pyrobaculum organotrophum*. *Gene* **121:** 103–110.

Dalgaard, J.Z., R.A. Garrett, and M. Belfort. 1993. A site-specific endonuclease encoded by a typical archaeal intron. *Proc. Natl. Acad. Sci.* **90:** 5414–5417.

Davis, E.O., S.G. Sedgwick, and M.J. Colsten. 1991. Novel structure of the *RecA* locus of *Mycobacterium tuberculosis* implies processing of the gene product. *J. Bacteriol.* **173:** 5653–5662.

Davis, E.O., P.J. Jenner, P.C. Brooks, M.J. Colston, and S.G. Sedgwick. 1992. Protein splicing in the maturation of the *M. tuberculosis* RecA protein: A mechanism for tolerating a novel class of intervening sequence. *Cell* **71:** 201–210.

Delahodde, A., V. Goguel, A.M. Becam, F. Creusot, J. Perea, J. Banroques, and C. Jacq. 1989. Site-specific DNA endonuclease and RNA maturase activities of two homologous intron-encoded proteins from yeast mitochondria. *Cell* **56:** 431–441.

Doolittle, R.F. 1993. The comings and goings of homing endonucleases and mobile introns. *Proc. Natl. Acad. Sci.* **90:** 5379–5381.

Dujon, B. 1980. Sequence of the intron and flanking exons of the mitochondrial 21S rRNA gene of yeast strains having different alleles at the *omega* and *rib*I loci. *Cell* **20:** 185–197.

Dujon, B. and A. Jacquier. 1983. Organization of the mitochondrial 21S rRNA gene in *Saccharomyces cerevisiae*: Mutants of peptidyl transferase center and nature of the omega locus. In *Mitochondria* (ed. W. deGruyter), pp. 389–403. deGruyter, Berlin.

Dujon, B. and F. Michel. 1976. Genetics and physical characterization of a segment of the mitochondrial DNA involved in the control of genetic recombination. In *The genetic function of mitochondrial DNA* (ed. C. Saccone and A.M. Kroon), pp. 175–184. North-Holland Biomedical, Amsterdam.

Dujon, B., M. Belfort, R.A. Butow, C. Jacq, C. Lemieux, P.S. Perlman, and V.M. Vogt.

1989. Mobile introns: Definition of terms and recommended nomenclature. *Gene* **82**: 115–118.

Durrenberger, F. and J.-D. Rochaix. 1991. Chloroplast ribosomal intron of *Chlamydomonas reinhardtii*: *In vitro* self-splicing, DNA endonuclease activity and *in vivo* mobility. *EMBO J.* **10**: 3495–3501.

―――. 1993. Characterization of the cleavage site and the recognition sequence of the I-*Cre*I DNA endonuclease encoded by the chloroplast ribosomal intron of *Chlamydomonas reinhardtii*. *Mol. Gen. Genet.* **236**: 409–414.

Eddy, S.R. and L. Gold. 1991. The phage T4 *nrd*B intron: A deletion mutant of a version found in the wild. *Genes Dev.* **5**: 1032–1041.

Ellison, E.L. and V.M. Vogt. 1993. Interaction of the intron-encoded mobility endonuclease I-PpoI with its target site. *Mol. Cell. Biol.* (in press).

Ellison, E.L., D.E. Muscarella, and V.M. Vogt. 1993. Characterization of I-*Ppo*I, an intron-encoded endonuclease from *Physarum polycephalum*. *J. Cell. Biochem.* **17c**: 177.

Gauthier, A., M. Turmel, and C. Lemieux. 1991. A group I intron in the chloroplast large subunit rRNA gene of *Chlamydomonas eugametos* encodes a double-strand endonuclease that cleaves the homing site of this intron. *Curr. Genet.* **19**: 43–47.

Gimble, F.S. and J. Thorner. 1992. Homing of a DNA endonuclease gene by meiotic gene conversion in *Saccharomyces cerevisiae*. *Nature* **357**: 301–306.

Goguel, V., A. Delahodde, and C. Jacq. 1992. Connections between RNA splicing and DNA intron mobility in yeast mitochondria. RNA maturase and DNA endonuclease switching experiments. *Mol. Cell Biol.* **12**: 696–705.

Goodrich-Blair, H. 1993. "Introns in HMU-bacteriophage intron structure and open reading frame function." Ph.D. thesis, State University of New York, Albany.

Goodrich-Blair, H. and D. A. Shub. 1993. A site-specific intron-endonuclease is required for marker exclusion. *J. Cell. Biochem.* **17c**:177.

Goodrich-Blair, H., V. Scarlato, J.M. Gott, M.-Q. Xu, and D.A. Shub. 1990. A self-splicing group I intron in the DNA polymerase gene of *Bacillus subtilis* bacteriophage SPO1. *Cell* **63**: 417–424.

Gott, J.M., A. Zeeh, D. Bell-Pedersen, K. Ehrenman, M. Belfort, and D.A. Shub. 1988. Genes within genes: Independent expression of phage T4 intron open reading frames and the genes in which they reside. *Genes Dev.* **2**: 1791–1799.

Grivell, L.A. 1992. Homing in on an endosymbiotic endonuclease. *Curr. Biol.* **2**: 450–452.

Gu, H.H., J. Xu, M. Gallagher, and G.E. Dean. 1993. Peptide splicing in the vacuolar ATPase subunit A from *Candida tropicalis*. *J. Biol. Chem.* **268**: 7372–7381.

Hanson, D.K., M.R. Lamb, H.R. Mahler, and P.S. Perlman. 1982. Evidence for translated intervening sequences in the mitochondrial genome of *Saccharomyces cerevisiae*. *J. Biol. Chem.* **257**: 3218–3224.

Heinemann, J.A. 1991. Genetics of gene transfer between species. *Trends Genet.* **7**: 181–185.

Heitman, J. 1992. How the *Eco*RI endonuclease recognizes and cleaves DNA. *BioEssays* **14**: 445–454.

―――. 1993. On the origins, structures, and functions of restriction-modification enzymes. *Genet. Eng.* **15**: 57–108.

Hendrix, R. 1991. Protein carpentry. *Curr. Biol.* **1**: 71–73.

Hensgens, L.A.M., L. Bonen, M. da Haan, G. van der Horst, and L.A. Grivell. 1983. Two intron sequences in yeast mitochondrial *COX1* gene: Homology among URF-containing introns and strain dependent variation in flanking exons. *Cell* **32**: 379–389.

Heyting, C. and H.H. Menke. 1979. Fine structure of the 21S ribosomal RNA region in yeast mitochondrial DNA. III. Physical location of mitochondrial genetic markers and the molecular nature of *omega*. *Mol. Gen. Genet.* **168:** 279–291.

Hirata, R. and Y. Anraku. 1992. Mutations at the putative junction sites of the yeast VMA1 protein, the catalytic subunit of the vacuolar membrane H$^+$-ATPase inhibit its processing by protein splicing. *Biochem. Biophys. Res. Commun.* **188:** 40–47.

Hirata, R., Y. Ohsumi, A. Nakano, H. Kawasaki, K. Suzuki, and Y. Anraku. 1990. Molecular structure of a gene, *VMA1*, encoding the catalytic subunit of H$^+$-translocating adenosine triphosphatase from vacuolar membranes of *Saccharomyces cerevisiae*. *J. Biol. Chem.* **265:** 6726–6733.

Hodges, R.A., F.B. Perler, C.J. Noren, and W.E. Jack. 1992. Protein splicing removes intervening sequences in an archaea DNA polymerase. *Nucleic Acids Res.* **20:** 6153–6157.

Jacq, C., P. Pajot, J. Lazowska, G. Dujardin, M. Claisse, O. Groudinsky, H. delaSalle, C. Grandchamps, M. Labouesse, A. Gargouri, B. Guiard, A. Spyridakis, M. Dreyfus, and P. Slonimski. 1982. Role of introns in the yeast cytochrome-*b* gene: *cis*- and *trans*-acting signals, intron manipulation, expression and intergenic communications. In *Mitochondrial genes* (ed. P. Slonimski et al.), pp. 155–183. Cold Spring Harbor Laboratory, Cold Spring Harbor, New York.

Jacquier, A. and B. Dujon. 1985. An intron-encoded protein is active in a gene conversion process that spreads an intron into a mitochondrial gene. *Cell* **41:** 383–394.

Kane, P.M., C.T. Yamashiro, D.F. Wolczyk, N. Neff, M. Goebl, and T.H. Stevens. 1990. Protein splicing converts the yeast *TFP1* gene product to the 69-kD subunit of the vacuolar H$^+$-adenosine triphosphatase. *Science* **250:** 651–657.

Kjems, J. and R.A. Garrett. 1985. An intron in the 23S RNA gene of the archaebacterium *Desulfurococcus mobilis*. *Nature* **318:** 675–677.

Klug, A. and D. Rhodes. 1987. "Zinc fingers": A novel protein motif for nucleic acid recognition. *Trends Biochem. Sci.* **12:** 464–469.

Lambowitz, A.M. 1989. Infectious introns. *Cell* **56:** 323–326.

Lambowitz, A.M. and M. Belfort. 1993. Introns as mobile genetic elements. *Annu. Rev. Genet.* **62:** 587–622.

Lazowska, J., C. Jacq, and P.P. Slonimski. 1980. Sequence of introns and flanking exons in wild-type and *box3* mutants of cytochrome *b* reveals an interlaced splicing protein coded by an intron. *Cell* **22:** 333–348.

Lowery, R., L. Hung, K. Knoche, and R. Bandziulis. 1992. Properties of I-*Ppo*I: A rare cutting intron-encoded endonuclease. *Promega Notes* **38:** 8–12.

Macreadie, I.G., R.M. Scott, A.R. Zinn, and R.A. Butow. 1985. Transposition of an intron in yeast mitochondria requires a protein encoded by that intron. *Cell* **41:** 395–402.

Marshall, P. and C. Lemieux. 1991. Cleavage pattern of the homing endonuclease encoded by the fifth intron in the chloroplast large subunit rRNA-encoding gene of *Chlamydomonas eugametos*. *Gene* **104:** 241–245.

———. 1992. The I-*Ceu*I endonuclease recognizes a sequence of 19 base pairs and preferentially cleaves the coding strand of the *Chlamydomonas moewussi* chloroplast large subunit rRNA gene. *Nucleic Acids Res.* **20:** 6401–6407.

Michel, F. and D.J. Cummings. 1985. Analysis of class I introns in a mitochondrial plasmid associated with senescence of *Podospora anserina* reveals extraordinary resemblance in the *Tetrahymena* ribosomal intron. *Curr. Genet.* **10:** 69–79.

Michel, F. and B. Dujon. 1986. Genetic exchanges between bacteriophage T4 and filamentous fungi? *Cell* **46:** 323.

Michel, F. and E. Westhof. 1990. Modelling of the three-dimensional architecture of

group I catalytic introns based on comparative sequence analysis. *J. Mol. Biol.* **216:** 585-610.
Michel, F., A. Jacquier, and B. Dujon. 1982. Comparison of fungal mitochondrial introns reveals extensive homologies in RNA secondary structure. *Biochimie* **64:** 867-881.
Monteilhet, C., A. Perrin, A. Thierry, L. Colleaux, and B. Dujon. 1990. Purification and characterization of the *in vitro* activity of I-*Sce*I, a novel and highly specific endonuclease encoded by a group I intron. *Nucleic Acids Res.* **18:** 1407-1413.
Moran, J.V., C.M. Wernette, K.L. Mecklenburg, R.A. Butow, and P.S. Perlman. 1992. Intron 5α of the *COX1* gene of yeast mitochondrial DNA is a mobile group I intron. *Nucleic Acids Res.* **20:** 4069-4076.
Mota, E.M. and R.A. Collins. 1988. Independent evolution of structural and coding regions in a *Neurospora* mitochondrial intron. *Nature* **332:** 654-656.
Muscarella, D.E. and V.M. Vogt. 1989. A mobile group I intron in the nuclear rDNA of *Physarum polycephalum. Cell* **56:** 443-454.
―――. 1993. A mobile group I intron from *Physarum polycephalum* can insert itself and induce point mutations in the nuclear ribosomal DNA of *Saccharomyces cerevisiae. Mol. Cell Biol.* **13:** 1023-1033.
Muscarella, D.E., E.L. Ellison, B.M. Ruoff, and V.M. Vogt. 1990. Characterization of I-*Ppo*, an intron-encoded endonuclease that mediates homing of a group I intron in the rDNA of *Physarum polycephalum. Mol. Cell Biol.* **10:** 3386-3396.
Nakagawa, K., N. Morishima, and T. Shibata. 1991. A maturase-like subunit of the sequence-specific endonuclease Endo.*Sce*I from yeast mitochondria. *J. Biol. Chem.* **266:** 1977-1984.
Orr-Weaver, T.L., J.W. Szostak, and R.J. Rothstein. 1981. Yeast transformation: A model system for the study of recombination. *Proc. Natl. Acad. Sci.* **78:** 6354-6358.
Pedersen-Lane, J. and M. Belfort. 1987. Variable occurrences of the *nrd*B intron in T-even phages suggests intron mobility. *Science* **237:** 182-184.
Perea, J., C. Desdouets, M. Schapria, and C. Jacq. 1993. I-*Sce*III: A novel group I intron-encoded endonuclease from the yeast mitochondria. *Nucleic Acids Res.* **21:** 358.
Perler, F.B., D.G. Comb, W.E. Jack, L.S. Moran, B. Quiang, R.B. Kucera, J. Benner, B.E. Slatko, D.O. Nwankwo, S.K. Hempstead, C.K.S. Carlow, and H. Jannasch. 1992. Intervening sequences in an Archaea DNA polymerase gene. *Proc. Natl. Acad. Sci.* **89:** 5577-5581.
Perlman, P.S. and R.A. Butow. 1989. Mobile introns and intron-encoded proteins. *Science* **246:** 1106-1109.
Perrin, A., M. Buckle, and B. Dujon. 1993. Asymmetrical recognition and activity of the I-*Sce*I endonuclease on its site and on intron-exon junctions. *EMBO J.* **12:** 2939-2947.
Plessis, A., A. Perrin, J.E. Haber, and B. Dujon. 1992. Site-specific recombination determined by I-*Sce*, a mitochondrial group I intron-encoded endonuclease expressed in the yeast nucleus. *Genetics* **130:** 451-460.
Quirk, S.M., D. Bell-Pedersen, and M. Belfort. 1989a. Intron mobility in the T-even phages: High frequency inheritance of group I introns promoted by intron open reading frames. *Cell* **56:** 455-465.
Quirk, S.M., D. Bell-Pedersen, J. Tomaschewski, W. Ruger, and M. Belfort. 1989b. The inconsistent distribution of introns in T-even phages indicates recent genetic exchanges. *Nucleic Acids Res.* **17:** 301-315.
Rochaix, J.D., M. Rahire, and F. Michel. 1985. The chloroplast ribosomal intron of *Chlamydomonas reinhardtii* codes for a polypeptide related to mitochondrial maturases. *Nucleic Acids Res.* **13:** 975-984.
Ruoff, B., S. Johansen, and V.M. Vogt. 1992. Characterization of the self-splicing prod-

ucts of a mobile intron from the nuclear rDNA of *Physarum polycephalum. Nucleic Acids Res.* **20:** 5899–5906.

Russell, D.W., R. Jensen, M.J. Zoller, J. Burke, B. Errede, M. Smith, and I. Herskowitz. 1986. Structure of the *Saccharomyces cerevisiae HO* gene and analysis of its upstream regulatory region. *Mol. Cell Biol.* **6:** 4281–4294.

Saldanha, R., G. Mohr, M. Belfort, and A.M. Lambowitz. 1993. Group I and group II introns. *FASEB J.* **7:** 15–24.

Sargueil, B., A. Delahodde, D. Hatat, G. L. Tian, J. Lazowska, and C. Jacq. 1991. A new specific DNA endonuclease activity in yeast mitochondria. *Mol. Gen. Genet.* **225:** 340–341.

Seeman, N.C., J.M. Rosenberg, and A. Rich. 1976. Sequence-specific recognition of double helical nucleic acids by proteins. *Proc. Natl. Acad. Sci.* **73:** 804–808.

Seraphin, B., G. Faye, D. Hatat, and C. Jacq. 1992. The yeast mitochondrial intron aI5α: Associated endonuclease activity and *in vivo* mobility. *Gene* **113:** 1–8.

Sharma, M., R.L. Ellis, and D.M. Hinton. 1992. Identification of a family of bacteriophage T4 genes encoding proteins similar to those present in group I introns of fungi and phage. *Proc. Natl. Acad. Sci.* **89:** 6658–6662.

Shub, D.A. and H. Goodrich-Blair. 1992. Protein introns: A new home for endonucleases. *Cell* **71:** 183–186.

Shub, D.A., J.M. Gott, M.-Q. Xu, B.F. Lang, M. Michel, J. Tomaschewski, J. Pedersen-Lane, and M. Belfort. 1988. Structural conversation among three homologous introns of bacteriophage T4 and the group I introns of eukaryotes. *Proc. Natl. Acad. Sci.* **85:** 1151–1155.

Szostak, J.W., T.L. Orr-Weaver, R.J. Rothstein, and F.W. Stahl. 1983. The double-strand-break repair model for recombination. *Cell* **33:** 25–35.

Thielking, V., U. Selent, E. Kohler, H. Wolfes, U. Pieper, R. Geiger, C. Urbanke, F.K. Winkler, and A. Pingoud. 1991. Site-directed mutagenesis studies with *Eco*RV restriction endonuclease to identify regions involved in recognition and catalysis. *Biochemistry* **30:** 6416–6422.

Thierry, A., A. Perrin, J. Boyer, C. Fairhead, B. Dujon, B. Frey, and G. Schmitz. 1991. Cleavage of yeast and bacteriophage T7 genomes at a single site using the rare cutter endonuclease I-*Sce*I. *Nucleic Acids Res.* **19:** 189.

Thompson, A.J., X. Yuan, W. Kudlicki, and D.L. Herrin. 1992. Cleavage and recognition pattern of a double-strand-specific endonuclease (I-*Cre*I) encoded by the chloroplast 23S rRNA intron in *Chlamydomonas reinhardtii. Gene* **119:** 247–251.

Tian, G.-L., C. Macadre, A. Kruszewska, B. Szczesniak, A. Ragnini, P. Grisanti, R. Rinaldi, C. Palleschi, L. Frontali, P.P. Slonimski, and J. Lazowska. 1991. Incipient mitochondrial evolution in yeasts. I. The physical map and gene order of *Saccharomyces douglasii* mitochondrial DNA discloses a translocation of a segment of 15,000 base-pairs and the presence of new introns in comparison with *Saccharomyces cerevisiae. J. Mol. Biol.* **218:** 735–746.

Tomaschewski, J. and W. Ruger. 1987. Nucleotide sequence and primary structures of gene products coded for by the T4 genome between map positions 48.266 kb and 39.166 kb. *Nucleic Acids Res.* **15:** 3632–3633.

Turmel, M., J. Boulanger, M.N. Schnare, M.W. Gray, and C. Lemieux. 1991. Six group I introns and three internal transcribed spacers in the chloroplast large subunit ribosomal RNA gene of the green algae *Chlamydomonas eugametos. J. Mol. Biol.* **217:** 293–311.

Vogel, A.M. and A. Das. 1992. Mutational analysis of *Agrobacterium tumefaciens* vir D2: Tyrosine 29 is essential for endonuclease activity. *J. Bacteriol.* **174:** 303–308.

Volbeda, A., A. Lahm, F. Sakiyama, and D. Suck. 1991. Crystal structure of *Penicillium*

citrinum P1 nuclease at 2.8 Å resolution. *EMBO J.* **10:** 1607–1618.
Wallace, C.J.A. 1993. The curious case of protein splicing: Mechanistic insights suggested by protein semisynthesis. *Protein Sci.* **2:** 697–705.
Waring, R.B., R.W. Davies, C. Scazzocchio, and T.A. Brown. 1982. Internal structure of a mitochondrial intron of *Aspergillus nidulans*. *Proc. Natl. Acad. Sci.* **79:** 6332–6336.
Wenzlau, J.M., R.J. Saldanha, R.A. Butow, and P.S. Perlman. 1989. A latent intron-encoded maturase is also an endonuclease needed for intron mobility. *Cell* **56:** 421–430.
Wernette, C.M., R. Saldanha, P.S. Perlman, and R.A. Butow. 1990. Purification of a site-specific enconuclease. I-*Sce*II, encoded by intron 4α of the mitochondrial *coxI* gene of *Saccharomyces cerevisiae*. *J. Biol. Chem.* **265:** 18976–18982.
Wernette, C., R. Saldanha, D. Smith, D. Ming, P.S. Perlman, and R.A. Butow. 1992. Complex recognition site for the group I intron-encoded endonuclease I-*Sce*II. *Mol. Cell Biol.* **12:** 716–723.
Woese, C.R., O. Kandler, and M.L. Wheelis. 1990. Towards a natural system of organisms: Proposal for the domains Archaea, Bacteria, Eucarya. *Proc. Natl. Acad. Sci.* **87:** 4576–4579.
Zhu, H., I.G. Macreadie, and R.A. Butow. 1987. RNA processing and expression of an intron-encoded protein in yeast mitochondria: Role of a conserved dodecamer sequence. *Mol. Cell. Biol.* **7:** 2530–2537.
Zhu, H., H. Conrad-Webb, X.S. Liao, P.S. Perlman, and R.A. Butow. 1989. Functional expression of yeast mitochondrial intron-encoded protein requires RNA processing at a conserved dodecamer sequence at the 3′ end of the gene. *Mol. Cell. Biol.* **9:** 1507–1512.
Zinn, A.R. and R.A. Butow. 1985. Nonreciprocal exchange between alleles of the yeast mitochondrial 21S rRNA gene: Kinetics and involvement of a double-strand break. *Cell* **40:** 887–895.

5
The Nucleases of Genetic Recombination

Stephen C. West
Imperial Cancer Research Fund
Clare Hall Laboratories
South Mimms, Herts EN6 3LD, United Kingdom

I. Introduction
II. Recombination Models
    A. Strand Assimilation Model
    B. Single-strand Reannealing Model
III. Nucleases That Initiate Recombination
    A. *E. coli* RecBCD Enzyme
    B. *E. coli* RecJ Protein
    C. *E. coli* Exonuclease VIII (RecE Protein)
    D. λ Exonuclease
    E. T7 Exonuclease
    F. *S. cerevisiae* Sep1 Protein
    G. *S. pombe* Exonuclease I and Exonuclease II
    H. Higher Eukaryotic Proteins
IV. Nucleases That Resolve Recombination Intermediates
    A. *E. coli* RuvC Protein
    B. T4 Endonuclease VII
    C. T7 Endonuclease I
    D. *S. cerevisiae* Holliday Junction Resolvases
    E. Mammalian Holliday Junction Resolvase
V. Concluding Remarks

I. INTRODUCTION

Genetic recombination plays two fundamental roles in the cell: It provides (1) a mechanism for the generation of genetic diversity and (2) a route for the repair of DNA lesions caused by irradiation or chemical damage. Nucleases play important roles in genetic recombination, and this chapter focuses on those enzymes that are directly implicated in the process. Although recombination can take two forms, generalized and site-specific, this chapter concentrates on general genetic recombination, i.e., recombination processes that occur between two homologous chromosomes. For excellent reviews on the mechanisms and enzymes of site-specific and transpositional recombination, see Craig (1988), Cox

(1988), Hatfull and Grindley (1988), Landy (1989), and Mizuuchi (1992).

Nucleases are required both for the initiation of crossover events between two interacting chromosomes and for the resolution of crossovers to allow the separation of recombinant DNA molecules. Because of the difficulty in directly ascribing many eukaryotic nucleases to a defined role in recombination (due to the lack of mutants), most of our knowledge has come from studies of bacterial and bacteriophage nucleases. However, even in bacteria, the situation is not totally clear, since mutants frequently have little or no phenotype due to the ability of one nuclease to take over in the absence of another. In this case, recombination-defective phenotypes are only observed when two or more nucleases of a given type have been inactivated by mutation. It is possible that the observed redundancy reflects the important role that nucleases play in the recombination process.

II. RECOMBINATION MODELS

A detailed review of the various possible mechanisms of recombination is beyond the scope of this chapter. However, a number of reviews, both classical (Holliday 1964; Hotchkiss 1974; Meselson and Radding 1975; Radding 1978) and more recent (Howard-Flanders et al. 1984; Cox and Lehman 1987; Radding 1988, 1991; Eggleston and Kowalczykowski 1991; Kowalczykowski 1991; West 1992), are available. Here, two simple models for recombination are presented. The models are mechanistically different and are in accord with the known enzymatic properties of several of the nucleases described below.

A. Strand Assimilation Model

Figure 1 shows an adaptation of the double-strand break repair model first suggested by Szostak et al., following their study of the recombinational repair of linearized plasmid DNA in yeast (Orr-Weaver et al. 1981; Szostak et al. 1983; Sun et al. 1989, 1991). The model proposes that an exonuclease activity attacks DNA following a double-strand break to generate extensive 3'-overhanging, single-stranded tails on both sides of the break (Fig. 1a,b). One 3' end then invades a homologous duplex to form heteroduplex DNA by strand displacement or D-loop formation (Fig. 1c). The D-loop may be enlarged by repair synthesis to enable pairing with the 3' end of the second single strand (Fig. 1d). The second 3' end then acts as a primer for DNA polymerase, and repair synthesis leads to the formation of an intermediate in which the DNA molecules are held

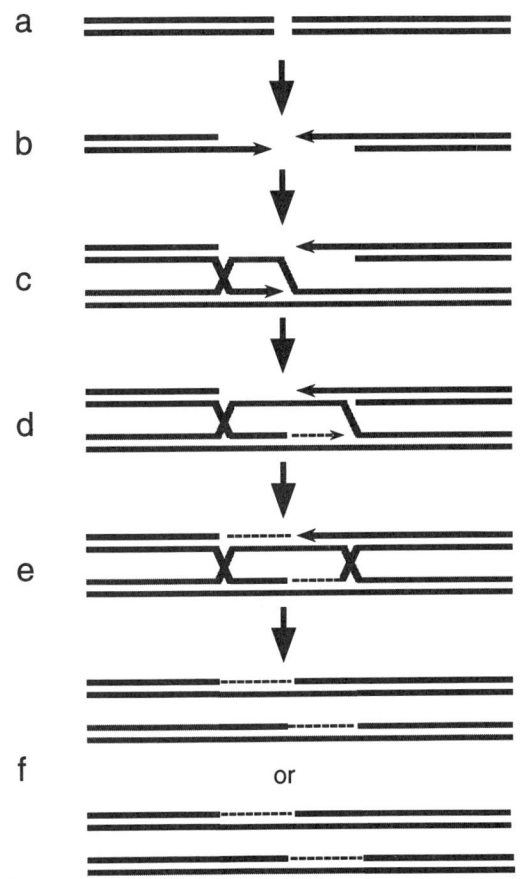

Figure 1 Double-strand break repair model for recombination. (*a*) Duplex DNA with a double-strand break; (*b*) digestion of 5′ termini by a 5′→3′ exonuclease produces overhanging 3′ termini; (*c*) one 3′ end invades homologous duplex DNA to form a D-loop; (*d*) D-loop is enlarged by repair synthesis and the second 3′ end pairs with the displaced strand; (*e*) repair synthesis is initiated at the second 3′ end to produce a joint in which the two DNA molecules are linked by a double Holliday junction; (*f*) junctions are resolved by a Holliday-junction-specific endonuclease to allow the chromosomes to separate. Resolution gives rise to recombinant products that are of either crossover or noncrossover (i.e., no exchange of outside markers) types.

together by two Holliday junctions (Fig. 1e). Finally, the junctions are resolved by endonucleolytic cleavage, resulting in the fomation of recombinant chromosomes (Fig. 1f), and the remaining nicks and gaps can be filled by the action of DNA polymerase and ligase.

This model requires a number of enzymatic activities, many of which have been purified and studied in vitro. For example, although the initiating double-strand break could be produced by X-ray damage, some cells are known to contain nucleases that are capable of introducing double-strand breaks at defined sequences. Enzymes of this type include the *Saccharomyces cerevisiae* HO endonuclease, which cuts DNA to initiate mating-type switching (Kostriken et al. 1983; Kostriken and Heffron 1984), and Endo.*Sce*I, which is thought to initiate the recombination of mitochondrial DNA in yeast (Kawasaki et al. 1991; Nakagawa et al. 1992). Double-strand breaks would then need to be processed by 5'→3' exonucleases to generate molecules with 3'-overhanging single-stranded tails, a role that can be carried out by a number of well-studied enzymes. Two bacteriophage proteins, λ exonuclease and T7 exonuclease, degrade duplex DNA with a 5'→3' polarity, as does *Escherichia coli* exonuclease VIII (the product of *recE*). However, 3' overhangs can also be produced by the action of a DNA helicase (e.g., *E. coli* RecQ helicase) in combination with a single-strand nuclease that exhibits a 5'→3' polarity (e.g., *E. coli* RecJ protein).

The 3' overhangs are thought to be substrates for proteins that catalyze homologous pairing and strand exchange. The *E. coli* RecA protein is the best characterized enzyme of this type (Roca and Cox 1990; Radding 1991; West 1992), and eukaryotic analogs of RecA (*S. cerevisiae* Rad51 and Dmc1) have recently been identified by sequence homology (Bishop et al. 1992; Shinohara et al. 1992). By catalyzing pairing and strand exchange, RecA promotes the formation of recombination intermediates that are linked by Holliday junctions. Recent studies indicate that a novel class of nucleases, the Holliday junction resolvases, specifically recognize Holliday junctions and introduce defined nicks into two strands at the site of the junction. Again, *E. coli* provides our best example of a Holliday junction resolvase, RuvC protein (described below). The resulting nicked duplex (recombinant) products are then repaired by the action of DNA ligases.

B. Single-strand Reannealing Model

Studies of the recombination events that occur following transfection of DNA molecules into mammalian somatic cells led to the proposal of an alternative type of recombination model (Lin et al. 1984). This scheme, developed from the Broker and Lehman model for recombination of T4 DNA (Broker and Lehman 1971), proposes that recombination occurs via the annealing of single-stranded tails rather than by strand invasion (Fig. 2). The mechanism might be particularly suitable for the recombination of repetitive sequence DNA. In this case, a 5'→3' exonuclease digests DNA to reveal complementary single-stranded sequences that are

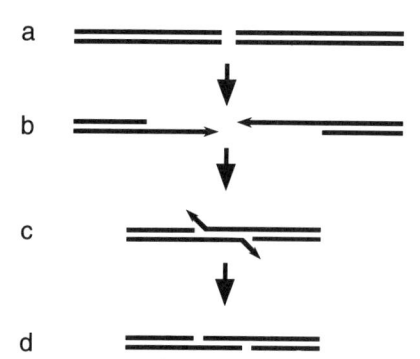

Figure 2 Single-strand reannealing model. A 5'→3' exonuclease activity degrades broken DNA (*b*) to produce complementary single strands that anneal to form a heteroduplex intermediate (*c*). The overlapping 3' ends are removed by a single-strand endonuclease or exonuclease (*d*) to produce DNA that can be repaired by polymerase and ligase.

able to reanneal to form a heteroduplex joint (Fig. 2c). The nonpaired 3' overhangs can then be removed (Fig. 2d) to allow repair by DNA polymerase and ligase. Like the first model, this scheme also requires 5'→3' exonuclease activity and is facilitated by an activity that reanneals DNA (although strand exchange is not required). Interestingly, Sep1 protein from *S. cerevisiae* (Johnson and Kolodner 1991), ExoII from *Schizosaccharomyces pombe* (Szankasi and Smith 1992b), Rrp1 protein from *Drosophila* (Sander et al. 1991a), and Hpp1 from human cells (Moore and Fishel 1990) all appear to contain both exonuclease and reannealing activity. These proteins may facilitate recombination by a mechanism similar to that of the λ *Red* pathway, which requires β protein and λ exonuclease (Muniyappa and Radding 1986).

III. NUCLEASES THAT INITIATE RECOMBINATION

A. *E. coli* RecBCD Enzyme

The RecBCD enzyme of *E. coli*, formally known as exonuclease V (Goldmark and Linn 1972), consists of three protein subunits (134,000, 129,000, and 67,000 molecular weight) encoded by the *recB*, *recC*, and *recD* genes, respectively (Finch et al. 1986a,b,c). This protein plays an important role in recombination and DNA repair in bacteria and has been the subject of numerous reviews (Smith 1987, 1988, 1990; Taylor 1988). In vitro, RecBCD enzyme exhibits a number of activities, including (1) ATP-dependent double-strand and single-strand exonuclease, (2) ATP-stimulated single-strand endonuclease, and (3) DNA helicase activity.

The protein is required for conjugal recombination but not for the recombination of circular DNA molecules.

Strains with null mutations in *recB* or *recC* are recombination-deficient, are sensitive to DNA-damaging agents such as UV light and ionizing irradiation, have reduced cell viability, and lack detectable RecBCD activities (Clark 1973; Chaudhury and Smith 1984a; Amundsen et al. 1990). In contrast, *recD* mutants are recombination-proficient and radiation-resistant, have normal cell viability, and lack nuclease activity, although they retain DNA unwinding activity (Chaudhury and Smith 1984b; Amundsen et al. 1986).

RecBCD enzyme, which functions as a heterotrimer, tracks along and unwinds linear duplex DNA in a highly processive manner (Rosamond et al. 1979; Taylor and Smith 1980, 1985). Indeed, approximately 30 kb of DNA can be unwound before the enzyme dissociates from DNA (Roman et al. 1992). In the absence of the *E. coli* single-strand-binding protein (SSB), unwound DNA reanneals behind the enzyme, such that single-stranded loops are formed at about 100 nucleotides per second as the enzyme tracks at 300 nucleotides per second (Taylor 1988; Roman and Kowalczykowski 1989b). ATP hydrolysis is required for the helicase activity, and two ATP molecules are hydrolyzed per base pair unwound (Roman and Kowalczykowski 1989a).

Early genetic studies showed that RecBCD-mediated recombination is enhanced by specific DNA sequences (Faulds et al. 1979). These recombination hot spots, Chi sites, increase the frequency of genetic exchange in their vicinity (Smith 1983). Stimulation of recombination results from a specific RecBCD-mediated nicking event that occurs five nucleotides to the 3' side of the Chi sequence (5'-GCTGGTGG-3'). Nicking only occurs when the enzyme approaches Chi from the right, as the sequence is written here (Taylor et al. 1985).

The properties of RecBCD have been assimilated into a model for recombination by Smith and colleagues (Smith 1981), and a recent adaptation of this model is shown in Figure 3 (Dixon and Kowalczykowski 1993). RecBCD binds the ends of linear duplex DNA and begins unwinding (Fig. 3a,b). The nuclease activity of RecBCD shows an asymmetry (Fig. 3b), with the 3'-terminal strand being degraded into small fragments and the 5'-terminal strand remaining intact (Dixon and Kowalczykowski 1991, 1993). RecBCD-mediated cleavage at Chi only occurs when a Chi site is encountered in the correct orientation (Taylor et al. 1985) and results in the introduction of a nick in the strand containing the Chi sequence (Fig. 3c). This interaction with Chi attenuates the nuclease activity of the RecBCD enzyme (Dixon and Kowalczykowski 1991, 1993; Taylor and Smith 1992), which continues to track along the

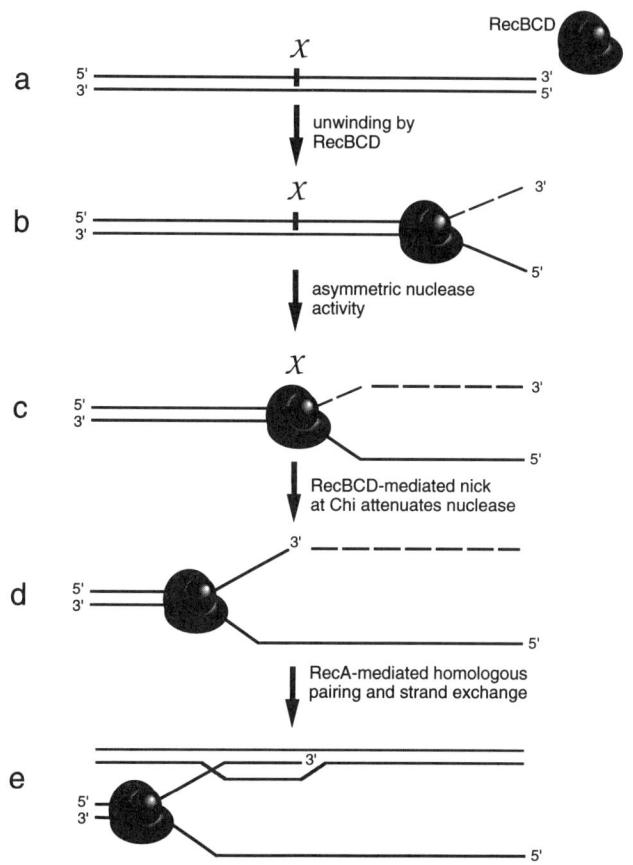

Figure 3 Model for the initiation of recombination by RecBCD enzyme. RecBCD enters DNA via a double-stranded end (*a*). The enzyme tracks along DNA and promotes its unwinding (*b*). The nuclease activity of RecBCD is asymmetric, with the 3'-ended strand being degraded and the 5'-ended strand remaining intact (*c*). Upon interaction with a Chi sequence (χ), the strand containing Chi is nicked and the nuclease activity of RecBCD is attenuated (*d*). RecBCD (or possibly RecBC) continues to move along the DNA, and single-stranded DNA is produced by the helicase. The strand nicked at Chi is paired into a homologous duplex by RecA protein to form a joint molecule (*e*). Subsequent steps, which include pairing of the intact unwound strand, Holliday junction resolution (by RuvC), and gap filling (by polymerase and ligase), are not shown. (Model adapted from Smith [1987] and Dixon and Kowalczykowski [1993].)

DNA as a DNA helicase (Fig. 3d). The single-stranded DNA produced by RecBCD is bound by RecA protein to form a helical nucleoprotein filament (for review, see Stasiak and Egelman 1988; West 1992), which initiates homologous pairing with intact duplex DNA (Fig. 3e). In recent studies, the combined action of RecBCD and RecA protein on Chi-containing DNA has been shown to result in the initiation of homologous pairing and provides support for this model (Dixon and Kowalczykowski 1991; Roman et al. 1991). The next step of the reaction is likely to involve pairing of the displaced strand of the duplex with the complementary unwound strand (not shown), followed by resolution of the Holliday junction by RuvC protein to give rise to recombinant products.

The mechanism by which the RecBCD nuclease activity is attenuated upon Chi cleavage is unknown. Studies of *recBCD*‡ mutants (which lack nuclease activity and are phenotypically "hyper-rec" in the absence of Chi) led to suggestions that RecBCD is physically altered by interaction with Chi, possibly by loss of the RecD subunit (Chaudhury and Smith 1984b; Amundsen et al. 1986; Lovett et al. 1988; Thaler et al. 1988, 1989). A comparison of the biochemical properties of RecBC and RecBCD argues in favor of this suggestion, since RecBC protein retained significant levels of DNA-dependent ATPase and helicase activity, yet exonuclease activity on single- and double-stranded DNAs was not detected (Palas and Kushner 1990).

To help define the roles of each of the three protein subunits, the *recB*, *recC*, and *recD* genes were cloned into overexpression vectors and the gene products were purified individually (Boehmer and Emmerson 1991; Masterson et al. 1992). The presence of the three subunits were required for reconstitution of all RecBCD activities. However, although the RecC and RecD subunits showed no activities on their own, RecB exhibited ATPase activity and RecBC showed a limited DNA helicase activity in the absence of RecD (Masterson et al. 1992).

B. *E. coli* RecJ Protein

The recombination proficiency and UV resistance of a *recD* mutant is dependent on a functional *recJ* gene (Lloyd et al. 1988; Lovett et al. 1988; Lloyd and Buckman 1991), indicating that RecD and RecJ can provide overlapping functions that compensate for one another in the single mutants. In addition, plasmid recombination, which is independent of RecBCD, is highly dependent on RecJ since *recJ* mutants show recombination frequencies that are reduced 1,000–10,000-fold (Kolodner et al. 1985).

The RecJ protein (63,000 molecular weight) has been overexpressed and purified (Lovett and Kolodner 1989, 1991). RecJ is an exonuclease that digests linear single-stranded DNA, with a preference for 5′ termini (Lovett and Kolodner 1989). Its ability to compensate for the nuclease deficiency of *recD* mutants indicates that it functions in the formation of duplex DNA molecules with 3'-overlapping termini. The formation of 3'ends would presumably require unwinding of the duplex DNA, and this could be provided by RecQ helicase or RecBC (in a *recD* mutant).

C. *E. coli* Exonuclease VIII (RecE Protein)

Mutations in a gene designated *sbcA* suppress the recombination deficiency and UV sensitivity of strains carrying *recB* or *recC* mutations (Barbour et al. 1970). The *sbcA* mutation results in the induction of exonuclease VIII, a highly processive 5'→3' exonuclease that acts upon double-stranded DNA (Kushner et al. 1974; Joseph and Kolodner 1983a,b). Exonuclease VIII is functionally similar to λ exonuclease and can act as its substitute in λ *Red* recombination (Gillen and Clark 1974).

The structural gene for exonuclease VIII is *recE* (Kushner et al. 1974). Both *recE* and *sbcA* map within a cryptic lambdoid prophage (Rac) that is positioned between 29.6 and 30.2 minutes of the *E. coli* chromosome. The *recE* gene has been cloned, and the 140-kD RecE protein (exonuclease VIII) has been overexpressed and purified (Joseph and Kolodner 1983a; Luisi-DeLuca et al. 1988). Gel filtration and sedimentation data indicate that the protein forms a multimer with a molecular weight of 480,000 (Joseph and Kolodner 1983a). The exonuclease acts in a highly processive manner and digests linear duplex DNA to produce 5' mononucleotides and short oligonucleotides (Joseph and Kolodner 1983b). The polarity of the exonuclease (5'→3') indicates that it acts to produce 3' overhangs that can be reannealed by RecT protein, an analog of λ β protein (Hall et al. 1993). The *recT* gene maps within *recE*, and the 33-kD RecT protein is encoded by the carboxy-terminal 33-kD portion of exonuclease VIII (Hall et al. 1993).

D. λ Exonuclease

λ exonuclease is a 24-kD protein that is required for recombination via the λ *Red* system. The protein has been purified to homogeneity (Radding 1966; Little et al. 1967) and is widely used as a reagent due to its commercial availability. Physical studies indicate that the protein is a tetramer with a relative molecular weight of 107,000 (Van Oostrum et al. 1985). λ exonuclease is the product of the α gene and interacts with the 28-kD β protein that promotes the reannealing of single-stranded DNA (Radding and Shreffler 1966; Muniyappa and Radding 1986). Like *E. coli* exonuclease VIII, λ exonuclease acts upon linear duplex DNA, which it digests with a 5'→3' polarity to produce mononucleotides

(Radding 1966; Little 1967). Again, it is likely that the biological role of the enzyme is to attack linear duplex DNA to produce 3′-overhanging tails, which in this case can be reannealed by β protein (Takahashi and Kobayashi 1990). This mechanism of recombination is similar to that in the single-strand reannealing model presented in Figure 2.

E. T7 Exonuclease

Genetic recombination of bacteriophage T7 DNA requires the products of genes *3, 4, 5,* and *6* (Kerr and Sadowski 1975). The product of gene *6* is T7 exonuclease that, like *E. coli* exonuclease VIII and λ exonuclease, is a 5′→3′ exonuclease (Kerr and Sadowski 1972a,b). However, unlike these highly processive nucleases, T7 exonuclease acts by a distributive mechanism and is less fastidious in its substrate requirements. It acts equally well upon nicks in DNA as well as at double-stranded ends. In combination with the T7 DNA-binding protein (the product of gene *2.5*), which reanneals single-stranded DNA, complex branched DNA intermediates are observed that resemble recombination intermediates extracted from T7-infected cells (Paetkau et al. 1977). Little work has been done on this enzyme in recent years, and further details of its properties are reviewed elsewhere (Sadowski 1985).

F. *S. cerevisiae* Sep1 Protein

The Sep1 protein of *S. cerevisiae* was initially identified on the basis of its ability to promote the pairing of linear duplex DNA with single-stranded DNA (Kolodner et al. 1987). It was therefore thought to be a protein that promoted recombination reactions by a mechanism analogous to those mediated by the *E. coli* RecA protein (Heyer et al. 1988). However, a number of observations now make this analogy untenable: (1) Unlike reactions catalyzed by RecA, pairing by Sep1 occurs without need for ATP (Kolodner et al. 1987), (2) the pairing reaction was found to involve nucleolytic digestion of the linear duplex DNA (Johnson and Kolodner 1991), (3) the exonuclease activity responsible for digestion is intrinsic to Sep1 protein (Johnson and Kolodner 1991), and (4) *S. cerevisiae* proteins that hydrolyze ATP and share extensive sequence homology with RecA have now been isolated (e.g., Rad51 protein).

The gene for Sep1 protein (*SEP1*) has been cloned and encodes a 175-kD protein (Dykstra et al. 1991; Tishkoff et al. 1991). The gene is identical to *XRN1*, which encodes a 5′→3′ exoribonuclease (Stevens 1980; Larimer and Stevens 1990). Mutants of the gene exhibit a growth defect and a small decrease in the rate of spontaneous mitotic recombination. Sep1 protein has been purified to homogeneity and shown to possess 5'→3' exonuclease activity on single- and double-stranded DNAs (Johnson and Kolodner 1991). The exonuclease activity can be functionally separated from the DNA reassociation activity by the substitution of Ca^{++} for Mg^{++}. In the presence of Ca^{++}, the nuclease is inactive, and reassociation activity requires the presence of an exogenous nuclease (Johnson and Kolodner 1991).

It is possible that Sep1 catalyzes single-strand reannealing reactions similar to those presented in the model of Figure 2 (i.e., by a mechanism related to that of λ exonuclease and β protein).

G. *S. pombe* Exonuclease I and Exonuclease II

Two nucleases, exonuclease I (ExoI) and exonuclease II (ExoII), that are implicated in genetic recombination have been isolated from *S. pombe*. The genes encoding the nucleases have been cloned (*exo1* and *exo2*), and null mutants have been generated (Szankasi and Smith 1992a,b). ExoI is a 36-kD protein that is induced fivefold during meiosis (Szankasi and Smith 1992a). It is a nonprocessive 5'→3' exonuclease that digests linear duplex DNA to produce 3' single-stranded tails. ExoI also acts, with the same affinity, at nicks in duplex DNA to release 5' mononucleotides. This protein may again be analogous to λ exonuclease.

The second *S. pombe* nuclease, ExoII, is not induced during meiosis. ExoII (134,000 molecular weight) is similar to *E. coli* RecJ protein and degrades single-stranded DNA with a 5'→3' polarity (Szankasi and Smith 1992b). The protein is known to act processively and releases 5' mononucleotides. ExoII may be the *S. pombe* analog of *S. cerevisiae* Sep1.

H. Higher Eukaryotic Proteins

A number of proteins have been isolated from higher organisms that appear to be functionally analogous to the yeast Sep1 protein. The best characterized enzyme of this type is the *Drosophila* Rrp1 protein (McCarthy et al. 1988; Lowenhaupt et al. 1989), but human proteins such as Hpp1 have also been detected (Fishel et al. 1988; Moore and Fishel 1990). These proteins were isolated on the basis of their ability to promote heteroduplex DNA formation, and it was subsequently noted that they possess exonuclease activity.

The *Drosophila* Rrp1 protein (*r*ecombination *r*epair *p*rotein 1) is quite unusual. This 105-kD protein has an amino-terminal 427-amino-acid region that is unrelated to known proteins and a 252-amino-acid carboxy-terminal region that shares sequence homology with *E. coli* exonuclease III and *Streptococcus pneumoniae* exonuclease A (Sander et al. 1991b). Like exonuclease III and exonuclease A, the Rrp1 protein exhibits 3'→5' exonuclease activity and apurinic endonuclease activity (Sander et al. 1991a). By analogy with *E. coli* exonuclease III, this protein may be involved in nucleotide excision repair (see also Chapter 9, this volume). To determine whether it also plays a role in cellular recombination, it will be necessary to characterize *Drosophila* mutants that lack Rrp1.

IV. NUCLEASES THAT RESOLVE RECOMBINATION INTERMEDIATES

A. *E. coli* RuvC Protein

The RuvC protein of *E. coli* is required for the resolution of Holliday junctions during recombination and the recombinational repair of DNA damage. Cells carrying *ruvC* mutations are sensitive to a wide range of DNA-damaging agents (including UV- and γ-irradiation and various mutagens) and are recombination-defective in combination with *recG*, *recBCsbcA*, or *recBCsbcB* mutations (Otsuji et al. 1974; Stacey and Lloyd 1976; Lloyd et al. 1984, 1987; Luisi-DeLuca et al. 1989; Benson et al. 1991). The *ruvC* gene has been cloned and encodes a 19-kD protein (Sharples and Lloyd 1991; Takahagi et al. 1991). The gene maps at minute 41 on the *E. coli* genetic map close to *ruvA* and *ruvB*, which together encode a protein (RuvAB protein) that catalyzes branch migration (i.e., the movement) of a Holliday junction. At the present time, it is not known whether RuvA, RuvB, and RuvC interact together.

The RuvC protein has been purified and shown to resolve (1) recombination intermediates made by RecA (Dunderdale et al. 1991), (2) synthetic Holliday junctions produced by annealing four complementary oligonucleotides (Dunderdale et al. 1991; Bennett et al. 1993), and (3) cruciform structures that extrude from supercoiled plasmids (Iwasaki et al. 1991). The RuvC protein is responsible for the Holliday junction resolution activity first detected in fractionated *E. coli* cell-free extracts (Connolly and West 1990; Connolly et al. 1991). Resolution occurs by a mechanism similar to that shown in Figure 4 (Bennett et al. 1993). RuvC protein recognizes junctions in DNA (Fig. 4a) and forms specific protein-DNA complexes (Fig. 4b) that have been visualized by band-shift assays (Dunderdale et al. 1991). In the presence of a divalent metal cofactor (Mn^{++} or Mg^{++}), the protein catalyzes cleavage by the introduction of symmetrically related nicks into two strands of the same polarity (Fig. 4b–c). For reasons that are not yet clear but may involve the align-

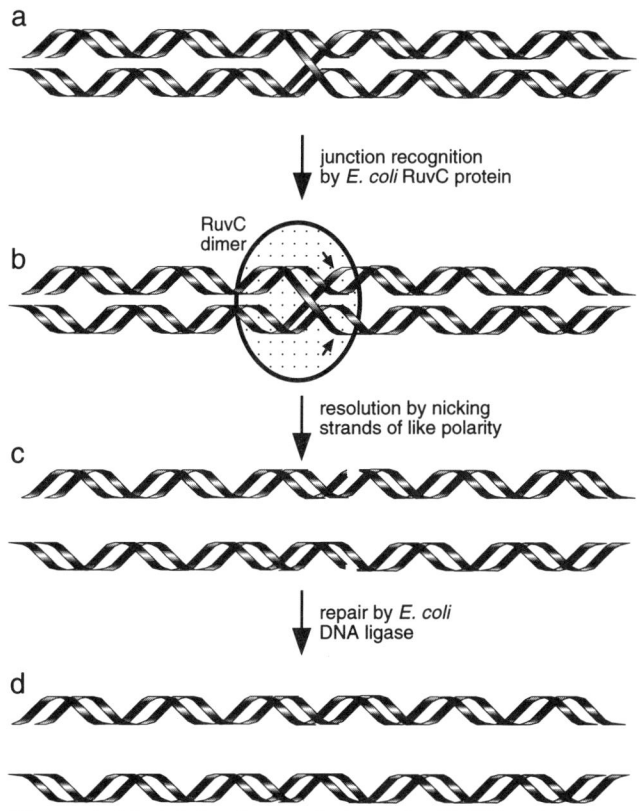

Figure 4 Model for the resolution of a Holliday junction by RuvC protein. A Holliday junction (*a*) is bound specifically by the RuvC dimer (*b*). The junction is resolved by the introduction of symmetrically related nicks into strands of like polarity (*c*). These occur at the 3′ side of thymine residues located close to the junction point. Nicking results in resolution of the intermediate structure to produce two nicked duplexes (*c*). The nicks are subsequently sealed by *E. coli* DNA ligase (*d*). In this figure, resolution occurs by cleavage of crossed strands. However, it is likely that the junction can isomerize to interconvert noncrossed and crossed strands. This isomerization step is genetically important since cleavage of alternate pairs of strands gives rise to recombinant products with or without an exchange of flanking genetic markers.

ment of homologous helices with the active site, the enzyme requires homologous DNA sequences (≥6 bp) to catalyze the cleavage reaction. In the absence of homology, binding complexes are observed, but the efficiency of cleavage is reduced more than 100-fold and is characteristic of an abortive resolution reaction. Although the nuclease binds junctions in a structure-specific manner, it also shows a sequence specificity in that the nicks are introduced at the 3′ side of thymine residues. The requirement for homology may be related to a need for symmetrically related thymines close to the junction. The resulting nicked duplexes are finally repaired by *E. coli* DNA ligase (Fig. 4d) (Bennett et al. 1993).

The RuvC protein is highly specific for natural or model recombination intermediates and has little or no activity on single-stranded or linear duplex DNA. It will not cut Y-junctions, mismatches, or loops in duplex DNA. However, the protein is active upon three-stranded junctions that resemble equivalent four-stranded Holliday junctions, and in this case, the enzyme again requires homologous DNA sequences at the junction point (F.E. Benson and S.C. West, unpubl.). This requirement for homology seems reasonable since the protein acts at a step in recombination that involves the alignment of homologous DNA sequences. It is possible that alignment within the active site of the enzyme involves bending the DNA, since RuvC–Holliday junction complexes have been found to be hyperreactive to minor-groove-specific agents such as hydroxyl radicals (Bennett et al. 1993). Alternatively, the homology requirement may be related to the sequence specificity of the protein and its requirement for symmetrically related thymines.

B. T4 Endonuclease VII

The product of bacteriophage T4 gene *49* is an 18-kD protein that is required during phage maturation. Mutants in gene *49* are defective in packaging, as observed by the formation of highly branched multimeric DNA and abortive infection. This accumulation of branched DNA indicates the important role that recombination plays in the life cycle of the T-even phages (Mosig 1987) and shows that T4 endonuclease VII is required to cleave branched DNA into unit-length phage DNA molecules. The gene encoding endonuclease VII has been cloned (Tomaschewski and Rüger 1987; Barth et al. 1988), and the protein has been purified to homogeneity from an overexpression vector (Kosak and Kemper 1990).

Following early studies which showed that T4 endonuclease VII could resolve plasmid recombination intermediates and cruciform DNA structures, the protein assumed the role of the prototypic "Holliday junction–resolvase" (Mizuuchi et al. 1982; Kemper et al. 1984; Lilley and Kemper 1984; Müller et al. 1990b). Using synthetic Holliday junctions, it was shown that cleavage occurred at the site of the junction and involved the introduction of nicks in strands of like polarity. The nicks occurred by a nonconcerted mechanism (Pottmeyer and Kemper 1992), and the strand breaks in the DNA products were repaired by addition of DNA ligase (Mizuuchi et al. 1982). Although the interaction with a junction appears to occur in a structure-specific manner, the enzyme shows a slight sequence specificity. In general, nicks are introduced at the 3' side of pyrimidine residues located within two to three nucleotides of the junction point (Kemper et al. 1984; Picksley et al. 1990; Pottmeyer and Kemper 1992).

Sedimentation studies have shown that T4 endonuclease VII forms a dimer in solution (Kemper and Garabett 1981; Kosak and Kemper 1990). To investigate the interaction of the protein with Holliday junctions, band-shift assays were used and specific junction-DNA complexes were observed in the absence of Mg^{++}. Attempts to define the binding site of the enzyme were somewhat inconclusive, but chemical protection experiments indicated the presence of a weak footprint that covered approximately five nucleotides close to the junction point (Parsons et al. 1990). Unlike RuvC protein, T4 endonuclease VII does not produce sites that are hypersensitive to hydroxyl radicals, suggesting that the molecular mechanisms of resolution may differ. Indeed, T4 endonuclease VII is quite different from RuvC protein in that it is a broad-range nuclease that can cleave distortions in duplex DNA (Fig. 5). Whereas RuvC protein is specific for Holliday junctions, T4 endonuclease VII cleaves at X and Y structures, heteroduplex loops, extended single-stranded termini, and sequence-induced bends in duplex DNA (Jensch and Kemper 1986; Duckett et al. 1988; Kleff and Kemper 1988; Mueller et al. 1988; Kemper et al. 1990; Bhattacharyya et al. 1991; Pottmeyer and Kemper 1992). As a general rule, cleavage occurs at the 3' side of the junction or site of secondary structure. This broad specificity is thought to be consistent with its role as a general debranching and repair enzyme that prepares the phage DNA for packaging.

C. T7 Endonuclease I

The product of T7 gene *3* (endonuclease I) is similar to T4 endonuclease VII and can complement T4 gene *49* mutants (de Massy et al. 1984, 1987; Panayotatos and Fontaine 1985; Pham and Coleman 1985). Endonuclease I has two essential roles during phage growth: (1) It degrades cellular host DNA following phage infection and (2) it cleaves junctions in replicating phage DNA to produce unbranched linear duplexes that may subsequently be packaged. Mutations in gene *3* result in mutant phage that are unable to degrade host DNA (Center et al. 1970) and are defective in phage maturation (Paetkau et al. 1977) and genetic recombination (Kerr and Sadowski 1975; Tsujimoto and Ogawa 1978; Lee and Sadowski 1981).

Endonuclease I degrades single-stranded DNA and cuts duplex DNA preferentially at X- and Y-junctions (Center and Richardson 1970;

a. Holliday junction

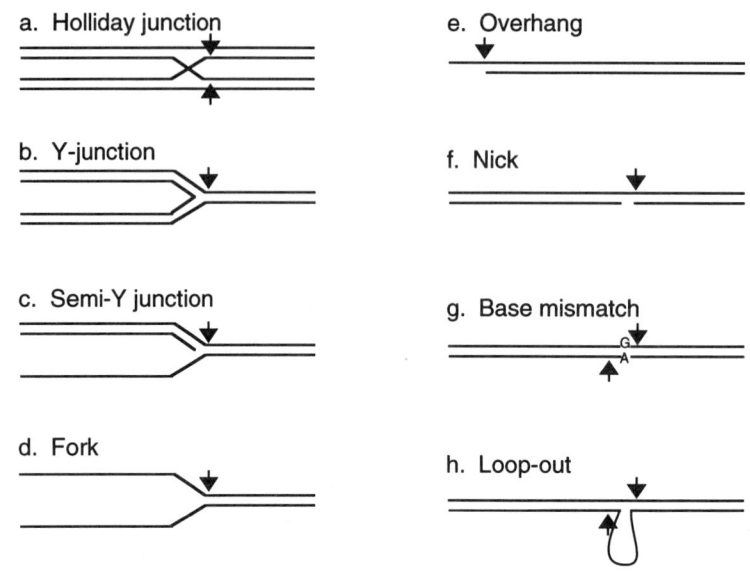

b. Y-junction

c. Semi-Y junction

d. Fork

e. Overhang

f. Nick

g. Base mismatch

h. Loop-out

Figure 5 Broad range of DNA substrates cut by T4 endonuclease VII. Sites of cutting, indicated by arrows, generally occur at the 3' side of the site of the secondary structure.

Sadowski 1971; de Massy et al. 1984, 1987; Dickie et al. 1987; Müller et al. 1990a; Parsons and West 1990; Lu et al. 1991). As observed with T4 endonuclease VII, the efficiency and site of cleavage are influenced by DNA sequence, and a preference for cleavage at the 5' side of pyrimidine residues has been observed (Dickie et al. 1987; Picksley et al. 1990). The protein appears to be dimeric and is composed of two identical 17-kD subunits. Like T4 endonuclease VII, the protein binds synthetic X junctions to form protein-DNA complexes that can be detected by band-shift assays (Parsons and West 1990).

D. *S. cerevisiae* Holliday Junction Resolvases

Two apparently distinct activities that cleave cruciforms or synthetic Holliday structures have been partially purified from *S. cerevisiae*. One activity, with a native molecular weight of approximately 200,000, was partially purified from vegetative yeast cells treated with the DNA-damaging agent mechlorethamine (West and Korner 1985; West et al. 1987). Studies with cruciform DNA structures showed that cleavage occurred by the introduction of symmetrically related nicks within the cruciform arms (Parsons and West 1988; Parsons et al. 1989). At the present time, the gene that encodes this activity is unknown.

A similar but unrelated 41-kD protein was shown to resolve cruciform DNA structures and figure-8 DNA molecules in vitro (Evans and Kolodner 1987, 1988; Symington and Kolodner 1985). Recently, its gene (*CCE1*) was identified and cloned (Kleff et al. 1992), and it was surprising to find that a *cce1* null mutant showed no defect in meiotic or mitotic recombination. Since *cce1* mutants showed a high proportion of petite cells, it was suggested that the Cce1 protein might be important for the maintenance of mitochondrial DNA (Kleff et al. 1992).

E. Mammalian Holliday Junction Resolvase

Using synthetic Holliday junctions, an activity that resolves Holliday junctions has been partially purified from calf thymus extracts (Elborough and West 1990). Subsequent studies indicate that this activity is functionally analogous to the *E. coli* RuvC protein (H. Hyde et al., unpubl.). Holliday junction resolution by the calf thymus protein occurs by the introduction of symmetrically related cuts across the junction, and the resulting nicked duplex products can be repaired by DNA ligase. Like RuvC protein, but unlike T4 endonuclease VII, this activity is specific for Holliday junctions and shows no activity on Y-junctions, mismatches, or heteroduplex loops. The protein resolves synthetic Holliday junctions and recombination intermediates made by RecA protein. Recent studies show that the protein can be partially purified from mammalian and rodent cells grown in culture.

V. CONCLUDING REMARKS

It is clear that nucleases play important roles in recombination and the recombinational repair of DNA damage. Nucleases provide the substrates that initiate recombination and finish the reaction by catalyzing the resolution of recombination intermediates. Interestingly, it seems that nucleases often play overlapping roles such that one nuclease can compensate for the loss (by mutation) of another.

Most of our knowledge of recombination nucleases comes from studies of bacterial and bacteriophage-encoded enzymes, and substantial progress has been made in recent years. In particular, the study of Holliday junction resolvases has progressed quickly following identification of the *E. coli* Holliday junction resolvase, RuvC protein. Further studies of protein-DNA interactions in this area of research will reveal how these enzymes recognize the junction point and manipulate four DNA strands during the resolution process. It will be of great interest to determine the three-dimensional structure of a Holliday junction resolvase by X-ray crystallography.

The detailed picture that now emerges from studies of bacterial recombination enzymes leads us to hope that similar activities can be isolated from eukaryotic sources and that their cellular roles can be defined. In this regard, it is to be hoped that the general principles of bacterial recombination will provide a framework for the basic biochemical reactions that occur during recombination in mammalian cells.

REFERENCES

Amundsen, S.K., A.M. Neiman, S. Thibodeaux, and G.R. Smith. 1990. Genetic dissection of the biochemical activities of RecBCD enzyme. *Genetics* **126:** 25-40.

Amundsen, S.K., A.F. Taylor, A.M. Chaudhury, and G.R. Smith. 1986. *recD*: The gene for an essential third subunit of exonuclease V. *Proc. Natl. Acad. Sci.* **83:** 5558-5562.

Barbour, S.D., H. Nagaishi, A. Templin, and A.J. Clark. 1970. Biochemical and genetic studies of recombination proficiency in *Escherichia coli*. II. Rec+ revertants caused by indirect suppression of Rec− mutations. *Proc. Natl. Acad. Sci.* **67:** 128-135.

Barth, K.A., D. Powell, M. Trupin, and G. Mosig. 1988. Regulation of two nested proteins from gene *49* (recombination endonuclease VII) and of a lambda RexA-like protein of bacteriophage T4. *Genetics* **120:** 329-343.

Bennett, R.J., H.J. Dunderdale, and S.C. West. 1993. Resolution of Holliday junctions by RuvC resolvase: Cleavage specificity and DNA distortion. *Cell* (in press).

Benson, F., S. Collier, and R.G. Lloyd. 1991. Evidence of abortive recombination in *ruv* mutants of *Escherichia coli* K-12. *Mol. Gen. Genet.* **225:** 266-272.

Bhattacharyya, A., A.I.H. Murchie, E. von Kitzing, S. Diekmann, B. Kemper, and D.M.J. Lilley. 1991. A model for the interaction of DNA junctions and resolving enzymes. *J. Mol. Biol.* **221:** 1191-1207.

Bishop, D.K., D. Park, L.Z. Xu, and N. Kleckner. 1992. DMC1—A meiosis-specific yeast homolog of *Escherichia coli* RecA required for recombination, synaptonemal complex-formation, and cell-cycle progression. *Cell* **69:** 439-456.

Boehmer, P.E. and P.T. Emmerson. 1991. Construction of over-expression vectors to facilitate the purification of the *Escherichia coli* RecBCD enzyme and its constituent subunits. *Gene* **102:** 1-6.

Broker, T.R. and I.R. Lehman. 1971. Branched DNA molecules: Intermediates in T4 recombination. *J. Mol. Biol.* **60:** 131-149.

Center, M. and C. Richardson. 1970. An endonuclease induced after infection of *Escherichia coli* with bacteriophage T7. *J. Biol. Chem.* **245:** 6285-6292.

Center, M., F. Studier, and C. Richardson. 1970. The structural gene for a T7 endonuclease essential for phage DNA synthesis. *Proc. Natl. Acad. Sci.* **65:** 242-248.

Craig, N.L. 1988. The mechanism of conservative site-specific recombination. *Annu. Rev. Genet.* **22:** 77-106.

Chaudhury, A.M. and G.R. Smith. 1984a. *Escherichia coli recBC* deletion mutants. *J. Bacteriol.* **160:** 788-791.

―――. 1984b. A new class of *Escherichia coli recBC* mutants: Implications for the role of RecBC enzyme in homologous recombination. *Proc. Natl. Acad. Sci.* **81:** 7850-7854.

Clark, A.J. 1973. Recombination deficient mutants of *E. coli* and other bacteria. *Annu. Rev. Genet.* **7:** 67–86.

Connolly, B. and S.C. West. 1990. Genetic recombination in *Escherichia coli*: Holliday junctions made by RecA protein are resolved by fractionated cell-free extracts. *Proc. Natl. Acad. Sci.* **87:** 8476–8480.

Connolly, B., C.A. Parsons, F.E. Benson, H.J. Dunderdale, G.J. Sharples, R.G. Lloyd, and S.C. West. 1991. Resolution of Holliday junctions *in vitro* requires *Escherichia coli ruvC* gene product. *Proc. Natl. Acad. Sci.* **88:** 6063–6067.

Cox, M.M. 1988. FLP site-specific recombination system of *Saccharomyces cerevisiae*. In *Genetic recombination* (ed. R. Kucherlapati and G.R. Smith), pp. 429–443. American Society for Microbiology, Washington, D.C.

Cox, M.M. and I.R. Lehman. 1987. Enzymes of genetic recombination. *Annu. Rev. Biochem.* **56:** 229–262.

de Massy, B., R.A. Weisberg, and F.W. Studier. 1987. Gene *3* endonuclease of bacteriophage T7 resolves conformationally branched structures in double-stranded DNA. *J. Mol. Biol.* **193:** 359–376.

de Massy, B., F.W. Studier, L. Dorgai, E. Appelbaum, and R.A. Weisberg. 1984. Enzymes and sites of genetic recombination: Studies with gene *3* endonuclease of phage T7 and with site affinity mutants of phage lambda. *Cold Spring Harbor Symp. Quant. Biol.* **49:** 715–726.

Dickie, P., G. McFadden, and A.R. Morgan. 1987. The site-specific cleavage of synthetic Holliday junction analogs and related branched DNA structures by bacteriophage T7 endonuclease I. *J. Biol. Chem.* **262:** 14826–14836.

Dixon, D.A. and S.C. Kowalczykowski. 1991. Homologous pairing *in vitro* stimulated by the recombination hotspot, Chi. *Cell* **66:** 361–371.

———. 1993. The recombination hotspot, Chi, is a regulatory sequence that acts by attenuating the nuclease activity of the *Escherichia coli* RecBCD enzyme. *Cell* **73:** 87–96.

Duckett, D.R., A.I.H. Murchie, S. Diekmann, E. von Kitzing, B. Kemper, and D.M.J. Lilley. 1988. The structure of the Holliday junction and its resolution. *Cell* **55:** 79–89.

Dunderdale, H.J., F.E. Benson, C.A. Parsons, G.J. Sharples, R.G. Lloyd, and S.C. West. 1991. Formation and resolution of recombination intermediates by *E. coli* RecA and RuvC proteins. *Nature* **354:** 506–510.

Dykstra, C.C., K. Kitada, A.B. Clark, R.K. Hamatake, and A. Sugino. 1991. Cloning and characterization of *DST2*, the gene for DNA strand transfer protein β from *Saccharomyces cerevisiae*. *Mol. Cell Biol.* **11:** 2583–2592.

Eggleston, A.K. and S.C. Kowalczykowski. 1991. An overview of homologous pairing and DNA strand exchange proteins. *Biochimie* **73:** 163–176.

Elborough, K.M. and S.C. West. 1990. Resolution of synthetic Holliday junctions in DNA by an endonuclease activity from calf thymus. *EMBO J.* **9:** 2931–2936.

Evans, D.H. and R. Kolodner. 1987. Construction of a synthetic Holliday junction analog and characterization of its interaction with a *Saccharomyces cerevisiae* endonuclease that cleaves Holliday junctions. *J. Biol. Chem.* **262:** 9160–9165.

———. 1988. Effect of DNA structure and nucleotide sequence on Holliday junction resolution by a *Saccharomyces cerevisiae* endonuclease. *J. Mol. Biol.* **201:** 69–80.

Faulds, D., N. Dower, M.M. Stahl, and F.W. Stahl. 1979. Orientation dependent recombination hotspot activity in bacteriophage lambda. *J. Mol. Biol.* **131:** 681–695.

Finch, P.W., A. Story, K. Brown, I.D. Hickson, and P.T. Emmerson. 1986a. Complete nucleotide sequence of *recD*, the structural gene for the alpha subunit of exonuclease V of *E. coli*. *Nucleic Acids Res.* **14:** 8583–8593.

Finch, P.W., R.E. Wilson, K. Brown, I.D. Hickson, and P.T. Emmerson. 1986b. Complete nucleotide sequence of the *E. coli recC* gene and of the *thyA-recC* intergenic region. *Nucleic Acids Res.* **14:** 4437–4451.

Finch, P.W., A. Story, K.E. Chapman, K. Brown, I.D. Hickson, and P.T. Emmerson. 1986c. Complete nucleotide sequence of the *E. coli recB* gene. *Nucleic Acids Res.* **14:** 8573–8582.

Fishel, R.A., K. Detmar, and A. Rich. 1988. Identification of homologous pairing and strand exchange activity from a human tumor cell line based on Z-DNA affinity chromatography. *Proc. Natl. Acad. Sci.* **85:** 36–40.

Gillen, J.R. and A.J. Clark. 1974. The RecE pathway of bacterial recombination. In *Mechanisms in recombination* (ed. R.E. Grell), pp. 123–136. Plenum Press, New York.

Goldmark, P.J. and S. Linn. 1972. Purification and properties of the *recBC* DNase of *Escherichia coli* K12. *J. Biol. Chem.* **247:** 1849–1860.

Hall, S.D., M.F. Kane, and R.D. Kolodner. 1993. Identification and characterization of the *Escherichia coli* RecT protein, a protein encoded by the *recE* region that promotes renaturation of homologous single-stranded DNA. *J. Bacteriol.* **175:** 277–287.

Hatfull, G.H. and N.D.F. Grindley. 1988. Resolvases and DNA-invertases: A family of enzymes active in site-specific recombination. In *Genetic recombination* (ed. R. Kucherlapati and G.R. Smith), pp. 357–396. American Society for Microbiology, Washington, D.C.

Heyer, W.D., D.H. Evans, and R.D. Kolodner. 1988. Renaturation of DNA by a *Saccharomyces cerevisiae* protein that catalyses homologous pairing and strand exchange. *J. Biol. Chem.* **263:** 15189–15195.

Holliday, R. 1964. A mechanism for gene conversion in fungi. *Genet. Res.* **5:** 282–304.

Hotchkiss, R.D. 1974. Models for genetic recombination. *Annu. Rev. Microbiol.* **28:** 445–468.

Howard-Flanders, P., S.C. West, and A.J. Stasiak. 1984. Role of RecA spiral filaments in genetic recombination. *Nature* **309:** 215–220.

Iwasaki, H., M. Takahagi, T. Shiba, A. Nakata, and H. Shinagawa. 1991. *Escherichia coli* RuvC protein is an endonuclease that resolves the Holliday structure. *EMBO J.* **10:** 4381–4389.

Jensch, F. and B. Kemper. 1986. Endonuclease VII resolves Y-junctions in branched DNA *in vitro*. *EMBO J.* **5:** 181–189.

Johnson, A.W. and R.D. Kolodner. 1991. Strand exchange protein 1 from *Saccharomyces cerevisiae*: A novel multifunctional protein that contains DNA strand exchange and exonuclease activities. *J. Biol. Chem.* **266:** 14046–14054.

Joseph, J.W. and R. Kolodner. 1983a. Exonuclease VIII of *Escherichia coli*. I. Purification and physical properties. *J. Biol. Chem.* **258:** 10411–10417.

———. 1983b. Exonuclease VIII of *Escherichia coli*. II. Mechanism of action. *J. Biol. Chem.* **258:** 10418–10424.

Kawasaki, K., M. Takahashi, M. Natori, and T. Shibata. 1991. DNA sequence recognition by a eukaryotic sequence-specific endonuclease, Endo.*Sce*I, from *Saccharomyces cerevisiae*. *J. Biol. Chem.* **266:** 5342–5347.

Kemper, B. and M. Garabett. 1981. Studies in T4-head maturation. 1. Purification and characterization of gene *49* controlled endonuclease. *Eur. J. Biochem.* **115:** 123–132.

Kemper, B., S. Pottmeyer, P. Solaro, and H. Kosak. 1990. Resolution of DNA secondary structures by endonuclease VII (endo VII) from phage T4. In *Structure and methods. I. Human genome initiative and DNA recombination* (ed. R.H. Sarma and M.H. Sarma), pp. 215–229. Adenine Press, New York.

Kemper, B., F. Jensch, M. Depka-Prondzynski, H.J. Fritz, U. Borgmeyer, and K.

Mizuuchi. 1984. Resolution of Holliday structures by endonuclease VII as observed in interactions with cruciform DNA. *Cold Spring Harbor Symp. Quant. Biol.* **49:** 815-825.

Kerr, C. and P.D. Sadowski. 1972a. Gene 6 exonuclease of bacteriophage T7. I. Purification and properties of the enzyme. *J. Biol. Chem.* **247:** 305-310.

———. 1972b. Gene 6 exonuclease of bacteriophage T7. II. Mechanism of the reaction. *J. Biol. Chem.* **247:** 311-318.

———. 1975. The involvement of genes 3, 4, 5 and 6 in genetic recombination in bacteriophage T7. *Virology* **65:** 281-285.

Kleff, S. and B. Kemper. 1988. Initiation of heteroduplex loop repair by T4-encoded endonuclease VII *in vitro. EMBO J.* **7:** 1527-1535.

Kleff, S., B. Kemper, and R. Sternglanz. 1992. Identification and characterization of yeast mutants and the gene for a cruciform cutting endonuclease. *EMBO J.* **11:** 699-704.

Kolodner, R., D.H. Evans, and P.T. Morrison. 1987. Purification and characterization of an activity from *S. cerevisiae* that catalyses homologous pairing and strand exchange. *Proc. Natl. Acad. Sci.* **84:** 5560-5564.

Kolodner, R., R.A. Fishel, and M. Howard. 1985. Genetic recombination of bacterial plasmid DNA: Effect of RecF pathway mutations on plasmid recombination in *Escherichia coli. J. Bacteriol.* **163:** 1060-1066.

Kosak, H.G. and B.W. Kemper. 1990. Large-scale preparation of T4 endonuclease VII from over-expressing bacteria. *Eur. J. Biochem.* **194:** 779-784.

Kostriken, R. and F. Heffron. 1984. The product of the *HO* gene is a nuclease: Purification and characterization of the enzyme. *Cold Spring Harbor Symp. Quant. Biol.* **49:** 89-96.

Kostriken, R., J.N. Strathern, A.J.S. Klar, J.B. Hicks, and F. Heffron. 1983. A site-specific endonuclease essential for mating-type switching in *S. cerevisiae. Cell* **35:** 167-174.

Kowalczykowski, S.C. 1991. Biochemistry of genetic recombination: Energetics and mechanism of DNA strand exchange. *Annu. Rev. Biophys. Biophys. Chem.* **20:** 539-575.

Kushner, S., H. Nagaishi, and A.J. Clark. 1974. Isolation of exonuclease VIII: The enzyme associated with the *sbcA* indirect suppressor. *Proc. Natl. Acad. Sci.* **71:** 3593-3597.

Landy, A. 1989. Dynamic, structural and regulatory aspects of lambda site-specific recombination. *Annu. Rev. Biochem.* **58:** 913-950.

Larimer, F.W. and A. Stevens. 1990. Disruption of the gene *XRN1*, coding for a 5'-3' exoribonuclease, restricts yeast cell growth. *Gene* **95:** 85-90.

Lee, D. and P. Sadowski. 1981. Genetic recombination of bacteriophage T7 *in vivo* studied by use of a simple physical assay. *J. Virol.* **40:** 839-847.

Lilley, D.M.J. and B. Kemper. 1984. Cruciform-resolvase interactions in supercoiled DNA. *Cell* **36:** 413-422.

Lin, F.L., K. Sperle, and N. Sternberg. 1984. Model for homologous recombination during transfer of DNA into mouse L cells: Role for DNA ends in the recombination process. *Mol. Cell. Biol.* **4:** 1020-1034.

Little, J.W. 1967. An exonuclease induced by bacteriophage lambda. II. Nature of the enzymatic reaction. *J. Biol. Chem.* **242:** 679-686.

Little, J.W., I.R. Lehman, and A.D. Kaiser. 1967. An exonuclease induced by bacteriophage lambda. I. Preparation of the crystalline enzyme. *J. Biol. Chem.* **242:** 672-678.

Lloyd, R.G. and C. Buckman. 1991. Overlapping functions of *recD*, *recJ* and *recN* provide evidence of three epistatic groups of genes in *Escherichia coli* recombination and DNA repair. *Biochimie* **73:** 313-320.

Lloyd, R.G., F.E. Benson, and C.E. Shurvinton. 1984. Effect of *ruv* mutations on recombination and DNA repair in *Escherichia coli* K12. *Mol. Gen. Genet.* **194:** 303-309.

Lloyd, R.G., C. Buckman, and F.E. Benson. 1987. Genetic analysis of conjugational recombination in *Escherichia coli* K12 strains deficient in RecBCD enzyme. *J. Gen. Microbiol.* **133:** 2531-2538.

Lloyd, R.G., M.C. Porton, and C. Buckman. 1988. Effect of *recF*, *recJ*, *recN*, *recO* and *ruv* mutations on ultraviolet survival and genetic recombination in a *recD* strain of *Escherichia coli*. *Mol. Gen. Genet.* **212:** 317-324.

Lovett, S.T. and R.D. Kolodner. 1989. Identification and purification of a single-stranded-DNA-specific exonuclease encoded by the *recJ* gene of *Escherichia coli*. *Proc. Natl. Acad. Sci.* **86:** 2627-2631.

―――. 1991. Nucleotide sequence of the *Escherichia coli recJ* chromosomal region and construction of RecJ-overexpression plasmids. *J. Bacteriol.* **173:** 353-364.

Lovett, S.T., C. Luisi-DeLuca, and R.D. Kolodner. 1988. The genetic dependence of recombination in *recD* mutants of *Escherichia coli*. *Genetics* **120:** 37-45.

Lowenhaupt, K., M. Sander, C. Hauser, and A. Rich. 1989. *Drosophila melanogaster* strand transferase. A protein that forms heteroduplex DNA in the absence of both ATP and single-strand binding protein. *J. Biol. Chem.* **264:** 20568-20575.

Lu, M., Q. Guo, F.W. Studier, and N.R. Kallenbach. 1991. Resolution of branched DNA substrates by T7 endonuclease I and its inhibition. *J. Biol. Chem.* **266:** 2531-2536.

Luisi-DeLuca, C., A.J. Clark, and R.D. Kolodner. 1988. Analysis of the *recE* locus of *Escherichia coli* K-12 by use of polyclonal antibodies to exonuclease VIII. *J. Bacteriol.* **170:** 5797-5805.

Luisi-DeLuca, C., S.T. Lovett, and R.D. Kolodner. 1989. Genetic and physical analysis of plasmid recombination in *recB recC sbcB* and *recB recC sbcA Escherichia coli* K-12 mutants. *Genetics* **122:** 269-278.

Masterson, C., P.E. Boehmer, F. McDonald, S. Chaudhuri, I.D. Hickson, and P.T. Emmerson. 1992. Reconstitution of the activities of the RecBCD holoenzyme of *Escherichia coli* from the purified subunits. *J. Biol. Chem.* **267:** 13564-13572.

McCarthy, J.G., M. Sander, K. Lowenhaupt, and A. Rich. 1988. Sensitive homologous recombination strand transfer assay: Partial purification of a *Drosophila melanogaster* enzyme. *Proc. Natl. Acad. Sci.* **85:** 5854-5858.

Meselson, M.M. and C.M. Radding. 1975. A general model for genetic recombination. *Proc. Natl. Acad. Sci.* **72:** 358-361.

Mizuuchi, K. 1992. Transpositional recombination: Mechanistic insights from studies of Mu and other elements. *Annu. Rev. Biochem.* **1051:** 1011-1051.

Mizuuchi, K., B. Kemper, J. Hays, and R.A. Weisberg. 1982. T4 endonuclease VII cleaves Holliday structures. *Cell* **29:** 357-365.

Moore, S.P. and R. Fishel. 1990. Purification and characterization of a protein from human cells which promotes homologous pairing of DNA. *J. Biol. Chem.* **265:** 11108-11117.

Mosig, G. 1987. The essential role of recombination in phage T4 growth. *Annu. Rev. Genet.* **21:** 347-372.

Mueller, J.E., B. Kemper, R.P. Cunningham, N.R. Kallenbach, and N.C. Seeman. 1988. T4 endonuclease VII cleaves the crossover strands of Holliday junction analogs. *Proc. Natl. Acad. Sci.* **85:** 9441-9445.

Müller, B., C. Jones, and S.C. West. 1990a. T7 endonuclease I resolves Holliday junctions formed *in vitro* by RecA protein. *Nucleic Acids Res.* **18:** 5633-5636.

Müller, B., C. Jones, B. Kemper, and S.C. West. 1990b. Enzymatic formation and resolution of Holliday junctions *in vitro*. *Cell* **60:** 329-336.

Muniyappa, K. and C.M. Radding. 1986. The homologous recombination system of phage λ. *J. Biol. Chem.* **261:** 7472-7478.

Nakagawa, K., N. Morishima, and T. Shibata. 1992. An endonuclease with multiple cutting sites, Endo.SceI, initiates genetic recombination at its cutting site in yeast mitochondria. *EMBO J.* **11:** 2707-2715.

Orr-Weaver, T., J.W. Szostak, and R.J. Rothstein. 1981. Yeast transformation: A model system for the study of recombination. *Proc. Natl. Acad. Sci.* **78:** 6354-6358.

Otsuji, N., H. Iyehara, and Y. Hideshima. 1974. Isolation and characterisation of an *Escherichia coli ruv* mutant which forms nonseptate filaments after low doses of ultraviolet irradiation. *J. Bacteriol.* **117:** 337-344.

Paetkau, V., L. Langman, P. Bradley, D. Scraba, and R. Miller. 1977. Folded, concatenated genomes as replication intermediates of bacteriophage T7 DNA. *J. Virol.* **22:** 130-141.

Palas, K.M. and S.R. Kushner. 1990. Biochemical and physical characterization of exonuclease V from *Escherichia coli*. Comparison of the catalytic activities of the RecBC and RecBCD enzymes. *J. Biol. Chem.* **265:** 3447-3454.

Panayotatos, N. and A. Fontaine. 1985. An endonuclease specific for single-stranded DNA selectively damages the genomic DNA and induces the SOS response. *J. Biol. Chem.* **260:** 3173-3177.

Parsons, C.A. and S.C. West. 1988. Resolution of model Holliday junctions by yeast endonuclease is dependent upon homologous DNA sequences. *Cell* **52:** 621-629.

―――. 1990. Specificity of binding to four-way junctions in DNA by bacteriophage T7 endonuclease I. *Nucleic Acids Res.* **18:** 4377-4384.

Parsons, C.A., B. Kemper, and S.C. West. 1990. Interaction of a four-way junction in DNA with T4 endonuclease VII. *J. Biol. Chem.* **265:** 9285-9289.

Parsons, C.A., A.I.H. Murchie, D.M.J. Lilley, and S.C. West. 1989. Resolution of model Holliday junctions by yeast endonuclease: Effect of DNA structure and sequence. *EMBO J.* **8:** 239-246.

Pham, T.T. and J.E. Coleman. 1985. Cloning, expression, and purification of gene *3* endonuclease from bacteriophage T7. *Biochemistry* **24:** 5672-5677.

Picksley, S.M., C.A. Parsons, B. Kemper, and S.C. West. 1990. Cleavage specificity of bacteriophage T4 endonuclease VII and bacteriophage T7 endonuclease I on synthetic branch migratable Holliday junctions. *J. Mol. Biol.* **212:** 723-735.

Pottmeyer, S. and B. Kemper. 1992. T4 endonuclease VII resolves cruciform DNA with nick and counter nick and its activity is directed by local nucleotide sequence. *J. Mol. Biol.* **223:** 607-615.

Radding, C.M. 1966. Regulation of λ exonuclease. I. Properties of λ exonuclease purified from lysogens of λT11 and wild-type. *J. Mol. Biol.* **18:** 235-250.

―――. 1978. Genetic recombination: Strand transfer and mismatch repair. *Annu. Rev. Biochem.* **47:** 847-880.

―――. 1988. Homologous pairing and strand exchange promoted by *Escherichia coli* RecA protein. In *Genetic recombination* (ed. R. Kucherlapati and G.R. Smith), pp. 193-230. American Society for Microbiology, Washington, D.C.

―――. 1991). Helical interactions in homologous pairing and strand exchange driven by RecA protein. *J. Biol. Chem.* **266:** 5355-5358.

Radding, C.M. and D.C. Shreffler. 1966. Regulation of λ exonuclease. II. Joint regulation of exonuclease and a new λ antigen. *J. Mol. Biol.* **18:** 251-261.

Roca, A.I. and M.M. Cox. 1990. The RecA protein: Structure and function. *Crit. Rev. Biochem. Mol. Biol.* **25:** 415–456.

Roman, L.J. and S.C. Kowalczykowski. 1989a. Characterization of the adenosine triphosphatase activity of the *Escherichia coli* RecBCD enzyme: Relationship of ATP hydrolysis to the unwinding of duplex DNA. *Biochemistry* **28:** 2873–2880.

———. 1989b. Characterization of the helicase activity of the *Escherichia coli* RecBCD enzyme using a novel helicase assay. *Biochemistry* **28:** 2863–2872.

Roman, L.J., D.A. Dixon, and S.C. Kowalczykowski. 1991. RecBCD-dependent joint-molecule formation promoted by the *Escherichia coli* RecA and SSB proteins. *Proc. Natl. Acad. Sci.* **88:** 3367–3371.

Roman, L.J., A.K. Eggleston, and S.C. Kowalczykowski. 1992. Processivity of the DNA helicase activity of *Escherichia coli* RecBCD enzyme. *J. Biol. Chem.* **267:** 4207–4214.

Rosamond, J., K.M. Telander, and S. Linn. 1979. Modulation of the action of the RecBC enzyme of *Escherichia coli* K12 by Ca^{++}. *J. Biol. Chem.* **254:** 8646–8652.

Sadowski, P.D. 1971. Bacteriophage T7 endonuclease I: Properties of the enzyme purified from T7 phage infected *Escherichia coli* B. *J. Biol. Chem.* **246:** 209–216.

———. 1985. Role of nucleases in genetic recombination. In *Nucleases* (ed. S.M. Linn and R.J. Roberts), pp. 23–40. Cold Spring Harbor Laboratories, Cold Spring Harbor, New York.

Sander, M., K. Lowenhaupt, and A. Rich. 1991a. *Drosophila* Rrp1 protein: An apurinic endonuclease with homologous recombination activities. *Proc. Natl. Acad. Sci.* **88:** 6780–6784.

Sander, M., K. Lowenhaupt, W.S. Lane, and A. Rich. 1991b. Cloning and characterization of Rrp1, the gene encoding *Drosophila* strand transferase: Carboxy-terminal homology to DNA-repair endo exonucleases. *Nucleic Acids Res.* **19:** 4523–4529.

Sharples, G.J. and R.G. Lloyd. 1991. Resolution of Holliday junctions in *E. coli:* Identification of the *ruvC* gene product as a 19 kDa protein. *J. Bacteriol.* **173:** 7711–7715.

Shinohara, A., H. Ogawa, and R. Ogawa. 1992. Rad51 protein involved in repair and recombination in *Saccharomyces cerevisiae* is a RecA-like protein. *Cell* **69:** 457–470.

Smith, G.R. 1981. Chi sites, RecBC enzyme, and generalized recombination. *Stadler Genet. Symp.* **13:** 25–37.

———. 1983. Chi hotspots of generalized recombination. *Cell* **34:** 709–710.

———. 1987. Mechanism and control of homologous recombination in *Escherichia coli*. *Annu. Rev. Genet.* **21:** 179–201.

———. 1988. Homologous recombination in procaryotes. *Microbiol. Rev.* **52:** 1–28.

———. 1990. RecBCD enzyme. *Nucleic Acids Mol. Biol.* **4:** 78–98.

Stacey, K.A. and R.G. Lloyd. 1976. Isolation of rec⁻ mutants from an F′-merodiploid strain of *Escherichia coli* K-12. *Mol. Gen. Genet.* **143:** 223–232.

Stasiak, A. and E.H. Egelman. 1988. Visualization of recombination reactions. In *Genetic recombination* (ed. R. Kucherlapati and G.R. Smith), pp. 265–308. American Society for Microbiology, Washington, D.C.

Stevens, A. 1980. Purification and characterization of a *Saccharomyces cerevisiae* exoribonuclease which yields 5′-mononucleotides by a 5′-3′ mode of hydrolysis. *J. Biol. Chem.* **255:** 3080–3085.

Sun, H., D. Treco, and J.W. Szostak. 1991. Extensive 3′-overhanging, single-stranded DNA associated with the meiosis-specific double-strand breaks at the ARG4 recombination initiation site. *Cell* **64:** 1155–1162.

Sun, H., D. Treco, N.P. Schultes, and J.W. Szostak. 1989. Double-strand breaks at an initiation site for meiotic gene conversion. *Nature* **338:** 87–90.

Symington, L.S. and R. Kolodner. 1985. Partial purification of an enzyme from *Sac-*

charomyces cerevisiae that cleaves Holliday junctions. *Proc. Natl. Acad. Sci.* **82:** 7247–7251.

Szankasi, P. and G.R. Smith. 1992a. A DNA exonuclease induced during meiosis of *Schizosaccharomyces pombe. J. Biol. Chem.* **267:** 3014–3023.

———. 1992b. A single-stranded DNA exonuclease from *Schizosaccharomyces pombe. Biochemistry* **31:** 6769–6773.

Szostak, J.W., T.L. Orr-Weaver, R.J. Rothstein, and F.W. Stahl. 1983. The double-strand break repair model for recombination. *Cell* **33:** 25–35.

Takahagi, M., H. Iwasaki, A. Nakata, and H. Shinagawa. 1991. Molecular analysis of the *Escherichia coli ruvC* gene, which encodes a Holliday junction specific endonuclease. *J. Bacteriol.* **173:** 5747–5753.

Takahashi, N. and I. Kobayashi. 1990. Evidence for the double-strand break repair model of bacteriophage lambda recombination. *Proc. Natl. Acad. Sci.* **87:** 2790–2794.

Taylor, A.F. 1988. RecBCD enzyme of *Escherichia coli.* In *Genetic recombination* (ed. R. Kucherlapati and G.R. Smith), pp. 231–264. American Society for Microbiology, Washington, D.C.

Taylor, A.F. and G.R. Smith. 1980. Unwinding and rewinding of DNA by the RecBC enzyme. *Cell* **22:** 447–457.

———. 1985. Substrate specificity of the DNA unwinding activity of the RecBC enzyme of *E. coli. J. Mol. Biol.* **185:** 431–444.

———. 1992. RecBCD enzyme is altered upon cutting DNA at a Chi recombination hotspot. *Proc. Natl. Acad. Sci.* **89:** 5226–5230.

Taylor, A.F., D.W. Schultz, A.S. Ponticelli, and G.R. Smith. 1985. RecBC enzyme nicking at Chi sites during DNA unwinding: Location and orientation-dependence of the cutting. *Cell* **41:** 153–163.

Thaler, D.S., E. Sampson, I. Siddiqi, S.M. Rosenberg, F.W. Stahl, and M. Stahl. 1988. A hypothesis: Chi-activation of RecBCD enzyme involves removal of the RecD subunit. In *Mechanisms and consequences of DNA damage processing* (ed. E. Friedberg and P. Hanawalt), pp. 413–422. A.R. Liss, New York.

Thaler, D.S., E. Sampson, I. Siddiqi, S.M. Rosenberg, L.C. Thomason, F.W. Stahl, and M.M. Stahl. 1989. Recombination of bacteriophage λ in *recD* mutants of *Escherichia coli. Genome* **31:** 53–67.

Tishkoff, D.X., A.W. Johnson, and R.D. Kolodner. 1991. Molecular and genetic analysis of the gene encoding the *Saccharomyces cerevisiae* strand exchange protein Sep1. *Mol. Cell Biol.* **11:** 2593–2608.

Tomaschewski, J. and W. Rüger. 1987. Nucleotide sequence and primary structures of gene products coded for by the T4 genome between map positions 48.266 kb and 39.166 kb. *Nucleic Acids Res.* **15:** 3632–3633.

Tsujimoto, Y. and H. Ogawa. 1978. Intermediates in genetic recombination of bacteriophage T7 DNA. Biological activity and the roles of gene *3* and gene *5. J. Mol. Biol.* **125:** 255–273.

Van Oostrum, J., J.L. White, and R.M. Burnett. 1985. Isolation and crystallization of lambda exonuclease. *Arch. Biochem. Biophys.* **243:** 332–337.

West, S.C. 1992. Enzymes and molecular mechanisms of homologous recombination. *Annu. Rev. Biochem.* **61:** 603–640.

West, S.C. and A. Korner. 1985. Cleavage of cruciform DNA structures by an activity purified from *Saccharomyces cerevisiae. Proc. Natl. Acad. Sci.* **82:** 6445–6449.

West, S.C., C.A. Parsons, and S.M. Picksley. 1987. Purification and properties of a nuclease from *Saccharomyces cerevisiae* that cleaves cruciform junctions in DNA. *J. Biol. Chem.* **262:** 12752–12758.

6
Fungal and Mitochondrial Nucleases

Murray J. Fraser
Children's Leukaemia and Cancer Research Centre
Prince of Wales Children's Hospital
Randwick, N.S.W., Australia 2031

Robert L. Low
Department of Pathology
University of Colorado Health Sciences Center
Denver, Colorado 80262-0216

I. Introduction
II. Extracellular (Secreted) Fungal Nucleases
    A. Nucleases S1 and P1 of *Aspergillus oryzae* and *Penicillium citrinum*
        1. Properties
        2. Amino Acid Sequences
        3. Three-dimensional Structures and Proposed Mechanisms of Action
        4. Biological Role and Possible Relationships to Nucleases of Other Organisms
    B. Other Secreted Fungal Nucleases
III. Intracellular Fungal Nucleases
    A. Endo-exonucleases of *Neurospora crassa*, *Aspergillus nidulans*, and *Saccharomyces cerevisiae*
        1. Active and Trypsin-activated Forms of the Endo-exonuclease of *Neurospora* Are Both Present in the Nuclei and Mitochondria
        2. Evidence for Roles of Endo-exonucleases in DNA Repair and Recombination
        3. Properties of Purified Endo-exonucleases
            a. Intact Enzyme and Active Fragments
            b. Substrates
            c. Modes of Action
            d. Divalent Metal Ion Requirements
            e. Inhibitors
        4. Predicted Amino Acid Sequences of the Yeast Endo-exonucleases
    B. Other Fungal Nucleases Implicated in DNA Repair and Recombination
        1. Nuclease α of *Ustilago maydis*
        2. A 3´ Endo-exonuclease of *Coprinus cinereus*
        3. Exonuclease I of *Schizosaccharomyces pombe*
        4. Resolvase of *Saccharomyces cerevisiae*
        5. HO Endonuclease of *Saccharomyces cerevisiae*
        6. Double-strand-specific Endonucleases Encoded by Mitochondrial DNA of *Saccharomyces cerevisiae*
            a. Endonucleases I-*Sce*I, II, III, and IV
            b. Endo.*Sce*I
    C. Other Intracellular Fungal Nucleases
        1. Nuclease β of *Ustilago maydis*
        2. Nuclease γ of *Ustilago maydis*

    3. Yeast Exonucleases
IV. **Mitochondrial Nucleases of Higher Eukaryotes**
   A. Background: The Major Nuclease Activity of Mitochondria
   B. The Purified and Characterized Eukaryotic Mitochondrial Endonucleases
      1. Endonuclease of *Drosophila melanogaster*
      2. The Single-strand Endonuclease of Mouse Plasmacytoma Cells
      3. Bovine Mitochondrial Endonuclease
   C. Biological Role of the Major Mitochondrial Nuclease
   D. Minor Mitochondrial Endonucleases Specific for Damaged Sites in DNA
V. **Perspectives**

I. INTRODUCTION

This chapter deals mainly with new developments in understanding structures, functions, and biological roles of three distinct but related classes of fungal and mitochondrial nucleases: the secreted single-strand-specific endonucleases, the intracellular endo-exonucleases, and the major mitochondrial nucleases. These enzymes act on both DNA and RNA to release 5'-phosphoryl or 3'-phosphoryl-terminated products. The extracellular single-strand endonucleases act in conjunction with nucleotide-metabolizing enzymes to scavenge phosphate and nucleosides for cell growth. Endo-exonucleases in nuclei likely have roles in DNA repair, recombination, and possibly DNA replication. The major mitochondrial nucleases, both endo-exonucleases and endonucleases, probably have roles in DNA repair and replication of mitochondrial DNA (mtDNA). In addition, the endo-exonucleases of fungi may also play a role in recombination of mtDNA. On the other hand, the intracellular single-strand-specific endonuclease isolated previously from *Neurospora* is probably derived from endo-exonucleases via limited proteolysis. Many of the sugar nonspecific endonucleases isolated earlier from mammalian cells may also have been derived in like manner or from mitochondrial endonuclease. Finally, the endonucleases isolated from mammalian mitochondria have been found to be directly related to the endo-exonucleases of fungal mitochondria.

II. EXTRACELLULAR (SECRETED) FUNGAL NUCLEASES

A. **Nucleases S1 and P1 of *Aspergillus oryzae* and *Penicillium citrinum***

Much of the previous interest in the single-strand-specific endonucleases stemmed from their useful applications in recognizing and cleaving or in eliminating single-strand regions in DNA duplexes or DNA-RNA hybrids (Shishido and Ando 1982). With the recent publications of the amino acid sequences of nuclease S1 (Iwamatsu et al. 1991) and nuclease P1 (Maekawa et al. 1991) and the three-dimensional structures of both nucleases (Volbeda et al. 1991; Sück et al. 1993), it is now possible to begin to consider their structure-function relationships.

1. Properties

Nucleases S1 and P1 are heat-stable glycoproteins that show Zn^{++}-dependent activity on both single-stranded (ss)DNA and ssRNA. These activities are very single-strand-specific at relatively high-ionic-strength reaction conditions (0.1–0.4 M NaCl). However, the single-strand specificity is not sufficient to permit recognition of single-base mismatches in duplex DNA (Silber and Loeb 1981), but it will allow action opposite nicks such as those generated in duplex DNA by ionizing radiations (Martin-Bertram 1981).

The enzymatic properties of nucleases S1 and P1 were outlined previously (Shishido and Ando 1982) and are summarized only briefly here. The actions are concerted endo- and exonucleolytic, releasing 5'-mononucleotides even at early times of digestion. Both enzymes also exhibit 3' (or 2') nucleotidase activity. The pH optimum for polynucleotide degradation by nuclease P1 (pH 5.3) is significantly higher than that for nuclease S1 (pH 4.2). Both enzymes are inhibited by phosphate ion (50% inhibition at 2 mM) and even more effectively by nucleotides and nucleoside triphosphates. Nuclease S1 is inhibited about 50% by 1 mM dATP. Both enzymes contain three Zn^{++}/mole and show optimal activity with 1 mM Zn^{++}. They are quite sensitive to inhibition by EDTA; no activities are seen in buffers containing 0.1 mM EDTA in the absence of metal ions. Nuclease S1 is at least 70% reactivated under these conditions with Zn^{++} and Co^{++} but not at all with Mg^{++} or Ca^{++}. The above properties provide a basis for comparison with the single-strand-specific endonuclease activities of endo-exonucleases (see below).

2. Amino Acid Sequences

Direct sequencing of nuclease S1 by Iwamatsu et al. (1991) has shown that it contains 267 amino acids with a calculated molecular weight of 29,029. The higher molecular weight of 32,000 observed by SDS-gel electrophoresis is accounted for by two N-glycosylations at Asn-92 and Asn-228. The protein contains two disulfide bridges, presumably between Cys-80 and Cys-85 and Cys-72 and Cys-216 and one free cysteine sulfhydryl (Cys-25). The locations of the two disulfide bridges are based on the positioning of the two bridges in nuclease P1. Direct sequencing of nuclease P1 by Maekawa et al. (1991) has shown that it contains 270 amino acids with a calculated molecular weight of 29,221. The considerably higher molecular weight of 36,000–37,000 observed on SDS gels (see Maekawa et al. 1991) is likely accounted for by four N-glycosylations, at asparagine residues 92, 138, 184, and 197. These posttranslational modifications produce a 19% increase in mass as carbohydrate. There is no free sulfhydryl in nuclease P1, but two disulfide bridges exist between Cys-80 and Cys-85 and Cys-72 and Cys-217. The excess of acidic aspartic acid and glutamic acid residues (39) over the basic lysine, arginine, and histidine residues (20) in nuclease S1 compared with 31 acidic residues versus 24 basic residues in nuclease P1 helps account for the lower pI of nuclease S1, 3.7 versus 4.5 for nuclease P1. It also helps explain the lower pH optimum in degrading polynucleotides, pH 4.2 for nuclease S1 versus pH 5.3 for nuclease P1 (Sück et al. 1993). Alignment of the two amino acid sequences (Maekawa et al. 1991) shows only a one-residue deletion in nuclease S1 and high homology with nuclease P1, 50% identical and 66% highly conserved residues.

3. Three-dimensional Structures and Proposed Mechanisms of Action

The three-dimensional structure of nuclease P1 has been determined at 2.8 Å resolution from tetragonal crystals (Volbeda et al. 1991) and more recently refined at 2.2 Å resolution from orthorhombic crystals (Sück et al. 1993). The three-dimensional structure of nuclease S1 has been derived by comparative modeling based on the refined P1 structure (Sück et al. 1993). In nuclease P1, 56% of the amino acid residues reside in 14 α helices. The α-helix content is almost twice that determined earlier by circular dichroism (CD)-spectrum analysis (see Shishido and Ando 1982). The folding of the polypeptide, especially in the active site region around the three Zn^{++} cluster is very similar to that found in the Zn^{++}-dependent phospholipase C from *Bacillus cereus* despite only very limited sequence similarity with the phospholipase. The Zn^{++} cluster is at the bottom of a hydrophobic cleft that is inaccessible to duplex DNA and RNA structures, but it is fully accessible to single-strand polynucleotides. This explains the strand specificity of these enzymes. Unlike nuclease S1, nuclease P1 has a second nucleotide-binding site, 20 Å away from the active site as detected by crystallographic analysis of nuclease P1 saturated with a thiophosphorylated dinucleotide (Volbeda et al. 1991). More recent studies on the binding of the dithiophosphorylated derivative of the tetranucleotide ATTT show that the two sites form a contiguous channel (Sück et al. 1993), i.e., an extended binding site.

The binding of the dinucleotide analog at the bottom of the active site cleft shows that the phosphate of the phosphodiester linkage is placed near the Zn^{++} cluster. Together with the neighboring arrangement of the amino acid residues and bound water molecules, this structure suggests that the mechanism of catalysis proceeds via an attack of Zn^{++}-activated water (hydroxyl ion). It is possible (Volbeda et al. 1991) that the side-chain carboxylate of Asp-153 assists in the water activation and that the side-chain guanidinium of Arg-48 interacts with the phosphate to stabilize the pentacoordinate transition state. Such a mechanism is similar to that proposed earlier (Cotton et al. 1979) for staphylococcal (micrococcal) nuclease, but, in the case of nuclease P1, cleavage occurs on the other side of the phosphate to release 5'-phosphoryl-terminated products. Site-directed mutagenesis of the nuclease P1 and S1 genes will provide insight on this proposed mechanism.

4. Biological Role and Possible Relationships to Nucleases of Other Organisms

Secretion of nucleases P1 and S1 into the culture media likely indicates that a main biological role of the enzymes released from the fungi in their native habitats is to provide nucleosides and phosphate for growth. In this respect and in view of their enzymatic properties, nucleases P1 and S1 are very much like nuclease I of many plant species (mung bean, wheat seedlings, barley seedlings, tobacco). These nucleases are all glycoproteins with Zn^{++}-dependent endonuclease and 3'-nucleotidase activities, which have pH optima in the range 5.0–6.5, molecular weights in the range 31,000–39,000, and all release 5'-phosphoryl-terminated products (see Wilson 1982). The first 16 amino acid residues from the amino terminus of barley nuclease I have been determined (Brown and Ho 1987). This sequence shows eight residues that are identical or highly conserved when aligned with the amino termini of nucleases S1 and P1 (Iwamatsu et al. 1991), indicating significant sequence similarity among these enzymes. Since secreted enzymes are processed, at least to remove amino-terminal signal sequences, it will be of great interest to determine their entire predicted amino acid sequences by sequencing the cloned genes.

In some cases, sugar-nonspecific, single-strand-specific endonuclease activities have been purified and characterized without reference to their intracellular locations or whether they are secreted and without tests for nucleotidase activities. Consequently, ambiguity remains as to whether they are more closely related to nucleases S1 and P1 than to the intracellular endo-exonucleases. Two cases in point are mentioned, the nuclease isolated from the basidiomycete fungus, *Schizophyllum commune* (Martin et al. 1986), and nuclease SP of spinach (Strickland et al. 1991).

If the possession of both sugar-nonspecific single-stranded endonuclease and 3'-nucleotidase activities and Zn^{++} dependence can be taken as diagnostic of the S1/P1 class of nuclease, then it is clear that this class of enzyme is not confined to fungi and plants. Trypanosomes such as *Crithidia luciliae* (Neubert and Gottlieb 1990) and *Leishmania donovani* (Campbell et al. 1991) have such nucleases localized on their surface membranes. These are phosphate-repressible enzymes that help the trypanosomes survive purine and/or phosphate starvation.

B. Other Secreted Fungal Nucleases

Secretion of nucleases from fungi is common and not confined to single-strand-specific endonucleases that act on DNA and RNA. However, only a few of these enzymes have been well characterized, and only those with DNase activity are mentioned here. The mycelia of both *Neurospora crassa* and *Aspergillus nidulans* secrete a Ca^{++}, Mg^{++}- or Ca^{++}, Mn^{++}-dependent, DNase-I-like endonuclease activity. The enzyme from *Neurospora*, referred to as DNase A, shows no strand specificity (Fraser 1979). It is derepressed 200-fold when mycelia are grown on phosphate-free medium (Käfer and Witchell 1984). The endonuclease from *Aspergillus*, known as DNase-4, is double-strand-specific (Campbell and Winder 1983). These two DNases both generate $5'$-phosphoryl-terminated products and appear to be dimeric proteins with native molecular masses of 60–65 kD. A third enzyme, also called DNase A, has been isolated from *A. nidulans* as well, but it differs from DNase-4 in acting on both ssDNA and double-stranded (ds)DNA and in terms of other properties (Käfer et al. 1989). Although very similar in enzymatic properties, the two DNase A endonucleases of *Neurospora* and *Aspergillus* were found to be immunochemically distinct from each other and from pancreatic DNase I (Fraser et al. 1986; Käfer et al. 1989). DNase-4 was not tested in these experiments. It seems very probable that these three fungal endonucleases are closely related, but more work is needed to clarify the situation.

N. crassa mycelia also secrete a phosphate-repressible, Mg^{++}-dependent, single-strand-specific exonuclease (DNase B) that acts on both DNA and RNA (Fraser 1979; Käfer and Witchell 1984). This enzyme is released in large amounts from the periplasmic space of frozen and thawed conidia. The conidial exonuclease was purified to homogeneity and extensively characterized earlier (Rabin et al. 1972; Tenenhouse and Fraser 1973). It is composed of a large (72–78 kD) single polypeptide. The exonucleolytic action was found to move in the $5' \rightarrow 3'$ direction and to release $5'$ mononucleotides. No immunochemical cross-reaction was found between DNase B and the 76-kD intra-cellular endo-exonuclease of *Neurospora* described in the next section.

III. INTRACELLULAR FUNGAL NUCLEASES

A. Endo-exonucleases of *Neurospora crassa, Aspergillus nidulans*, and *Saccharomyces cerevisiae*

Endo-exonucleases have single-strand-specific endonuclease activity like nucleases P1 and S1, but they are distinct in lacking 3'-nucleotidase activity and in having a processive 5'→3'-exonuclease activity that acts on either ssDNA or dsDNA to release 5'-phosphoryl-terminated oligo- or mononucleotides (Chow and Fraser 1983; Chow and Resnick 1987; Dake et al. 1988; Koa et al. 1990). They act optimally on polynucleotides in the neutral pH range. Related endo-exonucleases have recently been isolated from *Drosophila melanogaster* (Shuai et al. 1992), monkey CV-1 cells (Couture and Chow 1992), and human leukemic cells (Fraser et al. 1993). Inactive (precursor?) forms of endo-exonuclease have been detected in *N. crassa, A. nidulans*, and mammalian cells but not in *Saccharomyces* or *Drosophila*. So far, only one gene for an endo-exonuclease has been detected in *Neurospora*, but in *Saccharomyces*, two are known, *NUC1* on chromosome X encoding the mitochondrial enzyme (Vincent et al. 1988) and *RNC1* on chromosome XI encoding a different, but related, nuclear enzyme (Chow et al. 1992). The expression of *RCN1* is regulated by the DNA-repair/recombination *RAD52* gene, less than 10% of the immunoprecipitable activity being detected in *rad52* mutants. Gene disruption experiments have indicated that in yeast, neither gene is essential (Zassenhaus et al. 1988; Chow et al. 1992). All endo-exonucleases identified so far in this group cross-react immunochemically with specific antibodies raised to purified *Neurospora* endo-exonuclease. Studies of these endo-exonucleases have been greatly facilitated by the use of specific antibodies (which show no cross-reactions with nucleases S1 and P1 or with a variety of other commercially available nucleases) and by the use of activity gel analysis (Fraser et al. 1986). Using the latter technique, active polypeptides derived from endo-exonucleases can be renatured after SDS-gel electrophoresis and will then degrade DNA in situ.

1. Active and Trypsin-activated Forms of the Endo-exonuclease of Neurospora Are Both Present in the Nuclei and Mitochondria

Studies on the localization of the endo-exonuclease in mycelia of *Neurospora* (Fraser and Cohen 1983; Ramotar et al. 1987) showed that there are active as well as protease-activated forms of the enzyme in both nuclei and mitochondria. The inactive form, presumably the precursor of the endo-exonuclease, could be quantitated using an assay for ssDNase activity after first treating the protein extract with trypsin. In both nuclei and mitochondria, the ratio of inactive to active enzyme was about 2:1. However, unless caution was used to minimize vacuolar contamination in the mitochondrial preparations and protease inhibitors were included in the extractions, the inactive form was readily activated. Although only the active enzyme was found in the extracts of the lysosome-like vacuoles of the mycelia, the inactive form was predominant in the cytosol. The vacuolar activity was rapidly lost, even in the presence of various protease inhibitors, so it could not be characterized extensively or purified by the methods used for the nuclear and mitochondrial endo-exonucleases.

The endo-exonuclease in the nucleus is mainly bound to chromatin, but about 15–20% of the activity is tightly bound to the nuclear matrix fraction where it may have a function different from that of the soluble and chromatin-bound forms. The endo-exonuclease accounts for more than 90% of the total nuclear ssDNase activity as deduced from specific inhibition by antibody raised to the purified endo-exonuclease (Ramotar et al. 1987). Several different-sized active polypeptides of 37, 43, and 76 kD were found by activity gel analysis of nuclear fractions and in immunoprecipitates of the soluble nuclear fractions. The largest, a 76-kD polypeptide, was associated with the nuclear matrix fraction. Since this polypeptide was also detected in fresh sonicates of rapidly growing mycelia (Fraser et al. 1986), it seems likely that the smaller active species have been generated from this species by limited proteolysis. In fact, the endo-exonuclease that has been purified from *Neurospora* being only 31 kD in size (Chow and Fraser 1983) is likely to be a proteolyzed fragment of this 76-kD polypeptide. Even the purified 31-kD *Neurospora* endo-exonuclease is not protease-resistant, since a brief treatment of the enzyme with trypsin in vitro first abolished the processivity of its exonuclease activity and then the exonuclease activity itself, leaving a single-strand-specific endonuclease with somewhat diminished acid-solubilizing activity (Chow and Fraser 1983). This activity strongly resembled the single-strand-specific endonuclease that was first isolated from *Neurospora* by Linn and Lehman (1965a,b) and later shown to be immunologically related to the 31-kD endo-exonuclease (Fraser et al. 1986). A similar sensitivity to proteolysis of the exonuclease compared to endonuclease activity has also been observed for the extracellular bacterial BAL31 nuclease (Hauser and Gray 1990). This periplasmic enzyme is processed in the culture medium from a larger (120-kD) active protein.

The mitochondrial form of the endo-exonuclease of *Neurospora* is immunologically related to the purified nuclear enzyme and is also altered in vitro by proteolysis. Within the mitochondrion, the endo-exonuclease is localized to the inner membrane from which it can be partially (60–70%) released by sonication and fully solubilized when membranes are disrupted using Triton X-100 (Fraser and Cohen 1983). The ssDNase activity in fresh extracts of mitochondria was completely inhibited by antibody to the purified enzyme, indicating that it is a major activity of this organelle as well as of the nucleus. Activity gel analysis indicated that the largest active polypeptide in fresh extracts of mitochondria (66 kD) was rapidly lost on aging, whereas shorter polypeptides (43 kD and especially one of 28 kD) accumulated (Fraser et al. 1986).

The intracellular distribution of inactive and active forms of endo-exonuclease in *A. nidulans* was found to be the same as in *N. crassa*, but suborganellar localizations were not done (Koa et al. 1990). Unfortunately, no systematic localization studies of the endo-exonucleases have been made as yet in *Saccharomyces*, although it is clear that the yeast mitochondrial enzyme is also localized to the inner mitochondrial membrane (Dake et al. 1988). No inactive form of the endo-exonuclease has been detected in *Saccharomyces*. Antibody to the *Neurospora* endo-exonuclease cross-reacts with both the yeast nuclear and mitochondrial endo-exonucleases (Chow and Resnick 1987; Dake et al. 1988), even though they are distinct proteins encoded by different genes. The antibody also inhibits all detectable ssDNase activity in yeast mitochondrial extracts, and furthermore, disruption of the *NUC1* gene eliminates all detectable ssDNase and nonspecific RNase activities (Dake et al. 1988). These observations strongly indicate that the several apparently different yeast mitochondrial endonucleases reported previously (Jacquemin-Sablon et al. 1979; Morosoli and Lusena 1980; Rosamond 1981; Foury 1982; von Tigerstrom 1982) were all derived from the 38-kD *NUC1*-encoded endo-exonuclease. It seems likely that limited proteolysis was mainly responsible for generating the smaller endonucleases from endo-exonuclease. The protease problem in the isolation of yeast mitochondrial nuclease has also been discussed by other investigators (Jacquemin-Sablon and Jacquemin-Sablon 1984). In both yeast and *Neurospora*, the problem stems mainly from contamination of the mitochondrial preparations by protease-laden vacuoles.

Early attempts to purify and study the inactive precursor form of the *Neurospora* endo-exonuclease (Kwong and Fraser 1978) led to the isolation of only small amounts of a partially purified 93-kD protein. The assignment of this protein as the precursor to the endo-exonuclease was confirmed when a polypeptide of this size was subsequently identified on Western blots using an antibody probe for the endo-exonuclease (Hatahet 1989). The conversion of the precursor into the active endo-exonuclease may be regulated in part by a heat-resistant-specific inhibitor of the endo-exonuclease that has been identified (Hatahet and Fraser 1989) and found to be induced 20-fold by heat shock (Ramotar and Fraser 1989). This inhibitor appears to form a stable complex with the enzyme. Proteolytic inactivation of the inhibitor itself may therefore be essential in order for the endo-exonuclease to be fully activated. Although the most abundant form of inhibitor purified was 24 kD, activity gel analysis of the inhibitor clearly indicated that there are three larger heat-stable inhibitory polypeptides. Remarkably, one of these was 93 kD. These results raise the possibility that both the inhibitor and the endo-exonuclease are derived from the same precursor. Thus, a large precursor exquisitely sensitive to proteolysis might give rise to inactive complexes of inhibitor and endo-exonuclease, active endo-exonuclease, or free inhibitor, depending on where in the precursor the proteolytic breaks were made.

It is likely very relevant that the intracellular nuclease O of *Aspergillus oryzae* also has a specific heat-stable 22-kD inhibitor (Uozumi et al. 1976). This sugar-nonspecific endonuclease has some specificity for ssDNA and releases small 5'-phosphoryl-terminated oligonucleotides like endo-exonuclease. It is likely also derived from an endo-exonuclease by limited proteolysis (Koa et al. 1990).

2. Evidence for Roles of Endo-exonucleases in DNA Repair and Recombination

In extracts of logarithmically growing diploid cells of *S. cerevisiae*, about 30-40% of the total ssDNase activity measured at pH 8 was immunoprecipitable with antibody to *Neurospora* endo-exonuclease. However, no activity was immunoprecipitated from stationary-phase cells, and less than 10% of the total activity was immunoprecipitated from logarithmically growing diploid cells of a recombination-deficient and double-strand break repair-deficient *rad52* mutant (Chow and Resnick 1988). In each case, the presence or absence of immunoprecipitable activity correlated quantitatively with the presence or absence of a 72-kD polypeptide corresponding to the purified endo-exonuclease as detected by Western blotting using the same antibody. The results indicate that this recombination-deficient and double-strand break DNA-repair-deficient mutant contained only about 10% of the normal level of this endo-exonuclease and that the enzyme may also have some role in normal growth. The results indicate that the expression of the endo-exonuclease gene (*RNC1*) is regulated by the *RAD52* gene. Using the same assay, a tenfold increase in endo-exonuclease activity was detected in wild-type cells synchronized during meiosis (Resnick et al. 1984). Since the wild-type *RAD52* gene is essential for meiotic recombination, endo-exonuclease is implicated in the high levels of recombination observed during meiosis. A recent study of in vitro recombination in extracts of *rad52* and *rnc1* mutants (Moore et al. 1993) has shown that the endo-exonuclease is essential for recombination when both substrates are double-stranded. The addition of antibody raised against the *N. crassa* endo-exonuclease to the extracts abolished this recombination.

In *Neurospora*, the endo-exonuclease levels found in mitochondria and nuclei of the *uvs-3* mutant were 10–12% of those in the wild type (Fraser and Cohen 1983; Ramotar et al. 1987). This mutant is DNA-repair-deficient, exhibiting sensitivities to several mutagens, and may have some defect in recombination. Thus, in both *Saccharomyces* and *Neurospora*, the nuclear endo-exonuclease is implicated in DNA repair and recombination. A further indirect piece of evidence to support this possibility is the finding that antibody to the *Neurospora* endo-exonuclease cross-reacts specifically with one of the three polypeptides (RecC) of the major recombination nuclease of *Escherichia coli* (Fraser et al. 1990).

3. Properties of Purified Endo-exonucleases

a. Intact Enzyme and Active Fragments. Both intact and active endo-exonuclease fragments have been purified and characterized. Even through the *Neurospora* endo-exonuclease was purified rapidly in two steps in the presence of protease inhibitors, an active polypeptide of only 31–33 kD was recovered (Chow and Fraser 1983), in contrast to the expected one of 76 kD (see above). The same methods yielded an even smaller (28-kD) purified polypeptide from *A. nidulans* (Koa et al. 1990). On the other hand, three-step affinity chromatography did yield the intact 38-kD endo-exonuclease from yeast mitochondria (Dake et al. 1988), and a one-step immunoaffinity chromatography using antibody to the *Neurospora* endo-exonuclease yielded a nearly intact 72-kD yeast nuclear endo-exonuclease lacking only the predicted amino-terminal 37-amino-acid residues (Chow and Resnick 1987; Chow et al. 1992). The same immunoaffinity method was also used recently to isolate a 65-kD endo-exonuclease from monkey CV-1 cells (Couture and Chow 1992).

b. Substrates. Despite differences in size and primary structure (see below), the purified endo-exonucleases of *Neurospora*, *Aspergillus*, and nuclei and mitochondria of *Saccharomyces* have the same basic enzymological properties. They have true single-strand-specific endonuclease activity with linear and circular ssDNAs and ssRNA. They convert superhelical DNAs rapidly to nicked forms and then act slowly in the region of the nick to generate linear dsDNAs that, in turn, are good substrates for the exonuclease action (see below). *Neurospora* endo-exonuclease has little or no action on covalently closed, relaxed circular dsDNA (Fraser et al. 1989). Polyriboadenylic acid is a good RNA substrate for endo-exonuclease, but transfer RNA, ribosomal RNA, polyriboguanylic acid, and other RNAs with a high content of secondary structure are poor substrates.

c. Modes of Action. Endo-exonucleases have true single-strand-specific endonuclease activity as seen by cleavage of single-stranded circular phage DNAs (Chow and Resnick 1987; Dake et al. 1988) and nicking of superhelical, but not relaxed, covalently closed circular dsDNA (Fraser et al. 1989). The exonuclease actions on linear dsDNAs of the endo-exonucleases proceed in the 5'→3' direction, releasing short 5'-phosphoryl-terminated oligonucleotides (Chow and Fraser 1983), but differ in processivity from one enzyme to another. For example, the yeast mitochondrial enzyme is nonprocessive, whereas the *Neurospora* enzyme is highly processive. At limiting enzyme concentrations, *Neurospora* endo-exonuclease generates single-stranded tails and gaps thousands of nucleotides long in duplex DNA (Chow and Fraser 1983). A further study (Fraser et al. 1989) showed that endo-exonuclease actions on linear dsDNAs have an absolute requirement for the 5'-phosphoryl termini. In the presence of a molar excess of enzyme over double-stranded termini, linear dsDNAs are not only rapidly shortened, but also incised and slowly acquire double-strand breaks that show some site specificity of cleavage. Both of these internal actions are dependent on the 5'-phosphoryl termini. The latter double-strand breaks result in the appearance of "ladders" of dsDNA fragments when the treated DNA is subjected to electrophoresis in agarose gels. The double-strand breaks occurred most frequently in the middle of AGCACT and related sequences and likely result from nicks within two such closely apposed sites on opposite strands. This site preference for making double-strand breaks in linear DNAs is not as stringent as that of the RecBCD nuclease of *E. coli*, which recognizes octanucleotide Chi sites unrelated to AGCACT (Smith 1988). The fungal endo-exonucleases resemble the RecBCD nuclease not only in nicking single-stranded circular DNA and degrading linear dsDNA in a processive exonucleolytic manner, but also in having a scanning endonucleolytic activity that nicks dsDNA at preferred sites. This latter capability has only been tested for the *Neurospora* and *Aspergillus* endo-exonucleases (Fraser et al. 1989; Koa et al. 1990).

d. Divalent Metal Ion Requirements. All of the endo-exonucleases studied show optimal activities in the presence of 2–10 mM Mg^{++} and no activity in the presence of EDTA. Mn^{++} is also an efficient cofactor for all of these enzymes, but differences in responses to other divalent metal ions have been reported. For example, only the yeast mitochondrial and monkey CV-1 cell endo-exonucleases are also activated by Ca^{++} (Dake et al. 1988; Couture and Chow 1992). The yeast mitochondrial endo-exonuclease is activated by 2–10 mM Zn^{++}, which also fully activates the extracellular single-strand-specific endonucleases at millimolar levels (see above). However, 1 mM Zn^{++} completely inhibits the other three endo-exonucleases. It has since been found, however, that much lower concentrations of Zn^{++} (10–50 μM) activate the *Neurospora* endo-exonuclease (M.J. Fraser, unpubl.). The reasons for these differences in metal ion activation are not understood at present.

e. Inhibitors. Endo-exonucleases are very sensitive to inhibition by aurin tricarboxylic acid (ATA), widely considered to be a "general" nuclease inhibitor (Hallick et al. 1977). The purified *Neurospora* enzyme is completely inhibited by 1–2 μg/ml ATA (M.J. Fraser, unpubl.), whereas the yeast mitochondrial endo-exonuclease was reported to be inhibited 94% at 0.04 μg/ml ATA (Dake et al. 1988). No inhibition was detected for the extracellular *Neurospora* DNase A or DNase B enzymes discussed above, even at high concentrations of ATA (200 μg/ml). Both the *Neurospora* and yeast mitochondrial enzymes are also inactivated by ethoxyformic anhydride (EFA), a reagent fairly specific for modifying histidine residues in proteins (Chow and Fraser 1983; Dake et al. 1988).

4. Predicted Amino Acid Sequences of the Yeast Endo-exonucleases

Polyclonal, but nearly monospecific, antibody raised to the purified yeast mitochondrial endo-exonuclease was used to screen a λgt11 library of yeast genomic DNA fragments in order to isolate the *NUC1* gene (Zassenhaus et al. 1988). The same approach, but using antibody to the purified *Neurospora* endo-exonuclease, has been used to clone the *RNC1* gene for the yeast nuclear endo-exonuclease from a genomic library (Chow et al. 1992).

The amino acid sequence predicted from the cloned *NUC1* gene is 329 amino acids long with a calculated molecular weight of 37,200, very close to that observed for the purified yeast mitochondrial endo-exonuclease (Vincent et al. 1988). Although the initial search of the NBRF protein database failed to detect any homology with *NUC1* and other proteins, when it was found that antibody to the *Neurospora* endo-exonuclease also cross-reacted specifically with the *E. coli* RecC polypeptide, alignment of the mitochondrial endo-exonuclease sequence with the predicted RecC polypeptide sequence showed low sequence similarity between the carboxy-terminal 306-amino-acid residues of both polypeptides (Fraser et al. 1990). Five blocks of 24–56 residues contained 22% identical residues and 40% conserved residues. Much greater sequence similarity has now been reported (Ruiz-Carrillo and Côté 1993) with bovine endonuclease G (43% identity in a region of 226 amino acids) and with another sugar-nonspecific endonuclease (Ball et al. 1987) secreted from the bacterium, *Serratia marcesens* (43% identity in a region of 65 amino acids). Endonuclease G is likely identical to the major bovine mitochondrial endonuclease (see Section IV) and does not show cross-reaction with antibodies raised to *Neurospora* endo-exonuclease.

The amino acid sequence predicted from the cloned *RNC1* gene for the yeast nuclear endo-exonuclease contains 485 residues with a predicted molecular mass of 57 kD (Chow et al. 1992). Surprisingly, it shows little similarity to the *NUC1* sequence. A detailed analysis awaits the availability of other nuclear endo-exonuclease sequences. The difference between this and the observed molecular mass of the isolated enzyme of 72 kD has been attributed to posttranslational modifications, in particular N-glycosylation. The amino-terminal five-amino-acid residues of the purified mature enzyme were determined to be DEKNL. These were identical to residues 38–42 of the sequence predicted from the genomic clone, indicating that the purified enzyme had also been proteolytically processed at the amino-terminal end either in vivo or in vitro during extraction and purification.

A striking sequence similarity between residues 67 and 253 of the *RNC1* sequence and that of the human *rhoB* oncogene product was found (Chow et al. 1992). It showed 47% identity and 66% similarity, hence the name *RNC1* (*R*ho *N*u*C*lease) for the gene. Most of this sequence similarity is confined to a highly conserved four-section GTP-binding consensus (contained in residues 73–187 of the yeast sequence) in common with both the yeast and human ρ proteins (see Chow et al. 1992). The yeast nuclease has recently been shown to bind to a GTP-affinity column (T.Y.-K. Chow, pers. comm.), but no GTPase activity has been detected as yet. The inactive, trypsin-activatable forms of endo-exonucleases from *Neurospora* (T.Y.K. Chow, unpubl.) and human leukemic CEM cells (M.J. Fraser, unpubl.), but not their active forms, have now also been identified as GTP-binding proteins. Possibly, these GTP-binding sites will be found to be involved in the regulation of endo-exonucleases.

B. Other Fungal Nucleases Implicated in DNA Repair and Recombination

1. Nuclease α of Ustilago maydis

The isolation and characterization of a single-strand-specific endonuclease from *Ustilago maydis* was originally reported by Holloman and Holliday (1973) and Holloman (1973). Certain mutant strains of the fungus having a low frequency of mitotic gene conversion were found to have a reduced level of this enzyme. The properties of the enzyme resembled strongly those of the single-strand-specific nuclease of *N. crassa* described by Linn and Lehman (1965a,b) and now believed to be a derivative of *Neurospora* endo-exonuclease (see above). A reinvestigation of nuclease α (Holloman et al. 1981) resulted in the isolation of a larger form of the enzyme (a 55-kD vs. a 42-kD single polypeptide) that exhibited most of the same properties as the enzyme isolated originally, except that the enzyme progressively shortened linear dsDNA from both termini ("nibbling" of the duplex). In addition, endonucleolytic action was carefully demonstrated to occur at different damaged sites in duplex structures caused by depurination, deamination, and UV and ionizing radiations. Like endo-exonucleases, nuclease α releases 5'-phosphoryl-terminated products, including both mono- and small oligonucleotides. Although no exonuclease activity on linear dsDNA was reported, the "nibbling" reaction observed may be explained by the successive actions of a nonprocessive 5'→3' exonuclease and a single-strand-specific endonuclease that attacks the short 3' single-stranded tail generated.

2. A 3' Endo-exonuclease of Coprinus cinereus

An endo-exonuclease that appears in large amounts during synchronous meiosis in the basidiomycete *Coprinus cinereus* has been extensively purified and characterized by Lu and Sakaguchi (1991). It is a monomeric 43-kD protein and acts optimally on ssDNA at pH 8.3 in a Mg^{++}-dependent reaction. However, it differs from the other endo-exonucleases in having a single-strand-specific endonuclease activity with both circular ssDNA and supercoiled circular dsDNA that generates breaks having 3'-phosphoryl and 5'-hydroxyl termini. Its exonuclease activity is also single-strand-specific, and it hydrolyzes linear DNA in the 3'→5' direction. Both activities were shown to be associated with the same polypeptide. Curiously, although ssDNA, prepared by heat denaturation of restriction-enzyme-digested plasmid DNA, was a good substrate for the enzyme, single-stranded circular M13 bacteriophage DNA was a very poor substrate. However, when the phage DNA was linearized by treatment with a limiting amount of pancreatic DNase I, the resulting product was a good substrate. In the presence of alkaline phosphatase, the M13 DNA was extensively degraded by the endo-exonuclease in a short time. Thus, it seems that the 3′-phosphoryl-terminated linear product of the initial endonucleolytic cleavage is not a substrate for the exonuclease, but becomes a substrate on removal of the 3′-phosphoryl group. It is thus difficult to understand how this enzyme acts exonucleolytically in the absence of phosphatase. The nature of the acid-soluble products of exonuclease action and the possible action of this endo-exonuclease on RNA substrates have not been reported.

3. Exonuclease I of Schizosaccharomyces pombe

A double-strand-specific 5′ exonuclease activity in meiotic cells of the fission yeast *Schizosaccharomyces pombe* increased fivefold after premeiotic S phase and decreased to the initial level before meiotic divisions, the period in which meiotic recombination is believed to occur (Szankasi and Smith 1992a). Isolation and characterization of this enzyme have shown that it is an Mg^{++}- or Mn^{++}-dependent 36-kD monomeric nonprocessive exonuclease. It releases 5′ mononucleotides from dsDNA at one tenth of the rate seen with ssDNA. The optimal pH for action on dsDNA is 7.0–7.5. However, possible actions on ssRNA and dsRNA have not been rigorously eliminated. Supercoiled plasmid DNA was not a substrate, but the enzyme did act at nicks with 5′-phosphoryl termini in relaxed DNA and linear dsDNA but not with 5′-hydroxyl termini. No associated endonucleolytic activity was detected. Recent cloning of the exonuclease I gene has shown that it is encoded by an open reading frame that predicts a 53-kD protein. Disruption of this gene does not prevent meiosis (G.R. Smith, pers. comm.). The latter observation indicates that exonuclease I is not essential for meiosis and may be replaced by another enzyme with similar activity.

4. Resolvase of Saccharomyces cerevisiae

Nicks and double-strand breaks in DNA stimulate recombination, and exonucleases are thought to act at the nicks or double-strand breaks in conjunction with ssDNA-binding proteins to generate single-stranded tails for switching by RecA-like proteins into homologous duplexes in order to initiate the formation of the covalent Holliday intermediate. Resolvases are proposed to act on this intermediate to resolve the recombinant products (see Chapter 5, this volume). Resolution of the Holliday junction requires cleavage at or near the crossed single strands, realignment, and ligation to generate intact recombinant duplexes. Artificially constructed Holliday structures (cruciform structures in plasmids) have been used to identify such endonucleases (see West and Korner 1985;

Symington and Kolodner 1985). Supercoiled plasmids that contain inverted repeat sequences extrude cruciform structures in vitro. The configuration of the DNA strands at the base of the junction closely resembles a Holliday junction.

One difficulty is to distinguish endonuclease activities that specifically recognize Holliday intermediates from activities that cleave all ssDNAs. It is possible, however, that endo-exonucleases may play a dual role in recombination, i.e., the exonuclease activity serves in the initiation phase, whereas the endonuclease activity acts in the resolution phase. In this regard, two purified bacteriophage-encoded single-strand-specific endonucleases, T4 endonuclease VII and T7 endonuclease I, act at the site of a cruciform to introduce nicks into regions that are symmetrically apposed at the base of the Holliday junction. These endonucleases are believed to act in recombination in conjunction with phage-encoded exonucleases and to have other roles such as in host genomic DNA degradation. However, an endonuclease specific for Holliday junctions has been identified and partially purified from *S. cerevisiae* (Symington and Kolodner 1985; West and Korner 1985; West et al. 1987). The enzyme is Mg^{++}-dependent and shows no action on ssDNA, linear dsDNA, or relaxed circular dsDNA. The enzyme has an apparent native molecular weight of about 200,000. Specific binding to a synthetic Holliday junction has been demonstrated (Evans and Kolodner 1987), and cleavage dependence on homologous DNA sequences has been demonstrated in strands of the same polarity approximately four to eight nucleotides from the base of the Holliday junction (Parsons and West 1988). The cleavages within homologous sequences occurred with precise symmetry across the junction, but they occurred asymmetrically in a junction containing four arms of unrelated sequence. The cleavages appear to occur in a concerted manner by nicking the two homologous duplexes when held in alignment. This likely requires the enzyme to have a dimeric (or tetrameric) structure in order to form the two duplex binding sites. It will be interesting to determine whether this highly specialized endonuclease has any homology with the strand-switching enzymes that can also bind two DNA duplexes.

5. HO Endonuclease of Saccharomyces cerevisiae

HO endonuclease is a Mg^{++}-dependent site-specific endonuclease of 66 kD. It is found in haploid *S. cerevisiae* undergoing mating-type switching, a mitotic gene conversion event (Kostriken and Heffron 1984). It shares amino acid sequence similarities with other double-strand endonucleases and mitochondrial intron-encoded double-strand endonucleases which also stimulate site-specific gene conversions (see Section III.B.6), especially at two dodecapeptide motifs (Shub and Goodrich-Blair 1992). The enzyme initiates mating-type interconversion by making a double-strand break with 3'-hydroxyl and 5'-phosphoryl termini and with four nucleotide 3'-overhanging ends within the *MAT* locus (Kostriken et al. 1983). The minimal recognition site for HO endonuclease action in vitro is 18 bp long (Nickoloff et al. 1986). Likely intermediates have been detected in the interconversion following the double-strand break, one of them being the generation of a much longer 3'-overhanging single-strand region on one side of the break (White and Haber 1990; Haber et al. 1993) presumably required for efficient strand switching in the recombination reaction. The 5'→3' recission is more extensive in *rad52* mutant cells, indicating that the *RAD52*-controlled *RNC1* endo-exonuclease is not involved in generating the long single-strand region.

Insertion of the HO endonuclease recognition site into yeast genes unrelated to the *MAT* locus has been found to stimulate homologous recombination within and near these genes (Nickoloff et al. 1986; Haber et al. 1993). In addition to gene conversion at these loci, a highly efficient alternative pathway of double-strand break repair involving the *RAD50* group functions has been detected (Haber et al. 1993).

6. Double-strand-specific Endonucleases Encoded by Mitochondrial DNA of Saccharomyces cerevisiae

a. Endonucleases I-SceI, II, III, and IV. Five DNA endonucleases have now been identified in *Saccharomyces* mitochondria that are either encoded entirely or in part by intron sequences in mtDNA. These activities help mediate specific types of sequence-directed recombination in the mitochondrial genome. Such endonucleases are apparently absent from metazoan mitochondria whose DNAs lack intron sequences. Although these activities are likely to be present in a variety of fungal and plant mitochondria, only the yeast DNA endonucleases, being better characterized, are reviewed (see also Chapter 4).

Four of these five DNA endonucleases are members of a class of mitochondrial endonucleases that each function to promote the homing of specific, mobile group I intron sequences (in which the endonuclease is encoded) into an intronless copy of the same gene present in a second mtDNA circle (Table 1; see Chapter 4) (Dujon 1989; Lambowitz 1989). These four enzymes are each denoted first by the prefix "I" for their intron origin, followed by "Sce" for *Saccharomyces cerevisiae* and the roman numeral indicating the order of their discovery (Dujon et al.

Table 1 DNA Endonucleases of *Saccharomyces cerevisiae* That Are Encoded by Group I Introns in mtDNA

Endonuclease designation	Function	Gene locus	Intron inserted	Sequence of cleavage site
I-*Sce*I	intron homing	21S rRNA	omega	↓ TAGGGATAACAGGGT ATCCCTATTGTCCCA ↑
I-*Sce*II	intron homing	*COXI*	I4α	↓ TTGGTCATCCAGAAGTAT AACCAGTAGGTCTTCATA ↑
I-*Sce*III	intron homing	*COXI*	aI3	not reported
I-*Sce*IV	intron homing	*COXI*	aI5α	↓ TCATGATTAGCTCTAATC AGTACTAATCGAGATTAG ↑

See text for details and references.

1989). The first enzyme of this type is the omega-encoded endonuclease called I-*Sce*I (Jacquier and Dujon 1985; Macreadie et al. 1985; Colleaux et al. 1986). The I-*Sce*I endonuclease is a 26-kD polypeptide encoded by an open reading frame within the 1.1-kb so-called omega intron of the yeast 21S rRNA gene of the recipient DNA. The break is a four-base staggered cut that leaves 3′-hydroxyl termini. The endonucleolytic cleavage is then followed by duplication of the intron sequence from the donor DNA by replication (Zinn and Butow 1985). Mutational analysis around the cleavage site has defined a recognition sequence of at least 18 bp (Colleaux et al. 1988).

The I-*Sce*II, I-*Sce*III, and I-*Sce*IV DNA endonucleases are each encoded by separate introns in the cytochrome *c* oxidase subunit I (*COXI*) gene. The I-*Sce*II DNA endonuclease is encoded in the group I4α intron and introduces a staggered double-strand break at the junction of exon A4 and exon A5. This endonuclease is a 30-kD polypeptide that is processed from a 62-kD polypeptide precursor encoded by both exon (*COXI* exons 1–4) and intron sequences. The amino acid sequence of the endonuclease shows a conserved sequence motif (LAGLI-DADG) identified in I-*Sce*I and in other group I introns (Hensgens et al. 1983). Recently, the I-*Sce*II protein has been purified and partially characterized (Wernette et al. 1990). It requires Mg^{++} or Mn^{++} for activity, is sensitive to salt inhibition, and introduces the same specific double-strand break into recipient DNA at the I-*Sce*II cleavage site identified from in vivo studies. The properties of this purified enzyme are similar to the I-*Sce*II activity previously expressed in *E. coli* from a genetically engineered copy of the gene that contains codons compatible with the translation program of *E. coli* (Banroques et al. 1987; Delahodde et al. 1989). Mutational analyses have identified several mutations in the 18-bp I-*Sce*II recognition sequence that significantly reduce endonucleolytic cleavage (Sargueil et al. 1990; Wernette et al. 1992).

DNA endonucleases I-*Sce*III and I-*Sce*IV are encoded by *COXI* intron aI3 (Sargueil et al. 1991) and intron aI5α (Moran et al. 1992; Séraphin et al. 1992), respectively. The I-*Sce*III enzyme apparently makes a site-specific cleavage at the exon A3–exon A4 junction homing site for aI3 in the *COXI* gene sequence, whereas the I-*Sce*IV DNA endonuclease cleaves the aI5α homing site, in a *COXI* gene that lacks the aI5α intron. I-*Sce*IV is apparently encoded by both the aI5α intron and upstream exon sequences. The I-*Sce*IV endonuclease requires Mg^{++} for activity, is inhibited by moderate levels of monovalent salts, and shows activity between pH 6.5 and pH 9.5 (Moran et al. 1992). Similar to the I-*Sce*I and I-*Sce*II enzymes, I-*Sce*IV creates a four-base staggered cut with 3′-hydroxyl termini (Moran et al. 1992; Séraphin et al. 1992).

b. *Endo.SceI*. The fifth DNA endonuclease encoded in yeast mtDNA likely assumes a more general role in genetic recombination of the mitochondrial genome. This enzyme, Endo.SceI, is composed of a 75-kD polypeptide encoded by a nuclear gene *ENS1* and transported into the mitochondrion and a 50-kD subunit encoded by a mitochondrial gene *ENS2* in the RF3 open reading frame (Nakagawa et al. 1988, 1991). The *ENS1* gene has recently been found to be identical to a gene for a *HSP70*-related heat-shock protein (Morishima et al. 1990). The Endo.SceI enzyme, first discovered from its ability to fragment phage dsDNA into discrete-sized fragments (Watabe et al. 1983), has been shown to target specific DNA sequences that are found in natural DNA once every few thousand base pairs and that are defined by an asymmetric 26-bp consensus sequence (Kawasaki et al. 1991). It is estimated that more than 30 such sites exist in the yeast mitochondrial genome. Similar to the I-*Sce* DNA endonucleases, Endo.SceI creates a staggered cut with four bases protruding at the 3′ ends (Kawasaki et al. 1991). In vivo, the Endo.SceI activity presumably acts to create double-strand breaks required to initiate recombination reactions. This hypothesis is supported by recent experiments which show that there is enhanced segregation of antibiotic resistance markers in mtDNA near Endo.SceI cutting sites, which depends on the presence of a functional Endo.SceI activity in vivo (Nakagawa et al. 1992).

C. Other Intracellular Fungal Nucleases

1. Nuclease β of Ustilago maydis

This unique single-strand-specific, 5'→3' exonuclease releases 3'-phosphoryl-terminated mono- and small oligonucleotides from ssDNA 200 times faster than from dsDNA (Rusche et al. 1980). The 68-kD monomeric enzyme acts optimally at pH 6 in a nonprocessive mode. It has true endonucleolytic activity with single-stranded circular bacteriophage DNAs and a low endonucleolytic (nicking) activity with superhelical DNA. In addition, it hydrolyzes rRNA at one tenth of the rate that is seen with ssDNA. The enzyme also has intrinsic 5'-nucleotidase activity. Nuclease β is 50% inhibited by 1 mM ATP or dATP or with 4 mM dAMP. Another unique feature of this enzyme is a lack of requirement for divalent metal ions even in the presence of 10 mM EDTA or 1 mM 1,10-phenanthroline. Unfortunately, no function is known for this well-characterized enzyme.

2. Nuclease γ of Ustilago maydis

Yarnall et al. (1984) have isolated a third nuclease activity from *Ustilago*, an endonuclease of about 20 kD that acts preferentially on dsDNA, optimally at about pH 8, and requires divalent metal ions. In the presence of either Mg++ or Ca++, the enzyme nicked dsDNA, but in the presence of either Mn++, Co++, or Zn++, the enzyme made double-strand breaks in the DNA duplex. In each case, the products were oligonucleotides terminated with 5'-phosphoryl and 3'-hydroxyl groups. The endonuclease has the unique property of being soluble in 5% perchloric acid. A 22-kD endonuclease from HeLa cell nuclei shows the same nicking and breaking activities in the presence of the various divalent metal ions (Fischman et al. 1979) and may be a mammalian counterpart of nuclease γ. The functions of these endonucleases are as yet unknown.

3. Yeast Exonucleases

Two proofreading 3'→5' exonucleases of *S. cerevisiae* associated with DNA polymerases I and III, respectively, have been assigned the names exonuclease I and exonuclease III (see Burgers et al. 1988). These are not discussed further in this review, but it should be pointed out that the assignment of the term "exonuclease I" to a proofreading function is inconsistent with the terminology currently in use for *E. coli* and *S. pombe* (see Section III.B.3 above).

Two single-strand-specific 5'→3' exonucleases that release 5' mononucleotides from DNA and RNA, both designated exonuclease II, have been purified and characterized, respectively, from *S. cerevisiae* (Villadsen et al. 1982) and *S. pombe* (Szankasi and Smith 1992b). They require Mg^{++} for activity and are inhibited by Ca^{++} and Zn^{++}. The *S. cerevisiae* enzyme, a tetramer of 120-kD subunits, has a pH optimum of 7.6–8.2, whereas the *S. pombe* enzyme, a monomeric 134-kD protein, is maximally active at pH 9.3. The *S. pombe* enzyme has been demonstrated to act processively, but the *S. cerevisiae* enzyme has not been tested for processivity. The latter enzyme was also shown to lack endonucleolytic action on circular ssDNA. The *S. pombe* enzyme has also now been shown to have 50% identity with the Sep1 protein of *Saccharomyces* (Kolodner et al. 1993), which has both strand exchange and 5'→3' exonuclease activities (G.R. Smith, pers. comm.).

Exonucleases IV and V have been partially purified and characterized from an *S. cerevisiae* strain that has triple deficiencies in the major vacuolar proteases (Bauer et al. 1988; Burgers et al. 1988). Coupled with the use of protease inhibitors during extraction and purification in order to minimize the actions of minor proteases, this approach should lead to the isolation of intact nucleases. Exonuclease IV (Bauer et al. 1988) acts in the 5'→3' direction on ssDNA and dsDNA with little strand specificity and releases 5' mononucleotides. It also has RNase activity, but this activity has not been further characterized. Exonuclease V of *S. cerevisiae* (Burgers et al. 1988) has also been found to have single-strand-specific 5'→3' exonuclease activity like exonuclease II, but it acts processively on ssDNA to release mainly dinucleotides, and also a small proportion (10% of the total) of trinucleotides terminated with 5'-phosphoryl groups. No endonuclease activity has been detected using single-stranded circular and superhelical RFI forms of bacteriophage M13 DNA. Polyriboadenylic acid was not significantly degraded by the enzyme. The native molecular mass of the enzyme was determined to be 99 kD, but the purest preparation contained polypeptides of predominantly 55 and 57 kD in size. Unfortunately, these were not successfully renatured during attempted activity gel analysis, so they have not been confirmed to be associated with the exonuclease activity. It is possible that the two polypeptides comprise a heterodimer. This enzyme has been localized to the nucleus, but its function has not been elucidated. Given its intracellular location and large size, it would be interesting to know whether brief treatment of this enzyme with low concentrations of trypsin would unmask a sugar-nonspecific endonuclease activity and double-strand exonuclease activity like that of endo-exonuclease of *S. cerevisiae* nuclei and whether the levels of exonuclease V are affected by the *rad52* mutation.

IV. MITOCHONDRIAL NUCLEASES OF HIGHER EUKARYOTES

A. Background: The Major Nuclease Activity of Mitochondria

The presence of endonuclease activity in mitochondria of higher eukaryotes was first recognized more than 30 years ago (Beaufay et al. 1959), before the discovery of mtDNA. First identified in crude preparations of rat liver mitochondria, this endonuclease activity was active at alkaline pH in vitro and distinct from the DNase previously discovered in lysosomes (DNase II), an enzyme active at acidic pH (Bernardi 1967). Although the finding of a Mg^{++}-dependent endonuclease activity in rat liver mitochondria was later confirmed by other investigators (de Lamirande et al. 1967; Baudhuin et al. 1975), the most compelling evidence that mitochondria in fact contain an endonuclease activity came from the initial work on the mitochondrial nuclease of *N. crassa* (Linn and Lehman 1966). In that study, nuclease activity was first shown to copurify with mitochondria that were sedimented through a linear sucrose gradient. This procedure removed lysosomal and nuclear DNA contaminants.

Further purification and characterization of the mitochondrial endonucleases of rat liver (Curtis and Smellie 1966; Curtis et al. 1966) and the identification and partial purification of endonucleases with similar activities in mitochondria of *S. cerevisiae* (see Section III.A.1 above) suggested that the mitochondria of all eukaryotic cells contain a similar Mg^{++} (or Mn^{++})-dependent endonuclease activity with a capacity to degrade ssDNA, RNA, and dsDNA in vitro. When it was subsequently shown that the *Neurospora* and *Saccharomyces* mitochondrial nucleases were in fact endo-exonucleases (Chow and Fraser 1983; Fraser and Cohen 1983; Dake et al. 1988) and that a single 37-kD single-strand-specific endonuclease purified from mitochondria of mouse plasmacytoma cells as well as the 38-kD *Saccharomyces* endo-exonuciease showed weak immunochemical cross-reaction with antibody raised to the single *Neurospora* endo-exonuclease, it seemed even more likely that the mitochondrial nucleases of all eukaryotes were related. This conclusion is now further supported by the finding of 43% amino acid sequence identity between bovine endonuclease G and the *Saccharomyces NUC1*-encoded mitochondrial endo-exonuclease (Ruiz-Carrillo and Côté 1993), since endonuclease G is now believed (see Section IV.B.3 below) to be identical to the single endonuclease purified and characterized from bovine heart mitochondria (Cummings et al. 1987). The endonuclease G sequence predicted from the cloned cDNA encodes a polypeptide of 299 amino acids with a predicted mass of about 34 kD, which is significantly larger than that of the isolated endonuclease (26 kD) and indicates that some processing of the enzyme may have occurred. As for the *Neuro-*

spora mitochondrial activity, such processing could result in the loss of exonuclease activity and strand specificity (see Section III.A.1). Similar processing may account for the previous detection of as many as five apparently distinct activities in crude mitochondrial lysates from mammalian cells (Durphy et al. 1974).

B. The Purified and Characterized Eukaryotic Mitochondrial Endonucleases

1. Endonuclease of Drosophila melanogaster

This mitochondrial endonuclease has been purified and partially characterized from *D. melanogaster* embryos (Harosh et al. 1992). The *Drosophila* enzyme activity is partially inhibited by polyclonal antibodies raised to the mitochondrial nucleases of *S. cerevisiae* and bovine heart and exhibits activities on dsDNA and ssDNA similar to those of the *Neurospora* and *Saccharomyces* mitochondrial endo-exonucleases. The *Drosophila* embryo endonuclease is a monomeric protein of 44 kD that is stimulated by ATP and resistant to inhibition by N-ethylmaleimide (Sakaguchi et al. 1990). The Mg^{++} (Mn^{++})-dependent endo-exonuclease purified and characterized from adult flies of *D. melanogaster* (Shuai et al. 1992) also resembles the *Neurospora* endo-exonuclease but surprisingly differs in some respects from the *Drosophila* embryo enzyme. The enzyme isolated from adult flies is a monomeric protein of 33 kD, rather than 44 kD, is unaffected by ATP, but is sensitive to inhibition by N-ethylmaleimide. In addition, the activity with dsDNA is quite sensitive to inhibition by NaCl. This endo-exonuclease, however, has not been clearly demonstrated to originate from mitochondria.

2. The Single-strand Endonuclease of Mouse Plasmacytoma Cells

A single-strand-specific endonuclease has been extensively purified from the mitochondria of cultured mouse plasmacytoma cells (Tomkinson and Linn 1986). It was shown to copurify with mitochondria sedimented through a linear sucrose gradient. This procedure removed contaminants of nuclear DNA and lysosomes that are invariably present in standard preparations of mitochondria and demonstrated the mitochondrial origin of this endonuclease. The enzyme requires a divalent cation for activity, degrades ssDNA endonucleolytically, and nicks dsDNA in the neutral pH range, showing a remarkable degree of similarity to the *Saccharomyces* and *Neurospora* endo-exonucleases. The activity is associated with a 37-kD polypeptide as shown from activity gels and has a homodimeric subunit structure. In addition, the enzyme cross-reacts, albeit weakly, with the polyclonal antibody raised against the *Neurospora* endo-exonuclease.

3. Bovine Mitochondrial Endonuclease

The endonuclease identified and purified from bovine heart mitochondria requires a divalent cation for activity, is sensitive to inhibition by N-ethylmaleimide, and degrades undamaged DNA substrates in vitro producing 5'-phosphoryl-terminated oligonucleotides, not unlike the fungal endo-exonucleases (Cummings et al. 1987). This endonuclease, however, does exhibit some clear differences from the mouse cell enzyme. The bovine heart endonuclease, purified 2500-fold to near homogeneity, is a homodimer of a 29-kD, not a 37-kD, polypeptide. Unlike the mouse cell endonuclease, the bovine enzyme extensively degrades, not just nicks, dsDNA templates and maintains a high level of activity toward ssDNA and dsDNA at low pH (pH 5). Interestingly, one preparation of the yeast enzyme has been reported to degrade dsDNA at pH 5 (Rosamond 1981), whereas another did not degrade (Dake et al. 1988).

The bovine mitochondrial endonuclease is probably associated with the inner membrane of the mitochondrion. Treatment of mitochondria with 1–2% (w/v) Triton X-100 or N-octylglucoside is required to maximize release of the endonuclease activity, and the use of these detergents during purification of the enzyme seems to improve recovery of enzyme activity. The enzyme, although soluble in aqueous buffers, seems to be somewhat hydrophobic, as expected for a membrane-associated activity. For example, it binds avidly to the hydrophobic chromatographic resin phenyl-Sepharose. Sequencing analysis of the purified protein has so far revealed two peptide segments that appear to be quite hydrophobic (K.L. Houmiel and R.L. Low, unpubl.). Furthermore, previous studies found that the purified endonuclease is stimulated 10–15-fold by the major phospholipids of the mitochondrion, phosphatidylcholine and phosphatidylethanolamine, a property characteristic of enzyme activities of the inner-membrane compartment (Parks et al. 1990). Although the endonuclease likely associates with the inner membrane in vivo, recent studies show that the endonuclease is also stably associated with an insoluble complex of mtDNA and mtDNA-binding proteins that can be gently isolated from disrupted mitochondria through a series of differential centrifugation steps (Low et al. 1993). Several years ago, an electron microscopic study revealed a single large complex of protein and lipid bound to the mtDNA in the region of the displacement loop (Albring et al. 1977) and suggested that this complex could provide a site-specific attachment of the mtDNA to the inner membrane.

Although the bovine heart mitochondrial endonuclease will extensively degrade both ssDNA and dsDNA in vitro when the ionic strength of the assay is low, at more physiologic ionic strength, the enzyme nicks dsDNA in a highly nonrandom fashion, with a strong preference for attack between guanine residues in a specific sequence in bovine mtDNA that resides just upstream of the origin of DNA replication, at the border of the noncoding D-loop region and gene for tRNA^{phe} (Low et al. 1987). The site is an evolutionarily conserved 12-bp GC tract representing the longest run of G residues in the mtDNA sequence. The same strong preference for GC tracts is also observed with the partially purified mitochondrial endonucleases of human and rat heart in vitro, indicating that the nucleotide specificity of the endonuclease is conserved at least through mammalian evolution (Low et al. 1988). The conservation of a long tract of G residues near the origin of mtDNA replication during evolution implicates some role for this sequence in vivo, possibly as a special recognition site for the endonuclease. Of interest, there are two nuclear endonucleases, one called endonuclease G (Ruiz-Carrillo and Renaud 1987) and another called endonuclease R (Gottlieb and Muzyczka 1990), that have been reported to similarly nick within tracts of G residues in vitro. Endonuclease G, extensively purified from calf thymus nuclei, is a homodimer of a 26-kD polypeptide (Côté et al. 1989) and appears to be remarkably similar to bovine mitochondrial endonuclease (Low et al. 1988). A direct comparison of the enzymatic properties of the two enzymes confirms that these enzymes are very similar, if not identical (M. Gerschenson and R.L. Low, unpubl.). These observations are strongly supported by the sequence similarity between endonuclease G and the yeast mitochondrial endo-exonuclease discussed above and the additional finding of significant sequence identity between the bovine heart mitochondrial endonuclease and the yeast mitochondrial endo-exonuclease (K.L. Houmiel and R.L. Low, unpubl.).

C. Biological Role of the Major Mitochondrial Nuclease

Although it is still not known how the mitochondrial endonuclease actually participates in the metabolism of mtDNA, several lines of evidence indicate that the enzyme might serve in a pathway that corrects DNA damage. Although a role for the enzyme in mtDNA repair seems plausible, two obstacles make it difficult to define. The first problem is genetic. Inactivation of the *NUC1* gene in *Saccharomyces* has no apparent phenotype. Neither the complete loss nor the overproduction of this endo-exonuclease has any apparent effect on the survival and maintenance of mtDNA in vivo (Zassenhaus et al. 1988). The other problem is a biochemical one, namely, the difficulty in identifying what factor(s) normally prevents the nuclease from attacking undamaged mtDNA in the intact organelle and learning how such factors presumably activate the endonuclease if DNA damage is incurred. Although the *nuc1* mutant has not yet been informative, mutants of *Neurospora* (*uvs-3*) and *Drosophila* (*mus308*) have been identified that are deficient in the repair of specific types of DNA damage (i.e., damage induced by chemical mutagens) and that have deficiencies in the mitochondrial nuclease activities in vitro (Fraser and Cohen 1983; Boyd et al. 1990). The mutants seem to have a defect in posttranslational processing required for activation of the endo-exonuclease in vivo. In the case of the *Neurospora* mutant, there is a failure to convert the precursor polypeptide proteolytically into the active form. In the *Drosophila* mutant *mus308*, an alteration of the pI of the enzyme presumably occurs because of a change in some posttranslational processing step important for enzyme activation (Sakaguchi et al. 1990). Although the participation of the mammalian endonuclease in the repair of damage in mtDNA is less certain, there are unusually large variations in the level of the endonuclease among the mitochondria of different rat organs. This suggests that the enzyme could play some role in the removal of oxidative damage in mtDNA (Houmiel et al. 1991). The level of mitochondrial endonuclease per milligram of mitochondrial protein or per mtDNA genome is about 200 times higher in the heart than in the liver or spleen, whereas levels intermediate to these values are found in the brain and kidney. These variations seem to correlate with relative rates of oxidative metabolism among tissues and presumably with rates of oxidative injury to mtDNA that occurs from the leakage of electrons at points along the respiratory chain (Imlay and Linn 1988). In vivo, the mitochondrial endonuclease could serve to initiate the degradation and elimination of the damaged genome from the mtDNA pool. A selective degradation of oxidized mtDNA could be critical in mammalian cells where the rate of oxidative injury is high and the preservation of undamaged mtDNA essential for the assembly of a functional respiratory chain and long-term viability.

Despite evidence that fungal endo-exonucleases probably have roles in recombination and recombination double-strand break repair (see Section III.A.2) and that the yeast *NUC1*-encoded mitochondrial endo-exonuclease shows amino acid sequence similarity to one of the components (RecC) of the major recombination nuclease of *E. coli*, there is still no compelling evidence that the yeast *NUC1* gene product is involved in recombination of yeast mtDNA. It might be expected that it could participate in 5′→3′ recission of the DNA at the double-strand breaks generated by Endo.ScaI (Section III.B.6), which stimulate recombination of yeast mtDNA. On the other hand, there is evidence that the yeast nuclear endo-exonuclease (*RNC1* gene) can promote the intramolecular DNA recombination that occurs in the induction of "*petite*" mutations in mtDNA of *S. cerevisiae* (Chow and Kunz 1991). It is not known whether the *RNC1* gene product promotes the recombination stimulated by Endo.ScaI, since the appropriate mutants have not yet been constructed for testing this possibility. It seems unlikely that the mammalian mitochondrial endonucleases participate in recombination of mammalian mtDNA as these mtDNAs do not encode Endo.ScaI-like DNases and are not known to undergo recombination (Wallace 1992). However, the possibility of DNA-damage-induced recombination of mammalian mtDNA remains to be tested.

Mitochondrial endonuclease may additionally provide a role in mtDNA replication. This possibility has been suggested from the striking preference of the mammalian enzyme to nick within the conserved sequence tract of G residues that resides just upstream of the origin of DNA replication (Low et al. 1987). Nicking at this site in vivo could provide a swivelase activity that functions to relieve the torsional constraint imposed on the mtDNA template during DNA replication and facilitate resolution of the DNA circles once replication is complete. This proposed role for the mammalian endonuclease in mtDNA replication is quite different from that proposed for a novel dsDNA-specific endonuclease of the mitochondrion of the trypanosomatid *Crithidia fasciculata* (Shlomai and Linial 1986). The *Crithidia* enzyme introduces single nicks into catenated, postreplicated DNA minicircles of the kinetoplast DNA network. Nicking by the endonuclease in vivo apparently serves to prevent decatenation of the kinetoplast minicircles by DNA topoisomerase and to block further replication since only free minicircles are replicated. This ensures that each kinetoplast minicircle is only replicated once in each DNA replication cycle (Linial and Shlomai 1988). This novel mitochondrial endonuclease, a homodimer of a 60-kD polypeptide, requires Mg^{++} for activity and nicks at one of many potential sites near a sequence-directed bend in the DNA helix (Linial and Shlomai 1987).

D. Minor Mitochondrial Endonucleases Specific for Damaged Sites in DNA

Even though mitochondria lack a capacity to excise pyrimidine dimers (Clayton et al. 1974; Prakash 1975), three forms of UV-specific endonuclease with apurinic/apyrimidinic endonuclease activities have recently been identified and partially purified from the mitochondria of a mouse plasmacytoma cell line (Tomkinson et al. 1990). The three endonuclease activities can be separated from one another by chromatography on phosphocellulose. There is no absolute requirement for a divalent cation for activity. The endonuclease activities are active on DNA that has been UV-irradiated or depurinated at low pH but not on undamaged DNA. In vivo, these activities likely serve to help recognize and remove oxidative damage in mtDNA. It will be surprising if similar activities are not soon found in other mitochondria as well.

V. PERSPECTIVES

During the past dozen years, it has become apparent that at least three distinct, but distantly related, groups of major sugar-nonspecific nucleases are being recognized: (1) the secreted fungal single-strand-specific endonucleases represented by nucleases S1 and P1; (2) the large protease-sensitive multifunctional endo-exonucleases that are probably involved normally in recombination and recombinational double-strand break repair, represented by the *Neurospora* and *Saccharomyces RNC1* endo-exonucleases and now also found in mammalian cells; and (3) the mitochondrial nucleases that are more closely related to the secreted bacterial nuclease of *Serratia marcesens*, represented by the *Saccharomyces NUC1* mitochondrial endo-exonuclease and the bovine mitochondrial endonuclease. There is fertile ground for further work. More examples of each class need to be studied before the structures and functions of these activities are understood. For example, what are the structures and relatedness of endo-exonucleases of *Coprinus* and mammalian cells and nuclease α of *Ustilago* to those of *Neurospora* and *Saccharomyces*? Are the nuclear and mitochondrial endo-exonucleases processed in vivo or does the processing result only from artifactual limited proteolysis during extraction and isolation? Are the large endo-exonucleases regulated during the cell cycle and does this regulation involve GTP? How are these enzymes activated in vivo to carry out their functions in nuclei and in mitochondria and how are they then inactivated when not required? There are already indications that inactive precursors and inhibitors may be involved. Does *Neurospora* and possibly other ascomycetes lack a *NUC1*-like gene? What are the roles of the RNase activity of endo-exonucleases and mitochondrial endonuclease?

A number of minor fungal and mitochondrial DNase activities, both extracellular and intracellular, have also been discussed in this chapter. The extracellular enzymes are phosphate-repressible scavenger enzymes, whereas the intracellular DNases have been implicated in specialized roles in DNA repair and recombination. More information is needed at the molecular and genetic levels to understand their structure and biological roles and their relationships to nucleases with comparable functions in other organisms.

The new biochemical findings on sugar-nonspecific nucleases and specific DNases reported here, when combined with genetic and sequencing data, provide a basis for analyzing the structures and functions of similar enzymes in mammalian cells. Sugar-nonspecific nuclease activity, in particular, abounds in mammalian cells. Thus, it is to be regretted that publications still appear in this field that do not properly delineate the substrate specificities or the endonuclease and exonuclease activities of newly isolated enzymes. Adequate assays of both DNase and RNase activities with a variety of substrates should be a minimum requirement for such studies.

ACKNOWLEDGMENTS

We thank D. Sück, T.Y.-K. Chow, A. Ruiz-Carrillo, K. Houmiel, M. Gerschenson, and G.R. Smith for sharing unpublished data. M.J.F. wishes to acknowledge the support of the National Health and Medical Research Council of Australia, the great patience of his colleagues at the CLCRC, and the ever cheerful Cathryn Minahan in preparing the manuscript. R.L.L. thanks Clairene Mraz and Nancy Hart for the expert typing of the manuscript.

REFERENCES

Albring, M., J. Griffith, and G. Attardi. 1977. Association of a protein structure of probable membrane derivation with HeLa cell mitochondrial DNA near its origin of replication. *Proc. Natl. Acad. Sci.* **74:** 1348–1352.

Ball, T.K., P.N. Saurugger, and M.J. Benedik. 1987. The extracellular nuclease gene of *Serratia marcesens* and its secretion from *Escherichia coli. Gene* **57:** 183–192.

Banroques, J., J. Perea, and C. Jacq. 1987. Efficient splicing of two yeast mitochondrial introns controlled by a nuclear-encoded maturase. *EMBO J.* **6:** 1085–1091.

Baudhuin, P., C. Peeters-Joris, and J. Bartholeyns. 1975. Hepatic nucleases: Association of polyadenylase, alkaline ribonuclease, and deoxyribonuclease with rat-liver mitochondria. *Eur. J. Biochem.* **57:** 213–220.

Bauer, G.A., H.M. Heller, and P.M.J. Burgers. 1988. DNA polymerase III from *Saccharomyces cervisiae*. I. Purification and characterization. *J. Biol. Chem.* **263:** 917–924.

Beaufay, H., D.S. Bendall, P. Baudhuin, and C. deDuve. 1959. Tissue fractionation studies: Intracellular distribution of some dehydrogenases, alkaline deoxyribonuclease, and iron in rat-liver tissue. *Biochem. J.* **73:** 623–628.

Bernardi, G. 1967. Mechanism of action and structure of acid deoxyribonuclease. *Adv. Enzymol.* **31:** 1–49.

Boyd, J.B., K. Sakaguchi, and P.V. Harris. 1990. *mus308* mutants of *Drosophila* exhibit

hypersensitivity to DNA cross-linking agents and are defective in a deoxyribonuclease. *Genetics* **125:** 813-819.
Brown, P.H. and T.D. Ho. 1987. Biochemical properties and hormonal regulation of barley nuclease. *Eur. J. Biochem.* **168:** 357-364.
Burgers, P.M.J., G.A. Bauer, and L. Tam. 1988. Exonuclease V from *Saccharomyces cerevisiae*. A. 5′→3′-deoxyribonuclease that produces dinucleotides in a sequential fashion. *J. Biol. Chem.* **263:** 8099-8105.
Campbell, A.M. and F.G. Winder. 1983. Properties of deoxyribonuclease 4 from *Aspergillus nidulans*. *Biochim. Biophys. Acta* **746:** 125-132.
Campbell, T.A., G.W. Zlotnick, T.A. Neubert, J.B. Sacci, and M. Gottlieb. 1991. Purification and characterization of the 3′-nucleotidase/nuclease from pomastigotes of *Leishmania donovani*. *Mol. Biochem. Parasitol.* **47:** 109-118.
Chow, T.Y.-K. and M.J. Fraser. 1983. Purification and properties of single strand DNA-binding endo-exonuclease of *Neurospora crassa*. *J. Biol. Chem.* **258:** 12010-12018.
Chow, T.Y.-K. and B.A. Kunz. 1991. Evidence that an endo-exonuclease controlled by the *NUC2* gene functions in the induction of "petite" mutations in *Saccharomyces cerevisiae*. *Curr. Genet.* **20:** 39-44.
Chow, T.Y.-K. and M.A. Resnick. 1987. Purification and characterization of an endo-exonuclease from *Saccharomyces cerevisiae* that is influenced by the *RAD52* gene. *J. Biol. Chem.* **262:** 17659-17667.
———. 1988. An endo-exonuclease activity of yeast that requires a functional *RAD52* gene. *Mol. Gen. Genet.* **211:** 41-48.
Chow, T.Y.-K., E.L. Perkins, and M.A. Resnick. 1992. Yeast *RNC1* encodes a chimeric protein, RhoNUC, with a human rho motif and deoxyribonuclease activity. *Nucleic Acids Res.* **20:** 5215-5221.
Clayton, D.A., J.N. Doda, and E.C. Friedberg. 1974. The absence of a pyrimidine dimer repair mechanism in mammalian mitochondria. *Proc. Natl. Acad. Sci.* **71:** 2777-2781.
Colleaux, L., L. d'Auriol, F. Galibert, and B. Dujon. 1988. Recognition and cleavage site of the intron encoded omega transposase. *Proc. Natl. Acad. Sci.* **85:** 6022-6026.
Colleaux, L., L. d'Auriol, M. Betermier, G. Cottarel, A. Jacquier, F. Galibert, and B. Dujon. 1986. Universal code equivalent of a yeast mitochondrial intron reading frame is expressed into *E. coli* as a specific double strand endonuclease. *Cell* **44:** 521-533.
Côté, J., J. Renaud, and A. Ruiz-Carillo. 1989. Recognition of (dG)n•(dC)n sequences by endonuclease G. *J. Biol. Chem.* **264:** 3301-3310.
Cotton, F.A., E.E. Hazen, and M.J. Legg. 1979. *Staphylococcal* nuclease: Proposed mechanism of action based on structure of enzyme-thymidine 3′,5′-bisphosphate-calcium ion complex at 1.5 Å resolution. *Proc. Natl. Acad. Sci.* **76:** 2551-2555.
Couture, C. and T.Y.-K. Chow. 1992. Purification and characterization of a mammalian endo-exonuclease. *Nucleic Acids Res.* **20:** 4355-4361.
Cummings, O.W., T.C. King, J.A. Holden, and R.L. Low. 1987. Purification and characterization of the potent endonuclease in extracts of bovine heart mitochondria. *J. Biol. Chem.* **262:** 2005-2015.
Curtis. P.J. and R.M.S. Smellie. 1966. Properties of a purified rat liver nuclease. *Biochem. J.* **98:** 818-825.
Curtis, P.J., M.G. Burdon, and R.M.S. Smellie. 1966. The purification from rat liver of nuclease hydrolysing ribonucleic acid and deoxyribonucleic acid. *Biochem. J.* **98:** 813-825.
Dake, E., T.J. Hofmann, S. McIntire, A. Hudson, and H.P. Zassenhaus. 1988. Purification and properties of the major nuclease from mitochondria of *Saccharomyces cerevisiae*.

J. Biol. Chem. **263:** 7691-7702.

Delahodde, A., V. Goguel, A.M. Becam, F. Creusot, J. Perea, J. Banroques, and C. Jacq. 1989. Site-specific DNA endonuclease and RNA maturase activities of two homologous intron-encoded proteins from yeast mitochondria. *Cell* **56:** 431-441.

de Lamirande, G., R. Morais, and M. Blackstein. 1967. Intracellular distribution of 5'-ribonuclease and 5'-phosphodiesterase in rat liver. *Arch. Biochem. Biophys.* **118:** 347-351.

Dujon, B. 1989. Group I introns as mobile genetic elements: Facts and mechanistic speculations—A review. *Gene* **82:** 91-114.

Dujon, B., M. Belfort, R.A. Butow, C. Jacq, C. Lemieux, P.S. Perlman, and V.M. Vogt. 1989. Mobile introns: Definition of terms and recommended nomenclature. *Gene* **82:** 115-118.

Durphy, M., P.N. Manley, and E.C. Friedberg. 1974. A demonstration of several deoxyribonuclease activities in mammalian cell mitochondria. *J. Cell Biol.* **62:** 695-706.

Evans, D.H. and R. Kolodner. 1987. Construction of a synthetic Holliday junction analog and characterization of its interaction with a *Saccharomyces cervisiae* endonuclease that cleaves Holliday junctions. *J. Biol. Chem.* **262:** 9160-9165.

Fischman, G.J., M.W. Lambert, and G.P. Studzinski. 1979. Purification and properties of a nuclear DNA endonuclease from HeLa S_3 cells. *Biochim. Biophys. Acta* **567:** 464-471.

Foury, F. 1982. Endonucleases in yeast mitochondria: Apurinic and manganese-stimulated deoxyribonuclease activities in the inner mitochondrial membrane of *Saccharomyces cerevisiae*. *Eur. J. Biochem.* **124:** 253-259.

Fraser, M.J. 1979. Alkaline deoxyribonucleases released from *Neurospora crassa* mycelia: Two activities not released by mutants with multiple sensitivities to mutagens. *Nucleic Acids Res.* **6:** 231-246.

Fraser, M.J. and H. Cohen. 1983. Intracellular localization of *Neurospora crassa* endo-exonuclease and its putative precursor. *J. Bacteriol.* **154:** 460-470.

Fraser, M.J., Z. Hatahet, and X. Huang. 1989. The actions of *Neurospora* endo-exonuclease on double strand DNAs. *J. Biol. Chem.* **264:** 13093-13101.

Fraser, M.J., H. Koa, and T.Y.-K. Chow. 1990. *Neurospora* endo-exonuclease is immunochemically related to the *recC* gene product of *Escherichia coli*. *J. Bacteriol.* **172:** 507-510.

Fraser, M.J., T.Y.-K. Chow, H. Cohen, and H. Koa. 1986. An immunochemical study of *Neurospora* nucleases. *Biochem. Cell Biol.* **64:** 106-116.

Fraser, M.J., C.M. Ireland, S.J. Tynan, and A. Papaioannou. 1993. Evidence for the role of an endo-exonuclease in the chromatin DNA fragmentation which accompanies apoptosis. In *Programmed cell death: The cellular and molecular basis of apoptosis* (ed. M. Lavin and D. Watters), pp. 111-122. Harwood, Switzerland.

Gottlieb, J. and N. Muzyczka. 1990. Substrate specificity of *HeLa* endonuclease R. *J. Biol. Chem.* **265:** 10842-10850.

Haber, J., N. Sugawara, F. Fishman-Lobell, C. White, B. Ray, and E. Ivanov. 1993. *HO* endonuclease and other exo- and endonucleases involved in site-specific homologous recombination in *Saccharomyces cerevisiae*. *J. Cell. Biochem. Suppl.* **17C:** 146.

Hallick, R.B., B.K. Chelm, P.W. Gray, and E.M. Orozco. 1977. Use of aurin tricarboxylic acid as an inhibitor of nucleases during nucleic acid isolation. *Nucleic Acids Res.* **4:** 3055-3064.

Harosh, T., M. Mezzina, P.V. Harris, and J.B. Boyd. 1992. Purification and characterization of a mitochondrial endonuclease from *Drosophila melanogaster* embryos. *Eur. J.*

Biochem. **210:** 455-460.
Hatahet, Z. 1989. "*Neurospora* endo-exonuclease: Actions on double-strand DNAs and a specific inhibitor." Ph.D. Thesis, McGill University, Montreal, Canada.
Hatahet, Z. and M.J. Fraser. 1989. Specific inhibitors of *Neurospora* endo-exonuclease. *Biochem. Cell Biol.* **67:** 631-641.
Hauser, C.R. and H.B. Gray. 1990. Precursor-product relationship of larger to smaller molecular forms of the BAL31 nuclease from *Alteromonas espejiana*: Preferential removal of duplex exonuclease relative to endonuclease activity by proteolysis. *Arch. Biochem. Biophys.* **276:** 451-459.
Hensgens, L.A.M., L. Bonen. M. DeHaan, G. Van der Horst, and L.A. Grivell. 1983. Two intron sequences in yeast mitochondrial *COX1* gene: Homology among URF-containing introns and strain-dependent variation in flanking exons. *Cell* **32:** 379-389.
Holloman, W.K. 1973. Studies on a nuclease from *Ustilago maydis*. II. Substrate specificity and mode of action of the enzyme. *J. Biol. Chem.* **248:** 8114-8119.
Holloman, W.K. and R. Holliday. 1973. Studies on a nuclease from *Ustilago maydis*. I. Purification, properties and implication in recombination of the enzyme. *J. Biol. Chem.* **256:** 5087-5094.
Holloman, W.K., T.C. Rowe, and J.R. Rusche. 1981. Studies on nuclease α from *Ustilago maydis. J. Biol. Chem.* **256:** 5087-5094.
Houmiel, K.L., M. Gerschenson, and R.L. Low. 1991. Mitochondrial endonuclease activity in the rat varies markedly among tissues in relation to the rate of tissue metabolism. *Biochim. Biophys. Acta* **1079:** 197-202.
Imlay, J.A. and S. Linn. 1988. DNA damage and oxygen radical toxicity. *Science* **240:** 1302-1309.
Iwamatsu, A., H. Aoyama, G. Dibo, S. Tsunasawa, and F. Sakiyama. 1991. Amino acid sequence of nuclease S1 from *Aspergillus oryzae. J. Biochem.* **110:** 151-158.
Jacquemin-Sablon, H., A. Jacquemin-Sablon, and C. Paoletti. 1979. Yeast mitochondrial deoxyribonuclease stimulated by ethidium bromide. 1. Purification and properties. *Biochemistry* **18:** 119-127.
Jacquemin-Sablon, H. and A. Jacquemin-Sablon. 1984. Effect of proteolysis of the yeast mitochondrial deoxyribonucleases. *Biochim. Biophys. Acta* **786:** 2523-2526.
Jacquier, A. and B. Dujon. 1985. An intron-encoded protein is active in a gene conversion process that spreads an intron into a mitochondrial gene. *Cell* **41:** 383-394.
Käfer, E. and G.R. Witchell. 1984. Effects of *Neurospora* nuclease halo (*nuh*) mutants on secretion of two phosphate-repressible alkaline deoxyribonucleases. *Biochem. Genet.* **22:** 403-417.
Käfer, E., A. Tittler, and M.J. Fraser. 1989. A single, phosphate repressible deoxyribonuclease, DNaseA, secreted in *Aspergillus nidulans. Biochem. Genet.* **27:** 153-166.
Kawasaki, K., M. Takahashi, M. Natori, and T. Shibata. 1991. DNA sequence recognition by a eukaryotic sequence-specific endonuclease, *endo.SceI*, from *Saccharomyces cerevisiae. J. Biol. Chem.* **266:** 5342-5347.
Koa, H., M.J. Fraser, and E. Käfer. 1990. Endo-exonuclease of *Aspergillus nidulans. Biochem. Cell Biol.* **68:** 387-392.
Kolodner, R., S. Hall, A. Johnson, and D. Tishkoff. 1993. Novel homologous pairing proteins from *Saccharomyces cerevisiae* and *Escherichia coli. J. Cell. Biochem. Suppl.* **17C:** 146.
Kostriken, R. and F. Heffron. 1984. The product of the *HO* gene is a nuclease: Purification and characterization of the enzyme. *Cold Spring Harbor Symp. Quant. Biol.* **49:**

89-96.

Kostriken, R., J.N. Strathern, A.J.S. Klar, and F. Heffron. 1983. A site-specific endonuclease essential for mating-type switching in *Saccharomyces cerevisiae*. *Cell* **35:** 167-174.

Kwong, S. and M.J. Fraser. 1978. *Neurospora* endo-exonuclease and its inactive (precursor?) form. *Can. J. Biochem.* **56:** 370-377.

Lambowitz, A.M. 1989. Infectious introns. *Cell* **56:** 323-326.

Linial, M. and J. Shlomai. 1987. Sequence-directed bent DNA helix is the specific binding site for *Crithidia fasiculata* nicking enzyme. *Proc. Natl. Acad. Sci.* **84:** 8205-8209.

———. 1988. A unique endonuclease from *Crithidia fasiculata* which recognizes a bend in the DNA helix. *J. Biol. Chem.* **263:** 290-297.

Linn, S. and I.R. Lehman. 1965a. An endonuclease from *Neurospora crassa* specific for polynucleotides lacking an order structure. I. Purification and properties of the enzyme. *J. Biol. Chem.* **240:** 1287-1293.

———. 1965b. An endonuclease from *Neurospora crassa* specific for polynucleotides lacking an order structure. II. Studies of enzyme specificity. *J. Biol. Chem.* **240:** 1294-1304.

———. 1966. An endonuclease from mitochondria of *Neurospora crassa*. *J. Biol. Chem.* **241:** 2694-2699.

Low, R.L., J.M. Buzan, and C.L. Couper. 1988. The preference of the mitochondrial endonuclease for a conserved sequence block in mitochondrial DNA is highly conserved during mammalian evolution. *Nucleic Acids Res.* **16:** 6427-6445.

Low, R.L., O.W. Cummings, and T.C. King. 1987. The bovine mitochondrial endonuclease prefers a conserved sequence in the displacement loop region of mitochondrial DNA. *J. Biol. Chem.* **262:** 16164-16170.

Low, R.L., K. Houmiel, M. Gerschenson, and W. Parks. 1993. The mitochondrial endonuclease of bovine heart and its potential role in the metabolism of the mitochondrial genome. *J. Cell. Biochem. Suppl.* **17C:** 153.

Lu, B.C. and K. Sakaguchi. 1991. An endo-exonuclease from meiotic tissues of the basidiomycete *Coprinus cinereus*. Its purification and characterization. *J. Biol. Chem.* **266:** 21060-21066.

Macreadie, I.G., R.M. Scott, A.R. Zinn, and R.A. Butow. 1985. Transposition of an intron in yeast mitochondria requires a protein encoded by that intron. *Cell* **41:** 395-402.

Maekawa, K., S. Tsunasawa, G. Dibo, and F. Sakiyama. 1991. Primary structure of nuclease P1 from *Penicillium citrinum*. *Eur. J. Biochem.* **200:** 651-661.

Martin, S.A., R.C. Ullrich, and W.L. Meyer. 1986. Partial purification and properties of a nuclease from *Schizophyllum commune* with a preference toward single-stranded nucleic acid. *Biochim. Biophys. Acta* **867:** 67-75.

Martin-Bertram, H. 1981. S1-sensitive sites in DNA after γ-irradiation. *Biochim. Biophys. Acta* **652:** 261-265.

Moore, P.D., J.R. Simon, L.J. Wallace, and T.Y.-K. Chow. 1993. *In vitro* recombination in *rad* and *rnc* mutants of *Saccharomyces cerevisiae*. *Curr. Genet.* **23:** 1-8.

Moran, J.V., C.M. Wernette, K.L. Mecklenburg, R.A. Butow, and P.S. Perlman. 1992. Intron 5α of the *COX 1* gene of yeast mitochondrial DNA is a mobile group I intron. *Nucleic Acids Res.* **20:** 4069-4076.

Morishima, N., K.I. Nakagawa, E. Yamamoto, and T. Shibata. 1990. A subunit of yeast site-specific endonuclease *ScelI* is a mitochondrial version of the 70-kDa heat shock protein. *J. Biol. Chem.* **265:** 15189-15197.

Morosoli, R. and C.V. Lusena. 1980. An endonuclease from yeast mitochondrial fractions. *Eur. J. Biochem.* **110:** 431-437.

Nakagawa, K.I., N. Morishima, and T. Shibata. 1991. A maturase-like subunit of the sequence-specific endonuclease Endo.SceI from yeast mitochondria. *J. Biol. Chem.* **266:** 1977–1984.

———. 1992. An endonuclease with multiple cutting sites, endo.SceI, initiates genetic recombination at its cutting site in yeast mitochondria. *EMBO J.* **11:** 2707–2715.

Nakagawa, K., J. Hashikawa, O. Makino, T. Ando, and T. Shibata. 1988. Subunit structure of a yeast site-specific endodeoxyribonuclease, endoSceI. *Eur. J. Biochem.* **171:** 23–29.

Neubert, T.A. and M. Gottlieb. 1990. An inducible 3′ nucleotidase/nuclease from the trypanosomatid *Crithidia luciliae*. Purification and characterization. *J. Biol. Chem.* **265:** 7236–7242.

Nickoloff, J.A., E.Y. Chen, and F. Heffron. 1986. A 24-base-pair DNA sequence from the *MAT* locus stimulates intergenic recombination in yeast. *Proc. Natl. Acad. Sci.* **83:** 7831–7835.

Parks, W.A., C.L. Couper, and R.L. Low. 1990. Phosphatidylcholine and phosphatidylethanolamine enhance the activity of the mammalian mitochondrial endonuclease *in vitro*. *J. Biol. Chem.* **265:** 3436–3439.

Parsons, C.A. and S.C. West. 1988. Resolution of model Holliday junctions by yeast endonuclease is dependent upon homologous DNA sequences. *Cell* **52:** 621–629.

Prakash, L. 1975. Repair of pyrimidine dimers in nuclear and mitochondrial DNA of yeast irradiated with low doses of ultraviolet light. *J. Mol. Biol.* **98:** 781–795.

Rabin, E.Z., H. Tenenhouse, and M.J. Fraser. 1972. An exonuclease of *Neurospora crassa* specific for single-stranded nucleic acids. *Biochim. Biophys. Acta* **259:** 50–68.

Ramotar, D. and M.J. Fraser. 1989. *Neurospora* endo-exonuclease in heat-shocked mycelia: Evidence for a novel heat shock induced function. *Biochem. Cell Biol.* **67:** 642–652.

Ramotar, D., A.H. Auchincloss, and M.J. Fraser. 1987. Nuclear endo-exonuclease of *Neurospora crassa*. Evidence for a role in DNA repair. *J. Biol. Chem.* **262:** 425–431.

Resnick, M.A., A. Sugino, J. Nitiss, and T. Chow. 1984. DNA polymerase, deoxyribonucleases, and recombination during meiosis in *Saccharomyces cerevisiae*. *Mol. Cell Biol.* **4:** 2811–2817.

Rosamond, J. 1981. Purification and properties of an endonuclease from the mitochondrion of *Saccharomyces cerevisiae*. *Eur. J. Biochem.* **120:** 541–546.

Ruiz-Carrillo, A. and J. Côté. 1993. Cloning of bovine endonuclease G cDNA. A step closer towards understanding its function. *J. Cell. Biochem. Suppl.* **17C:** 175.

Ruiz-Carrillo, A. and J. Renaud. 1987. Endonuclease G: A $(dG)_n \cdot (dC)_n$-specific DNase from higher eukaryotes. *EMBO J.* **6:** 401–407.

Rusche, J.R., T.C. Rowe, and W.K. Holloman. 1980. Purification and characterization of nuclease β from *Ustilago maydis*. *J. Biol. Chem.* **255:** 9117–9123.

Sakaguchi, K., P.V. Harris, R. van Kuyk, A. Singson, and J.B. Boyd. 1990. A mitochondrial nuclease is modified in *Drosophila* mutants (*mus308*) that are hypersensitive to DNA crosslinking agents. *Mol. Gen. Genet.* **224:** 333–340.

Sargueil, B., A. Delahodde, D. Hatat, G.L. Tian, J. Lasowska, and C. Jacq. 1991. A new specific DNA endonuclease activity in yeast mitochondria. *Mol. Gen. Genet.* **225:** 340–341.

Sargueil, B., D. Hatat, A. Delahodde, and C. Jacq. 1990. *In vivo* and *in vitro* analyses of an intron-encoded DNA endonuclease from yeast mitochondria. Recognition site by site-directed mutagenesis. *Nucleic Acids Res.* **18:** 5659–5665.

Séraphin, B., G. Faye, D. Hatat, and C. Jacq. 1992. The yeast mitochondrial intron aI5α: Associated endonuclease activity and *in vivo* mobility. *Gene* **113:** 1–8.

Shishido, K. and T. Ando. 1982. Single-strand-specific nucleases. In *Nucleases* (ed. S.M. Linn and R.J. Roberts), pp. 155–185. Cold Spring Harbor Laboratory, Cold Spring Harbor, New York.

Shlomai, J. and M. Linial. 1986. A nicking enzyme from trypanosomatids which specifically affects the topological linking of duplex DNA circles. *J. Biol. Chem.* **261:** 16219–16225.

Shuai, K., C.K. Das Gupta, R.S. Hawley, J.W. Chase, K.L. Stone, and K.R. Williams. 1992. Purification and characterization of an endo-exonuclease from adult flies of *Drosophila melanogaster*. *Nucleic Acids Res.* **20:** 1379–1385.

Shub, D.A. and H. Goodrich-Blair. 1992. Protein introns: A new home for endonucleases. *Cell* **71:** 183–186.

Silber, J.R. and L.A. Loeb. 1981. S1 nuclease does not cleave DNA at single-base mismatches. *Biochim. Biophys. Acta* **656:** 256–264.

Smith, G.R. 1988. Homologous recombination in procaryotes. *Microbiol. Rev.* **52:** 1–28.

Strickland, J.A., L.G. Marzilli, J.M. Puckett, and P.W. Doetsch. 1991. Purification and properties of nuclease SP. *Biochemistry* **30:** 9749–9756.

Sück, D., R. Dominguez, A. Lahm, and A. Volbeda. 1993. The three-dimensional structures of *Penicillium* P1 and *Aspergillus* S1 nucleases. *J. Cell. Biochem. Suppl.* **17C:** 154.

Symington, L.S. and R. Kolodner. 1985. Partial purification of an enzyme from *Saccharomyces cerevisiae* that cleaves Holliday junctions. *Proc. Natl. Acad. Sci.* **82:** 7247–7251.

Szankasi, P. and G.R. Smith. 1992a. A DNA exonuclease induced during meiosis of *Schizosaccharomyces pombe*. *J. Biol. Chem.* **267:** 3014–3023.

———. 1992b. A single-stranded DNA exonuclease from *Schizosaccharomyces pombe*. *Biochemistry* **31:** 6769–6773.

Tenenhouse, H. and M.J. Fraser. 1973. The ribonuclease activities of the single-strand specific nucleases of *Neurospora crassa*. *Can. J. Biochem.* **51:** 569–580.

Tomkinson, A.E. and S. Linn. 1986. Purification and properties of a single-strand-specific endonuclease from mouse cell mitochondria. *Nucleic Acids Res.* **14:** 9579–9593.

Tomkinson, A.E., R.T. Bonk,. J. Kim, N. Bartfeld, and S. Linn. 1990. Mammalian mitochondrial endonuclease activities specific for ultraviolet-irradiated DNA. *Nucleic Acids Res.* **18:** 929–935.

Uozumi, T., K. Ishino, T. Beppu, and K. Arima. 1976. Purification and properties of the nuclease inhibitor of *Aspergillus orgyzae* and kinetics of its interaction with crystalline nuclease O. *J. Biol. Chem.* **251:** 2808–2813.

Villadsen, I.S., S.E. Bjorn, and A. Vrang. 1982. Exonuclease II from *Saccharomyces cerevisiae*. An enzyme which liberates 5'-deoxyribomononucleotides from single-stranded DNA by a 5'→3' mode of hydrolysis. *J. Biol. Chem.* **257:** 8177–8182.

Vincent, R.D., T.J. Hofmann, and H.P. Zassenhaus. 1988. Sequence and expression of *NUC1*, the gene encoding the mitochondrial nuclease in *Saccharomyces cerevisiae*. *Nucleic Acids Res.* **16:** 3297–3312.

Volbeda, A., A. Lahm, F. Sakiyama, and D. Sück. 1991. Crystal structure of *Penicillium citrinum* P1 nuclease at 2.8 Å resolution. *EMBO J.* **10:** 1607–1618.

von Tigerstrom, R.G. 1982. Purification and characteristics of a mitochondrial endonuclease from the yeast *Saccharomyces cerevisiae*. *Biochemistry* **21:** 6397–6403.

Wallace, D.C. 1992. Diseases of the mitochondrial DNA. *Annu. Rev. Biochem.* **61:** 1175–1212.

Watabe, H.-O., T. Iino, T. Kaneko, T. Shibata, and T. Ando. 1983. A new class of

site-specific endodeoxyribonucleases. *J. Biol. Chem.* **258:** 4663–4665.
Wernette, C.M., R. Saldahna, P.S. Perlman, and R.A. Butow. 1990. Purification of a site-specific endonuclease, I-*Sce*II, encoded by intron 4α of the mitochondrial *Cox1* gene of *Saccharomyces cerevisiae. J. Biol. Chem.* **265:** 18976–18982.
Wernette, C., R. Saldanha, D. Smith, D. Ming, P.S. Perlman, and R.A. Butow. 1992. Complex recognition site for the group I intron-encoded endonuclease I-*Sce*II. *Mol. Cell. Biol.* **12:** 716–723.
West, S.C. and A. Korner. 1985. Cleavage of cruciform DNA structures by an activity from *Saccharomyces cerevisiae. Proc. Natl. Acad. Sci.* **82:** 6445–6449.
West, S.C., C.A. Parsons, and S.M. Picksley. 1987. Purification and properties of a nuclease from *Saccharomyces cerevisiae* that cleaves DNA at cruciform junctions. *J. Biol. Chem.* **262:** 12752–12758.
White, C.I. and J.E. Haber. 1990. Intermediates of recombination during mating type switching in *Saccharomyces cerevisiae. EMBO J.* **9:** 663–673.
Wilson, C.M. 1982. Plant nucleases: Biochemistry and development of multiple molecular forms. *Curr. Top. Biol. Med. Res.* **6:** 33–54.
Yarnall, M., T.C. Rowe, and W.K. Holloman. 1984. Purification and properties of nuclease γ from *Ustilago maydis. J. Biol. Chem.* **259:** 3026–3032.
Zassenhaus, H.P., T.J. Hofmann, R. Uthayashanker, R.D. Vincent, and M. Zona. 1988. Construction of a yeast mutant lacking the mitochondrial nuclease. *Nucleic Acids Res.* **16:** 3283–3296.
Zinn, A.R. and R.A. Butow. 1985. Non-reciprocal exchange between alleles of the yeast mitochondrial 21S rRNA gene: Kinetics and the involvement of a double strand break. *Cell* **40:** 887–895.

7
DNA Topoisomerases

Tao-shih Hsieh
Department of Biochemistry
Duke University Medical Center
Durham, North Carolina 27710

I. Introduction
II. **Multiplicity of Topoisomerases**
III. **Mechanistic Aspects**
   A. Type II Topoisomerases
      1. DNA Binding
      2. Strand Cleavage and Religation
      3. ATP Hydrolysis and Strand Passage
   B. Type I Topoisomerases
      1. Specificity for Single/Double-stranded DNA
      2. Strand Cleavage and Transfer
IV. **Structural Homology among DNA Topoisomerases**
V. **Biological Functions of DNA Topoisomerases**
VI. **Targeting by Antimicrobial and Anticancer Drugs**
VII. **Concluding Remarks**

Recent developments in the analysis of DNA structures clearly demonstrate that there is a rich repertoire of various structural motifs that DNA can assume and that the biological functions of DNA require a dynamic aspect in the transition among these structures. DNA topoisomerases play a critical role in mediating such structural transitions of DNA and thus are important in regulating many aspects of its biological functions. Numerous reviews (Liu 1983; Maxwell and Gellert 1986; Osheroff 1989; Cozzarelli and Wang 1990; Reece and Maxwell 1991; Hsieh 1992) and a chapter on DNA topoisomerases in the predecessor of this current volume (Wang 1982) give comprehensive accounts of the exciting developments in this field. This chapter summarizes the background of topoisomerases and highlights recent literature.

I. INTRODUCTION

DNA topoisomerase activity was first discovered with the ω protein (topoisomerase I) from *Escherichia coli* by virtue of its ability to relax covalently closed, supercoiled DNA (Wang 1971). The ω protein belongs to the type I DNA topoisomerase family of enzymes that can break and rejoin one DNA strand at a time. In 1976, *E. coli* DNA gyrase was identified (Gellert et al. 1976a); it defined the type II DNA topoisomerases, which act by passing a segment of DNA through a reversible double-strand break. Other reactions that can be catalyzed by topoisomerases include relaxation of supercoils, catenation and decatenation, and knotting/unknotting of DNA rings. Some DNA topoisomerases can also introduce superhelical turns into DNA by utilizing the energy from ATP hydrolysis: Eubacterial DNA gyrase can generate negative supercoils and archebacterial reverse gyrase can introduce positive supercoils. A hallmark for all topoisomerase reactions is that under most circumstances, the chemical structure of DNA is not ultimately altered. Instead, the DNA topoisomerases mediate two coupled, sequential transesterification reactions that break and rejoin the DNA phosphodiester backbone bond so as to change only the higher-order structure of DNA.

II. MULTIPLICITY OF TOPOISOMERASES

DNA topoisomerases are ubiquitous in nature, and both type I and II enzymes are present in all organisms examined so far. Moreover, many organisms contain multiple forms of either type I or type II topoisomerases (topo I and topo II). Eubacteria, such as *E. coli*, appear to have duplicate copies of both type I and II topoisomerases. In addition to the original ω protein, *E. coli* contains a second type I enzyme, topoisomerase III (topo III) (Dean et al. 1983; DiGate and Marians 1988). These enzymes show some differences in biochemical properties and share limited, but significant, sequence homology (Tse-Dinh and Wang 1986; DiGate and Marians 1989). Neither enzyme appears to be essential (Sternglanz et al. 1981; DiGate and Marians 1989).

Eubacterial type I topoisomerases are implicated in many aspects of DNA metabolism. Topo I, in conjunction with DNA gyrase, is responsible for regulating a homeotic state of DNA supercoiling (Menzel and Gellert 1983; Tse-Dinh and Beran 1988). The enzymes act in concert to remove supercoiling stress generated during DNA replication or transcription (Liu and Wang 1987). Type I topoisomerases are also implicated in genetic recombination: A deletion of the gene for topo I affects homologous recombination (Fishel and Kolodner 1984), and the gene for topo III is identical to *mutR*, which enhances mutational frequency as a result of elevated deletion events caused by recombination between repeated sequence elements (Schofield et al. 1992).

In the budding yeast, there are two type I topoisomerases, topo I and topo III, that are not related by sequence. Similar to other eukaryotic topoisomerases I, yeast topo I accounts for the majority of topo I activity in these cells, whereas topo III is inefficient in relaxing either positive or negative DNA supercoils based on the analysis of plasmid DNA supercoiling in yeast cells carrying mutations in these topoisomerase genes (Giaever and Wang 1988). The gene encoding yeast topo III, *EDR1* or *TOP3*, was first identified by its effect on recombination between the terminal repeats of the yeast retrotransposon Ty1 (Rothstein et al. 1987; Wallis et al. 1989). There is limited, but significant, homology between yeast topo III and the bacterial type I topoisomerases (Wallis et al. 1989). Moreover, yeast topo III is mechanistically similar to both topo I and topo III of *E. coli* (Kim and Wang 1992).

In addition to DNA gyrase, eubacteria also harbor a second type II enzyme, topo IV (Kato et al. 1990, 1992). The subunits of topo IV, ParC and ParE, are related by amino acid sequence homology with the gyrase subunits, GyrA and GyrB, respectively (Kato et al. 1990; Luttinger et al. 1991). Despite their sequence homology, both enzymes are essential for viability, suggesting that each has a distinct and critical biological function. An essential function of topo IV is during chromosome segregation, since the genes for the topo IV subunits were first identified by mutations affecting bacterial nucleoid separation during cell division (Kato et al. 1988; Schmid 1990). Topo IV also participates in the partitioning of replicated plasmid DNAs that are topologically interlocked (Adams et al. 1992). However, since DNA gyrase also plays a role in *E. coli* chromosome segregation (Steck and Drlica 1984), the division of labor between gyrase and topo IV in chromosome dynamics remains a fascinating question.

In vertebrates, including *Xenopus* and humans, there are also duplicate type II DNA topoisomerases, topo IIα (170,000 M_r) and topo IIβ (180,000 M_r), which share extensive amino acid sequence homology (Chung et al. 1989; Jenkins et al. 1992; Hirano and Mitchison 1993). Topo IIα appears to be highly expressed in proliferating cells, whereas topo IIβ is predominantly present in noncycling G_0 cells (Woessner et al. 1991; Hirano and Mitchison 1993; for review, see Hsieh 1992).

III. MECHANISTIC ASPECTS

A. Type II Topoisomerases

1. DNA Binding

That DNA topo II[1] apparently displays preferences in its interactions with DNA at both the sequence and structural levels was originally ob-

---

[1] Unless specified otherwise, properties ascribed to "topo II" imply all type II topoisomerases irrespective of source.

served during the analysis of sequences surrounding the cleavage sites of topo II (for review, see Hsieh 1990). Other methods used to monitor enzyme/DNA interactions in further detail include nitrocellulose filter binding (Morrison et al. 1980; Sander et al. 1987), glass fiber filter binding (Roca and Wang 1992), retardation of mobility in agarose gel electrophoresis (Osheroff 1986), alteration of DNA writhe (Liu and Wang 1978a), nuclease protection (Liu and Wang 1978b; Fisher et al. 1981; Kirkegaard and Wang 1981; Morrison and Cozzarelli 1981; Lee et al. 1989b), electron microscopy (Kreuzer and Huang 1983; Kirschhausen et al. 1985; Moore et al. 1985; Zechiedrich and Osheroff 1990; Howard et al. 1991), and electric dichroism (Rau et al. 1987). These studies yielded several major conclusions. First, a DNA segment of 110–160 base pairs can wrap around the eubacterial DNA gyrase in one right-handed turn, whereas the eukaryotic enzyme makes contact with DNA over a region of only 20–30 base pairs. This difference in DNA contact is likely the mechanistic basis for the different topoisomerization reactions, i.e., supercoiling versus relaxation, catalyzed by bacterial gyrase and eukaryotic topo II, respectively. Second, the sequence preference in the DNA cleavage reaction is also reflected in the DNA-binding reaction. Finally, the presence of an ATP analog in the binding reaction may have a profound effect on the structure and stability of DNA/enzyme complexes.

DNA topo II also has a propensity to bind DNA having unusual structures: bent loci (Howard et al. 1991), left-handed Z-DNA (Glikin et al. 1991), and a crossover of two DNA segments (Kreuzer and Huang 1983; Osheroff and Brutlag 1983; Zechiedrich and Osheroff 1990; Roca et al. 1993). Binding of topo II to a node of a pair of DNA helices is anticipated from its reaction mechanism involving transport of one DNA segment through another via a transient break. The enzyme's ability to carry out both intermolecular condensation (catenation) and intramolecular condensation (knotting) is related to this characteristic of topo II. However, a quantitative analysis of the crossover binding reaction suggested that it may not be obligatory for strand passage events, and thus this property of topo II may be more related to its putative structural role in the nucleus (Roca et al. 1993). On the basis of the DNA cleavage reaction, it was also suggested that topo II may interact with parallel stranded tetraplex DNA (Chung et al. 1992). The biochemical basis for the interaction of topo II with noncanonical DNA structures is unclear. However, it is possible that a key step in the catalytic cycle of topo II reactions is a distortion of DNA helical structure due to its interactions with the enzyme, and a distorted DNA structure may therefore facilitate its binding to topo II.

2. Strand Cleavage and Religation

A unique feature of the topoisomerases is that upon the addition of a strong protein denaturant, such as alkali or SDS, a covalent intermediate can be generated in which the enzyme is linked to DNA through a phosphodiester bond between the phosphate group at a newly generated DNA end and a specific tyrosine residue in the topoisomerase. Earlier reviews dealt with this topic in more detail (Wang and Liu 1979; Wang 1985). For topo II, two identical subunits from the enzyme are joined to 5'-phosphoryl ends of a double-strand break that is staggered by four base pairs. This covalent complex, the "cleavage complex," is likely to be an intermediate in the catalytic cycle of topoisomerase reactions.

The apparent equilibrium between the DNA cleavage and religation reactions can be perturbed under various reaction conditions. This property has been exploited to analyze the kinetic parameters of the cleavage/religation reactions by topo II (Osheroff and Zechiedrich 1987; Robinson and Osheroff 1991). Furthermore, the preference between single- and double-strand cleavage can be altered if reaction conditions are changed by switching divalent cations, using acidic pH, or adding topo-II-targeting drugs (Gellert et al. 1977; Chen et al. 1984; Osheroff and Zechiedrich 1987; Muller et al. 1988). It has been suggested that double-strand cleavage by topo II may involve two sequential, but coordinated, single-strand cleavage events (Lee et al. 1989a; Zechiedrich et al. 1989).

The DNA cleavage reaction has provided a convenient means to analyze the sequence preference exhibited by topo II during the cleavage/religation reaction (Udvardy et al. 1985, 1986; Yang et al. 1985; Riou et al. 1986). Statistical analysis of cleavage site sequences has produced a cleavage consensus sequence that is characterized by numerous degeneracies (Morrison and Cozzarelli 1979; Lockshon and Morris 1985; Sander and Hsieh 1985; Spitzner and Muller 1988). Nevertheless, it is clear that nucleotide sequence is an important determinant for the topo II cleavage reaction, since analysis of cleavage sites in synthetic oligonucleotide substrates demonstrates that cleavage takes place as predicted from the consensus sequence (Fisher et al. 1986; Lee et al. 1989b; Andersen et al. 1991). This sequence preference in the cleavage/religation reaction is probably the basis of the sequence preference noted for the relaxation of supercoils by topo II (Sander et al. 1987).

An interrupted cycle of the cleavage/religation reaction could lead to DNA strand transfer, a reaction similar to the critical step of illegitimate recombination. Using a sensitive genetic assay, rare recombinant products can be detected from the in vitro strand transfer reaction mediated by several type II topoisomerases, including DNA gyrase (Ikeda and Shiozaki 1984), phage T4 topo II (Ikeda 1986), and the calf thymus enzyme (Bae et al. 1988). Intermolecular ligation events can also be detected using special DNA substrates for which DNA cleavage can occur spontaneously, and an invading DNA fragment can be ligated to the cleavage site (Andersen et al. 1991; Gale and Osheroff 1992).

3. ATP Hydrolysis and Strand Passage

The complete catalytic cycle of type II topoisomerases requires ATP, yet both DNA binding and cleavage/religation by topo II can proceed in its absence. It is likely that ATP is required for the strand passage step, although hydrolysis of the ATP seems not to be necessary at this point as the addition of a nonhydrolyzable ATP analog can support partial cycling of the topo II reaction (Sugino et al. 1978; Liu et al. 1979; Osheroff et al. 1983). Hence, it appears that the binding of ATP or an analog can suffice for the DNA strand passage reaction and that ATP hydrolysis is necessary at some subsequent point for the complete turnover of the catalytic cycle (Peebles et al. 1979; Wang et al. 1981; Osheroff 1986).

Because a preformed complex between an ATP analog and topo II cannot carry out any of the topoisomerization reactions (Roca and Wang 1992), a post-strand passage turnover is not the only role for ATP hydrolysis. There are two ATP-binding sites, and the binding of ATP to topo II appears to be cooperative, suggesting an allosteric interaction between the two sites (Tamura et al. 1992; Lindsley and Wang 1993b). This cooperativity in ATP binding may be the mechanistic means for ATP to regulate the concerted cleavage of two DNA strands by the two topo II subunits and to coordinate strand passage with the cleavage/religation steps. The allosteric transition brought about by ATP binding probably induces topo II to act as a clamp, which can trap a segment of DNA prior to its subsequent translocation through the reversible topo-II-mediated DNA gate (Roca and Wang 1992). The high-resolution crystal structure of a dimer of the amino-terminal domain of GyrB, corresponding to the ATP-binding/hydrolysis region, revealed an opening with a diameter of 20 Å formed between these two fragments, which may be part of the DNA clamp structure (Wigley et al. 1991).

The conformational transition elicited by ATP binding is also manifested by an ATP-induced protease-sensitive site located in the interdomain region between the GyrB and GyrA homologous domains in yeast topo II (Lindsley and Wang 1991). It is possible that the binding of a single ATP molecule to a protomer is sufficient to trigger such a conformational change and to close the clamp with an entrapped DNA segment (Lindsley and Wang 1993a). This is also consistent with the fact that at low ATP concentrations, a single ATP molecule allows completion of one cycle of strand passage (Lindsley and Wang 1993b).

B. Type I Topoisomerases

1. Specificity for Single/Double-stranded DNA

A unique feature of the eubacterial topo I is its specificity for negatively supercoiled DNA in the supercoil relaxation reaction (Wang 1971), possibly due to the enzyme's specificity for single-stranded regions that may be present in a highly negatively supercoiled DNA or that can be readily formed upon binding to the enzyme. Both DNA-binding and enzyme-mediated DNA cleavage assays showed that *E. coli* DNA topo I has a high affinity for single-stranded DNA (Depew et al. 1978). Furthermore, bacterial topo I can relax positively supercoiled DNA when it contains a single-stranded, heteroduplex loop (Kirkegaard and Wang 1985). The second type I DNA topoisomerase in *E. coli*, topo III, also appears to require the presence of single-stranded regions for its activities (Dean et al. 1983; DiGate and Marians 1988).

Eukaryotic topo I, which can readily relax both positively and negatively supercoiled DNAs (Champoux and Dulbecco 1972), has a specificity different from that of the bacterial enzyme in that it can bind to and generate a transient swivel in double-stranded DNA regardless of the extent of supercoiling. An interesting demonstration of this point is that eukaryotic topo I can generate a population of topoisomers from relaxed DNA with a unique linking number (Pulleyblank et al. 1975). Analysis of the enzyme-mediated DNA cleavage reaction indicates that eukaryotic topo I requires a double-stranded region for its activity (Been and Champoux 1984). However, supercoiled DNA is a better substrate than relaxed DNA for the reactions mediated by eukaryotic topo I (Camilloni et al. 1988). It is possible that the preference for a supercoiled molecule is due to the enzyme's affinity for bent DNA loci (Caserta et al. 1989; Krogh et al. 1991) or to a higher affinity for a pair of duplexes in proximity (Zechiedrich and Osheroff 1990). Analysis of the DNA sequences surrounding the topo I cleavage sites indicated that the enzyme may have an affinity for a DNA structure with a particular helical geometry (Shen and Shen 1990).

Whereas the major type I topoisomerase in eukaryotes, topo I, has an apparent specificity for double-stranded DNA, the second type I activity in yeast appears to be single-strand-specific (Kim and Wang 1992). The sequence preference for DNA cleavage by yeast topo III appears to be similar to the preference of *E. coli* topo III. The amino acid sequence of yeast topo III is apparently unrelated to that of yeast topo I, but it displays significant homology with the bacterial topo I and topo III (Wallis et al. 1989).

2. Strand Cleavage and Transfer

For both eubacterial and eukaryotic type I topoisomerases, it has been possible to uncouple the concerted action of cleavage and rejoining so as to mediate either an intra- or intermolecular strand transfer reaction. *E. coli* topo I can spontaneously cleave oligodeoxyribonucleotide homopolymers without the addition of any denaturant (Tse-Dinh et al. 1983), and the resulting cleavage complex is competent for an intermolecular strand transfer reaction to the 3'-hydroxyl end of a linear or nicked circular duplex DNA (Tse-Dinh 1986). Eukaryotic topo I can form the spontaneous cleavage complex with single-stranded DNA substrates and can carry out both intra- and intermolecular strand transfer reactions (Been and Champoux 1981; Halligan et al. 1982; Svejstrup et al. 1991; Shuman 1992). Cleavage by this enzyme presumably occurs within a duplex or a transient duplex loop in single-stranded DNA (Been and Champoux 1984). This cleavage event may result in either fraying or dissociation of the newly generated 5'-hydroxyl end from the site of the nick to produce a spontaneous cleavage product. This topo-I-linked DNA complex is competent for covalent joining to a different 5'-hydroxyl end. These reactions, which occur in the absence of strong denaturants, may be biologically significant, possibly playing a role in promoting genetic recombination, especially the illegitimate recombination involving specific but nonhomologous sequences (Bullock et al. 1985; Shuman 1991).

IV. STRUCTURAL HOMOLOGY AMONG DNA TOPOISOMERASES

On the basis of amino acid sequence homology, topoisomerases can be grouped into three families: eubacterial topo I (e.g., *topA*), eukaryotic topo I (e.g., *TOP1*), and topo II.

In addition to the eubacterial topo I, the first family also includes eubacterial topo III (*topB*) and yeast topo III (*TOP3*). The homology within these enzymes is largely limited to the amino-terminal half, centering around the active site tyrosine residue (DiGate and Marians 1989; Wallis et al. 1989). This residue, Tyr-319 in bacterial topo I (Lynn and Wang 1989), serves as the nucleophile in the transesterification reaction to generate the link between enzyme and DNA. The sequences are more divergent toward the carboxyl terminus, and it is interesting to note that for *E. coli* topo I, the carboxy-terminal quarter of the molecule is dispensable for both its catalytic and biological activities (Zumstein and Wang 1986). Recent sequencing of the archebacterial reverse gyrase, a type I enzyme capable of introducing positive DNA supercoils in an ATP-coupled reaction (Kikuchi and Asai 1984), indicates that it is a member of the *topA* family (Confalonieri et al. 1993). Reverse gyrase is bipartite in its structural domains: the carboxy-terminal half is homologous to eubacterial topo I and the amino terminus contains sequence motifs that are the signature of DNA helicases (Confalonieri et al. 1993). This structural information has important mechanistic implications. The positive supercoiling reaction could be a consequence of the coupled action of the helicase tracking along the DNA duplex and relaxation by topo I of the negative supercoils generated in the wake of the helicase fork, thus leaving behind positive supercoils (Liu and Wang 1987; Confalonieri et al. 1993).

Eukaryotic topo I share extensive homology among themselves and form a family distinct from the eubacterial topo I. On the basis of amino acid sequence comparisons, eukaryotic topo I can be divided into several domains (Fig. 1A; for a detailed description of the boundaries of these domains, see Hsieh et al. 1993). Neither the amino acid sequence nor their length is conserved in the nonhomologous domain at the amino terminus or in another domain near the carboxyl terminus. However, these sequences are characterized by an abundance of charged amino acid residues, especially at the amino terminus. Both a functional nuclear localization signal (Alsner et al. 1992) and a phosphate acceptor site (Durban et al. 1985) are located in the amino end of the molecule. Although this domain of eukaryotic topo I is more sensitive to proteolysis and is dispensable for in vitro catalytic activity (Liu and Miller 1981; Bjornsti and Wang 1987; D'Arpa et al. 1988; Alsner et al. 1992), it may serve critical roles for its in vivo functions such as nuclear targeting and the regulation of topoisomerase activities.

Genetic and biochemical data also suggest functions associated with the homologous domains in eukaryotic topo I. The active site tyrosine that initiates the transesterification step during the catalytic cycle has been mapped to the carboxy-proximal homologous domain (Eng et al. 1989; Lynn et al. 1989), suggesting that this region is involved in DNA breakage/rejoining. A D533G missense mutation was identified in the highly conserved GKDSI motif in a camptothecin-resistant human topo I gene (Tamura et al. 1991), implicating this domain in the interaction between the enzyme and the antitumor drug camptothecin, which targets topo I. This notion is also supported by the isolation of a vaccinia topo I mutation that converts the wild-type, drug-resistant enzyme to a sensitive form (Gupta et al. 1992). The mutant enzyme contains a mutation of

A. Eucaryotic topoisomerase I

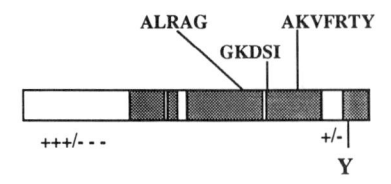

```
           ALRAG       AKVFRTY
                 GKDSI
    ┌────┬──┬──┬─────┬──┬──┐
    │    │  │  │     │  │  │
    └────┴──┴──┴─────┴──┴──┘
     +++/---            +/-
                         Y
```

B. Type II DNA topoisomerase

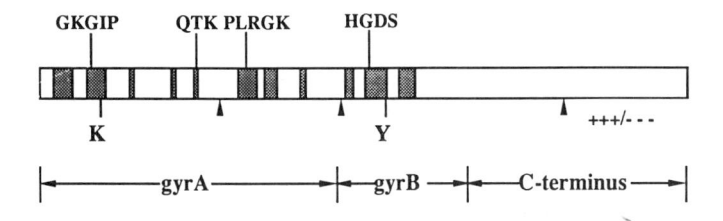

```
GKGIP    QTK PLRGK   HGDS
┌┬──┬─┬─┬─┬──┬──┬─┬──┬──────────────┐
│││  │ │ │ │  │  │ │  │              │
└┴──┴─┴─┴─┴──┴──┴─┴──┴──────────────┘
  K       ▲    ▲  Y       ▲    +++/---
 |←────gyrA────→|←gyrB→|←─C-terminus─→|
```

0 100 200 300 400 500 aa

Figure 1 Diagrammatical representation of structural homologies in eukaryotic topoisomerase I (*A*) and type II DNA topoisomerase (*B*). (*Shaded areas*) Sequence blocks with extensive homology; (*open areas*) sequence blocks with no homology. Arrowheads denote the mapped proteolytic-sensitive sites and +/- marks the charged and hydrophilic sequences. The amino acid residues marked underneath the homology diagram indicate those that are critical for the biochemical functions, whereas the residues shown above locate the landmark motifs that are discussed in the text. Homology comparison for type II enzymes includes the sequences from bacteria, bacteriophage, and eukaryotes, and thus their extent of homology is significantly less than that for eukaryotic topoisomerase I.

aspartate to valine in the homology motif AKDFRTYNAS (Gupta et al. 1992). However, since there are reports for camptothecin-resistant mutations mapped in the other regions of eukaryotic topo I (Kubota et al. 1992; Caron and Wang 1993), it is likely that determinants for camptothecin sensitivity are distributed over several regions of the linear sequence of topo I.

A recently isolated mutation in yeast topo I results in an enzyme with an enhanced level of DNA cleavage activities in the absence of camptothecin and elicits a number of interesting biological effects in yeast cells, including hyperrecombination and induction of DNA-repair genes (Levin et al. 1993). Analysis of this mutation, a change of arginine to lysine in the motif ALRAG, suggests that the biochemical function of coordinating the topo I cleavage/rejoining equilibrium resides in this domain (Levin et al. 1993).

Type II DNA topoisomerases of both eukaryotic and prokaryotic origins form a third family of homologous enzymes, which includes both topo II of the same organism, e.g., DNA gyrase and topo IV of eubacteria, and the vertebrate topo IIα and IIβ. In comparison to its bacterial counterpart, the structure of eukaryotic topo II is characterized by two features. It is encoded by a single gene that has extensive homologies with the twin subunits in bacterial DNA gyrase, and it has a unique carboxyl tail that is extremely hydrophilic and charged (Fig. 1B) (for review, see Caron and Wang 1993). The hydrophilic carboxyl terminus is dispensable for in vitro catalytic activity; however, it may confer some critical in vivo functions (Caron and Wang 1993; Crenshaw and Hsieh 1993; see also Shiozaki and Yanagida 1992). This domain is also exquisitely sensitive to proteolysis, suggesting that it has an extended and exposed structure (Lindsley and Wang 1991; Shiozaki and Yanagida 1991).

The carboxyl terminus of the bacterial GyrA subunit, which is less charged than its eukaryotic counterpart and is separated by a proteolytic-sensitive site from the amino-terminal half of GyrA, also appears to be dispensable for part of the gyrase activity (Reece and Maxwell 1991). Other protease-sensitive sites have been mapped and may delineate domain junctions as well. There is a protease-sensitive site for eukaryotic topo II at the junction between the GyrB and GyrA homology domains (Lindsley and Wang 1991; Shiozaki and Yanagida 1991). A protease-sensitive site mapped in the GyrB domain is conserved between DNA gyrase (Brown et al. 1979; Gellert et al. 1979; Adachi et al. 1987) and eukaryotic topo II (Lindsley and Wang 1991; Shiozaki and Yanagida 1991; Lee and Hsieh 1993). These partial proteolysis data suggest that some aspects of the overall domain structure are conserved in topo II across the phylogenies (Lee and Hsieh 1993).

Some of the conserved sequence motifs are known to be associated with certain aspects of the biochemical functions of topo II. Located in the GyrA domain is the active site tyrosine for covalent linkage to DNA during the breakage/religation cycle (Horowitz and Wang 1987; Worland and Wang 1989). Cross-linking with an ATP analog and site-specific mutagenesis data have established that a glycine-rich region (marked by GKGIP in Fig. 1B) near the amino terminus of the GyrB domain is part of the ATP-binding site (Tamura and Gellert 1990; Lindsley and Wang 1993b). Crystallographic data of the amino-terminal half of the GyrB protein confirm the role of this glycine-rich loop in ATP binding and implicate another sequence motif (QTK) in the middle of GyrB in the binding of the γ-phosphate of ATP (Wigley et al. 1991). Mapping data from some of the drug-resistant gyrase and eukaryotic topo II mutants were used to identify regions in the topo II structure that are involved in the catalytic function of DNA cleavage/religation (see Caron and Wang 1993).

V. BIOLOGICAL FUNCTIONS OF DNA TOPOISOMERASES

The biochemical properties of DNA topoisomerases suggest that they have critical roles in many, if not all, aspects of DNA metabolism and chromosome structure/function. The available genetic data from both eubacteria and yeast confirm this notion in that all of the known type II enzymes are absolutely required for the growth and replication of these organisms. Topo II in both fission and budding yeasts is involved in premitotic chromosome condensation (Uemura et al. 1987) and in the segregation of replicated chromosomes during both mitosis (DiNardo et al. 1984; Holm et al. 1985; Uemura and Yanagida 1986) and meiosis (Rose et al. 1990). However, topo II may have other functions outside of the mitotic phase that could be shared with topo I, because although the lethality of a *top2⁻* single mutant occurs uniquely at M phase, yeast cells that are *top1⁻* and *top2^{ts}* are arrested immediately at any point in the cell cycle if both topoisomerase activities are inactivated (Uemura and Yanagida 1984; Goto and Wang 1985; Brill et al. 1987; also for discussion, see Yanagida and Sternglanz 1990). It is plausible that at least one topoisomerase activity must be present during S phase to relieve the torsional stress generated during transcription and replication.

In *E. coli*, there is also an intimate interaction between topo I (ω protein) and topo II (DNA gyrase). Eubacterial topo I is not essential, but cells lacking topo I activity grow slowly and usually acquire compensatory mutations, some of which map to the gyrase loci (DiNardo et al. 1982; Pruss et al. 1982; Raji et al. 1985). DNA gyrase and topo I have diametrically opposed enzymatic activities with respect to DNA supercoiling, and it is necessary to have both activities present to remove the torsional stresses efficiently within the twin supercoiling domains produced by any DNA-tracking process (Liu and Wang 1987 and references therein). Similar supercoiling domains induced by transcription or other

DNA-tracking processes are likely to be present in eukaryotic cells as well (Giaever and Wang 1988).

In marked contrast to the essential nature of topo II in yeast, neither of the type I DNA topoisomerases, topo I and III, is essential in these organisms (Uemura and Yanagida 1984; Goto and Wang 1985; Thrash et al. 1985; Wallis et al. 1989). However, it is apparent that both enzymes are needed for normal growth of yeast since a mutation in either of them results in a slow-growth phenotype (Wallis et al. 1989; Yanagida and Sternglanz 1990). There are several functions attributed to type I DNA topoisomerases. Upon their discovery in bacterial cells (Wang 1971) and mammalian cells (Champoux and Dulbecco 1972), it was suggested that these enzymes serve as a DNA swivel during replication. The functional requirement of topo I as a swivel has been demonstrated in an in vitro DNA replication system (Yang et al. 1987), and analysis of the synthesis and processing of nascent DNA chains in yeast cells also revealed a functional role of topo I in the chain elongation step of DNA synthesis (Kim and Wang 1989a).

Transcription also requires DNA swivels, and biochemical and cytological data demonstrate the physical association of topo I with rDNA loci (Higashinakagawa et al. 1977; Fleischmann et al. 1984; Muller et al. 1985; Zhang et al. 1988) and other actively transcribing loci (Fleischmann et al. 1984; Gilmour et al. 1986; Gilmour and Elgin 1987). Another intriguing biological function associated with topo I is the suppression of DNA recombination. Yeast cells harboring a *top1* mutation or a temperature-sensitive *top2* mutation and grown under semipermissive conditions exhibit a 50- to 200-fold enhancement in mitotic recombination among the rDNA repeats (Christman et al. 1988). An observation likely to be related to this phenomenon is that in yeast cells that are *top1*$^-$ and *top2*^{ts}, grown at the permissive temperature, more than half of the rDNA repeats are present as extrachromosomal rings (Kim and Wang 1989b). Yeast topo III, which was originally identified by its effect on recombination between repeated sequences of the Ty δ element (Wallis et al. 1989), suppresses recombination between homologous sequences including the rDNA locus (Bailis et al. 1992). Topo III may be involved in gene conversion events as well, since the conversion tracts in *top3*$^-$ yeast are anomalous and are longer than those that occur in wild-type cells (Bailis et al. 1992). One plausible interpretation for the effect of topo I mutants on genetic recombination is that a type I topoisomerase is required for unlinking synapsed DNA molecules (Wang et al. 1990), a reaction very similar to the segregation of replicated plasmid molecules containing single-stranded gaps by *E. coli* topo III (DiGate and Marians 1988). As addressed earlier, both type I topoisomerases (topo I and topo

III) in eubacteria have effects on the level of genetic recombination (Fishel and Kolodner 1984; Schofield et al. 1992).

A genetic system has recently been established to analyze the functions of topo I in *Drosophila* (Lee et al. 1993). The cloned gene was used to identify the cytogenetic location of topo I, and mutants with deficiencies in *top1* were then isolated. Analysis of these mutants indicates that topo I has an essential zygotic function for growth and development in the fruit fly and has a maternal function as evidenced by the maternal storage of *top1* transcripts and protein in early embryos. The developmental regulation of topo I expression suggests a critical function of topo I in DNA replication (Lee et al. 1993). In mammalian tissue-culture cells, the expression of topo I is also linked to cell proliferation and DNA replication (Hwong et al. 1989; Romig and Richter 1990).

VI. TARGETING BY ANTIMICROBIAL AND ANTICANCER DRUGS

A major driving force for research on DNA topoisomerases is the discovery that these enzymes are the intracellular pharmacological targets of some clinically important antitumor and antibiotic drugs (for recent reviews, see Andoh et al. 1993; Liu 1993). Except for eubacterial topo I, these enzymes serve as targets for numerous cytotoxic agents. The two subunits of bacterial DNA gyrase, GyrA and GyrB, are targeted by antibiotics of the quinolone and coumarin families, respectively (for review, see Reece and Maxwell 1991). The coumarin drugs, including novobiocin and coumermycin, inhibit supercoiling and ATPase activities (Gellert et al. 1976b; Sugino et al. 1978). The cytotoxicity of the quinolone drugs appears to be mediated differently from that of the coumarins in that the effect of the quinolone is not due only to inhibition of DNA gyrase. Mutations exist in *gyrA* that can result in resistance to quinolone drugs like nalidixate, but the resistance phenotype is recessive to the sensitive phenotype when both resistant and wild-type *gyrA* genes are present in merodiploid cells (Hane and Wood 1969). The concentration of the quinolone drug necessary to stop bacterial growth is much lower than that needed to inhibit gyrase supercoiling in vitro (Gellert et al. 1977). The quinolone drugs have the unique property of being able to promote the formation of a gyrase-mediated cleavable DNA complex (Gellert et al. 1977; Sugino et al. 1977). These data, plus the observation that bacterial cells treated with quinolone drugs exhibit different phenotypes in comparison to cells containing thermolabile DNA gyrase, led to the hypothesis that in the presence of these drugs, gyrase was converted into a cytotoxic agent (Kreuzer and Cozzarelli 1979). In addition to the trapping of a gyrase/DNA cleavable complex, the bactericidal action of quinolone drugs also invokes the SOS response and cell-division control pathways (for review, see Liu 1989).

Most of the antitumor drugs targeting either eukaryotic topo I or II exert their cytotoxic action through trapping the cleavable DNA complex and possibly induction of the DNA-repair response, a course of action very similar to that proposed for the quinolone drugs in prokaryotes. Both drugs with potent DNA intercalating activity, such as acridines and anthracyclines (Nelson et al. 1984), and those that do not appear to intercalate into DNA, such as epipodophyllotoxins (Chen et al. 1984), can promote topo-II-mediated DNA cleavage in vitro and in vivo (for reviews, see Pommier and Kohn 1989; Beck and Danks 1991; Osheroff et al. 1991). A characteristic feature of DNA cleavage reactions by all topoisomerases is their apparent reversibility; after formation of the cleavable complex, reversal can be readily accomplished by changing the reaction conditions by, for example, chelating the divalent cations, adding high salt, or heating or cooling the reaction mixtures (for review, see Osheroff 1989; Hsieh 1990). However, DNA cleavage promoted by some of the topo-II-targeting drugs cannot be completely reversed, implicating a possible mechanism for the irreversible cytotoxicity of these drugs (Lee and Hsieh 1992). There is also a group of potent antitumor agents, with diverse chemical structures, that target eukaryotic topo I and enhance enzyme-induced DNA cleavages. These drugs include compounds without notable DNA-binding affinity such as camptothecin (Hsiang et al. 1985), DNA intercalators such as actinomycin D (Trask and Muller 1988), and DNA minor-groove-binding ligands, including bisbenzimidazole and the related Hoechst dye series (Chen et al. 1993a,b).

Although a large body of biochemical, pharmacological, and cytological evidence strongly indicates that topoisomerases are the critical targets for the cell-killing action of these antitumor drugs, the most rigorous demonstration comes from yeast genetics. Yeast with a null mutation in *top1* is resistant to camptothecin, and the presence of a functional topo I, from either a homologous or heterologous gene, is requisite for camptothecin sensitivity (Eng et al. 1988; Nitiss and Wang 1988; Bjornsti et al. 1989). Since topo II is essential for the viability of yeast cells, a less direct approach has been taken to identify the target of topo II drugs. First, overexpression of topo II in yeast results in hypersensitivity to amsacrine and etoposide (Nitiss et al. 1992). Second, for *top2*^{ts} yeast cells growing at a semipermissive temperature where yeast can survive on a limiting amount of topo II activity, there is a marked enhancement in resistance to amsacrine and etoposide (Nitiss et al. 1993). A correlation also exists between drug resistance and a diminished level of topo II in mammalian tissue-culture cells (Beck et al. 1987; Potmesil et al. 1988). Besides the use of yeast as a model system for drug-targeting studies, bacteriophage T4 has been applied to analyze actions of drugs targeting topo II (Huff et al. 1989).

A group of new drugs can inhibit the function of topoisomerases at points in the mechanistic pathway other than cleavage and rejoining. Not only are these drugs useful reagents for investigating the mechanistic aspects of the enzyme, but they can potentially complement the other topoisomerase drugs in cancer chemotherapy. For instance, both merbarone (Drake et al. 1989) and bis-(2,6-dioxopiperazine) derivatives such as ICRF-193, -154, and -159 (Tanabe et al. 1991) can inhibit the catalytic activities of topo II and can furthermore abolish the stimulation of DNA cleavage by other topo II drugs. The target of these agents is presumably at a point prior to the cleavage/rejoining of DNA, although the details remain to be defined biochemically.

VII. CONCLUDING REMARKS

Studies of the mechanism of action of DNA topoisomerases have resulted in a large body of information about their biochemical and biophysical properties. The application of cell biological and genetic techniques has demonstrated that these enzymes are essential for many aspects of DNA metabolism and chromosome function. The discovery that clinically important antibiotic and antineoplastic drugs target DNA topoisomerases further heightens interest in understanding these enzymes. The molecular tools available to dissect their structure-function relationships should facilitate our progress in studying these enzymes in the future.

ACKNOWLEDGMENTS

I am grateful to my colleagues who generously provided preprints, reprints, and stimulating discussions. The work carried out in my laboratory is supported by a grant from the National Institutes of Health (GM-29006).

REFERENCES

Adachi, T., M. Mizuuchi, E.A. Robinson, E. Appella, M.H. O'Dea, M. Gellert, and K. Mizuuchi. 1987. DNA sequence of the *E. coli gyrB* gene: Application of a new sequencing strategy. *Nucleic Acids Res.* **15:** 771–783.

Adams, D., E.M. Shektman, E.L. Zechiedrich, M.B. Schmid, and N.R. Cozzarelli. 1992.

The role of topoisomerase IV in partitioning bacterial replicons and the structure of catenated intermediates in DNA replication. *Cell* **71:** 277-288.

Alsner, J., J.Q. Svejstrup, E. Kjeldsen, B.S. Soerensen, and O. Westergaard. 1992. Identification of an N-terminal domain of eucaryotic DNA topoisomerase I dispensable for catalytic activity but essential for *in vivo* function. *J. Biol. Chem.* **267:** 12408-12411.

Andersen, A.H., B.S. Soerensen, K. Christiansen, J.Q. Svesjstrup, K. Lund, and O. Westergaard. 1991. Studies of the topoisomerase II-mediated cleavage and religation reactions by use of a suicidal double-stranded DNA substrate. *J. Biol. Chem.* **266:** 9203-9210.

Andoh, T., H. Ikeda, and M. Oguro, eds. 1993. *Molecular biology of DNA topoisomerases and its application to chemotherapy*. CRC Press, Boca Raton, Florida.

Bae, Y.S., I. Kawasaki, H. Ikeda, and L.F. Liu. 1988. Illegitimate recombination mediated by calf thymus DNA topoisomerase II *in vitro*. *Proc. Natl. Acad. Sci.* **85:** 2076-2080.

Bailis, A.M., L. Arthur, and R. Rothstein. 1992. Genome rearrangement in *top3* mutants of *Saccharomyces cerevisiae* requires a functional RAD1 excision repair gene. *Mol. Cell. Biol.* **12:** 4988-4993.

Beck, W.T. and M.K. Danks. 1991. Mechanism of resistance to drugs that inhibit DNA topoisomerases. *Semin. Cancer Biol.* **2:** 235-244.

Beck, W.T., M.C. Cirtain, M.K. Danks, R.L. Felsted, A.R. Safa, J.S. Wolverton, D.P. Suttle, and J.M. Trent. 1987. Pharmacological, molecular, and cytogenetic analysis of "atypical" multidrug-resistant human leukemia cells. *Cancer Res.* **47:** 5455-5460.

Been, M.D. and J.J. Champoux. 1981. DNA breakage and closure by rat liver type I topoisomerase: Separation of the half-reactions by using a single-stranded DNA substrate. *Proc. Natl. Acad. Sci.* **78:** 2883-2887.

———. 1984. Breakage of single-stranded DNA by eukaryotic type I topoisomerase occurs only at regions with the potential for base-pairing. *J. Mol. Biol.* **180:** 515-531.

Bjornsti, M.-A. and J.C. Wang. 1987. Expression of yeast DNA topoisomerase I can complement a conditional-lethal DNA topoisomerase I mutation in *Escherichia coli*. *Proc. Natl. Acad. Sci.* **84:** 9871-9875.

Bjornsti, M.-A., P. Benedetti, G.A. Viglianti, and J.C. Wang. 1989. Expression of human DNA topoisomerase I in yeast cells lacking yeast DNA topoisomerase I: Restoration of sensitivity of the cells to the antitumor drug camptothecin. *Cancer Res.* **49:** 6318-6323.

Brill, S.J., S. DiNardo, K. Voelkel-Meiman, and R. Sternglanz. 1987. Need for DNA topoisomerase activity as a swivel for DNA replication and for transcription of ribosomal RNA. *Nature* **326:** 414-416.

Brown, P.O., C.L. Peebles, and N.R. Cozzarelli. 1979. A topoisomerase from *Escherichia coli* related to DNA gyrase. *Proc. Natl. Acad. Sci.* **76:** 6110-6114.

Bullock, P., J.J. Champoux, and M. Botchan. 1985. Association of crossover points with topoisomerase I cleavage sites: A model for nonhomologous recombination. *Science* **230:** 954-958.

Camilloni, G., E. Di Mautino, M. Caserta, and E. Di Mauro. 1988. Eukaryotic DNA topoisomerase I reaction is topology dependent. *Nucleic Acids Res.* **16:** 7071-7085.

Caron, P.R. and J.C. Wang. 1993. DNA topoisomerase as targets of therapeutics: A structural overview. In *Molecular biology of DNA topoisomerases and its application to chemotherapy* (ed. T. Andoh et al.), pp. 1-18. CRC Press, Boca Raton, Florida.

Caserta, M., A. Amadei, E. Di Mauro, and G. Camilloni. 1989. *In vitro* preferential topoisomerization of bent DNA. *Nucleic Acids Res.* **17:** 8463-8474.

Champoux, J.J. and R. Dulbecco. 1972. An activity from mammalian cells that untwists superhelical DNA-A possible swivel for DNA replication. *Proc. Natl. Acad. Sci.* **69:**

143-146.
Chen, A., C. Yu, B. Gatto, and L.F. Liu. 1993a. DNA minor groove binding ligands: A new class of mammalian DNA topoisomerase I inhibitors. *Proc. Natl. Acad. Sci.* **90:** 8131-8135.
Chen, A., C. Yu, A. Bodley, L.F. Peng, and L.F. Liu. 1993b. A new mammalian DNA topoisomerase I poison Hoechst 33342: Cytotoxicity and drug resistance in human cell cultures. *Cancer Res.* **53:** 1332-1336.
Chen, G.L., L. Yang, T.C. Rowe, B.D. Halligan, K.L. Tewey, and L.F. Liu. 1984. Nonintercalative antitumor drugs interfere with the breakage-reunion reaction of mammalian DNA topoisomerase II. *J. Biol. Chem.* **259:** 13560-13566.
Christman, M.F., F.S. Dietrich, and G.R. Fink. 1988. Mitotic recombination in the rDNA of *S. cerevisiae* is suppressed by the combined action of DNA topoisomerases I and II. *Cell* **55:** 413-425.
Chung, I.K., V.B. Mehta, J.R. Spitzner, and M.T. Muller. 1992. Eucaryotic topoisomerase II cleavage of parallel stranded DNA tetraplexes. *Nucleic Acids Res.* **20:** 1973-1977.
Chung, T.D.Y., F.H. Drake, K.B. Tan, S.R. Per, S.T.R. Crooke, and C.K. Mirabelli. 1989. Characterization and immunological identification of cDNA clones encoding two human DNA topoisomerase II isozymes. *Proc. Natl. Acad. Sci.* **86:** 9431-9435.
Confalonieri, F., C. Elie, M. Nadal, C. Bou de la Tour, P. Forterre, and M. Duguet. 1993. Reverse gyrase: A helicase-like domain and a type I topoisomerase in the same polypeptide. *Proc. Natl. Acad. Sci.* **90:** 4753-4757.
Cozzarelli, N.R. and J.C. Wang, eds. 1990. *DNA topology and its biological effects.* Cold Spring Harbor Laboratory Press, Cold Spring Harbor, New York.
Crenshaw, D.G. and T. Hsieh. 1993. Function of hydrophilic carboxyl terminus of type II DNA topoisomerase from *Drosophila melanogaster*. II. *In vivo* studies. *J. Biol. Chem.* **268:** 21328-21343.
D'Arpa, P., P.S. Machlin, H. Ratrie III, N.F. Rothfield, D.W. Cleveland, and W.C. Earnshaw. 1988. cDNA cloning of human DNA topoisomerase I: Catalytic activity of a 67.7-kDA carboxyl-terminal fragment. *Proc. Natl. Acad. Sci.* **85:** 2543-2547.
Dean, F., M.A. Krasnow, R. Otter, M.M. Matzuk, S.J. Spengler, and N.R. Cozzarelli. 1983. *Escherichia coli* type-I topoisomerases: Identification, mechanism, and role in recombination. *Cold Spring Harbor Symp. Quant. Biol.* **47:** 769-777.
Depew, R.E., L.F. Liu, and J.C. Wang. 1978. Interaction between DNA and *Escherichia coli* protein ω. Formation of a complex between single-stranded DNA and ω protein. *J. Biol. Chem.* **253:** 511-518.
DiGate, R.J. and K.J. Marians. 1988. Identification of a potent decatenating enzyme from *Escherichia coli*. *J. Biol. Chem.* **263:** 13366-13373.
―――. 1989. Molecular cloning and DNA sequence analysis of *Escherichia coli topB*, the gene encoding topoisomerase III. *J. Biol. Chem.* **264:** 17924-17930.
DiNardo, S., K. Voelkel, and R. Sternglanz. 1984. DNA topoisomerase II mutant of *Saccharomyces cerevisiae*: Topoisomerase II is required for segregation of daughter molecules at the termination of DNA replication. *Proc. Natl. Acad. Sci.* **81:** 2616-2620.
DiNardo, S., K.A. Voelkel, R. Sternglanz, A.E. Reynolds, and A. Wright. 1982. *Escherichia coli* DNA topoisomerase I mutants have compensatory mutations in DNA gyrase genes. *Cell* **31:** 43-51.
Drake, F.H., G.A. Hofman, S.-M. Mong, J.O. Bartus, R.P. Hertzberg, R.K. Johnson, M.R. Mattern, and C.K. Mirabelli. 1989. *In vitro* and intracellular inhibition of topoisomerase II by the antitumor agent merbarone. *Cancer Res.* **49:** 2578-2583.

Durban, E., M. Goodenough, J. Mills, and H. Busch. 1985. Topoisomerase I phosphorylation *in vitro* and in rapidly growing Novikoff hepatoma cells. *EMBO J.* **4:** 2921-2926.

Eng, W.-K., S.D. Pandit, and R. Sternglanz. 1989. Mapping of the active site tyrosine of eukaryotic DNA topoisomerase I. *J. Biol. Chem.* **264:** 13373-13376.

Eng, W.-K., L. Faucette, R.K. Johnson, and R. Sternglanz. 1988. Evidence that DNA topoisomerase I is necessary for the cytotoxic effects of camptothecin. *Mol. Pharmacol.* **34:** 755-760.

Fishel, R.A. and R. Kolodner. 1984. *Escherichia coli* strains containing mutations in the structural gene for topoisomerase I are recombination deficient. *J. Bacteriol.* **160:** 1168-1170.

Fisher, L.M., H.A. Barot, and M.E. Cullen. 1986. DNA gyrase complex with DNA: Determinants for site-specific DNA breakage. *EMBO J.* **5:** 1411-1418.

Fisher, L.M., K. Mizuuchi, M.H. O'Dea, H. Ohmori, and M. Gellert. 1981. Site-specific interaction of DNA gyrase with DNA. *Proc. Natl. Acad. Sci.* **78:** 4165-4169.

Fleischmann, G., G. Pflugfelder, E.K. Steiner, K. Javaherian, G.C. Howard, J.C. Wang, and S.C.R. Elgin. 1984. *Drosophila* DNA topoisomerase I is associated with transcriptionally active regions of the genome. *Proc. Natl. Acad. Sci.* **81:** 6958-6962.

Gale, K.C. and N. Osheroff. 1992. Intrinsic intermolecular DNA ligation activity of eucaryotic topoisomerase II. Potential roles in recombination. *J. Biol. Chem.* **267:** 12090-12097.

Gellert, M., L.M. Fisher, and M.H. O'Dea. 1979. DNA gyrase: Purification and catalytic properties of a fragment of gyrase B protein. *Proc. Natl. Acad. Sci.* **76:** 6289-6293.

Gellert, M., K. Mizuuchi, M.H. O'Dea, and H.A. Nash. 1976a. DNA gyrase: An enzyme that introduces superhelical turns into DNA. *Proc. Natl. Acad. Sci.* **73:** 3872-3876.

Gellert, M., M.H. O'Dea, T. Itoh, and J.I. Tomizawa. 1976b. Novobiocin and coumermycin inhibit DNA supercoiling catalyzed by DNA gyrase. *Proc. Natl. Acad. Sci.* **73:** 4474-4478.

Gellert, M., K. Mizuuchi, M.H. O'Dea, T. Itoh, and J.I. Tomizawa. 1977. Nalidixic acid resistance: A second genetic character involved in DNA gyrase activity. *Proc. Natl. Acad. Sci.* **74:** 4772-4776.

Giaever, G.N. and J.C. Wang. 1988. Supercoiling of intracellular DNA can occur in eukaryotic cells. *Cell* **55:** 849-856.

Gilmour, D.S. and S.C.R. Elgin. 1987. Localization of specific topoisomerase I interactions within the transcribed region of active heat shock genes using the inhibitor camptothecin. *Mol. Cell. Biol.* **7:** 141-148.

Gilmour, D.S., G. Pflugfelder, J.C. Wang, and J.T. Lis. 1986. Topoisomerase I interacts with transcribed regions in *Drosophila* cells. *Cell* **44:** 401-407.

Glikin, G.C., T.M. Jovin, and D.J. Arndt-Jovin. 1991. Interactions of *Drosophila* DNA topoisomerase II with left-handed Z-DNA in supercoiled minicircles. *Nucleic Acids Res.* **19:** 7139-7144.

Goto, T. and J.C. Wang. 1985. Cloning of yeast *TOP1*, the gene encoding DNA topoisomerase I, and construction of mutants defective in both DNA topoisomerase I and DNA topoisomerase II. *Proc. Natl. Acad. Sci.* **82:** 7178-7182.

Gupta, M., C.-X. Zhu, and Y.-C. Tse-Dinh. 1992. An engineered mutant of vaccinia virus DNA topoisomerase I is sensitive to the anti-cancer drug camptothecin. *J. Biol. Chem.* **267:** 24177-24180.

Halligan, B., J. Davis, K. Edwards, and L. Liu. 1982. Intra- and intermolecular strand transfer by HeLa DNA topoisomerase I. *J. Biol. Chem.* **257:** 3995-4000.

Hane, M.W. and T.H. Wood. 1969. *Escherichia coli* K-12 mutants resistant to nalidixic acid: Genetic mapping and dominance studies. *J. Bacteriol.* **99:** 238-241.

Higashinakagawa, T., H. Wahn, and R.H. Reeder. 1977. Isolation of ribosomal gene chromatin. *Dev. Biol.* **55:** 375-386.

Hirano, T. and T.J. Mitchison. 1993. Topoisomerase II does not play a scaffolding role in the organization of mitotic chromosomes assembled in *Xenopus* egg extracts. *J. Cell Biol.* **120:** 601-612.

Holm, C., T. Goto, J.C. Wang, and D. Botstein. 1985. DNA topoisomerase II is required at the time of mitosis in yeast. *Cell* **41:** 553-563.

Horowitz, D.S. and J.C. Wang. 1987. Mapping the active site tyrosine of *Escherichia coli* DNA gyrase. *J. Biol. Chem.* **262:** 5339-5344.

Howard, M.T., M.P. Lee, T. Hsieh, and J.D. Griffith. 1991. *Drosophila* topoisomerase II-DNA interactions are affected by DNA structure. *J. Mol. Biol.* **217:** 53-62.

Hsiang, Y.-H., R. Hertzberg, S. Hecht, and L.F. Liu. 1985. Camptothecin induces protein-linked DNA breaks via mammalian DNA topoisomerase I. *J. Biol. Chem.* **260:** 14873-14878.

Hsieh, T. 1990. Mechanistic aspects of type-II topoisomerases. In *DNA topology and its biological effects* (ed. N.R. Cozzarelli and J.C. Wang), pp. 243-263, Cold Spring Harbor Laboratory Press, Cold Spring Harbor, New York.

———. 1992. DNA topoisomerases. *Curr. Opin. Cell Biol.* **4:** 396-400.

Hsieh, T., M.P. Lee, and S.D. Brown. 1993. Structure of eucaryotic type I DNA topoisomerase. *Adv. Pharmacol.* (in press).

Huff, A., J. Leatherwood, and K. Kreuzer. 1989. Bacteriophage T4 DNA topoisomerase is the target of antitumor agent 4'-(9-acridinylamino)methanesulfon-m-aniside(mAMSA) in T4-infected *Escherichia coli*. *Proc. Natl. Acad. Sci.* **86:** 1307-1311.

Hwong, C.-L., M.-S. Chen, and J. Hwang. 1989. Phorbol ester transiently increases topoisomerase I mRNA levels in human skin fibroblasts. *J. Biol. Chem.* **264:** 14923-14926.

Ikeda, H. 1986. Bacteriophage T4 DNA topoisomerase mediates illegitimate recombination *in vitro*. *Proc. Natl. Acad. Sci.* **83:** 922-926.

Ikeda, H. and M. Shiozaki. 1984. Nonhomologous recombination mediated by *Escherichia coli* DNA gyrase: Possible involvement of DNA replication. *Cold Spring Harbor Symp. Quant. Biol.* **49:** 401-409.

Jenkins, J.R., P. Ayton, T. Jones, S.L. Davies, D.L. Simmons, A.L. Harris, D. Sheer, and I.D. Hickson. 1992. Isolation of cDNA clones encoding the β isozyme of human DNA topoisomerase II and localisation of the gene to chromosome 3p24. *Nucleic Acids Res.* **20:** 5587-5592.

Kato, J.-I., Y. Nishimura, and M.Y. Hirota. 1988. Gene organization in a region containing a new gene involved in chromosome partition in *Escherichia coli*. *J. Bacteriol.* **170:** 3967-3977.

Kato, J.-I., H. Suzuki, and H. Ikeda. 1992. Purification and characterization of DNA topoisomerase IV in *Escherichia coli*. *J. Biol. Chem.* **267:** 25676-25684.

Kato, J.-I., Y. Nishimura, R. Imamura, H. Niki, S. Hiraga, and H. Suzuki. 1990. New topoisomerase essential for chromosome segregation in *E. coli*. *Cell* **63:** 393-404.

Kikuchi, A. and K. Asai. 1984. Reverse gyrase—A topoisomerase which introduces positive helical turns into DNA. *Nature* **309:** 677-681.

Kim, R.A. and J.C. Wang. 1989a. Function of DNA topoisomerases as replication swivels in *Saccharomyces cerevisiae*. *J. Mol. Biol.* **208:** 257-267.

———. 1989b. A subthreshold level of DNA topoisomerases leads to the excision of yeast rDNA as extrachromosomal rings. *Cell* **57:** 975-985.

———. 1992. Identification of the yeast *TOP3* gene product as a single strand-specific DNA topoisomerase. *J. Biol. Chem.* **267:** 17178-17185.
Kirkegaard, K. and J.C. Wang. 1981. Mapping the topography of DNA wrapped around gyrase by nucleolytic and chemical probing of complexes of unique DNA sequences. *Cell* **23:** 721-729.
———. 1985. Bacterial DNA topoisomerase I can relax positively supercoiled DNA containing a single-stranded loop. *J. Mol. Biol.* **185:** 625-637.
Kirschhausen, T., J.C. Wang, and S.C. Harrison. 1985. DNA gyrase and its complexes with DNA: Direct observation by electron microscopy. *Cell* **41:** 933-943.
Kreuzer, K.N. and N.R. Cozzarelli. 1979. *Escherichia coli* mutants thermosensitive for DNA gyrase subunit A: Effects on DNA replication, transcription, and bacteriophage growth. *J. Bacteriol.* **140:** 425-430.
Kreuzer, K. and W.M. Huang. 1983. T4 DNA topoisomerase. In *Bacteriophage T4* (ed. C.K. Mathews et al.), pp. 90-96. American Society for Microbiology, Washington, D.C.
Krogh, S., U. Mortensen, O. Westergaard, and B.J. Bonven. 1991. Eucaryotic topoisomerase I-DNA interaction is stabilized by helix curvature. *Nucleic Acids Res.* **19:** 1235-1241.
Kubota, N., F. Kanzawa, K. Nishio, Y. Takeda, T. Ohmori, Y. Fujiwara, Y. Terashima, and N. Saijo. 1992. Detection of topoisomerase I gene point mutation in CPT-11 resistant lung cancer cell line. *Biochem. Biophys. Res. Commun.* **188:** 571-577.
Lee, M.P. and T. Hsieh. 1992. Incomplete reversion of double stranded DNA cleavage mediated by *Drosophila* topoisomerase II: Formation of single stranded DNA cleavage complex in the presence of an anti-tumor drug VM26. *Nucleic Acids Res.* **20:** 5027-5033.
———. 1993. Linker insertion mutagenesis of *Drosophila* topoisomerase II. Probing the structure of eukaryotic topoisomerase II. *J. Mol. Biol.* (in press).
Lee, M.P., M. Sander, and T. Hsieh. 1989a. Single strand DNA cleavage reaction of duplex DNA by *Drosophila* topoisomerase II. *J. Biol. Chem.* **264:** 13510-13518.
———. 1989b. Nuclease protection by *Drosophila* DNA topoisomerase II. *J. Biol. Chem.* **264:** 21779-21787.
Lee, M.P., S.D. Brown, A. Chen, and T. Hsieh. 1993. DNA topoisomerase I is essential in *Drosophila melanogaster*. *Proc. Natl. Acad. Sci.* **90:** 6656-6660.
Levin, N.A., M.-A. Bjornsti, and G.R. Fink. 1993. A novel mutation in DNA topoisomerase I of yeast causes DNA damage and RAD9-dependent cell cycle arrest. *Mol. Cell. Biol.* **133:** 799-814.
Lindsley, J.E. and J.C. Wang. 1991. Proteolysis patterns of epitopically labeled yeast DNA topoisomerase II suggest an allosteric transition in the enzyme induced by ATP binding. *Proc. Natl. Acad. Sci.* **88:** 10485-10489.
———. 1993a. Study of allosteric communication between protomers by immunotagging. *Nature* **361:** 749-750.
———. 1993b. On the coupling between ATP usage and DNA transport by yeast DNA topoisomerase II. *J. Biol. Chem.* **268:** 8096-8104.
Liu, L.F. 1983. DNA topoisomerases-enzymes that catalyse the breaking and rejoining of DNA. *CRC Crit. Rev. Biochem.* **15:** 1-24.
———. 1989. DNA topoisomerase poisons as antitumor drugs. *Annu. Rev. Biochem.* **58:** 351-375.
———. 1993. DNA topoisomerases and their applications in pharmacology. *Adv. Pharmacol.* (in press).
Liu, L.F. and K.G. Miller. 1981. Eukaryotic DNA topoisomerases: Two forms of type I

DNA topoisomerases from HeLa cell nuclei. *Proc. Natl. Acad. Sci.* **78:** 3487–3491.

Liu, L.F. and J.C. Wang. 1978a. Micrococcus luteus DNA gyrase: Active components and a model for its supercoiling of DNA. *Proc. Natl. Acad. Sci.* **74:** 2098–2102.

———. 1978b. DNA-DNA gyrase complex: The wrapping of the DNA duplex outside the enzyme. *Cell* **15:** 979–984.

———. 1987. Supercoiling of the DNA template during transcription. *Proc. Natl. Acad. Sci.* **84:** 7024–7027.

Liu, L.F., C.C. Liu, and B.M. Alberts. 1979. T4 DNA topoisomerase: A new ATP-dependent enzyme essential for initiation of T4 bacteriophage DNA replication. *Nature* **281:** 456–461.

Lockshon, D. and D.R. Morris. 1985. Sites of reaction of *Escherichia coli* DNA gyrase on pBR322 *in vivo* as revealed by oxolinic acid-induced plasmid linearization. *J. Mol. Biol.* **181:** 63–74.

Luttinger, A.L., A.L. Springer, and M.B. Schmid. 1991. A cluster of genes that affect nucleoid segregation in *Salmonella typhimurium*. *New Biol.* **3:** 687–697.

Lynn, R.M. and J.C. Wang. 1989. Peptide sequencing and site-directed mutagenesis identify tyrosine 319 as the active-site tyrosine of *Escherichia coli* topoisomerase I. *Proteins* **6:** 231–239.

Lynn, R.M., M.-A. Bjornsti, P.R. Caron, and J.C. Wang. 1989. Peptide sequencing and site-directed mutagenesis identify tyrosine-727 as the active site tyrosine of *Saccharomyces cerevisiae* DNA topoisomerase I. *Proc. Natl. Acad. Sci.* **86:** 3559–3563.

Maxwell, A. and M. Gellert. 1986. Mechanistic aspects of DNA topoisomerases. *Adv. Protein Chem.* **38:** 69–107.

Menzel, R. and M. Gellert. 1983. Regulation of the genes for *E. coli* DNA gyrase: Homeostatic control of DNA supercoiling. *Cell* **34:** 105–113.

Moore, C.L., L. Klevan, J.C. Wang, and J.D. Griffith. 1985. Gyrase-DNA complex visualized as looped structure by electron microscopy. *J. Mol. Biol.* **258:** 4612–4617.

Morrison, A. and N.R. Cozzarelli. 1979. Site-specific cleavage of DNA by *E. coli* DNA gyrase. *Cell* **17:** 175–184.

———. 1981. Contacts between DNA gyrase and its binding site on DNA: Features of symmetry and asymmetry revealed by protection from nucleases. *Proc. Natl. Acad. Sci.* **78:** 1416–1420.

Morrison, A., N.P. Higgins, and N.R. Cozzarelli. 1980. Interaction between DNA gyrase and its cleavage site on DNA. *J. Biol. Chem.* **255:** 2211–2219.

Muller, M.T., W.P. Pfund, V.B. Mehta, and D.K. Trask. 1985. Eukaryotic type I topoisomerase is enriched in the nucleolus and catalytically active on ribosomal DNA. *EMBO J.* **4:** 1237–1243.

Muller, M.T., J.R. Spitzner, J.H. DiDonato, V.B. Mehta, K. Tsutsui, and K. Tsutsui. 1988. Single-strand DNA cleavages by eucaryotic topoisomerase II. *Biochemistry* **27:** 8369–8379.

Nelson, E.M., K.M. Tewey, and L.F. Liu. 1984. Mechanism of antitumor drug action: Poisoning of mammalian DNA topoisomerase II on DNA by 4'-(9-acridinylamino)-methanesulfon-*m*-aniside. *Proc. Natl. Acad. Sci.* **81:** 1361–1365.

Nitiss, J. and J.C. Wang. 1988. DNA topoisomerase-targeting antitumor drugs can be studied in yeast. *Proc. Natl. Acad. Sci.* **85:** 7501–7505.

Nitiss, J., Y.-X. Liu, and Y. Hsiung. 1993. A temperature sensitive topoisomerase II allele confers temperature dependent drug resistance on amsacrine and etoposide: A genetic system for determining the targets of topoisomerase II inhibitors. *Cancer Res.* **53:** 89–93.

Nitiss, J., Y.-X. Liu, P. Harbury, M. Jannatipour, R. Wasserman, and J.C. Wang. 1992.

Amasacrine and etoposide hypersensitivity in yeast cells overexpressing DNA topoisomerase II. *Cancer Res.* **52:** 4467–4472.
Osheroff, N. 1986. Eukaryotic topoisomerase II. Characterization of enzyme turnover. *J. Biol. Chem.* **261:** 9944–9950.
———. 1989. Biochemical basis for the interactions of type I and type II topoisomerase with DNA. *Pharmacol. Ther.* **41:** 223–241.
Osheroff, N. and D.L. Brutlag. 1983. Recognition of supercoiled DNA by *Drosophila* topoisomerase II. In *Mechanism of DNA replication and recombination* (ed. N.R. Cozzarelli), pp. 55–64, A.R. Liss, New York.
Osheroff, N. and E.L. Zechiedrich. 1987. Calcium promoted DNA cleavage by eucaryotic topoisomerase II: Trapping the covalent enzyme-DNA complex in an active form. *Biochemistry* **26:** 4303–4309.
Osheroff, N., E.R. Shelton, and D.L. Brutlag. 1983. DNA topoisomerase II from *Drosophila melanogaster*. *J. Biol. Chem.* **258:** 9536–9543.
Osheroff, N., E.L. Zechiedrich, and K.C. Gale. 1991. Catalytic function of DNA topoisomerase II. *BioEssays* **13:** 269–275.
Peebles, C.L., N.P. Higgins, K.N. Kreuzer, A. Morrison, P.O. Brown, A. Sugino, and N.R. Cozzarelli. 1979. Structure and activities of *Escherichia coli* DNA gyrase. *Cold Spring Harbor Symp. Quant. Biol.* **43:** 41–52.
Pommier, Y. and K. Kohn. 1989. Topoisomerase II inhibition by antitumor intercalators and demethylepipodophyllotoxins. In *Development in cancer chemotherapy* (ed. R. Glazer), pp. 175–181. CRC Press, Boca Raton, Florida.
Potmesil, M., Y.-H. Hsiang, L.F. Liu, B. Bank, H. Grossberg, S. Kirschenbaum, T.J. Forlenza, A. Penziner, D. Kanganis, D. Knowles, F. Traganos, and R. Silber. 1988. Resistance of human leukemic and normal lymphocytes to drug-induced DNA cleavage and low levels of topoisomerase II. *Cancer Res.* **48:** 3537–3543.
Pruss, G.J., S.H. Manes, and K. Drlica. 1982. *Escherichia coli* DNA topoisomerase mutants: Increased supercoiling is corrected by mutations near gyrase genes. *Cell* **31:** 35–42.
Pulleyblank, D.E., M. Shure, D. Tang, J. Vinograd, and H.-P. Vosberg. 1975. Action of nicking-closing enzyme on supercoiled and nonsupercoiled closed circular DNA: Formation of a Boltzmann distribution of topological isomers. *Proc. Natl. Acad. Sci.* **72:** 4280–4284.
Raji, A., D.J. Zabel, C.S. Laufer, and R.E. Depew. 1985. Genetic analysis of mutations that compensate for loss of *Escherichia coli* DNA topoisomerase I. *J. Bacteriol.* **162:** 1173–1179.
Rau, D.C., M. Gellert, F. Thoma, and A. Maxwell. 1987. Structure of the DNA gyrase-DNA complex as revealed by transient electric dichroism. *J. Mol. Biol.* **193:** 555–569.
Reece, R.J. and A. Maxwell. 1991. DNA gyrase: Structure and function. *Crit. Rev. Biochem. Mol. Biol.* **26:** 335–375.
Riou, J.-F., E. Multon, M.-J. Vilarem, C.-J. Larsen, and G. Riou. 1986. *In vivo* stimulation by antitumor drugs of the topoisomerase II induced cleavage site in C-*myc* protooncogene. *Biochem. Biophys. Res. Commun.* **137:** 154–160.
Robinson, M.J. and N. Osheroff. 1991. Effects of antineoplastic drugs in the post-strand-passage DNA cleavage/religation equilibrium of topoisomerase II. *Biochemistry* **30:** 1807–1813.
Roca, J. and J.C. Wang. 1992. The capture of a DNA double helix by an ATP-dependent protein clamp: A key step in DNA transport by type II DNA topoisomerases. *Cell* **71:** 833–840.
Roca, J., J.M. Berger, and J.C. Wang. 1993. On the simultaneous binding of eucaryotic

DNA topoisomerase II to a pair of double-stranded DNA helices. *J. Biol. Chem.* **268:** 14250-14255.

Romig, H. and A. Richter. 1990. Expression of the topoisomerase I gene in serum stimulated human fibroblasts. *Biochim. Biophys. Acta* **1048:** 274-280.

Rose, D., W. Thomas, and C. Holm, C. 1990. Segregation of recombined chromosomes in meiosis I requires DNA topoisomerase II. *Cell* **60:** 1009-1017.

Rothstein, R., C. Helms, and N. Rosenberg. 1987. Concerted deletions and inversions are caused by mitotic recombination between δ sequences in *Saccharomyces cerevisiae*. *Mol. Cell. Biol.* **7:** 1198-1207.

Sander, M. and T. Hsieh. 1985. *Drosophila* topoisomerase II double-strand DNA cleavage: Analysis of DNA sequence homology at the cleavage site. *Nucleic Acids Res.* **13:** 1057-1072.

Sander, M., T. Hsieh, A. Udvardy, and P. Schedl. 1987. Sequence dependence of *Drosophila* topoisomerase II in plasmid relaxation and DNA binding. *J. Mol. Biol.* **194:** 219-229.

Schmid, M.B. 1990. Locus affecting nucleoid segregation in *Salmonella typhimurium*. *J. Bacteriol.* **172:** 5416-5424.

Schofield, M.A., R. Agbunag, M.L. Michaels, and J.H. Miller. 1992. Cloning and sequencing of *Escherichia coli mutR* shows its identity to *topB*, encoding topoisomerase III. *J. Bacteriol.* **174:** 5168-5170.

Shen, C.C. and C.-K.J. Shen. 1990. Specificity and flexibility of the recognition of DNA helical structure by eukaryotic topoisomerase I. *J. Mol. Biol.* **212:** 67-78.

Shiozaki, K. and M. Yanagida. 1991. A functional 125-kDa core polypeptide of fission yeast DNA topoisomerase II. *Mol. Cell. Biol.* **11:** 6093-6102.

―――. 1992. Functional dissection of the phosphorylated termini of fission yeast DNA topoisomerase II. *J. Cell Biol.* **119:** 1023-1036.

Shuman, S. 1991. Recombination mediated by vaccinia viral DNA topoisomerase I in *Escherichia coli* is sequence specific. *Proc. Natl. Acad. Sci.* **88:** 10104-10108.

―――. 1992. DNA strand transfer reactions catalyzed by vaccinia topoisomerase I. *J. Biol. Chem.* **267:** 8620-8627.

Spitzner, J.R. and M.T. Muller. 1988. A consensus sequence for cleavage by vertebrate DNA topoisomerase II. *Nucleic Acids Res.* **16:** 5533-5556.

Steck, T.R. and K. Drlica. 1984. Bacterial chromosome segregation: Evidence for DNA gyrase involvement in decatenation. *Cell* **36:** 1081-1088.

Sternglanz, R., S. DiNardo, K.A. Voelkel, Y. Nishimura, Y. Hirota, K. Becherer, L. Zumstein, and J.C. Wang. 1981. Mutations in the gene coding for *Escherichia coli* DNA topoisomerase I affect transcription and transposition. *Proc. Natl. Acad. Sci.* **78:** 2747-2751.

Sugino, A., C.L. Peebles, K.N. Kreuzer, and N.R. Cozzarelli. 1977. Mechanism of action of nalidixic acid: Purification of *E. coli nalA* gene product and its relationship to DNA gyrase and a novel nicking-closing enzyme. *Proc. Natl. Acad. Sci.* **74:** 4767-4771.

Sugino, A., N.P. Higgins, P.O. Brown, C.L. Peebles, and N.R. Cozzarelli. 1978. Energy coupling in DNA gyrase and the mechanism of action of novobiocin. *Proc. Natl. Acad. Sci.* **75:** 4838-4842.

Svejstrup, J.Q., K. Christiansen, I.I. Gromova, A.H. Andersen, and O. Westergaard. 1991. New technique for uncoupling the cleavage and religation reaction of eucaryotic topoisomerase I. *J. Mol. Biol.* **222:** 669-678.

Tamura, J.K. and M. Gellert. 1990. Characterization of the ATP binding site on *Escherichia coli* DNA gyrase. Affinity labeling of Lys-103 and Lys-110 of the B subunit by pyridoxal 5′-diphospho-5′-adenosine. *J. Biol. Chem.* **265:** 21342-21349.

Tamura, J.K., A.D. Bates, and M. Gellert. 1992. Slow interaction of 5′-adenylyl-β,γ,-imidodiphosphate with *Escherichia coli* DNA gyrase. Evidence for cooperativity in nucleotide binding. *J. Biol. Chem.* **267:** 9214–9222.

Tamura, H., C. Kohchi, R. Yamada, T. Ikeda, O. Koiwai, E. Patterson, J.D. Keene, K. Okada, E. Kjeldsen, K. Nishikawa, and T. Andoh. 1991. Molecular cloning of a cDNA of a camptothecin-resistant human DNA topoisomerase I and identification of mutation sites. *Nucleic Acids Res.* **19:** 69–75.

Tanabe, K., Y. Ikegami, R. Ishida, and T. Andoh. 1991. Inhibition of topoisomerase II by antitumor agents bis(2,6-dioxopiperazine) derivatives. *Cancer Res.* **51:** 4903–4908.

Thrash, C., A.T. Bankier, B.G. Barrel, and R. Sternglanz. 1985. Cloning, characterization, and sequencing of the yeast DNA topoisomerase I gene. *Proc. Natl. Acad. Sci.* **82:** 4374–4378.

Trask, D.K. and M.T. Muller. 1988. Stabilization of type I topoisomerase-DNA covalent complexes by actinomycin D. *Proc. Natl. Acad. Sci.* **85:** 1417–1421.

Tse-Dinh, Y.-C. 1986. Uncoupling of the DNA breaking and rejoining steps of *Escherichia coli* type I topoisomerase *J. Biol. Chem.* **261:** 10931–10935.

Tse-Dinh, Y.-C. and R.K. Beran. 1988. Multiple promoters for transcription of the *Escherichia coli* DNA topoisomerase I gene and their regulation by DNA supercoiling. *J. Mol. Biol.* **202:** 735–742.

Tse-Dinh, Y.-C. and J.C. Wang. 1986. Complete nucleotide sequence of the *topA* gene encoding *Escherichia coli* DNA topoisomerase I. *J. Mol. Biol.* **191:** 321–331.

Tse-Dinh, Y.-C., B.G.H. McCarron, R. Arentzen, and V. Chowdhry. 1983. Mechanistic study of *E. coli* DNA topoisomerase I: Cleavage of oligonucleotides. *Nucleic Acids Res.* **11:** 8691–8701.

Udvardy, A., P. Schedl, M. Sander, and T. Hsieh. 1985. Novel partitioning of DNA cleavage sites for *Drosophila* topoisomerase II. *Cell* **40:** 933–941.

———. 1986. Topoisomerase II cleavage in chromatin. *J. Mol. Biol.* **191:** 231–246.

Uemura, T. and M. Yanagida. 1984. Isolation of type I and II DNA topoisomerase mutants from fission yeast: Single and double mutants show different phenotypes in cell growth and chromatin organization. *EMBO J.* **3:** 1737–1744.

———. 1986. Mitotic spindle pulls but fails to separate chromosomes in type II DNA topoisomerase mutants: Uncoordinated mitosis. *EMBO J.* **5:** 1003–1010.

Uemura, T., H. Ohkura, Y. Adachi, K. Morino, K. Shiozaki, and M. Yanagida. 1987. DNA topoisomerase II is required for condensation and separation of mitotic chromosomes in *S. pombe*. *Cell* **50:** 917–925.

Wallis, J.W., G. Chrebet, G. Brodsky, M. Rolfe, and R. Rothstein. 1989. A hyperrecombination mutation in *S. cerevisiae* identifies a novel eukaryotic topoisomerase. *Cell* **58:** 409–419.

Wang, J.C. 1971. Interaction between DNA and an *Escherichia coli* protein ω. *J. Mol. Biol.* **55:** 523–533.

———. 1982. DNA topoisomerases. In *Nucleases* (ed. S.M. Linn and R.J. Roberts), pp. 41–57. Cold Spring Harbor Laboratory Press, Cold Spring Harbor, New York.

———. 1985. DNA topoisomerases. *Annu. Rev. Biochem.* 54: 665–697.

Wang, J.C. and L.L. Liu. 1979. DNA topoisomerases: Enzymes that catalyze the concerted breaking and rejoining of DNA backbone bonds. In *Molecular genetics* (ed. J.H. Taylor), part 3, pp. 65–88. Academic Press, New York.

Wang, J.C., P.R. Caron, and R.A. Kim. 1990. The role of DNA topoisomerases in recombination and genome stability: A double-edged sword? *Cell* **62:** 403–406.

Wang, J.C., R.I. Gumport, K. Javaherian, K. Kirkegaard, L. Klevan, M.L. Kotewicz, and Y.-C. Tse. 1981. DNA topoisomerases. In *Mechanistic studies of DNA replication and genetic recombination* (ed. B.M. Alberts and C.F. Fox), pp. 769-784. Academic Press, New York.

Wigley, D.B., G.J. Davies, E.J. Dodson, A. Maxwell, and G. Dodson. 1991. Crystal structure of an N-terminal fragment of the DNA gyrase B protein. *Nature* **351:** 624-629.

Woessner, R.D., M.R. Mattern, C.K. Mirabelli, R.K. Johnson, and F.H. Drake. 1991. Proliferation- and cell cycle-dependent differences in expression of the 170 kilodalton and 180 kilodalton forms of topoisomerase II in NIH-3T3 cells. *Cell Growth Differ.* **2:** 209-214.

Worland, S.T. and J.C. Wang. 1989. Inducible overexpression, purification, and active site mapping of DNA topoisomerase II from the yeast *Saccharomyces cerevisiae*. *J. Biol. Chem.* **264:** 4412-4416.

Yanagida, M. and R. Sternglanz. 1990. Genetics of DNA topoisomerases. In *DNA topology and its biological effects* (ed. N.R. Cozzarelli and J.C. Wang), pp. 299-320. Cold Spring Harbor Laboratory Press, Cold Spring Harbor, New York.

Yang, L., T.C. Rowe, E.M. Nelson, and L.F. Liu. 1985. *In vivo* mapping of DNA topoisomerase II-specific cleavage sites on SV40 chromatin. *Cell* **41:** 127-132.

Yang, L., M.S. Wold, J.J. Li, T.J. Kelly, and L.F. Liu. 1987. Roles of DNA topoisomerases in simian virus 40 DNA replication *in vitro*. *Proc. Natl. Acad. Sci.* **84:** 950-954.

Zechiedrich, E.L. and N. Osheroff. 1990. Eucaryotic topoisomerases recognize nucleic acid topology by preferentially interacting with DNA crossovers. *EMBO J.* **9:** 4555-4562.

Zechiedrich, E.L., K. Christiansen, A.H. Andersen, O. Westergaard, and N. Osheroff. 1989. Double-strand DNA cleavage /religation reaction of eukaryotic topoisomerase II: Evidence for a nicked DNA intermediate. *Biochemistry* **28:** 6229-6236.

Zhang, H., J.C. Wang, and L.F. Liu. 1988. Involvement of DNA topoisomerase I in transcription of human ribosomal RNA genes. *Proc. Natl. Acad. Sci.* **85:** 1060-1064.

Zumstein, L. and J.C. Wang. 1986. Probing the structural domains and function *in vivo* of *Escherichia coli* DNA topoisomerase I by mutagenesis. *J. Mol. Biol.* **191:** 333-340.

8
Proofreading Exonucleases: Error Correction during DNA Replication

Myron F. Goodman and Linda B. Bloom
University of Southern California
Department of Biological Sciences
Los Angeles, California 90089-1340

I. Introduction
II. Genetic and Biochemical Basis of Fidelity
   A. Proofreading: 3′ →5′ Exonuclease Activity during DNA Synthesis
      1. Proofreading Specificity Dependence on the Ratio of Melted to Annealed Primer Termini
      2. Costs Associated with Attaining High Replication Accuracy
   B. Mutant Polymerase Phenotypes Reveal Novel Biological Insights
      1. Interaction of Methylation-directed Mismatch Repair and Proofreading in *E. coli*
      2. An "Antimutator" with Mutator Properties
      3. Communicating between Polymerase and Exonuclease Active Sites
   C. Rapid Flow Analysis of Changes in 2-Aminopurine Fluorescence as a Probe of the Effects of Sequence Context on 3′ →5′ Exonuclease Activity and Specificity
III. Conclusions and Future Prospects

I. INTRODUCTION

In elucidating the structure of DNA, Watson and Crick (1953a,b) proposed that during DNA replication, each strand could serve as a template for synthesizing its complementary partner strand, pairing A with T and G with C. They also suggested that spontaneous transition mutations might be accounted for by the presence of disfavored imino and enol base tautomers that could form A·C and G·T mispaired replication intermediates, stabilized by two and three H bonds, respectively. A large number of DNA synthesis fidelity studies have employed DNA polymerases purified from numerous prokaryotic and eukaryotic cellular and viral sources (for recent reviews, see Echols and Goodman 1991; Goodman et al. 1993). These in vitro fidelity studies have been complemented by physicochemical (Aboul-ela et al. 1985; Breslauer et al. 1986; Petruska et al. 1988; Gaffney and Jones 1989), nuclear magnetic resonance (Kan et al. 1983; Patel et al. 1984a,b; Sowers et al. 1986a,b,

1988, 1989a,b), and X-ray (Brown et al. 1986, 1989; Hunter et al. 1986a,b, 1987) structural studies using natural DNA and synthetic oligonucleotides containing matched and mismatched base pairs. "Stable" multiply H-bonded base mispairs in cationic (Sowers et al. 1986a,b), anionic (Driggers and Beattie 1988; Sowers et al. 1988, 1989a,b; Yu et al. 1993), and wobble structures (Kan et al. 1983; Patel et al. 1984b; Hunter et al. 1986a,b; Sowers et al. 1988, 1989a,b; Brown et al. 1989) have clearly been observed in B-form DNA, although the originally suggested disfavored tautomeric structures have not been detected.

DNA polymerases can catalyze every type of mismatched intermediate involving normal nucleotides (Loeb and Kunkel 1982; Mendelman et al. 1989; Echols and Goodman 1991), leading to transition and transversion mutations. Polymerases also incorporate modified nucleotides (some of which can be quite bulky) onto growing primer-template termini (Singer et al. 1989), and to copy DNA template lesions, at greatly reduced rates relative to undamaged DNA template strands (Moore et al. 1981; Ide et al. 1985; Banerjee et al. 1988; Taylor and O'Day 1990; Dosanjh et al. 1991). In the case of errors involving normal nucleotides, polymerase misinsertion frequencies generally range from about 10^{-3} to 10^{-5}, G·T mispairs toward the high end and G·A and C·C mispairs at the low end. Spontaneous mutation rates have been most extensively characterized in *Escherichia coli*, where 10^{-9} to 10^{-10} per replicated base pair seem to be a reasonable estimate. Thus, despite the exceptionally high nucleotide insertion specificity exhibited by prokaryotic and eukaryotic polymerases, an additional 10^5-fold reduction in errors is required just to attain the levels estimated for *E. coli*.

Exonucleolytic proofreading provides a means to correct a significant fraction of insertion errors at the replication fork (Echols and Goodman 1991). *E. coli* polymerase I and bacteriophage T4 DNA polymerase are examples of repair and replicative prokaryotic enzymes, respectively, that contain $5'\rightarrow3'$ polymerase and $3'\rightarrow5'$ exonuclease activities on a single polypeptide chain; polymerase I also has an associated $5'\rightarrow3'$ exonuclease activity used for removal of RNA primers on lagging-strand Okazaki fragments (Kornberg and Baker 1992a). DNA polymerase III, the replicative polymerase in *E. coli*, has at least ten separate subunits (Kornberg 1988; McHenry 1991), the $3'\rightarrow5'$ exonuclease activity residing on the ε-subunit (Scheuermann et al. 1983; DiFrancesco et al. 1984; Scheuermann and Echols 1984). For eukaryotic polymerases (Linn 1991; Wang 1991), δ, ε, and mitochondrial γ polymerases have associated exonuclease activities, whereas β and α polymerases do not. Polymerases α (Skarnes et al. 1986; Wang 1991) and β (Mosbaugh and Meyer 1980) have been reported to associate with separate exonucleases. Recently,

Bialek and Grosse (1993) described the extensive purification of a calf thymus enzyme having both exonuclease and polymerase activities that copurified with the polymerase α–primase complex. The exonuclease behaves as a proofreading enzyme, excising mononucleotides in a 3′→5′ direction and showing a 20-fold specificity for excision of mismatched primer termini.

Presumably, DNA replication and repair have been optimized by balancing the spontaneous mutational load in the cell, thus permitting evolution to proceed without threatening fitness or survival. Is the action of associated or separate 3′→5′ exonucleases sufficient to reduce polymerase errors to the spontaneous levels observed in vivo? The answer is no. For reasons discussed in Sections II.A.1, II.A.2, and II.B.1, exonucleases may be directly responsible for up to about a several hundredfold reduction in errors (Sinha and Haimes 1981; Fersht et al. 1982; Fersht and Knill-Jones 1983; Sinha and Goodman 1983; Kunkel et al. 1984). An additional 10- to 100-fold elimination of errors is carried out by mismatch-specific postreplication repair enzymes (for review, see Radman and Wagner 1986; Modrich 1987, 1989) (Chapter 9, this volume).

What is the evidence that 3′→5′ exonucleases function as "proofreading" enzymes? The following section summarizes genetic studies in vivo, carried out with T4 bacteriophage and with *E. coli*, and in vitro biochemical analyses that demonstrate the importance of 3′ exonucleases in excising polymerase-catalyzed insertion errors. Enzymatic mechanisms ensuring selective removal of incorrectly formed base pairs are discussed in Section II. X-ray data on the *E. coli* polymerase I Klenow fragment (KF contains 5′→3′ polymerase and 3′→5′ exonuclease activities but is lacking the 5′→3′ exonuclease activity of wild-type polymerase I) have provided our first physical picture of the polymerase and exonuclease active clefts (Derbyshire et al. 1988). The structural analysis of polymerase I has stimulated investigations into enzymatic mechanisms for switching between synthetic and degradative modes (Cowart et al. 1989; Joyce 1989; Reddy et al. 1992) and mechanisms of distinguishing primer-terminal mismatches from correct matches. Recently, pre-steady-state fluorescent quench and depolarization techniques have been used to measure nuclease-catalyzed excision and primer termini and dNTP substrate mobility in the exonuclease active cleft (L.B. Bloom et al., in prep.) (Section II.C).

II. GENETIC AND BIOCHEMICAL BASIS OF FIDELITY

The *E. coli* bacteriophage T4 has a single DNA polymerase, gp43, that is required for DNA replication and repair (de Waard et al. 1965; Warner and Barnes 1966; Kornberg and Baker 1992b). Conditionally lethal temperature-sensitive mutants in T4 gene *43* cannot synthesize DNA at nonpermissive temperatures. In a series of seminal experiments carried out in the mid to late 1960s, Speyer (1965; Speyer et al. 1967), Freese (Freese and Freese 1967), and Drake (Drake and Allen 1968; Drake et al. 1969) showed that mutations in gene *43* strongly affect mutagenesis in the *r*II gene, a nonessential region of T4 used as a mutagenic target. The gene *43* mutants fall into two classes: those that increase *r*II reversion rates (Speyer 1965; Speyer et al. 1967; Freese and Freese 1967), called mutators, and those that decrease reversion rates (Drake 1969; Drake and Allen 1969; Drake et al. 1969), called antimutators. For some of the gene *43* alleles (e.g., *tsL56*), an increase of about 100-fold was observed at a specific *r*II locus reverting by an A·T→G·C mutation. Similarly, about a two order of magnitude decrease in an A·T→G·C transition was observed for the antimutator, *tsL141*. The primary conclusion from these studies was that although base selection might be governed to a very large extent by Watson-Crick base-pairing rules, DNA polymerases also had a significant role to play in base selection. However, it seemed surprising that mutations in gene *43* could lead to the "better," more accurate, antimutator enzyme.

A 3'→5' exonuclease acting in conjunction with DNA polymerase, either as part of a single polypeptide or as a separate component in a holoenzyme complex, can degrade DNA at a 3' terminus (Fig. 1a). When the exonuclease excises a primer-3'-terminal nucleotide during DNA synthesis, the net reaction of nucleotide insertion followed by excision leads to the conversion or "turnover" of a substrate dNTP→dNMP (Fig. 1b). In an in vitro assay using the 3' exonuclease of *E. coli* polymerase I to degrade DNA containing either correctly matched or mismatched primer-3' termini, Brutlag and Kornberg (1972) found that the mismatched termini were removed much more rapidly.

During the same period, Bessman and colleagues (Muzyczka et al. 1972; Bessman et al. 1974) reported an important correlation between the relative rates of exonucleolytic removal and forward polymerization using purified T4 mutator and antimutator polymerases corresponding to those used in the genetic studies of Speyer, Freese, and Drake. In an assay using *equal* polymerase units to measure rates for removal of nucleotide mismatches at primer-3' termini, antimutators were found to hydrolyze mismatches much more rapidly and mutators were found to hydrolyze much more slowly than wild-type polymerase (Fig. 2, top) (Muzyczka et al. 1972). Correctly paired primer termini were also excised most efficiently by an antimutator and least efficiently by a mutator T4 polymerase (Fig. 2, bottom) (Muzyczka et al. 1972). These data also

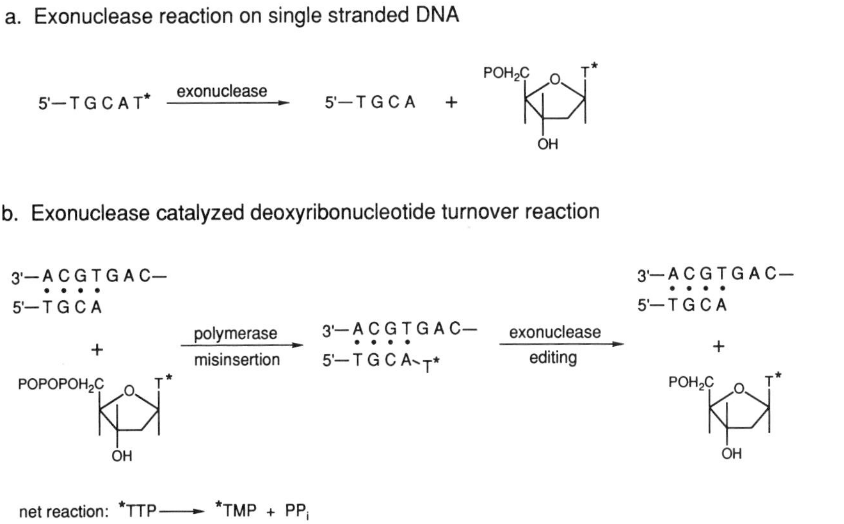

a. Exonuclease reaction on single stranded DNA

5'—TGCAT* —exonuclease→ 5'—TGCA + [deoxyribose-T*]

b. Exonuclease catalyzed deoxyribonucleotide turnover reaction

3'—ACGTGAC—
5'—TGCA
+
POPOPOH₂C—[sugar]—T*

—polymerase misinsertion→ 3'—ACGTGAC—
5'—TGCA-T*

—exonuclease editing→ 3'—ACGTGAC—
5'—TGCA
+
POH₂C—[sugar]—T*

net reaction: *TTP → *TMP + PP$_i$

Figure 1 Reactions catalyzed by the 3'→5' exonuclease activity associated with most DNA polymerases. (*a*) Hydrolysis of a 3'-terminal nucleotide of single-stranded DNA to release a deoxyribonucleoside monophosphate. (*b*) Polymerase-catalyzed misinsertion followed by exonucleolytic proofreading resulting in a net turnover of dNTP to dNMP.

reveal a significant reduction in nucleotide excision when DNA synthesis is permitted to occur by addition of dNTP substrates (Fig. 2, bottom, open circles). Nuclease/polymerase (N/P) ratios were observed to be large for antimutators and small for mutators, small N/P ratios giving rise to low replication fidelity in vitro and high mutation rates in vivo and large N/P ratios giving rise to accurate replication and low mutation rates (Muzyczka et al. 1972).

Genetic and biochemical studies in *E. coli* reveal a similar picture. Mutations in the *DnaE* gene, coding for the α subunit of the DNA polymerase III holoenzyme complex, give rise to mutator and antimutator phenotypes (Fijalkowska et al. 1993). Mutations in the *DnaQ* (*mutD*) gene, coding for the ε proofreading subunit polymerase III (Scheuermann et al. 1983; DiFrancesco et al. 1984), resulted in strong mutator phenotypes (Degnan and Cox 1974; Cox 1976; Cox and Horner 1982).

A. Proofreading: 3'→5' Exonuclease Activity during DNA Synthesis

As illustrated in Figure 2, a 3'→5' proofreading exonuclease can catalyze release of 3'-dNMP either under nonsynthesizing conditions,

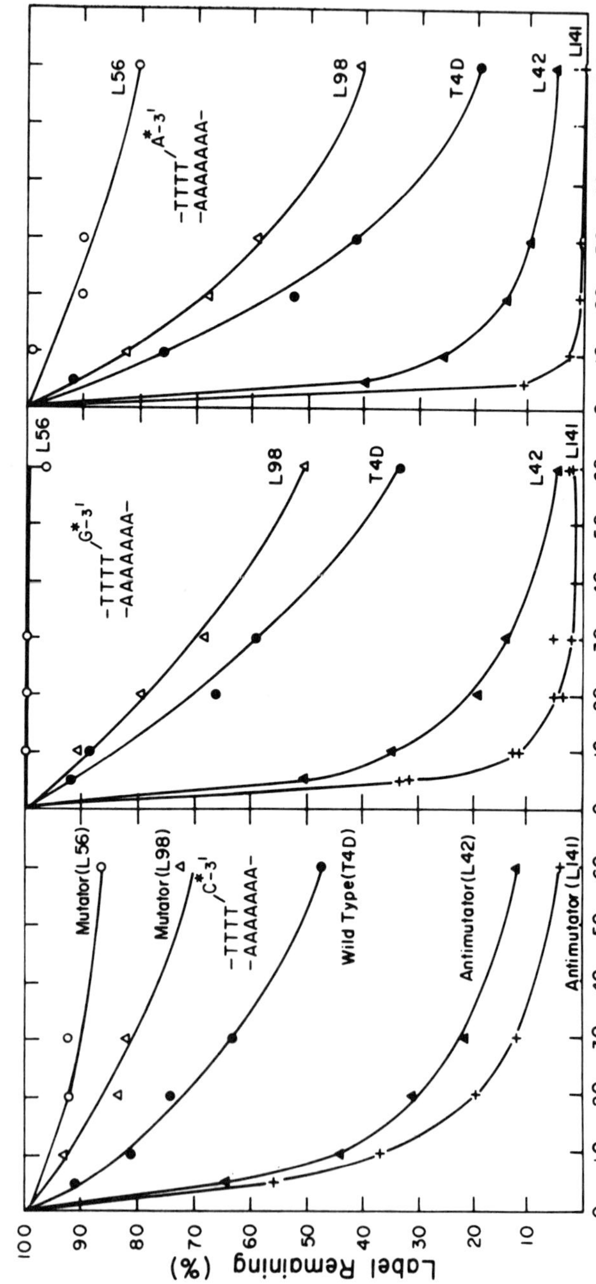

Label Remaining (%)
Minutes
Mutator (L56)
Mutator (L98)
Wild Type (T4D)
Antimutator (L42)
Antimutator (L141)
-TTTT C*-3'
-AAAAAAA-
-TTTT G*-3'
-AAAAAAA-
-TTTT A*-3'
-AAAAAAA-
L56
L98
T4D
L42
L141

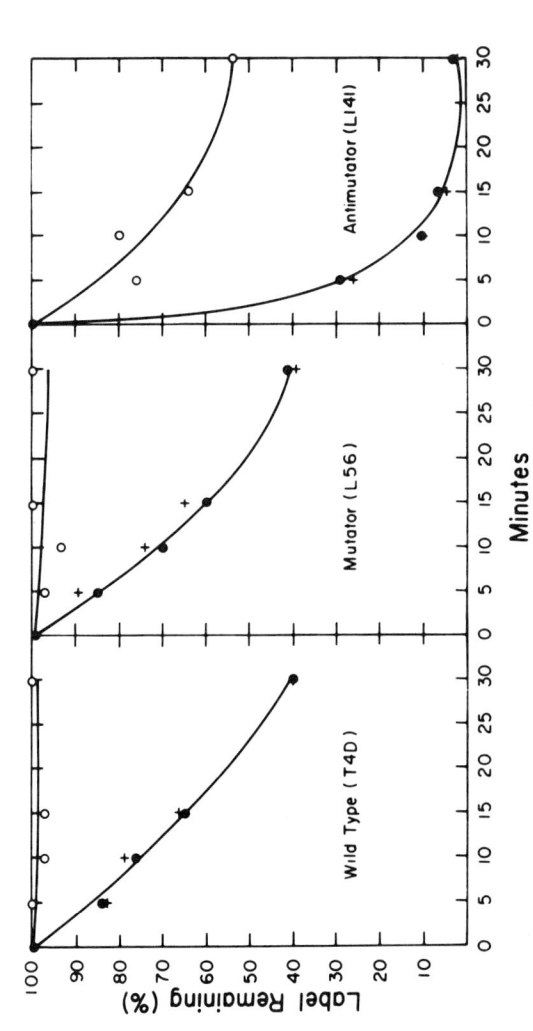

Figure 2 Time courses for removal of mismatched and matched nucleotides from 3′-primer termini by wild-type T4 DNA polymerase and T4 mutator and antimutator polymerases. (*Top*) Hydrolysis of mispaired nucleotides in reactions containing equal concentrations of DNA and equal units of polymerase activity. (*Bottom*) Hydrolysis of matched terminal nucleotides from poly(dA)$_{500}$:oligo(dT)$_{47}$ in the absence of added nucleotides (*closed circles*), in the presence of the next correct nucleotide, 250 μM dTTP (*open circles*), and in the presence of an incorrect nucleotide, 250 μM dCTP (+). Polymerase units were adjusted to give similar reaction rates. (Reprinted, with permission, from Muzyczka et al. 1972.)

i.e., in the absence of dNTP, or during DNA synthesis (Hershfield and Nossal 1972; Muzyczka et al. 1972; Hershfield 1973; Bessman et al. 1974). The dNTP→dNMP turnover reaction, measuring nucleotide excision during DNA synthesis, is commonly referred to as "proofreading." The specificity of proofreading pertains to the ability of an exonuclease to distinguish between correctly and incorrectly matched nucleotides for removal at primer-3' termini.

1. Proofreading Specificity Dependence on the Ratio of Melted to Annealed Primer Termini

A simple model proposed to explain preferential exonucleolytic removal of misincorporated nucleotides is that the 3'→5' exonuclease activities of polymerases are much more active in hydrolyzing single-stranded DNA compared to double-stranded DNA (Fig. 3). Misinserted nucleotides at a replication fork are much less likely than correctly inserted nucleotides to form H bonds with the template strand, and therefore form single-stranded (i.e., melted out) termini (Fig. 3, state M).

When carried out in the presence of the four dNTP substrates, competition takes place between the forward synthetic reaction (Fig. 3, annealed configuration, state A), involving nucleotide insertion and translocation steps, and the exonucleolytic proofreading reaction (Fig. 3, melted configuration, state M). The *next nucleotide effect* (Clayton et al. 1979; Fersht 1979)—in which exonucleolytic proofreading at template position i is inhibited as the concentration of the "next correct" dNTP, complementary to template position i +1, is increased—is a manifestation of this competition. In other words, conditions that favor the forward reaction will reduce the backward reaction and vice versa. Although the equilibrium ratio of melted/annealed primer termini may be different in the polymerase active cleft compared to aqueous solution, the ratio should not depend on the dNTP concentration. Experiments consistent with this point showed that proofreading of mismatched base pairs cannot be totally suppressed, even by the presence of saturating concentrations of the next correct dNTP (see Clayton et al. 1979; Fersht et al. 1982).

Polymerases that have no proofreading exonuclease activity have considerable difficulty extending mismatched primer termini. This difficulty is not caused by a decrease in polymerase binding, because recent experiments demonstrated that polymerases bind with roughly equal affinity to matched and mismatched primer termini (Mendelman et al. 1990; Wong et al. 1991; Creighton et al. 1992; Huang et al. 1992). The

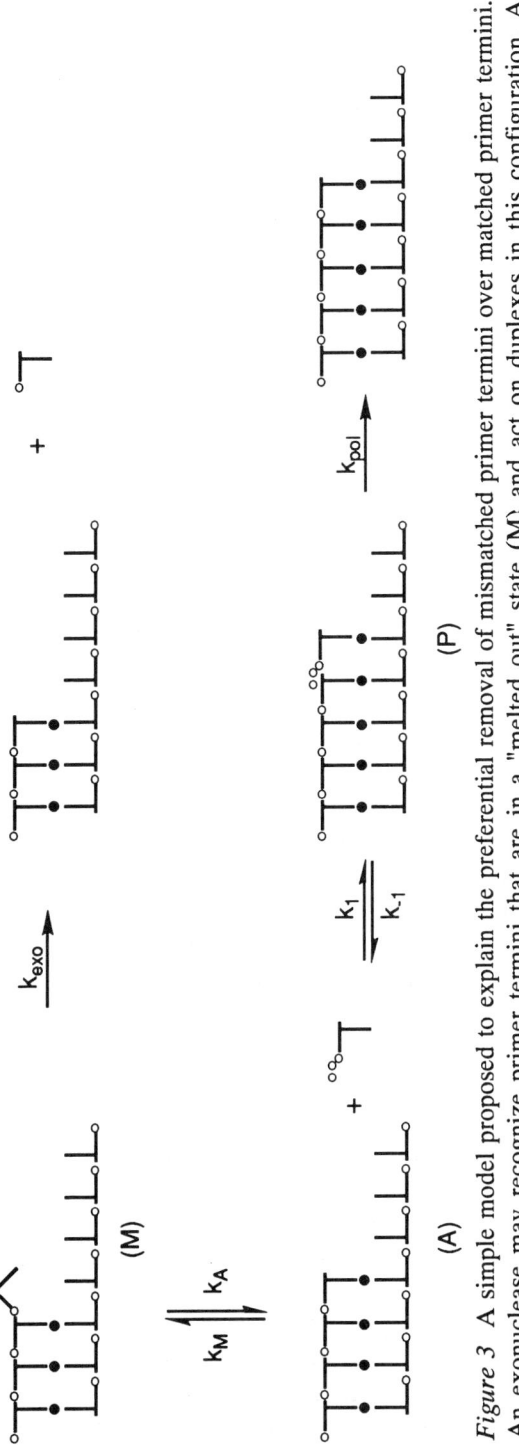

Figure 3 A simple model proposed to explain the preferential removal of mismatched primer termini over matched primer termini. An exonuclease may recognize primer termini that are in a "melted out" state (M) and act on duplexes in this configuration. A polymerase may only be able to extend primer termini that are in an annealed state (A). Selective hydrolysis of mismatches may result from a larger ratio of k_M/k_A for mismatches than for correct matches.

ratios of extending mismatched termini compared with matched termini are generally in the range of 10^{-3} to 10^{-6} (Kuchta et al. 1988; Petruska et al. 1988; Perrino and Loeb 1989; Perrino et al. 1989; Mendelman et al. 1990; Wong et al. 1991; Creighton et al. 1992; Huang et al. 1992), depending on the identity and location of the mismatch. Extension of a mismatch may require a transiently annealed configuration (Fig. 3, state M) and conformational change in the polymerase. It seems likely that a conformational change would at least be required to switch from polymerase to exonuclease when making a transition between synthesis and editing.

The following three observations imply that correctly matched primer termini are not always competent for elongation and that mismatched primer termini are not always competent for removal: (1) Proofreading of mismatched base pairs can be reduced but not eliminated by forcing the polymerase to go forward in the presence of saturating dNTP levels; (2) a significant fraction (~5–20%) of correct dNTPs are also turned over to dNMPs during normal DNA synthesis (Muzyczka et al. 1972; Bessman et al. 1974; Clayton et al. 1979; Fersht et al. 1982); (3) proofreading of mismatches is less than 100% efficient (its efficiency range is 90–99.5%). The simple "two state" model (Fig. 3) assumes that polymerases synthesize from annealed primer-template configurations (state A) and excise from melted out configurations (state M). Note that either continued polymerization or proofreading depends only on the "state" of the primer-template terminus, irrespective of whether it contains a "right" or "wrong" base pair. Wrong primer termini have a small probability of being in an annealed state, competent for elongation, and right termini can sometimes be melted out and hence be subject to excision. If it is necessary to melt several base pairs to gain access to the exonuclease active cleft, as is thought to be the case for *E. coli* polymerase I (Derbyshire et al. 1988), then the ratio of the melted/annealed rate constants, k_M/k_A, may change, but the basic suppositions of the simple model would still be valid.

It is interesting to note that there were no significant differences observed in the proofreading specificities of T4 mutator (*tsL56*), *43+*, and antimutator (*tsL141*) activities (Bessman et al. 1974; Clayton et al. 1979). Instead, the differences in accuracy for the three enzymes were attributed to differences in their nucleotide turnover activities (Fig. 1b). The highly active exonuclease associated with the antimutator polymerase succeeded in eliminating most of the mismatched base pairs along with a sizable fraction of correctly matched pairs, whereas the inactive mutator exonuclease excised a smaller number of wrong (and right) base pairs. Since both mutator and antimutator polymerases tend to replace the excised mismatched and correctly matched nucleotides with correctly matched nucleotides, a highly active nuclease will excise a larger absolute number of mispaired termini, leading to product DNA containing fewer misincorporations. Excision of either matched or mismatched primer termini in the absence of dNTPs (nonsynthesizing conditions) also occurred much more efficiently for T4 L141 antimutator polymerase and much less efficiently for L56 mutator polymerase compared to wild type. Similarly, Brenowitz et al. (1991) showed that the source of specificity for the ε subunit of the *E. coli* polymerase III holoenzyme complex is the greater melting capacity of a mispaired 3′ terminus.

It may be necessary to destabilize several adjacent normal base pairs to remove a single base mispair at a primer-3′ terminus (Cowart et al. 1989). An analysis of the X-ray crystal structure for the Klenow fragment suggests that melting of about four base pairs is required to position the growing primer at the exonuclease active site (Derbyshire et al. 1988). Klenow fragment contains a relatively inactive proofreading activity compared to DNA polymerases II and III from *E. coli* and bacteriophage T4 polymerase (Echols and Goodman 1991; Kornberg and Baker 1992b). Although there is no X-ray structural information available for T4 polymerase, kinetic data and cross-linking studies suggest that melting of as few as two primer-terminal base pairs is sufficient for exonucleolytic removal by the T4 enzyme (Cowart et al. 1989; Reddy et al. 1992). Although one might speculate that inefficient editing of the Klenow fragment may be attributed to a requirement for extensive primer melting, it is important to keep in mind that compared to T4 and the other two *E. coli* polymerases, the exonuclease associated with *E. coli* polymerase I is relatively inactive on single-stranded DNA (Kornberg and Baker 1992b).

2. Costs Associated with Attaining High Replication Accuracy

During DNA synthesis using *E. coli* polymerase III (Fersht et al. 1982) and T4 polymerase (Muzyczka et al. 1972; Bessman et al. 1974; Clayton et al. 1979) in vitro, the fraction of *correct* dNTP→dNMP turnover relative to incorporation is about 5–10% for G·C and 10–20% for A·T. For the case of an antimutator T4 polymerase, *tsL141*, removal of dAMP inserted opposite template T was observed to be about 40% using dNTP substrate concentrations required to approach V_{max} values for DNA synthesis (Muzyczka et al. 1972; Bessman et al. 1974; Clayton et al. 1979). Since it appears from the data that a significant fraction of correct base pairs are excised associated with wild-type polymerase exonuclease, even at maximal rates of DNA synthesis, it seems likely that a requirement to achieve high proofreading specificities may be limited by the ease with which correctly paired primer termini can melt out. Perhaps proofreading specificity is governed in large part by the ratio of melted to annealed primer-template ends (Clayton et al. 1979). Estimates of proofreading specificity for the *E. coli* polymerase III enzyme (ε subunit) are about 10–300 (Schaaper 1988; Echols and Goodman 1991); i.e., an active proofreading exonuclease reduces polymerase insertion errors by roughly one to two orders of magnitude.

The use of proofreading during DNA synthesis imposes a tradeoff on the cell. On the one hand, it is advantageous to optimize DNA synthesis fidelity, which is subject to evolutionary and mutational load constraints. On the other hand, it is also advantageous to minimize the energy required to resynthesize dNTP substrates from dNMP generated by removal of correctly formed base pairs. Since bacteriophage T4 devotes a considerable portion of its genome to carrying out dNTP metabolism, it is conceivable that the T4 replication enzymes have evolved to maintain a balanced ratio of cost (dNMP generation) to accuracy (Clayton et al. 1979; Fersht et al. 1982).

The T4 genome contains 1.66×10^5 base pairs (Kim and Davidson 1974; Wood and Revel 1976). It was estimated from in vitro data that the combined polymerase proofreading activities of the wild-type polymerase make roughly one base substitution error per 10^7 base pairs (Kunkel et al. 1984). If this estimate is also approximately valid as well in vivo, then about 49 out of 50 progeny viruses will undergo no DNA sequence changes. Any further improvement in replication accuracy to better maintain the T4 genome would seem superfluous, especially if, as in the case of *tsL141* antimutator, the costs involved destruction of a large number of correctly made base pairs. The burst size for *43+* (T4 wild type) is in the range of 100–200 progeny phage, whereas the antimutator *tsL141* burst size was reduced by a factor of between 2 and 3 (Drake and Allen 1969; Allen et al. 1970). Because no significant differences in dNTP pool sizes were observed in *E. coli* infected with T4 wild-type or *tsL141* antimutator phage (Hopkins and Goodman 1985), the lower *tsL141* burst could be a reflection of inefficient DNA synthesis caused by inefficient translocation of the mutant polymerase (Gillen and Nossal 1976).

B. Mutant Polymerase Phenotypes Reveal Novel Biological Insights

Simple biochemical explanations of genetic properties of mutator and antimutator polymerase alleles in *E. coli* and T4-infected *E. coli* describe only part of a more complex and interesting story.

1. *Interaction of Methylation-directed Mismatch Repair and Proofreading In* E. coli

Mutations in *mutD* (*DnaQ*), the structural gene coding for the ε subunit of the polymerase III holoenzyme complex (Scheuermann et al. 1983), led to the strongest mutator phenotype ever observed (Degnan and Cox 1974; Cox and Horner 1982). Spontaneous mutation rates in *E. coli* are generally on the order of about 10^{-10} per base pair replicated (Drake 1969). The spontaneous mutation frequencies in *mutD* backgrounds were roughly 10^{-5}, an increase of about 10^5-fold! This result contrasts with proofreading specificities measured in vitro that were about 95-99% and could therefore account for at most a 50-100-fold increase in mutagenesis. Until the recent experiments of Schaaper (1988; Schaaper and Radman 1989; Schaaper et al. 1989), it was extremely difficult to rationalize these excessively large differences in mutation frequencies observed in vivo from those predicted from the properties of the exonucleases measured in vitro.

Schaaper observed that *mutD* mutagenesis frequencies differed significantly for cells grown in rich medium compared to those grown in minimal medium (Schaaper 1988, 1989; Schaaper and Radman 1989). In rich medium, the mutation frequencies were roughly 10^5-fold greater than those of wild type, whereas in minimal medium, the mutation frequencies were increased by "only" 10^2-fold. The spectrum of mutations also differed significantly, transition mutations occurring in rich medium, whereas transversion and small frameshift mutational hot spots were predominant when the cells were grown in minimal medium.

Schaaper and Radman (Schaaper 1988; Schaaper and Radman 1989) proposed that the differences in mutational spectra and frequencies were caused by the contributions of methylation-directed postreplication mismatch repair (for a review on mismatch repair, see Modrich 1989; Chapter 9, this volume). In rich medium, cells grow rapidly and contain multiple chromosomes. Under rapid growth conditions, mismatch repair becomes saturated, permitting errors normally corrected by this system to survive as mutations; in minimal medium, methylation-directed mismatch repair can operate normally. The 10^3-fold difference in the mutation rates in rich versus minimal medium for *E. coli mutD-DnaQ* alleles suggests that methylation-directed mismatch repair corrects 99.9% of the errors that escape proofreading. Indeed, based on an analysis of the mutational spectra in minimal medium, most transversion and frameshift mutational hot spots can be attributed to primer-template slippage events, where primer-template strands undergo transient misalignment in the vicinity of the replication fork (Streisinger et al. 1967; Fowler et al. 1974; Kunkel and Soni 1988; Schaaper 1988). The suggestion that saturation of mismatch repair was responsible for excessively high mutation frequencies in rich medium was strongly supported by the observation that *mutD* mutation frequencies could be significantly reduced using an episome to provide the cell with additional MutH and MutL gene products, MutL being most effective. The presence of excess MutS or MutU proteins had no effect on mutation frequencies (Schaaper and Radman 1989).

The 10^2-fold increase in mutation rate in minimal medium provides a rough estimate of the contribution of proofreading to spontaneous mutation rates. The original expectation that proofreading should reduce spontaneous base substitutions by roughly 100-fold was based on (1) the assumption that incorrect base pairs can exist in transiently "annealed" conformations competent for elongation, and correct base pairs can assume melted conformations and undergo excision, and (2) more convincingly, the observation that proofreading specificities measured in vitro were generally no greater than 95-99% (Clayton et al. 1979; Sinha and Haimes 1981; Fersht et al. 1982; Sinha and Goodman 1983; Kunkel et al. 1984). Current experimental data appear to have validated these early expectations (Schaaper 1988).

2. An "Antimutator" with Mutator Properties

The T4 L141 polymerase, which had been shown to be antimutagenic, i.e., it exhibited a considerably smaller spontaneous mutation rate than wild type in an A·T→G·C transition pathway (Drake and Allen 1969; Drake et al. 1969), was also shown to mutate *more* than wild type with respect to small frameshifts (Ripley and Shoemaker 1983). Thus, a mutant polymerase that was found to behave as an antimutator in one mutagenic pathway behaved as a mutator in a completely different mutagenic pathway. Perhaps there are several changes in polymerase properties accompanying an increase in exonuclease activity caused by inefficient translocation. For example, perturbations affecting polymerase translocation might affect synthesis on transiently misaligned primer-template molecules. Insertions and deletions have been shown to result from DNA synthesis on misaligned templates (Streisinger et al. 1967; de Boer and Ripley 1984; Kunkel 1986; Kunkel and Bebenek 1988). Deletions occur following template strand slippage and insertions following primer strand slippage. How might polymerases cope with distorted DNA structures? A mutant enzyme might cause an increase in small frameshifts by synthesizing DNA from primer-templates containing internal loopout structures adjacent to perfectly matched termini. Maybe wild-type polymerase dissociates rapidly from DNA containing loopout regions near 3'-primer ends and so is unlikely to cause frameshift errors.

These simple speculations concerning the biochemical and structural bases for enhanced frameshift mutagenesis by the antimutator polymerase should be directly testable in vitro.

3. Communicating between Polymerase and Exonuclease Active Sites

There is a subtle point concerning the macroscopically measured N/P ratio: high for antimutator, intermediate for wild-type, and low for mutator polymerase. Antimutator, wild-type, and mutator T4 polymerases exhibited high, intermediate, and low exonuclease activities under both nonsynthesizing and synthesizing conditions (Muzyczka et al. 1972; Bessman et al. 1974; Clayton et al. 1979). As mentioned above, although saturating levels of the next correct nucleotide strongly inhibit 3' exonuclease activity, significant proofreading activity (dNTP→dNMP turnover activity) still persists (Clayton et al. 1979; Fersht et al. 1982).

A possible factor contributing to the high exonuclease/polymerase ratio of L141 antimutator polymerase may have to do with a diminished capacity for this enzyme to "slide" along the DNA to switch back and forth between polymerase and exonuclease active sites. If forward translocation were inhibited relative to wild type, then a mutant polymerase could show a high N/P ratio in vitro coupled with an antimutator phenotype in vivo. For the antimutator allele *CB120*, which was recently shown to be identical to *tsL141* (Reha-Krantz 1988, 1989), Gillen and Nossal (1976) found that the antimutator polymerase had "very low polymerase and high turnover activities...with poly[d(A-T)]...perhaps because of interference by intrastrand base pairing in the template." It is possible that an enhancement in small frameshift mutations observed for *tsL141* (Ripley and Shoemaker 1983) could be related to the enzyme's inefficient translocation during DNA synthesis, since "stuttering" backward and forward motions might encourage formation of transient loopout structures. It is also possible that there are abnormal interactions between L141 polymerase and the T4 polymerase accessory proteins, gp41, gp61, gp44-62, gp45, gp32, that could lead to frameshifts. It should be possible to address the question of primer-template conformation directly using a pre-steady-state fluorescence method designed to probe the structural properties of the primer-template DNA and dNTP substrates during DNA synthesis (see Section II.C).

An "opposite" class of translocation T4 mutant polymerases, mapping in the amino-terminal region of gene *43* (*mel62, L412M*), has been described by Reha-Krantz (1987, 1988). Although the purified polymerases have *normal* (wild-type) levels of proofreading exonuclease and polymerase activities, the phage exhibit mutator phenotypic behavior in vivo.

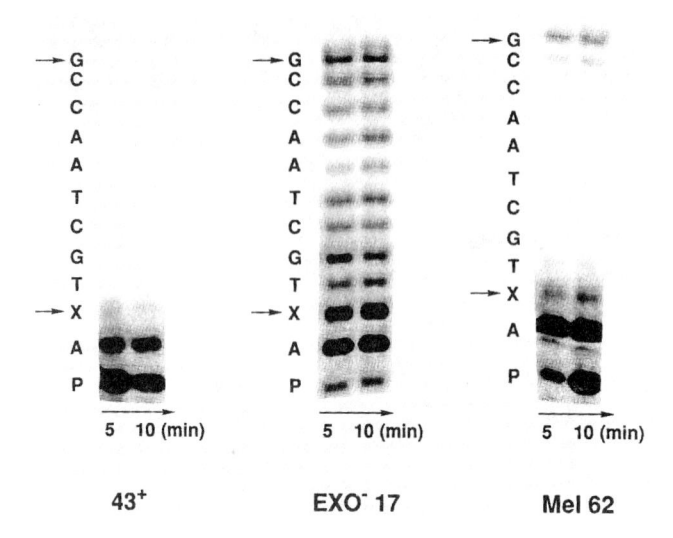

Figure 4 Nucleotide insertion and extension at an abasic site (X) by wild-type and mutant T4 DNA polymerases. 43⁺ is wild-type T4 DNA polymerase. EXO⁻ 17 is a mutant with less than 0.1% of the exonuclease activity of wild type. Mel 62 is a mutant that has polymerase and exonuclease activities indistinguishable from wild type.

These mutator polymerases appear to be more processive than wild type in vitro and show a much greater ability than wild-type polymerase to copy and bypass abasic (apurinic/apyrimidinic) template lesions in vitro. These two interesting properties are illustrated in Figure 4 for the translocation mutator T4 mel62 polymerase, relative to wild-type and exonuclease-deficient T4 polymerases (H. Cai et al., unpubl.). Each polymerase was used to extend a 5′^{32}P-labeled primer by addition of one nucleotide to reach a site-directed abasic template lesion, X. As anticipated, owing to its active exonuclease, the wild-type enzyme excised nucleotides inserted opposite X (Fig. 4, 43⁺). Elimination of the proofreading exonuclease allowed the exo⁻ polymerase to copy past the lesion efficiently (Fig. 4, EXO⁻ 17). The mel62 mutator polymerase, containing wild-type exonuclease and polymerase levels, was also able to copy the lesion (Fig. 4, Mel 62), although less efficiently than the exo⁻ mutant. However, in contrast with the exo⁻ mutant, once past the lesion, mel62 continued synthesis to the end of the template strand without dissociating. A significant fraction of exo⁻ polymerases dissociate at each template position beyond the lesion, as indicated by the presence of "pausing bands" at every template position.

A key conclusion to be drawn from the data in Figure 4 is that although proofreading is critically important in impeding lesion bypass, it is equally important that there be efficient coupling of polymerase and exonuclease activities. In other words, to excise nucleotides misinserted opposite template lesions effectively, the T4 polymerase must be able to switch back and forth efficiently between polymerase and exonuclease modes. Mutations in the amino-terminal region of gene *43* appear to modulate the degree of "communication" between the two active sites, and thereby also affect the enzyme's ability to translocate along the DNA.

C. Rapid Flow Analysis of Changes in 2-Aminopurine Fluorescence as a Probe of the Effects of Sequence Context on 3'→5' Exonuclease Activity and Specificity

The fluorescent base analog, 2-aminopurine (AP), can be used as a probe to measure the kinetics of nucleotide insertion and excision on the millisecond time scale on which these reactions occur. The dynamics within DNA duplexes (Nordlund et al. 1989; Guest et al. 1991) and DNA·polymerase interactions (L.B. Bloom et al., in prep.) can be measured on a nanosecond time scale also by using AP. AP serves as an excellent fluorescent probe due to its ability to base pair with thymine and because its fluorescence properties are sensitive to its environment (Ward et al. 1969). AP has been shown to form a Watson-Crick-type base pair with thymine (Fig. 5) without significant distortion of B-DNA structure (Sowers et al. 1986a; Nordlund et al. 1989). Because of its base pairing ability with thymine, AP would not be expected to cause severe steric interactions within an enzyme active site as some other more bulky fluorescent labels might cause. When AP is present in solution in the form of a free nucleotide, dAPMP, its steady-state fluorescence emission can be 25–125 times greater depending on sequence than when present at the 3'terminus of synthetic primer-templates (L.B. Bloom et al., in prep.). The kinetics of nucleotide insertion can therefore be measured by following the decrease in fluorescence when dAPMP is inserted into DNA, whereas the kinetics of nucleotide removal can be measured by following the increase in fluorescence as dAPMP is removed.

AP was used as a probe to investigate the effects of local DNA sequence on the proofreading efficiency of T4 DNA polymerase. The removal of AP from a "correct" AP·T base pair and incorrect AP·A, AP·C, and AP·G base pairs at the 3'-primer terminus of identical primer templates was measured to determine how the relative stability of a base pair affected the ability of T4 DNA polymerase to excise a nucleotide. Pre-steady-state kinetics of dAPMP removal were measured using rapid-mixing stopped-flow techniques by following the increase in AP fluorescence over the time course of the reaction. Removal of dAPMP was

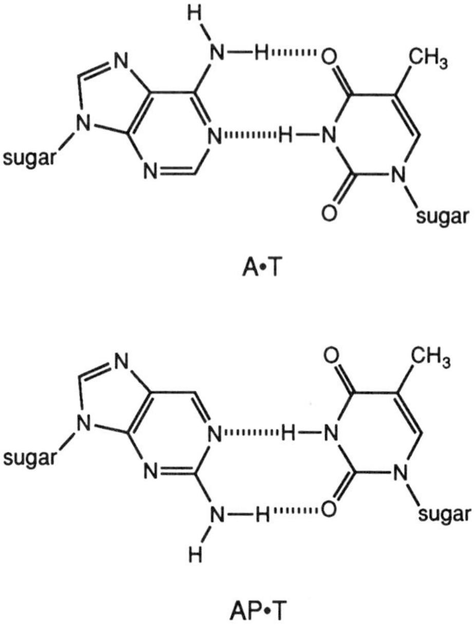

Figure 5 Watson-Crick type base pairs between adenine and thymine and AP and thymine.

slowest when paired opposite T. Rates of removal of dAPMP increase in the following order: AP·T (k_{exo} = 3.0 s^{-1}) < AP·A (k_{exo} = 7.1 s^{-1}) < AP·C (k_{exo} = 12 s^{-1}) < AP·G (k_{exo} = 18 s^{-1}) (L.B. Bloom et al., in prep.). The rates of removal of dAPMP from base pairs are inversely correlated with the melting temperatures for duplexes containing these base pairs. Melting temperatures were reported for four heptamer duplexes of identical sequence except for the base that was paired opposite a centrally located AP (Erijta et al. 1986). Because the heptamers were of identical sequence except for the AP base pair, differences in T_m values largely reflect differences in the stabilities of the AP base pairs. T_m values were 31.4, 28.0, 24.9, and 20.4°C for duplexes containing AP·T, AP·A, AP·C, and AP·G, respectively (Erijta et al. 1986), suggesting that the relative stabilities of the base pairs are AP·T>AP·A>AP·C>AP·G. The results of these experiments suggest that the less stable the base pair, the more efficiently it is removed by the proofreading exonuclease of T4 DNA polymerase.

Regions of DNA with a relatively high proportion of G·C base pairs tend to melt at higher temperatures and to be more stable than regions with a higher proportion of A·T base pairs. As a result, proofreading in G·C-rich regions of DNA may be less efficient than in more A·T-rich regions of DNA (Bessman and Reha-Krantz 1977; Petruska and Good-

A:
5'—C G G C P
3'—G C C G T A A—5'

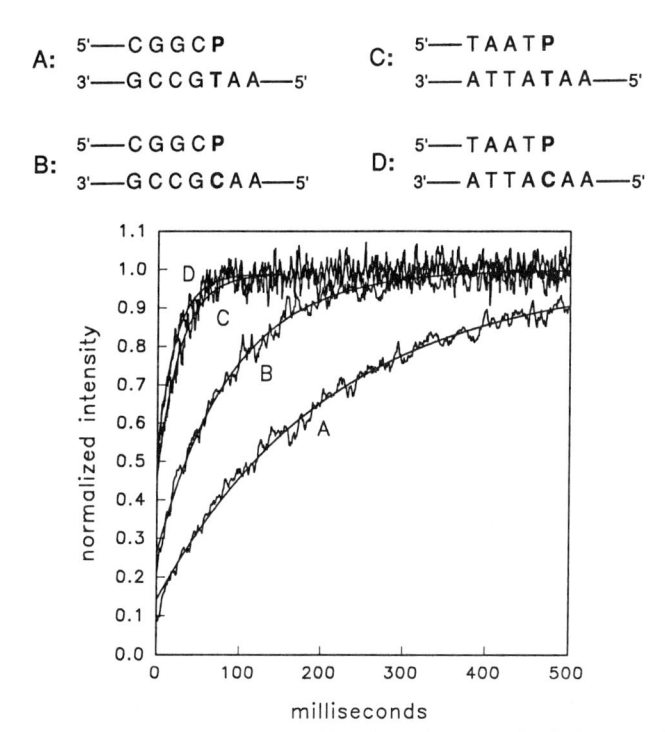

B:
5'—C G G C P
3'—G C C G C A A—5'

C:
5'—T A A T P
3'—A T T A T A A—5'

D:
5'—T A A T P
3'—A T T A C A A—5'

Figure 6 Pre-steady-state kinetics of removal of dAPMP from the 3'-primer terminus of synthetic primer-templates by T4 DNA polymerase at 20°C. 2-Aminopurine is designated as P at the primer-3' termini. Plots show the increase in fluorescence emission of AP over the time course of reactions. Reactions contained an excess of polymerase (400 nM) over 17mer/30mer DNA primer-templates (200 nM).

man 1985). The effect of the G·C content of the region upstream of the primer-3' terminus was examined by measuring the pre-steady-state kinetics of removal of dAPMP from AP·T and AP·C base pairs. Primer templates were of identical sequence except for the four base pairs immediately upstream of the AP base pair (Fig. 6). In one case, four G·C base pairs were immediately upstream of AP (Fig. 6A,B) and in the other, four A·T base pairs were immediately upstream of AP (Fig. 6C,D). As might be expected, dAPMP was excised fastest ($k_{exo} = 47$ s^{-1}) when mispaired opposite C in the A·T-rich sequence (Fig. 6D) and slowest ($k_{exo} = 4.7$ s^{-1}) when paired opposite T in the G·C-rich sequence (Fig. 6A) (L.B. Bloom et al., in prep.). A significant result was that AP was removed about three times faster when paired opposite T ($k_{exo} = 37$ s^{-1}) in the A·T-rich sequence (Fig. 6C) than when mispaired opposite C

(k_{exo} = 11 s^{-1}) in the G·C-rich sequence (Fig. 6B) (L.B. Bloom et al., in prep.). In contrast, AP was removed about four times slower when paired opposite T than when mispaired opposite C within identical sequence contexts as described above. This means that a "correct" base (AP·T) in the "wrong place" (A·T-rich sequence) can be removed by T4 DNA polymerase more efficiently than an incorrect nucleotide (AP·C) in the "right place" (G·C-rich sequence).

AP can also be used as a reporter of its environment to give information on a nanosecond time scale about dynamic interactions within a DNA duplex (Nordlund et al. 1989; Guest et al. 1991) and between polymerase and DNA. When AP is incorporated at the primer-3' terminus of a primer-template, its fluorescence is quenched relative to free dAPMP, whereas the fluorescence of AP in single-stranded DNA is intermediate between these two (L.B. Bloom et al., in prep.). The time-resolved fluorescence intensity and anisotropy decays are also sensitive to the environment of AP.

The intensity of fluorescence of free dAPTP decays as a single exponential with a lifetime of 7.85 ns, whereas the intensity of AP fluorescence at the primer terminus decays much faster. Four lifetime components are required to fit the data for duplex DNA, suggesting that AP may be in at least four resolvable states (L.B. Bloom et al., in prep.). A single rotational correlation time (0.149 ns) was observed for the smaller triphosphate molecule in solution, whereas long (~3-6 ns) and short (<1 ns) rotational correlation times were observed for AP at the 3' terminus of synthetic primer-templates. The shorter correlation time observed for AP in duplex DNA is most likely due to the internal motion of the base, whereas the longer correlation time is due to the motion of the large primer-template in solution. An important observation was that when a primer-template containing AP at the 3' terminus bound to T4 DNA polymerase in the presence of EDTA (i.e., no Mg^{++} present for catalysis), the steady-state fluorescence emission of AP increased (L.B. Bloom et al., in prep.). This increase in AP fluorescence was accompanied by a slower rate of decay of the fluorescence intensity.

This increase in the steady-state fluorescence of AP can be attributed to binding to the polymerase because of the dramatic increase in the rotational correlation time for the AP-containing DNA. Our interpretation of the increase in fluorescence on binding is that when the primer-template binds to the polymerase, it is held in such a way that the interactions between AP and the primer-template are decreased so that its fluorescence increases. It may be that the DNA is held in such a way that the primer terminus is separated from the template and possibly located in the exonuclease active site. Although this highly fluorescent bound state may

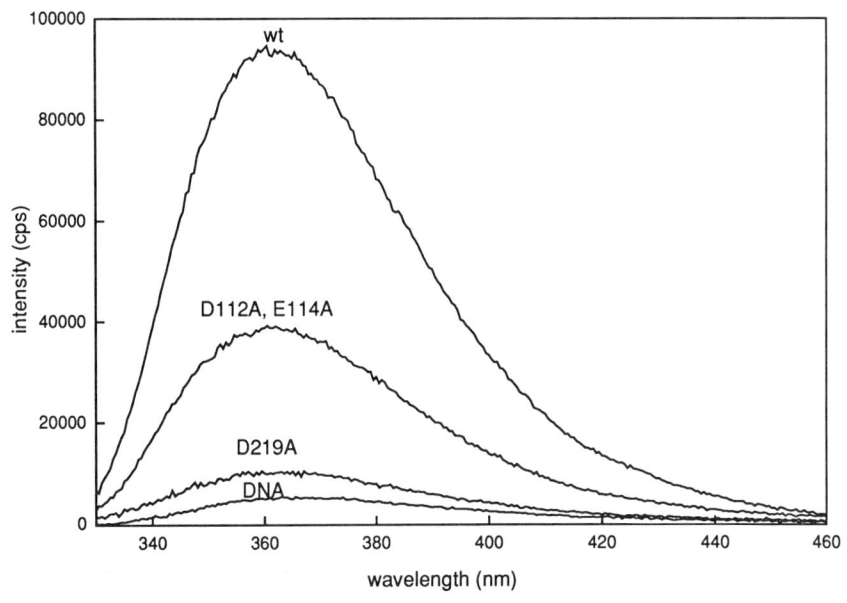

Figure 7 Steady-state fluorescence emission spectra for AP in different environments. AP is located at the 3'-primer terminus of a 17mer/30mer duplex (DNA). The AP-containing primer-template is bound to wild-type T4 DNA polymerase (wt) and bound to two mutant T4 DNA polymerases with reduced exonuclease activities. One mutant enzyme contains two amino acid substitutions (D112A, E114A), and the other contains a single-amino-acid substitution (D219A).

not be biologically significant because it occurs in the absence of Mg^{++}, which is required for catalysis and possibly for proper binding geometry of the primer-template, interesting differences in AP fluorescence occur when the AP-containing DNA is bound to wild-type T4 DNA polymerase and exonuclease-deficient mutant enzymes. When two different exonuclease-deficient mutant enzymes bind to a primer-template containing AP at the 3'terminus, the steady-state fluorescence of AP is not enhanced as much as with wild-type enzyme (Fig. 7). This result might suggest that part of the decrease in exonuclease activity of these mutants is a result of decreased binding in the exonuclease active site.

III. CONCLUSIONS AND FUTURE PROSPECTS

The potentially catastrophic effects of point mutations in genomic DNA has led to the development of sophisticated error-correcting machinery, including 3'→5'proofreading exonucleases that excise mismatches at the replication fork and endo- and exonucleases that can recognize and correct errors after replication has been completed. Genetic loci giving rise to "mutator" and "antimutator" phenotypes include genes coding for DNA polymerases and $3' \rightarrow 5'$ exonucleases. Biochemical studies comparing the properties of wild-type and mutant DNA polymerases of bacteriophage T4 and its host *E. coli* have helped to elucidate detailed mechanisms of error induction and correction at the replication fork.

The rapid increase in the availability of cloned and overexpressed polymerases, proofreading exonucleases, and polymerase accessory proteins has provided large quantities of purified proteins suitable for use in structural, physical, and biochemical studies. Experiments are currently under way to investigate interactions between DNA and dNTP substrates with individual replication enzymes and "holoenzyme" replication complexes. These studies involve structural analysis by X-ray diffraction and nuclear magnetic resonance, along with a variety of steady-state and pre-steady-state enzyme kinetic techniques, including fluorescence quench and depolarization. Key goals in the analysis of fidelity are to identify the individual enzymatic steps in synthetic and proofreading pathways and to determine the effects of local DNA sequence on the rates and specificities of nucleotide insertion and excision. The data should yield novel insights into the biochemical basis for mutational hot and cold spots and, more generally, in elucidating the balance between the requirement for high fidelity, to ensure genomic integrity, and the necessity to generate mutations so that organisms can evolve.

ACKNOWLEDGMENTS

Research support is gratefully acknowledged from National Institutes of Health grants GM-21422, GM-42554, AG-11398, and an NIH Postdoctoral Research Fellowship GM-15034. We also thank Dr. Joseph M. Beechem and Michael R. Otto (Vanderbilt University), Dr. Linda J. Reha-Krantz (University of Alberta), Dr. Ramon Eritja (CSIC, Barcelona), and Ms. Hong Cai (University of Southern California) for collaborating on the fluorescence and gel fidelity studies using bacteriophage T4 polymerases. Our special gratitude is reserved for Dr. John W. Drake (NIEHS), whose genetic studies helped to usher in the field of fidelity, and to Drs. Maurice J. Bessman (Johns Hopkins University) and Harrison Echols (University of California, Berkeley) whose elegant biochemical analyses provided a sound intellectual framework for relating in vitro biochemical data to in vivo genetic phenotypes.

This chapter is dedicated to Hatch Echols whose recent passing has taken from us a very close friend and immensely valued colleague.

REFERENCES

Aboul-ela, F., D. Koh, I.J. Tinoco, and F.H. Martin. 1985. Base-base mismatches. Thermodynamics of double helix formation for $dCA_3XA_3G + dCT_3YT_3G$ (X, Y = A,C,G,T). *Nucleic Acids Res.* **13:** 4811–4825.

Allen, E.F., I. Albrecht, and J.W. Drake. 1970. Properties of bacteriophage T4 mutants defective in DNA polymerase. *Genetics* **65:** 187–200.

Banerjee, S.K., R.B. Christensen, C.W. Lawrence, and J.E. LeClerc. 1988. Frequency and spectrum of mutations produced by a single *cis-syn* thymine-thymine cyclobutane dimer in a single-stranded vector. *Proc. Natl. Acad. Sci.* **85:** 8141–8145.

Bessman, M.J. and L.J. Reha-Krantz. 1977. Studies on the biochemical basis of spontaneous mutation. V. Effects of temperature on mutation frequency. *J. Mol. Biol.* **116:** 115–123.

Bessman, M.J., N. Muzyczka, M.F. Goodman, and R.L. Schnaar. 1974. Studies on the biochemical basis of spontaneous mutation. II. The incorporation of a base and its analogue into DNA by wild-type, mutator, and antimutator DNA polymerases. *J. Mol. Biol.* **88:** 409–421.

Bialek, G. and F. Grosse. 1993. An error-correcting proofreading exonuclease-polymerase that copurifies with DNA-polymerase-α-primase. *J. Biol. Chem.* **268:** 6024–6033.

Brenowitz, S., S. Kwack, M.F. Goodman, M. O'Donnell, and H. Echols. 1991. Specificity and enzymatic mechanism of the editing exonuclease of *Escherichia coli* DNA polymerase III. *J. Biol. Chem.* **266:** 7888–7892.

Breslauer, K.J., R. Frank, H. Blocker, and L.A. Marky. 1986. Predicting DNA duplex stability from the base sequence. *Proc. Natl. Acad. Sci.* **83:** 3746–3750.

Brown, T., W.N. Hunter, G. Kneale, and O. Kennard. 1986. Molecular structure of the G•A base pair in DNA and its implications for the mechanism of transversion mutations. *Proc. Natl. Acad. Sci.* **83:** 2402–2406.

Brown, T., G.A. Leonard, E.D. Booth, and J. Chambers. 1989. Crystal structure and stability of a DNA duplex containing A(anti)•G(syn) base-pairs. *J. Mol. Biol.* **207:** 455–457.

Brutlag, D. and A. Kornberg. 1972. Enzymatic synthesis of deoxyribonucleic acid, XXXVI. A proofreading function for the 3'→5' exonuclease activity in deoxyribonucleic acid polymerases. *J. Biol. Chem.* **247:** 241–248.

Clayton, L.K., M.F. Goodman, E.W. Branscomb, and D.J. Galas. 1979. Error induction and correction by mutant and wild type T4 DNA polymerases: Kinetic error discrimination mechanisms. *J. Biol. Chem.* **254:** 1902–1912.

Cowart, M.C., K.J. Gibson, D.J. Allen, and S.J. Benkovic. 1989. DNA substrate structural requirements for the exonuclease and polymerase activities of procaryotic and phage DNA polymerases. *Biochemistry* **28:** 1975–1983.

Cox, E.C. 1976. Bacterial mutator genes and the control of spontaneous mutagenesis. *Annu. Rev. Genet.* **10:** 135–156.

Cox, E.C. and D.L. Horner. 1982. Dominant mutators in *Escherichia coli*. *Genetics* **100:** 7–18.

Creighton, S., M.-M. Huang, H. Cai, N. Arnheim, and M.F. Goodman. 1992. Base mispair extension kinetics: Binding of avian myeloblastosis reverse transcriptase to matched and mismatched base pair termini. *J. Biol. Chem.* **267:** 2633–2639.

de Boer, J.G. and L.S. Ripley. 1984. Demonstration of the production of frameshift and base-substitution mutations by quasipalindromic DNA sequences. *Proc. Natl. Acad. Sci.* **81:** 5528–5531.

de Waard, A., A.V. Paul, and I.R. Lehman. 1965. The structural gene for deoxyribonucleic acid polymerase in bacteriophage T4 and T5. *Proc. Natl. Acad. Sci.* **54:** 1241-1248.

Degnan, G.E. and E.C. Cox. 1974. Conditional mutator gene in *Escherichia coli*: Isolation, mapping, and effector studies. *J. Bacteriol.* **117:** 477-487.

Derbyshire, V., P.S. Freemont, M.R. Sanderson, L. Beese, J.M. Friedman, C.M. Joyce, and T.A. Steitz. 1988. Genetic and crystallographic studies of the 3',5'-exonucleolytic site of DNA polymerase I. *Science* **240:** 199-201.

DiFrancesco, R., S.K. Bhatnagar, A. Brown, and M.J. Bessman. 1984. The interaction of DNA polymerase III and the product of the *Escherichia coli* mutator gene, *mutD*. *J. Biol. Chem.* **259:** 5567-5573.

Dosanjh, M.K., G. Galeros, M.F. Goodman, and B. Singer. 1991. Kinetics of extension of O^6-methylguanine paired with cytosine or thymine in defined oligonucleotide sequences. *Biochemistry* **30:** 11595-11599.

Drake, J.W. 1969. Comparative rates of spontaneous mutation. *Nature* **221:** 1132.

Drake, J.W. and E.F. Allen. 1969. Antimutagenic DNA polymerases of bacteriophage T4. *Cold Spring Harbor Symp. Quant. Biol.* **33:** 339-344.

Drake, J.W., E.F. Allen, S.A. Forsberg, R. Preparata, and E.O. Greening. 1969. Spontaneous mutation. Genetic control of mutation rates in bacteriophage T4. *Nature* **221:** 1128-1131.

Driggers, P.H. and K.L. Beattie. 1988. Effect of pH on base-mispairing properties of 5-bromouracil during DNA synthesis. *Biochemistry* **27:** 1729-1735.

Echols, H. and M.F. Goodman. 1991. Fidelity mechanisms in DNA replication. *Annu. Rev. Biochem.* **60:** 477-511.

Erijta, R.E., B.E. Kaplan, D. Mhaskar, L.C. Sowers, J. Petruska, and M.F. Goodman. 1986. Synthesis and properties of defined DNA oligomers containing base mispairs involving 2-aminopurine. *Nucleic Acids Res.* **14:** 5869-5884.

Fersht, A.R. 1979. Fidelity of replication of phage φX174 DNA by DNA polymerase III holoenzyme; spontaneous mutation by misincorporation. *Proc. Natl. Acad. Sci.* **76:** 4946-4950.

Fersht, A.R. and J.W. Knill-Jones. 1983. Contribution of 3'→5' exonuclease activity of DNA polymerase III holoenzyme from *Escherichia coli* to specificity. *J. Mol. Biol.* **165:** 669-682.

Fersht, A.R., J.W. Knill-Jones, and W.C. Tsui. 1982. Kinetic basis of spontaneous mutation. Misinsertion frequencies, proofreading specificities and cost of proofreading by DNA polymerases of *Escherichia coli*. *J. Mol. Biol.* **156:** 37-51.

Fijalkowska, I.J., R.L. Dunn, and R.M. Schaaper. 1993. Mutants of *Escherichia coli* with increased fidelity of DNA replication. *Genetics* (in press).

Fowler, R.G., G.E. Degnan, and E.C. Cox. 1974. Mutational specificity of conditional *Escherichia coli* mutator, *mutD5*. *Mol. Gen. Genet.* **133:** 179-191.

Freese, E.B. and E.F. Freese. 1967. On the specificity of DNA polymerase. *Proc. Natl. Acad. Sci.* **57:** 650-657.

Gaffney, B.L. and R.A. Jones. 1989. Thermodynamic comparison of the base pairs formed by the carcinogenic lesion O^6-methylguanine with reference both to Watson-Crick pairs and to mismatched pairs. *Biochemistry* **28:** 5881-5889.

Gillen, F.D. and N.G. Nossal. 1976. Control of mutation frequency by bacteriophage T4 DNA polymerase I. The CB120 antimutator DNA polymerase is defective in strand displacement. *J. Biol. Chem.* **251:** 5219-5224.

Goodman, M.F., S. Creighton, L.B. Bloom, and J. Petruska. 1993. Biochemical basis of DNA replication fidelity. *Crit. Rev. Biochem. Mol. Biol.* **28:** 83-126.

Guest, C.R., R.A. Hochstrasser, L.C. Sowers, and D.P. Millar. 1991. Dynamics of mismatched base pairs in DNA. *Biochemistry* **30:** 3271–3279.

Hershfield, M.S. 1973. On the role of deoxyribonucleic acid polymerase in determining mutation rates: Characterization of the defect in the T4 deoxyribonucleic acid polymerase caused by the *ts*L88 mutation. *J. Biol. Chem.* **248:** 1417–1423.

Hershfield, M.S. and N.G. Nossal. 1972. Hydrolysis of template and newly synthesized deoxyribonucleic acid by the 3′ to 5′ exonuclease activity of the T4 deoxyribonucleic acid polymerase. *J. Biol. Chem.* **247:** 3393–3404.

Hopkins, R.L. and M.F. Goodman. 1985. Ribonucleoside and deoxyribonucleoside triphosphate pools during 2-aminopurine mutagenesis in T4 mutator-, wild type-, and antimutator-infected *Escherichia coli*. *J. Biol. Chem.* **260:** 6618–6622.

Huang, M.-M., N. Arnheim, and M.F. Goodman. 1992. Extension of base mispairs by *Taq* DNA polymerase: Implications for single nucleotide discrimination in PCR. *Nucleic Acids Res.* **20:** 4567–4573.

Hunter, W.N., T. Brown, N.N. Anand, and O. Kennard. 1986a. Structure of an adenine-cytosine base pair in DNA and its implications for mismatch repair. *Nature* **320:** 552–555.

Hunter, W.N., G. Kneale, T. Brown, D. Rabinovich, and O. Kennard. 1986b. Refined crystal structure of an octanucleotide duplex with G·T mismatched base-pairs. *J. Mol. Biol.* **190:** 605–618.

Hunter, W.N., T. Brown, G. Kneale, N.N. Anand, D. Rabinovich, and O. Kennard. 1987. The structure of guanosine-thymidine mismatches in B-DNA at 2.5-Å resolution. *J. Biol. Chem.* **262:** 9962–9970.

Ide, H., Y.W. Kow, and S.S. Wallace. 1985. Thymine glycols and urea residues in M13 DNA constitute replicative blocks in vitro. *Nucleic Acids Res.* **13:** 8035–8052.

Joyce, C.M. 1989. How DNA travels between the separate polymerase and 3′→5′-exonuclease sites of DNA polymerase I (Klenow fragment). *J. Biol. Chem.* **264:** 10858–10866.

Kan, L.S., S. Chandrasegaran, S.M. Pulford, and P.S. Miller. 1983. Detection of a guanine X adenine base pair in a decadeoxyribonucleotide by proton magnetic resonance spectroscopy. *Proc. Natl. Acad. Sci.* **80:** 4263–4265.

Kim, J.-S. and N. Davidson. 1974. Electron microscope heteroduplex studies of sequence relations of T2, T4, and T6 bacteriophage DNAs. *Virology* **57:** 93–111.

Kornberg, A. 1988. DNA replication. *J. Biol. Chem.* **263:** 1–4.

Kornberg, A. and T.A. Baker. 1992a. DNA polymerase I of *E. coli*. In *DNA replication*, ch. 4, pp. 113–164. W.H. Freeman, New York.

——— . 1992b. Prokaryotic DNA polymerase other than *E. coli*. In *DNA replication*, ch. 5, pp. 165–196. W.H. Freeman, New York.

Kuchta, R.D., P. Benkovic, and S.J. Benkovic. 1988. Kinetic mechanism whereby DNA polymerase I (Klenow) replicates DNA with high fidelity. *Biochemistry* **27:** 6716–6725.

Kunkel, T.A. 1986. Frameshift mutagenesis by eucaryotic DNA polymerases in vitro. *J. Biol. Chem.* **261:** 13581–13587.

Kunkel, T.A. and K. Bebenek. 1988. Recent studies of the fidelity of DNA synthesis. *Biochim. Biophys. Acta* **951:** 1–15.

Kunkel, T.A. and A. Soni. 1988. Mutagenesis by transient misalignment. *J. Biol. Chem.* **263:** 14784–14789.

Kunkel, T.A., L.A. Loeb, and M.F. Goodman. 1984. On the fidelity of DNA replication. The accuracy of T4 DNA polymerases in copying φX174 DNA in vitro. *J. Biol. Chem.* **259:** 1539–1545.

Linn, S. 1991. How many pols does it take to replicate DNA? *Cell* **66:** 185-187.

Loeb, L.A. and T.A. Kunkel. 1982. Fidelity of DNA synthesis. *Annu. Rev. Biochem.* **52:** 429-457.

McHenry, C.S. 1991. DNA polymerase III holoenzyme: Components, structure, and mechanism of a true replicative complex. *J. Biol. Chem.* **266:** 19127-19130.

Mendelman, L.V., J. Petruska, and M.F. Goodman. 1990. Base mispair extension kinetics: Comparison of DNA polymerase α and reverse transcriptase. *J. Biol. Chem.* **265:** 2338-2346.

Mendelman, L.V., M.S. Boosalis, J. Petruska, and M.F. Goodman. 1989. Nearest neighbor influences on DNA polymerase insertion fidelity. *J. Biol. Chem.* **264:** 14415-14423.

Modrich, P. 1987. DNA mismatch correction. *Annu. Rev. Biochem.* **56:** 435-466.

———. 1989. Methyl-directed DNA mismatch correction. *J. Biol. Chem.* **264:** 6597-600.

Moore, P.D., K.K. Bose, S.D. Rabkin, and B.S. Strauss. 1981. Sites of termination of in vitro DNA synthesis on ultraviolet- and N-acetylaminofluorene-treated φX174 templates by prokaryotic and eukaryotic DNA polymerases. *Proc. Natl. Acad. Sci.* **78:** 110-114.

Mosbaugh, D.W. and R.R. Meyer. 1980. Interaction of mammalian deoxyribonuclease V, a double strand 3′→5′ and 5′→3′ exonuclease, with deoxyribonucleic acid polymerase β from the Novikoff hepatoma. *J. Biol. Chem.* **255:** 10239-10247.

Muzyczka, N., R.L. Poland, and M.J. Bessman. 1972. Studies on the biochemical basis of spontaneous mutation. I. A comparison of the deoxyribonucleic acid polymerase of mutator, antimutator, and wild type strains of bacteriophage T4. *J. Biol. Chem.* **247:** 7116-7122.

Nordlund, T.M., S. Andersson, L. Nilsson, R. Rigler, A. Gråslund, and L.W. McLaughlin. 1989. Structure and dynamics of a fluorescent DNA oligomer containing the *Eco*RI recognition sequence: Fluorescence, molecular dynamics, and NMR studies. *Biochemistry* **28:** 9095-9103.

Patel, D.J., S.A. Kozlowski, S. Ikuta, and K. Itakura. 1984a. Deoxyadenosine-deoxycytidine pairing in the d(C-G-C-G-A-A-T-T-C-A-C-G) duplex: Conformation and dynamics at and adjacent to the dA x dC mismatch site. *Biochemistry* **23:** 3218-3226.

———. 1984b. Dynamics of DNA duplexes containing internal G•T, G•A, A•C, and T•C pairs: Hydrogen exchange at and adjacent to mismatch sites. *Fed. Proc.* **43:** 2663-2670.

Perrino, F.W. and L.A. Loeb. 1989. Differential extension of 3′ mispairs is a major contribution to the high fidelity of calf thymus DNA polymerase-α. *J. Biol. Chem.* **264:** 2898-2905.

Perrino, F.W., B.D. Preston, L.L. Sandell, and L.A. Loeb. 1989. Extension of mismatched 3′ termini of DNA is a major determinant of the infidelity of human immunodeficiency virus type 1 reverse transcriptase. *Proc. Natl. Acad. Sci.* **86:** 8343-8347.

Petruska, J. and M.F. Goodman. 1985. Influence of neighboring bases on DNA polymerase insertion and proofreading fidelity. *J. Biol. Chem.* **260:** 7533-7539.

Petruska, J., M.F. Goodman, M.S. Boosalis, L.C. Sowers, C. Cheong, and I. Tinoco, Jr. 1988. Comparison between DNA melting thermodynamics and DNA polymerase fidelity. *Proc. Natl. Acad. Sci.* **85:** 6252-6256.

Radman, M. and R. Wagner. 1986. Mismatch repair in *Escherichia coli*. *Annu. Rev. Genet.* **20:** 523-538.

Reddy, M.K., S.E. Weitzel, and P.H. von Hippel. 1992. Processive proofreading is intrinsic to T4 DNA polymerase. *J. Biol. Chem.* **267:** 14157–14166.

Reha-Krantz, L.J. 1987. Genetic and biochemical studies of the bacteriophage T4 DNA polymerase. *UCLA Symp. Mol. Cell. Biol. New Ser.* **47:** 501–509.

———. 1988. Amino acid changes coded by bacteriophage T4 DNA polymerase mutator mutants: Relating structure to function. *J. Mol. Biol.* **202:** 711–724.

———. 1989. Locations of amino acid substitutions in bacteriophage T4 *ts*L56 DNA polymerase predict an *N*-terminal exonuclease domain. *J. Virol.* **63:** 4762–4766.

Ripley, L.S. and N.B. Shoemaker. 1983. A major role for bacteriophage T4 DNA polymerase in frameshift mutagenesis. *Genetics* **103:** 353–366.

Schaaper, R.M. 1988. Mechanisms of mutagenesis in the *Escherichia coli* mutator mutD5: Role of DNA mismatch repair. *Proc. Natl. Acad. Sci.* **85:** 8126–8130.

———. 1989. *Escherichia coli* mutator *mutD5* is defective in the *mutHLS* pathway of DNA mismatch repair. *Genetics* **121:** 205–212.

Schaaper, R.M. and M. Radman. 1989. The extreme mutator effect of *Escherichia coli mutD5* results from saturation of mismatch repair by excessive DNA replication errors. *EMBO J.* **8:** 3511–3516.

Schaaper, R.M., B.I. Bond, and R.G. Fowler. 1989. A·T→C·G transversions and their prevention by the *Escherichia coli mutT* and *mutHLS* pathways. *Mol. Gen. Genet.* **219:** 256–262.

Scheuermann, R.H. and H. Echols. 1984. A separate editing exonuclease for DNA replication: The ε-subunit of *Escherichia coli* DNA polymerase holoenzyme. *Proc. Natl. Acad. Sci.* **81:** 7747–7751.

Scheuermann, R., S. Tam, P.M. Burgers, C. Lu, and H. Echols. 1983. Identification of the ε-subunit of *Escherichia coli* DNA polymerase III holoenzyme as the dnaQ gene product: A fidelity subunit for DNA replication. *Proc. Natl. Acad. Sci.* **80:** 7085–7089.

Singer, B., F. Chavez, S.J. Spengler, J.T. Kusmierek, L. Mendelman, and M.F. Goodman. 1989. Comparison of polymerase insertion and extension kinetics of a series of O^2-alkyldeoxythymidine triphosphates and O^4-methyldeoxythymidine triphosphate. *Biochemistry* **28:** 1478–1483.

Sinha, N.K. and M.F. Goodman. 1983. Fidelity of DNA replication. In *Bacteriophage T4* (ed. C.K. Mathews et al.), pp. 131–137. American Society for Microbiology, Washington, D.C.

Sinha, N.K. and M.D. Haimes. 1981. Molecular mechanisms of substitution mutagenesis. An experimental test of the Watson-Crick and Topal-Fresco models of base mispairings. *J. Biol. Chem.* **256:** 10671–10683.

Skarnes, W., P. Bonin, and E. Baril. 1986. Exonuclease activity associated with a multiprotein form of HeLa DNA polymerase α. *J. Biol. Chem.* **261:** 6629–6636.

Sowers, L.C., R. Eritja, B. Kaplan, M.F. Goodman, and G.V. Fazakerly. 1988. Equilibrium between a wobble and ionized base pair formed between fluorouracil and guanine in DNA as studied by proton and fluorine NMR. *J. Biol. Chem.* **263:** 14794–14801.

Sowers, L.C., G.V. Fazakerley, R. Eritja, B.E. Kaplan, and M.F. Goodman. 1986a. Base pairing and mutagenesis: Observation of a protonated base pair between 2-aminopurine and cytosine in an oligonucleotide by proton NMR. *Proc. Natl. Acad. Sci.* **83:** 5434–5438.

Sowers, L.C., G.V. Fazakerley, H. Kim, L. Dalton, and M.F. Goodman. 1986b. Variation of nonexchangeable proton resonance chemical shifts as a probe of aberrant base pair formation in DNA. *Biochemistry* **25:** 3983–3988.

Sowers, L.C., M.F. Goodman, R. Eritja, B. Kaplan, and G.V. Fazakerley. 1989a. Ionized and wobble base-pairing for bromouracil-guanine in equilibrium under physiological conditions. A nuclear magnetic resonance study on an oligonucleotide containing a bromouracil-guanine base-pair as a function of pH. *J. Mol. Biol.* **205:** 437–447.

Sowers, L.C., R. Eritja, F.M. Chen, T. Khwaja, B.E. Kaplan, M.F. Goodman, and G.V. Fazakerley. 1989b. Characterization of the high pH wobble structure of the 2-aminopurine.cytosine mismatch by N-15 NMR spectroscopy. *Biochem. Biophys. Res. Commun.* **165:** 89–92.

Speyer, J.F. 1965. Mutagenic DNA polymerase. *Biochem. Biophys. Res. Commun.* **21:** 6–8.

Speyer, J.F., J.D. Karam, and A.B. Lenny. 1967. On the role of DNA polymerase in base selection. *Cold Spring Harbor Symp. Quant. Biol.* **31:** 693–697.

Streisinger, G., Y. Okada, J. Emrich, J. Newton, A. Tsugita, E. Terzaghi, and M. Inouye. 1967. Frameshift mutations and the genetic code. *Cold Spring Harbor Symp. Quant. Biol.* **31:** 77–84.

Taylor, J.S. and C.L. O' Day. 1990. *cis-syn* Thymine dimers are not absolute blocks to replication by DNA polymerase I of *Escherichia coli* in vitro. *Biochemistry* **29:** 1624–1632.

Wang, T.S. 1991. Eukaryotic DNA polymerases. *Annu. Rev. Biochem.* **60:** 513–552.

Ward, D.C., E. Reich, and L. Stryer. 1969. Fluorescence studies of nucleotides and polynucleotides. I. Formycin, 2-aminopurine riboside, 2,6-diaminopurine riboside, and their derivatives. *J. Biol. Chem.* **244:** 1228–1237.

Warner, H.R. and J.E. Barnes. 1966. Deoxyribonucleic acid synthesis in *Escherichia coli* infected with some deoxyribonucleic acid polymerase-less mutants of bacteriophage T4. *J. Virol.* **28:** 100–107.

Watson, J.D. and F.H.C. Crick. 1953a. Genetical implications of the structure of deoxyribonucleic acid. *Nature* **171:** 964–967.

———. 1953b. The structure of DNA. *Cold Spring Harbor Symp. Quant. Biol.* **18:** 123–131.

Wong, I., S.S. Patel, and K.A. Johnson. 1991. An induced-fit kinetic mechanism for DNA replication fidelity: Direct measurement by single-turnover kinetics. *Biochemistry* **30:** 526–537.

Wood, W.B. and H.R. Revel. 1976. The genome of bacteriophage T4. *Bacteriol. Rev.* **40:** 847–868.

Yu, H., R. Eritja, L.B. Bloom, and M.F. Goodman. 1993. Ionization of bromouracil and fluorouracil stimulates base mispairing frequencies with guanine. *J. Biol. Chem.* **268:** 15935–15943.

9
Nucleases Involved in DNA Repair

R. Stephen Lloyd
Sealy Center for Molecular Science
University of Texas Medical Branch
Galveston, Texas 77555-0852

Stuart Linn
Division of Biochemistry and Molecular Biology
University of California
Berkeley, California 94720

I. **A Word on Nomenclature**
II. **Types of DNA Damage and Causative Agents**
   A. Base Changes and Mismatched Base Pairs
   B. Baseless Sites
   C. Cross-links
   D. Strand Breaks, Deletions, and Duplications
   E. Alkylation and Bulky Adducts
   F. Oxidative Damage
III. **Base Excision Repair**
   A. General Considerations Concerning DNA Glycosylases and DNA Glycosylase/AP Lyases
   B. Oxidized Pyrimidines
     1. *E. coli* Endonuclease III and Closely Related Enzymes
     2. Mammalian UV Endonucleases I and II
   C. Oxidized Purines: FPG Glycosylase (8-Oxoguanine DNA Glycosylase: Mut *M* Gene Product)
   D. Deaminated Bases
     1. Uracil DNA Glycosylase
     2. G/T Mismatch-specific Thymine DNA Glycosylase (Mammalian) and Endonuclease (*E. coli*)
       a. Mammalian cell studies
       b. Studies in *E. coli*
     3. 5-Hydroxymethyluracil DNA Glycosylase
     4. Hypoxanthine DNA Glycosylase
   E. Alkylated Bases: N-Methylpurine DNA Glycosylases
   F. Cyclobutane Pyrimidine Dimers
     1. T4 Endonuclease V
     2. *M. luteus* (*M. lysodeikticus*) UV Endonuclease
     3. Yeast Pyrimidine Dimer Endonuclease
   G. Mismatched Bases: *MUTY*—An Adenine DNA Glycosylase for A/G or A/C Matches
   H. Other DNA Glycosylases: 1,N6-Ethenoadenine DNA Glycosylase

Nucleases, 2nd Edition
© 1993 Cold Spring Harbor Laboratory Press 0-87969-426-2/93 $5 + .00

I. Endonucleases
      1. Classification of Apurinic/Apyrimidinic Endonucleases
      2. Prokaryotic
         a. *E. coli* Exonuclease III
         b. *E. coli* Endonuclease IV
      3. Eukaryotic
         a. Yeast Apurinic Endonuclease, APN1
         b. *Drosophila* Apurinic Endonucleases
         c. Mammalian Apurinic Endonucleases
  IV. **Structure/Function Analyses of DNA Glycosylases/AP Lyases**
      A. T4 Endonuclease V
         1. X-ray Crystal Structure
         2. Identification of the Active Site Residue
         3. Amino Acids Involved in Nontarget DNA Binding
      B. *E. coli* Endonuclease III
   V. **End-trimming Nucleases**
  VI. **Nucleotide Excision Repair**
      A. UvrABC
         1. Overview
         2. Functional Analyses of Domains within the Proteins That Comprise the *E. coli* UvrABC Complex
            a. ATP-binding Domains
            b. Zinc Finger Motifs
            c. Helix-Turn-Helix Motif
            d. Catalytically Important Residues
      B. The Mammalian Counterpart to UvrABC
      C. The RAD1/RAD2 Incision Complex
      D. Other Incision Endonucleases
 VII. **Transcriptional Regulation of Repair Nucleases during Stress Responses**
      A. The SOS Response
      B. The Adaptive (ada) Response
      C. Responses to Oxidative Stress
VIII. **Methyl-directed Mismatch Repair**
      A. MutH
      B. Mismatch Repair in Other Prokaryotes and Eukaryotes
  IX. **Perspectives**

I. A WORD ON NOMENCLATURE

The problem of nomenclature is especially acute in the case of repair nucleases. The most critical problem arises from the discovery of "AP endonucleases" that act through a β-elimination (lyase), rather than by a hydrolytic mechanism (see Chapter 1). Because these enzymes are generally thought of as "nucleases" by investigators outside the field of repair, and because they clearly fall into the purview of "DNA-repair nucleases," they are included in the list of "repair nucleases" noted here; i.e., they are proteins that catalytically cleave phosphodiester bonds within a polynucleotide.

A second problem of nomenclature arises when proteins have several names due to their having been independently isolated by several laboratories, possibly assaying separate activities, or when homologous enzymes have been isolated from several organisms. In this chapter, we refer to them by their accepted gene name (e.g., *Escherichia coli* Fpg protein) or by their *E. coli* designation where several bacterial homologs have been described (e.g., exonuclease III).

II. TYPES OF DNA DAMAGE AND CAUSATIVE AGENTS

Virtually all DNA-damaging agents cause multiple types of lesions, and most lesions can be formed by several agents. In addition, the major biological response to a DNA-damaging agent might not be elicited by the most abundant lesion produced. For these reasons, it is not always possible to prepare specific substrates for particular nucleases or to identify specific nucleases involved in processing these particular lesions. The following types of DNA damage are intended to exemplify the situation and to illustrate the need for a number of DNA-repair systems; it is not an exhaustive catalog of each of the types of DNA damage.

A. Base Changes and Mismatched Base Pairs

Alterations in DNA sequence arise from errors in replication, by recombination between nonidentical molecules to produce heteroduplex mismatches and by spontaneous or chemically induced changes, such as those formed by deamination.

B. Baseless Sites

These apurinic or apyrimidinic (AP) sites arise spontaneously due to acid-catalyzed hydrolysis, by the action of DNA glycosylases (see below), by exposure to chemicals (particularly alkylating agents), or by radiation. Due to spontaneous hydrolysis, the DNA of a mammalian cell probably loses approximately 5,000–10,000 purines and 200–500 pyrimidines per 20-hour generation time (Lindahl 1979). Damaging agents increase this burden.

C. Cross-links

Interstrand cross-links arise after exposure to radiation, or to active oxygen species, probably through the action of free radicals on baseless sites that are formed. Baseless sites yield cross-links by the formation of a Schiff's base between the free aldehyde group of the baseless sugar and an amino group present on a base in the opposite strand. Finally, interstrand cross-links are also formed by bifunctional alkylating agents such as nitrogen mustard or by the light-activated psoralen compounds.

Intrastrand cross-links are exemplified by pyrimidine dimers, damage by bifunctional alkylating agents, and protein-DNA cross-links induced by radiation. Generally, intrastrand cross-links are presumed to be less difficult to repair than interstrand cross-links.

D. Strand Breaks, Deletions, and Duplications

DNA-damaging agents cause single- and double-strand breaks by a variety of mechanisms. Often, the break has an adjacent lesion, e.g., a baseless sugar. Deletions or duplications are presumed to be formed during replication or recombination.

E. Alkylation and Bulky Adducts

Electrophiles that react with the DNA bases can be simple alkyl groups or bulky ring compounds. Each type of electrophile has its own pattern of base addition and is dependent to a large extent not only on whether SN_1 or SN_2 reactions occur, but also on steric and other factors. Therefore, each agent has its peculiar biological responses. The multitude of possible addition sites for each alkylating and bulky agent, some of which, like S-adenosylmethionine, normally occur in the cell, calls for a number of specific repair systems.

F. Oxidative Damage

Oxidative DNA damage initiated by radiation, active oxygen species, and various other oxidizing agents is a major source of DNA damage. The precise nature of the damage is usually complex and includes about 50 characterized base damages as well as damaged sugar residues remaining at the termini of strand breaks that must be trimmed off.

III. BASE EXCISION REPAIR

A. General Considerations Concerning DNA Glycosylases and DNA Glycosylase/AP Lyases

The structural integrity of the DNA bases within cells is continuously being challenged by both spontaneous decomposition and a variety of damaging agents, including chemicals, ionizing radiation, and UV irradiation. Cells have elaborate nucleotide excision repair systems as well as recombinational pathways to enhance cell survival, but it is the base excision repair pathway that provides a major line of defense against mutagenesis and cell killing by the more common base lesions. The enzymes that initiate the base excision repair pathway consist of a large array of DNA glycosylases which may also contain a concomitant AP lyase activity. These enzymes are responsible for monitoring DNA for the presence of damaged bases, binding to those sites and catalyzing the breakage of the N1-C1′ glycosyl bond. If the glycosylase has an associated AP lyase activity, a break is introduced into the sugar-phosphate backbone leaving a 3′-α,β-unsaturated aldehyde and a 5′-phosphate (see Chapter 1). Once the altered base has been released, and independent of whether there is an associated AP lyase activity, these abasic sites must be processed by an endonuclease to create a 3′-hydroxyl, which is the necessary substrate for subsequent polymerization and ligation.

The DNA glycosylases are low-molecular-mass proteins, generally ranging from 16 kD to 42 kD, and do not require metal cofactors or exogenous energy sources for activity. Except for uracil DNA glycosylase and the *Micrococcus luteus* UV correndonuclease I, all DNA glycosylases examined show a strong preference for incising bases from adducted double-stranded DNA over single-stranded DNA. An additional feature of these enzymes that should be considered in structure/function analyses is their relative affinity for undamaged DNA. A high degree of nontarget DNA-binding affinity has been shown to facilitate the enzyme in locating its target DNA base within large domains of unmodified DNA by reducing the dimensionality of its search from three-dimensional to one-dimensional. This property has been demonstrated for T4 endonuclease V, *M. luteus* UV endonuclease, and *E. coli* uracil DNA glycosylase (Gruskin and Lloyd 1986; Hamilton and Lloyd 1989; Higley and Lloyd 1993).

In this chapter, we will not give an extensive review of all of the available literature on any specific glycosylase, glycosylase/AP lyase, or AP endonuclease, but rather we emphasize major forms of base damage and the enzymes that initiate the repair process at those altered sites, i.e., oxidized pyrimidines, oxidized purines, deaminated bases, alkylated bases, UV-induced cyclobutane pyrimidines, and mismatched bases.

B. Oxidized Pyrimidines

1. E. coli *Endonuclease III and Closely Related Enzymes*

Exposure of DNA to ionizing radiation or oxygen radicals produces a large number of base lesions, the pyrimidines being particularly susceptible to damage. The DNA glycosylase/AP lyase in *E. coli* responsible for the initiation of repair at many of these lesions is endonuclease III, the product of the *nth* gene (Gates and Linn 1977; Demple and Linn 1980; Warner et al. 1980; Katcher and Wallace 1983). This 23.5-kD protein recognizes thymine glycols (Gates and Linn 1977; Demple and Linn 1980; Katcher and Wallace 1983), 5,6-dihydrothymine (Demple and Linn 1980), 5-hydroxy-5-methylhydantoin (Breimer and Lindahl 1984), methyltartronylurea (Breimer and Lindahl 1980; Katcher and Wallace 1983), and at least one cytosine photoproduct (Doetsch et al. 1986; Helland et al. 1986). The rate of the glycosylase reaction is dependent on the substrate used and the topology of the DNA containing the lesion (Kow and Wallace 1987). For example, endonuclease III shows a preference for supercoiled DNAs containing thymine glycols over nicked or relaxed DNA substrates, whereas the superhelical density of the substrate does not affect reactivities with urea-containing or abasic substrates. Interestingly, the glycosylase plus AP lyase activity is ten- and twofold greater than AP lyase activity alone when comparing urea- and thymine-glycol-containing substrates, respectively, to abasic DNA.

Following the cloning of the *nth* gene (Asahara et al. 1989), the overexpressed endonuclease III was demonstrated to contain a 4Fe-4S cluster in the 2+ oxidation state with a zero net spin (Cunningham et al. 1989) by biophysical techniques, including Mössbauer and electron paramagnetic resonance (EPR) spectroscopy and elemental analysis. Although there was some speculation that the 4Fe-4S cluster could play a role in catalysis, subsequent studies of the stereochemical course of the reaction (Mazumder et al. 1991) (see Chapter 1) and the X-ray crystal structure (Kuo et al. 1992a,b) showed that this was not likely. In addition, resonance Raman spectroscopic analyses rule out a catalytic role of the 4Fe-4S cluster (Fu et al. 1992). Endonuclease III catalyzes the DNA strand cleavage reaction on the 3'side of the aldehyde abasic site by a *syn*-β-elimination, with the abstraction of the 2'-*pro*-S proton, to form a 3'-α,β-unsaturated aldehyde (Mazumder et al. 1991) (see Chapter 1).

Enzymes very similar to *E. coli* endonuclease III appear to be ubiquitous and have been described in calf thymus (Bacchetti and Benne 1975; Doetsch et al. 1986; Helland et al. 1986), mouse cells (Nes and Nissen-Mayer 1978; Nes 1980; Helland et al. 1985; Kim and Linn 1989), human lymphoblasts (Brent 1973, 1976, 1983), *M. luteus* (Jorgensen et al. 1987), and yeast (Gossett et al. 1988).

2. Mammalian UV Endonucleases I and II

Kim and Linn (1989) purified and characterized two UV endonucleases (I and II) from murine plasmacytoma cells. A similar activity had been previously reported for a 28-kD enzyme that recognized and incised DNA at abasic sites as well as DNA that had been damaged with UV light, γ-rays and OsO_4 (Nes 1980). These data suggested that the substrate might be a saturated pyrimidine, 5,6-dihydroxy-dihydrothymine. Kim and Linn demonstrated that neither enzyme appeared to recognize the two major UV photoproducts, the cyclobutane pyrimidine dimer or the 6-4 bipyrimidine photoproduct. Rather, both of these enzymes recognized thymine glycol damage. The mechanism of incision was demonstrated to proceed by way of a coupled DNA glycosylase/AP lyase reaction. Although these enzymes share substrate specificity, UV endonuclease I is larger (43 kD) and requires different parameters for optimal activity (e.g., salt, pH, and detergent effects) relative to UV endonuclease II (28 kD). UV endonuclease I acts on both superhelical and relaxed DNAs, whereas UV endonuclease II only incises superhelical DNAs. A mitochondrial UV endonuclease was also described with similar activities (Tomkinson et al. 1990).

C. Oxidized Purines: FPG Glycosylase (8-Oxoguanine DNA Glycosylase; MutM Gene Product)

Chetsanga and Lindahl (1979) and Laval and co-workers (Boiteux et al. 1984, O'Connor et al. 1988) first identified the Fpg gene product as an activity that catalyzes the removal of the imidazole ring-opened form of N7-methylguanine, 2,6-diamino-4-hydroxy-5-N-methyl-formamido-pyrimidine and 4,6-diamino-5-N-methyl-formamidopyrimidine. The same enzyme was independently isolated while investigating an activity that acts upon 8-oxo-7,8-dihydro-2'-deoxyguanosine (8-oxodG), a base lesion formed in DNA by ionizing radiation or oxygen radicals (Chung et al. 1991). Repair of 8-oxodG has been reviewed recently (Tchou and Grollman 1993).

Tchou et al. (1991) and Michaels and Miller (1992) demonstrated that 8-oxoguanine was released from synthetic duplexes containing 8-oxodG across from dC, dG, or dT but not across from dA, whereas 8-oxodA was not removed from duplex DNA irrespective of the nucleotide opposite it. The inability of the Fpg protein to catalyze the removal of the adducted guanine when it is opposite a dA can be rationalized since in vitro replication past 8-oxodG can result in the incorporation of dA, a mutagenic event (Shibutani et al. 1991). Thus, the removal of 8-oxodG across from a misincorporated dA would result in a G to T transversion following repair synthesis. Indeed, this has been shown to occur in vivo during phage and plasmid transformation assays (Wood et al. 1990; Moriya et al. 1991). Moreover, Michaels et al. (1991, 1992) showed that the MutM protein, which prevents G to T transversions in *E. coli*, is identical to the Fpg protein.

The cloning and overexpression of the *fpg* gene (Boiteux et al. 1987, 1990) have greatly facilitated biochemical and site-directed mutagenesis studies on the protein. In addition to its glycosylase activity, Fpg was also shown to contain an associated AP lyase activity, catalyzing a β-elimination reaction (Bailly et al. 1989a,b; Boiteux et al. 1990), and following β-elimination, an associated δ-elimination yields both 3' and 5' phosphomonoesters at a single nucleotide gap (Bailly et al. 1989a,b; Graves et al. 1992a,b). Although this scheme is believed to proceed by a concerted three-step reaction, glycosylase activity is not a mandatory first step in the reaction scheme since Fpg protein will also catalyze the removal of sugar residues at abasic sites.

The Fpg protein acts as a monomer and based on atomic absorption data contains one zinc atom/protein monomer, consistent with the presence of a zinc finger motif (Boiteux et al. 1990). To investigate the requirement for the zinc finger in enzyme catalysis and the role of other cysteine residues within the enzyme, biochemical analyses were performed on the Fpg protein and site-directed mutagenesis was carried out on each of the six cysteine residues (O'Connor et al. 1993). When denatured Fpg protein was renatured in the presence of $^{65}Zn^{++}$, one zinc atom was bound per enzyme molecule, and this binding could be competed with Cu^{++}, Cd^{++}, Hg^{++}, and Zn^{++} but not other divalent cations, a result consistent with the characteristics of other zinc finger proteins. In addition, cysteine-specific reagents, hydroxymercuriphenylsulfonic acid and methylmethanethiosulfonate, inhibit both DNA glycosylase and AP lyase activities, suggesting either a catalytic or structural role for cysteine residues within the protein.

Finally, single cysteine to glycine mutations within the Cys-X_2-Cys-X_{16}-Cys-X_2-Cys- consensus zinc-binding domain greatly reduced the binding affinity for Zn^{++}. These mutants retained only very low levels of all three catalytic activities, but mutations at cysteine residues outside the zinc finger consensus sequence retained the majority of their glycosylase/AP lyase activities, although the C195G mutant had lost most of its ability to release the α,β-unsaturated aldehyde containing deoxyribose (δ-elimination). Together, these results strongly suggest that the zinc finger motif within the Fpg protein plays a critical role in the structure and function of this enzyme.

D. Deaminated Bases

1. Uracil DNA Glycosylase

Uracil can arise in double-stranded DNA by two mechanisms. One is the incorporation of dUTP in place of dTTP during replication. A second mechanism, which is mutagenic, by which uracil can arise in DNA is the in situ deamination of cytosine to uracil (Lindahl and Nyberg 1974; Lindahl 1979). If these uracils are not removed, transition mutations

(GC→AT) will be fixed during the next round of DNA replication. Uracil DNA glycosylase (UDG) is responsible for the removal of uracil from DNA. The activity appears to be ubiquitous and has been well characterized in many organisms following its initial discovery in *E. coli* by Lindahl (1974). For a review, see Tomilin and Aprelikova (1989).

The UDGs appear to function as monomeric proteins, not requiring cofactors or divalent cations, and unlike the majority of other DNA glycosylases, many UDGs remove uracils effectively from single-stranded DNA (Delort et al. 1985). In no case is there evidence for an associated AP lyase activity with UDG. The enzyme is also inhibited by high concentrations of free uracil and abasic sites (Lindahl et al. 1977; Domena et al. 1988). Mammalian cells appear to contain both nuclear and mitochondrial forms of UDG (Domena and Mosbaugh 1985; Wittwer and Krokan 1985; Domena et al. 1988).

The gene encoding UDG has been cloned from a variety of organisms: *E. coli* (Duncan and Chambers 1984; Varshney et al. 1988), *Streptococcus pneumoniae* (Méjean et al. 1990), herpes simplex virus types 1 and 2 (Caradonna et al. 1987; Worrad and Caradonna 1988), Shope fibroma virus (Upton et al. 1993), *Saccharomyces cerevisiae* (Percival et al. 1989), and humans (Olsen et al. 1989; Meyer-Siegler et al. 1991). All of these genes show a high degree of sequence conservation except for one of the human genes (Meyer-Siegler et al. 1991) where UDG activity is associated with the 37-kD subunit of glyceraldehyde-3-phosphate dehydrogenase (G3PD).

2. G/T Mismatch-specific Thymine DNA Glycosylase (Mammalian) and Endonuclease (E. coli)

The spontaneous deamination of DNA cytosine or adenine produces the abnormal bases, uracil and hypoxanthine, respectively. However, the deamination of 5-methylcytosine within DNA produces thymine and yields a G/T mismatch. In *E. coli*, the inner cytosine in the sequence CCWGG is methylated by the action of the *Dcm* methyltransferase, and in many bacteria, cytosine is methylated due to the action of a methyltransferase that forms part of a restriction system (Chapter 2). In most eukaryotic cells, 5-methylcytosine occurs at frequencies of 2–8%, often in CpG islands (Doerfler 1983; Riggs and Jones 1983). In addition, the presence of 5-methylcytosine in the 5′ region of genes and in the inactive X chromosome in mammalian cells is associated with the loss of or very low levels of gene expression (Bird 1986). Thus, the correct repair of such spontaneous deaminations, especially deamination of 5-methylcytosine, is extremely important in both prokaryotic and eukaryotic cells.

a. Mammalian Cell Studies. The in vivo correction of G/T mismatches was first demonstrated by Brown and Jiricny (1987), who prepared recombinant SV40 molecules with a single G/T mismatch such that in one, a G/T to G/C correction created a new *Bam*HI site, whereas in the other, a G/T to A/T correction created a new *Bcl*I site. A lack of in vivo correction resulted in a mixture of *Bam*HI and *Bcl*I sites. It was found that in African green monkey kidney cells, greater than 90% of the correction was G/T to G/C, regardless of the strand orientation of the mismatch. Efficient repair does not appear to be sequence-context-dependent, since the spontaneous deamination products of ^{m}CpG or ^{m}CpA were repaired with equal efficiency (Brown and Jiricny 1987).

Wiebauer and Jiricny (1989) demonstrated in vitro correction of G/T mismatches. Using a 90-mer containing a single G/T mismatch, incubation of HeLa cell extracts plus ATP, Mg^{++}, and the four dNTPs resulted in the restoration of the G/T mismatch to G/C. Analyses of the incision of the G/T-containing 90-mer demonstrated that phosphodiester bond scission had occurred both 5' and 3' relative to the T of the G/T mispair to leave a one-nucleotide gap opposite the G residue. The initial catalytic action at the G/T mispair is a glycosylase activity as demonstrated by the release of free thymine from mismatched oligonucleotides (Wiebauer and Jiricny 1990). These authors also showed that the single-nucleotide gap in the DNA is filled by DNA polymerase β.

b. Studies in E. coli. On the basis of genetic recombination data between two very closely spaced markers in the *cI* gene of bacteriophage λ, a novel mismatch repair system in *E. coli* was postulated that processed G/T mismatches within the *Dcm* methyltransferase recognition sequence CCWGG (Lieb 1983, 1985). This was subsequently shown to involve a very short patch (VSP) repair mechanism (Jones et al. 1987; Zell and Fritz 1987). In these studies, heteroduplexes of λ DNA or M13 RF DNA, respectively, were constructed with a G/T mismatch at the site of a G/5-methylcytosine base pair and used to transfect *E. coli*. The data were interpreted to imply that this VSP repair was (1) *sequence specific* in that it occurred only at the CCWGG sequence, (2) *site specific* such that it occurred only when the T was derived from deamination of the internal C residue in the above sequence, (3) *mismatch specific* in that only G/T mispairs were repaired, and (4) *base specific* in that the correction was G/T to G/C. This repair pathway required the *mutL*, *mutS*, and *dcm* genes, but not the *mutH* or *mutU* genes.

The endonuclease that is responsible for the initiation of VSP repair has been identified and named the Vsr endonuclease (Hennecke et al. 1991). The Vsr gene product is an 18-kD protein that catalyzes the incision of the phosphodiester backbone 5' to the mismatched T of the G/T mispair. There was no evidence for glycosylase activity as reported for the human cell-free system, and heteroduplexes other than G/T were not recognized.

3. 5-Hydroxymethyluracil DNA Glycosylase

5-Hydroxymethyluracil DNA glycosylase (HmUDG) is one of at least three DNA glycosylases that can be found in eukaryotic cells but not in prokaryotic cells (Hollstein et al. 1984; Boorstein et al. 1987; Cannon-Carlson et al. 1989); the other glycosylases are 5-hydroxymethylcytosine DNA glycosylase (Cannon et al. 1988) and 1,N6-ethenoadenine DNA glycosylase. HmUDG has been purified as a 38-kD protein with no activity toward uracil-containing DNA, unlike the crude activity described in earlier reports (Hollstein et al. 1984; Cannon-Carlson et al. 1989). Like UDG, this enzyme has activity on both single-stranded and double-stranded DNAs. HmUDG may have an associated AP lyase activity since the most purified preparations were capable of incising the DNA phosphodiester backbone following glycosylase action. This DNA incising activity copurified with the glycosylase activity and displayed similar thermal inactivation properties.

Cannon-Carlson et al. (1989) speculated that mammalian cells contain HmUDG activity to repair 5-methylcytosine residues that have undergone a sequential deamination to thymine and oxidation to hydroxymethyluracil or a sequential oxidation to 5-hydroxymethylcytosine and deamination to hydroxymethyluracil. The necessity for such an enzyme can be easily justified in view of the key role that 5-methylcytosine plays in the regulation of gene expression.

4. Hypoxanthine DNA Glycosylase

The deamination of adenine in DNA results in the formation of hypoxanthine (HX), and when replicated, an HX:T base pair gives rise to an HX:C base pair and is thus mutagenic. A specific DNA glycosylase that initiates the removal of hypoxanthine from DNA has been isolated from *E. coli* (Karran and Lindahl 1978; Harosh and Sperling 1988) and mammalian cells (Karran and Lindahl 1980; Myrnes et al. 1982). The calf thymus activity has been shown to remove hypoxanthine more efficiently from an HX:T base pair than from an HX:C base pair (Dianov and Lindahl 1991). Other mismatched base pairs were not recognized.

E. Alkylated Bases: N-Methylpurine DNA Glycosylases

Alkylation damage to DNA includes a variety of modifications of all four bases as well as the phosphodiester backbone (for review, see Singer and Kusmierek 1982), and cells contain a number of DNA-repair mechanisms to initiate the removal or direct reversal of these lesions. In both prokaryotes and eukaryotes, the repair of O6-alkylguanine proceeds by a direct reversal mechanism in which the alkyl group is transferred onto a protein, O6-methylguanine DNA methyltransferase (Foote et al. 1980; Olsson and Lindahl 1980; for review, see Mitra and Kaina 1993). In *E. coli*, the UvrABC nucleotide excision repair system also removes a substantial number of different alkylated nucleotides (Voigt et al. 1989; Snowden et al. 1990; Snowden and Van Houten 1991).

With the exception of O6-alkylguanine, the primary mechanism for removing base alkylation damage is through the action of N-methylpurine DNA glycosylases. *E. coli* has two distinct genes that encode enzymes with N-methylpurine DNA glycosylase activities, *tag* and *alkA*, which encode 3-methyladenine DNA glycosylases I and II, respectively (Riazuddin and Lindahl 1978; Karran et al. 1982; Thomas et al. 1982). Similar activities are found in *M. luteus* (Laval 1977; Shackleton et al. 1979).

3-Methyladenine DNA glycosylase II not only catalyzes the release of alkylpurines (3-methyladenine, 3-methylguanine, and 7-methylguanine), but also releases alkylpyrimidines (O2-methylthymine and O2 methylcytosine). However, 3-methyladenine DNA glycosylase I catalyzes only the removal of 3-methyladenine. No endonucleolytic or AP lyase activities appear to be associated with these enzymes. The *tag* gene product accounts for about 90% of the 3-methyladenine DNA glycosylase activity within *E. coli* and appears to be constitutively expressed. Mutants in *tag* are moderately sensitive to alkylating agents (Karran et al. 1980). In contrast, *alkA* is known to be a major component of the adaptive response since exposure of cells to low levels of N-methyl-N'-nitro-N-nitrosoguanidine (MNNG) strongly induces the expression not only of *alkA*, but also of a series of other genes: *ada*, *alkB*, and *ogt*. This adaptive response greatly reduces both cell killing and mutagenesis (Samson and Cairns 1977; for review, see Walker 1984; Myles and Sancar 1989).

Activities similar to those of *E. coli* 3-methyladenine DNA glycosylase II have been found in mammalian cells. For example, proteins of 42 and 27 kD have been isolated from calf thymus which catalyze the removal of 3-methyladenine, 3-methylguanine, 7-methylguanine, and O2-methylthymine (Male et al. 1985, 1987). Genes encoding these glycosylases have been cloned from *E. coli* (Clarke et al. 1984; Nakabeppu et al. 1984; Sakumi 1986), *M. luteus* (Pierre and Laval 1986), *S. cerevisiae* (Chen et al. 1989, 1990; Berdal et al. 1990), rat (O'Connor and Laval 1990), mouse (Engelward et al. 1993), and humans (Chakravarti et al. 1991; O'Connor and Laval 1991; Samson et al. 1991; Vickers et al. 1993). Although the mammalian genes show extensive homology, there is little conservation between the mammalian genes and the prokaryotic and yeast genes.

F. Cyclobutane Pyrimidine Dimers

1. T4 Endonuclease V

Following exposure of aqueous solutions of DNA to 254 nm UV light, several bipyrimidine photoproducts are produced, the most prevalent being the covalent joining of two adjacent pyrimidines through the saturation of the 5,6-double bond to form a cyclobutane pyrimidine dimer. The other biologically relevant pyrimidine photoproduct is the 6,4-adduct, created by the formation of an oxetane or azetidine 4-membered ring structure that subsequently decomposes into a stable covalent bonding of the 6-position of the 5'-pyrimidine to the 4-position of the 3'-pyrimidine. Phage T4 endonuclease V has a narrow substrate specificity for the *cis-syn* cyclobutane pyrimidine dimer, although recent studies have shown that it can incise *trans-syn* dimers at a rate approximately 1% that of the *cis-syn* dimer (J.S. Taylor, pers. comm.).

The recent X-ray crystal structure of endonuclease V (Morikawa et al. 1992) provides a firm model to interpret the extensive studies relating the biochemical properties of the enzyme to specific residues that had previously been made using chemical modification and site-directed mutagenesis (for a more complete discussion, see Section IV).

In vivo, and under low salt conditions in vitro, endonuclease V binds to DNA electrostatically. The enzyme then randomly diffuses along the DNA by a processive sliding or looping mechanism until it reaches its target site, a *cis-syn* pyrimidine dimer (Lloyd et al. 1980; Ganesan et al. 1986; Gruskin and Lloyd 1986). Initial evidence in favor of a looping mechanism was provided by electron microscopic examination of populations of irradiated plasmid DNA molecules taken from enzyme reaction mixtures. Distinct loops in circular and linear DNAs were observed, the size of which approximated the intra-pyrimidine dimer distance (Lloyd et al. 1987). This processive DNA scanning has been shown to be correlated with relative biological efficacy of the enzyme. This has been measured by enhancements in UV survival of repair-deficient *E. coli* cells in which the gene has been introduced (Dowd and Lloyd 1989a,b, 1990).

If the enzyme encounters a pyrimidine dimer in either strand during the scanning process, endonuclease V binds to that site and initiates the repair process by catalyzing the cleavage of the N-glycosyl bond of the 5'-most pyrimidine of the dimer (Gordon and Haseltine 1980; Radany and Friedberg 1980; Seawell et al. 1980). Following this glycosylic bond scission, the enzyme may dissociate from the cleaved product (leaving an abasic site) or it may catalyze phosphodiester bond scission by β-elimination. The relative rates of these two outcomes may depend on the pH of the medium. Abasic sites formed by this mechanism or by acid depurination also serve as substrates for endonuclease-V-catalyzed β-elimination reactions. With either substrate, the final 3′ product is an α,β-unsaturated aldehyde with *trans* geometry, indicating that the β-elimination reaction was initiated with the open chain rather than the ring form of the sugar. After completing the phosphodiester bond scission, the enzyme dissociates from that site and may continue its diffusion along the DNA molecule.

2. M. luteus (M. lysodeikticus) *UV Endonuclease*

M. luteus, previously known as *M. lysodeikticus*, is a gram-positive coccus that is extremely resistant to the killing effects of short-wavelength UV light. A part of this UV resistance is likely to be due to enzymatic activities that initiate repair of cyclobutane pyrimidine dimers. Activity was described in extracts of *M. lysodeikticus* cells that inactivated the transforming ability of UV-irradiated *Bacillus subtilis* DNA (Strauss 1962). However, *B. subtilis* DNA that had not been irradiated, but had been treated with this extract, did not show reduced transforming ability. It was later shown that extracts from *M. lysodeikticus* induced breaks in UV-irradiated DNA (Strauss et al. 1966). These studies were followed by analyses that showed that thymine-thymine dimers and cytosine-thymine dimers were incised, leaving a terminal phosphate (Carrier and Setlow 1966). This activity was specific for UV-irradiated DNA and did not require divalent cations (Nakayama et al. 1967; Shimada et al. 1967; Setlow et al. 1970).

Five chromatographic peaks of endonucleolytic activity from *M. luteus* showed substrate preference for UV-irradiated versus unirradiated DNAs (Riazuddin and Grossman 1977a,b). Two of these activities, Py-Py correndonucleases I and II, had dramatically different isoelectric points of 4.7 and 8.7, respectively. In addition, correndonuclease I incised both single- and double-stranded UV-irradiated DNAs, whereas correndonuclease II acted only on double-stranded DNA. These studies also suggested that these enzymes cleaved 5′ to the damaged site and that the products were 5′-phosphoryl and 3′-hydroxyl termini. The in vivo role of these enzymes in the initiation of DNA repair at cyclobutane dimer sites was established (Riazuddin and Grossman 1977c). Finally, it was shown that the *M. luteus* UV endonuclease (correndonuclease II) possessed both DNA glycosylase activity and phosphodiester bond scission activity (Gordon and Haseltine 1980; Haseltine et al. 1980). T4 endonuclease V and *M. luteus* UV endonuclease both gave identical products and both enzymes have a high affinity for unirradiated DNA and use a salt-sensitive linear diffusion along DNA to locate sites within DNA that contain pyrimidine dimers (Hamilton and Lloyd 1989).

3. Yeast Pyrimidine Dimer Endonuclease

Recently, an activity, yeast pyrimidine dimer endonuclease (YPDE), has been isolated from *S. cerevisiae* that closely resembles T4 endonuclease V (Hamilton et al. 1992). YPDE is estimated to be between 16 and 20 kD and has a pI of 4.5 (P. Doetsch, pers. comm.). Both enzymes incise apurinic DNA by similar mechanisms and both enzymes have poor, but detectable, activity on UV-irradiated single-stranded DNAs, have no requirement for divalent cations, and are sensitive to inactivation by N-ethylmaleimide (P. Doetsch, pers. comm.).

G. Mismatched Bases: *MUTY*—An Adenine DNA Glycosylase for A/G or A/C Mismatches

Although the major pathway for the repair of mismatched bases in *E. coli* is the methyl-directed mismatch repair pathway (discussed later in this chapter), there is a specific DNA glycosylase for the recognition and incision of adenine bases from A/G, A/8-oxoG, and A/C mispairs (Lu and Chang 1988 a,b). The *mutY* (or *micA*) gene encodes a 39.1-kD protein that is responsible for preventing endogenous C/G to A/T transversions (Au et al. 1988, 1989; Tsai-Wu et al. 1991) and for repairing A/C mismatches in heteroduplex bacteriophage λ DNA (Radicella et al. 1988). However, the major function of MutY is likely to be the removal of adenine opposite 8-oxoG, a mismatch that can arise in DNA following the oxidation of dG to 8-oxodG and the incorporation of dA opposite 8-oxodG (Michaels et al. 1992; Michaels and Miller 1992).

The MutY protein was originally described as a DNA glycosylase lacking an associated abasic endonuclease/abasic lyase activity (Au et al. 1988). However, more recent studies suggest that MutY also catalyzes phosphodiester bond scission immediately 3' to the newly created abasic site (Tsai-Wu et al. 1992). Although this reaction was described as an endonuclease, it is more likely to be that of an AP lyase, with a mechansim similar to that of T4 endonuclease V, *M. luteus* UV endonuclease, or *E. coli* endonuclease III. In contrast to these enzymes, the MutY protein does not appear to incise depurinated DNA, suggesting an absolute requirement for sequential reactions (Tsai-Wu et al. 1992). The discrepancy between the findings of the Modrich and Lu laboratories has not been completely resolved.

The *E. coli mutY* gene has been cloned (Michaels et al. 1990; Tsai-Wu et al. 1991) and was shown to have strong sequence homology with the *E. coli* endonuclease III gene, *nth*. On the basis of sequence identity around the 4Fe-4S cluster, it was predicted that the MutY enzyme is an iron-sulfur protein (Michaels et al. 1990). This was further corroborated by renaturation experiments of MutY in which both ferrous ion and sulfide must be present during renaturation to produce the functional glycosylase and abasic lyase activities (Tsai-Wu et al. 1992).

A similar genetic system has been reported in *Salmonella typhimurium* (Lu et al. 1990), and the *S. typhimurium* homolog with *E. coli mutY* (*mutB*) has been cloned and sequenced (Desiraju et al. 1993). The amino acid sequence identity is 91%, including the 4Fe-4S cluster region.

H. Other DNA Glycosylases: 1,N6-Ethenoadenine DNA Glycosylase

Exposure of animals to carcinogenic agents such as ethyl carbamate, vinyl carbamate, and vinyl halides results in the formation of 1,N6-ethenodeoxyadenosine (εA) in DNA (Laib et al. 1981; Eberle et al. 1989; Leithauser et al. 1990). Rydberg et al. (1991) have identified a human protein (35 kD) that was able to bind specifically to a synthetic 25-bp oligonucleotide containing a single εA. In a follow-up study, the Singer laboratory demonstrated that this binding protein is a DNA glycosylase, specific for εA (Rydberg et al. 1992). Using crude extracts on synthetic double-stranded DNA containing a site-specific adduct, an incision was made 5' to the εA adduct site. However, in significantly more pure extracts, the release of the free base could be detected and quantitated with no evidence of the release of 1,N6-etheno-2'-deoxy-adenosine or 1,N6-etheno-2'-deoxyadenosine-5'-monophosphate. These data suggest that the εA DNA-binding protein and the εA DNA glycosylase are the same protein, but this point remains to be directly demonstrated.

I. Endonucleases

1. Classification of Apurinic/Apyrimidinic Endonucleases

Theoretically, phosphodiester bonds around an abasic site can be cleaved at four positions. A general nomenclature for classifying enzymes (class I–IV) with these activities was proposed by Mosbaugh and Linn (1980) (Fig. 1).

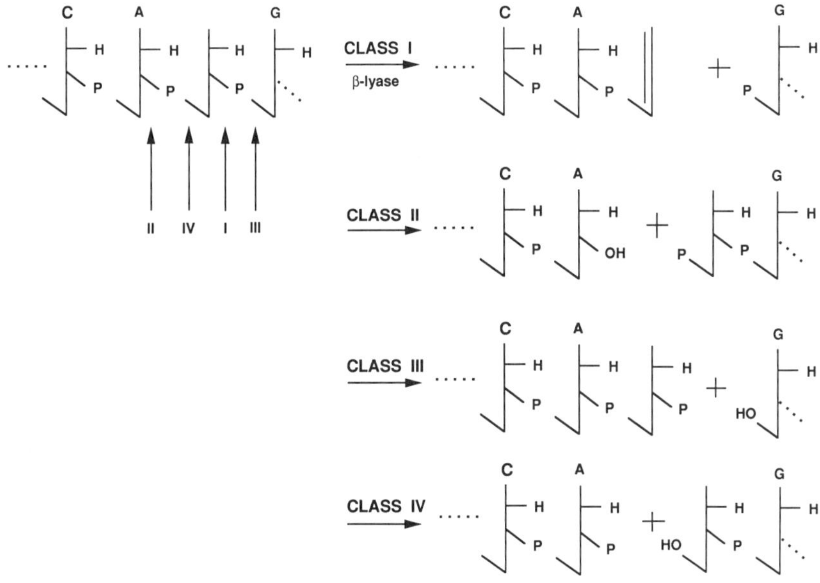

Figure 1 Classification of apurinic/apyrimidinic endonucleases.

As discussed in the opening remarks of this chapter, when the nomenclature was originally proposed, it was not recognized that the class I endonucleases were not hydrolyases, but rather acted as β-elimination catalysts and should have been referred to as apurinic/apyrimidinic (AP) lyases (Bailly and Verly 1987). These enzymes cleave 3' to abasic sites producing a 3'-terminal α,β-unsaturated aldehyde and a 5'-terminal phosphate. To date, all AP lyases are associated with base-specific DNA glycosylases, including T4 endonuclease V, *M. luteus* UV endonuclease, *E. coli* endonuclease III, *E. coli* formamido-pyrimidine DNA glycosylase, and possibly *E. coli* MutY protein (A·G mismatch adenine DNA glycosylase), as discussed above. It is likely that the chemistry of the glycosylase activity dictates the subsequent β-elimination reaction (Dodson et al. 1993). The 3' terminus generated by these enzymes is not effectively recognized as a substrate for the 3'→5' exonuclease activity of DNA polymerase and will not serve as a primer for DNA synthesis. To complete DNA-repair synthesis, the α,β-unsaturated aldehyde must be removed to yield a 3'-hydroxyl.

Class II endonucleases hydrolytically cleave DNA 5' to the abasic site, producing a 3'-terminal hydroxyl nucleotide and a 5'-terminal phosphate. The class II endonucleases are also capable of removing the 3'-α,β-unsaturated aldehyde product generated by the class I enzymes. Although the 3'-hydroxyl serves as an excellent substrate for DNA-repair synthesis, the 5'-deoxyribose-5-phosphate needs to be removed by a deoxyribophosphodiesterase to yield a single nucleotide gap prior to repair synthesis (Franklin and Lindahl 1988).

A class III enzyme cleaves 3' to an abasic site, yielding an abasic deoxyribose with a 5'-hydroxyl and a 3'-phosphate. Only one example of a class III endonuclease has been reported (Spiering and Deutsch 1986). No examples of class IV endonucleases have been reported.

2. Prokaryotic

a. E. coli Exonuclease III. Although the term exonuclease III implies a strict exonucleolytic activity, *E. coli* exonuclease III has a wide variety of catalytic activities: (1) 3'→5'-exonucleolytic activity on double-stranded DNA, (2) 3'-DNA phosphomonoesterase activity (Warner et al. 1980), (3) 3'-phosphodiesterase activity that removes 3'- phosphoglycoaldehyde esters (Demple et al. 1986) and 3'-phosphoglycolate esters (Niwa and Moses 1981; Henner et al. 1983), (4) endonucleolytic activity on double-stranded DNA that hydrolyzes 5' to urea residues (Kow and Wallace 1987), (5) 3'-phosphodiesterase activity that hydrolyzes double-stranded DNA 3' to alkylhydroxylamine-modified residues (Kow 1989), and (6) RNase H activity (Keller and Crouch 1972). The activity of *E. coli* exonuclease III was originally described as a contaminating side fraction during the purification of DNA polymerase I in which it was able to activate DNA templates for in vitro synthesis. Lindahl (1982) demonstrated that exonuclease III accounts for greater than 85% of the apurinic endonucleolytic activity of *E. coli*. In addition, the gene encoding exonuclease III, *xth*, has been cloned and sequenced (Saporito et al. 1988).

Exonuclease-III-deficient cells (*xth*) are exquisitely sensitive to the lethal effects of H_2O_2 (Demple et al. 1986). It was demonstrated that following H_2O_2 exposure, *xth* cells accumulate single-strand breaks in their genome at a 20-fold enhanced rate relative to wild-type cells. When DNA was isolated from H_2O_2-treated wild-type and *xth*⁻ cells and directly used as a template for DNA synthesis, even though the chromosomal DNA from the exonuclease-III-deficient cells contained 20 times the number of single-strand breaks (possible sites for the initiation of DNA synthesis), this DNA supported far less synthesis relative to DNA from the wild-type cells. However, if the DNA from *xth*⁻ cells was treated in vitro with purified extracts of exonuclease III, a tenfold stimulation in DNA polymerase activity was observed. A comparable result could also be obtained by the addition of purified preparations of endonuclease IV, which contains a 3' phosphatase and 3' repair activity but no 3'→5' exonucleolytic function. However, in vitro overexpression of endonuclease IV cannot substitute for exonuclease III in xth^- mutants that have been challenged with H_2O_2. Collectively, these data suggest that at least one important role for exonuclease III is associated with the processing of DNA termini following oxidative challenge such that these termini are converted into appropriate substrates for DNA synthesis (Demple et al. 1986).

b. E. coli Endonuclease IV. Endonuclease IV was originally isolated as an EDTA-resistant, 32-kD apurinic endonuclease (Ljungquist et al. 1976; Ljungquist 1977; Ljungquist and Lindahl 1977). Subsequently, Levin et al. (1988) found that endonuclease IV releases 3'-phosphoglycoaldehyde, hydrolyzes 3'-phosphates or 3'-(2,3-didehydro-2,3-dideoxy) ribose-5-phosphate esters, and incises intact apurinic sites.

Although endonuclease IV can be assayed in the presence of EDTA, prolonged storage of the enzyme in the presence of EDTA inactivates it; atomic absorption spectrometry revealed the presence of 2.4 atoms of zinc and 0.7 atoms of manganese per endonuclease IV molecule (Levin et al. 1991). Interestingly, EDTA-inactivated endonuclease IV could be restored to an active form by the addition of $CoCl_2$ and to some extent by $MnCl_2$ but not by $ZnCl_2$. The exact role that metals play in endonuclease IV activity remains to be elucidated.

The gene encoding endonuclease IV, *nfo*, has been identified (Cunningham et al. 1986; Saporito and Cunningham 1988), and *nfo* mutants are sensitive to killing by bleomycin, *t*-butyl hydroperoxide, and mitomycin C (Cunningham et al. 1986; Foster and Davis 1987). The *nfo* gene is inducible by agents that affect the redox status of *E. coli* (e.g., paraquat) (Chan and Weiss 1987; Greenberg et al. 1990).

3. Eukaryotic

a. Yeast Apurinic Endonuclease, APN1. At least five endonucleolytic activities were described in *S. cerevisiae* (D1, D2, D3, D4, and E) (Armel and Wallace 1984; Chang et al. 1987). These enzymes ranged in size from 10 to 37 kD and all required the presence of Mg^{++} for activity. It is possible that several of the lower-molecular-mass species resulted from proteolysis of the E protein. Endonuclease E effectively released urea from extensively oxidized DNA templates and thus appears to have activities similar to those of the *E. coli* exonuclease III (Chang et al. 1987). It also appears to be the same enzyme described by Johnson and Demple (1988a,b) as the Apn1 protein. This enzyme is both the major apurinic endonuclease and 3'-phosphoglycoaldehyde diesterase of *S. cerevisiae*. In addition to its other activities, Apn1 has a potent 3'-phosphatase activity, making it very similar in activity to *E. coli* endonuclease IV.

Like *E. coli* endonuclease IV, Apn1 can be inactivated by the metal chelating agents, EDTA and 1,10-phenanthroline (Levin et al. 1991), and atomic absorption spectrometry has revealed the presence of 3.3 zinc atoms and no manganese atoms per Apn1 molecule. EDTA-inactivated enzyme could be restored to an active form by the addition of $CoCl_2$, $MnCl_2$, or $ZnCl_2$.

The gene encoding Apn1 has been cloned (Popoff et al. 1990), and its first 280 amino acids show 55% identity with *E. coli nfo*. Gene disruptions were made to replace the wild-type *APN1* gene in haploid yeast. Relative to wild-type yeast cells, the *APN1::URA3* mutants lacked greater than 97% of the apurinic endonucleolytic activity and 3'-phosphoglycoaldehyde diesterase activity but were viable. They were hypersensitive to hydrogen peroxide, butyl hydroperoxide and methyl and ethylmethane sulfonate (Ramotar et al. 1991b). These cells also accumulate single-strand breaks in their genomic DNA in response to oxidative challenge in a manner very similar to that of *E. coli* exonuclease-III-deficient cells. The *APN1* gene can complement exonuclease III and endonuclease IV mutants in *E. coli*, when those cells are challenged with H_2O_2 (Ramotar et al. 1991a).

b. Drosophila *Apurinic Endonucleases.* A number of apurinic endonucleases have been reported in *Drosophila melanogaster*. Two chromatographically distinct apurinic endonucleases (I and II) were isolated from embryos, both of which are large (63 and 66 kD) relative to most other apurinic endonucleases (Spiering and Deutsch 1981, 1986; Margulies and Wallace 1984; Venugopal et al. 1990). Although the DNA cleavage products of AP endonuclease II were consistent with a class I enzyme, AP endonuclease I was demonstrated to use a class III mechanism, cleaving 3' to the apurinic site to leave a 3'-deoxyribose phosphate and a 5'-hydroxyl (Spiering and Deutsch 1986).

A class II apurinic endonuclease has also been identified from *Drosophila*, the Rrp1 protein (*r*ecombination *r*epair *p*rotein 1) (Sander et al. 1991). In addition to its apurinic endonucleolytic activity, Rrp1 was reported to have a 3'-exonuclease activity and to carry out a Mg^{++}-dependent renaturation of single-stranded DNA. Analysis of the sequence of the gene encoding Rrp1 reveals significant sequence homology with exonuclease III of *E. coli* and exonuclease A of *Streptococcus pneumoniae*.

In addition, Kelley et al. (1989) used antibodies directed against a HeLa cell apurinic endonuclease to screen a λgt11 library and identified a gene encoding a 35-kD protein (AP3). Grabowski et al. (1991) demonstrated that the AP3 protein showed a 66% identity with the human apurinic endonuclease, P0. The *P0* gene encodes a ribosomal protein and was cloned by Rich and Steitz (1987). Thus, it appears that the P0 gene product and the AP3 gene product have dual functions, serving both as ribosomal proteins and as DNA-repair endonucleases.

c. Mammalian Apurinic Endonucleases. Although too numerous to review in depth here, apurinic endonucleases have been isolated from a very large number of cell and tissue types from a variety of mammals (for review, see Doetsch and Cunningham 1990), including human placenta (Linsley et al. 1977; Grafstrom et al. 1982; Shaper et al. 1982), human cultured cells (Kane and Linn 1981; Kuhnlein 1985), calf thymus (Henner et al. 1987), and rat liver (Cesar and Verly 1983). Both class I and II enzymes are found.

The HeLa-cell-derived apurinic endonuclease is a 41-kD, Mg^{++}-dependent class II endonuclease that was shown to be inhibited by adenine, hypoxanthine, adenosine, and NAD^+ (Kane and Linn 1981). This endonuclease is capable of removing 3′-blocking groups from oxidatively damaged DNAs (Chen et al. 1991). The calf thymus enzyme shares many of the same properties with the HeLa-cell-derived enzyme (Henner et al. 1987; Sanderson et al. 1989). The substrate range of the calf thymus enzyme has been examined and shown to cleave interphosphate linkages containing ethylene glycol, propanediol, and tetrahydrofuran.

A significant number of genes that encode mammalian apurinic endonucleases have been cloned. Human apurinic endonucleases have been cloned from placenta (APE) (Demple et al. 1991) and tumor (HAP2) cells (Robson and Hickson 1991; Robson et al. 1991). The gene encoding human APE has been localized to human chromosome 14q11.2-12 (Harrison et al. 1992; Robson et al. 1992).

IV. STRUCTURE/FUNCTION ANALYSES OF DNA GLYCOSYLASES/AP LYASES

Since all their catalytic activities are contained within a single polypeptide chain, the DNA glycosylases are an attractive group of proteins for structure/function analyses of protein–nucleic acid interactions. Two enzymes are examined in detail, T4 endonuclease V and the *E. coli* endonuclease III. The crystallographic structures of both have been solved, and structure/function relationships have been examined by chemical modification and site-directed mutagenesis (see also Chapter 1).

A. T4 Endonuclease V

1. X-ray Crystal Structure

The first three-dimensional structure of a DNA-repair enzyme was for T4 endonuclease V in the absence of substrate (Morikawa et al. 1992). The

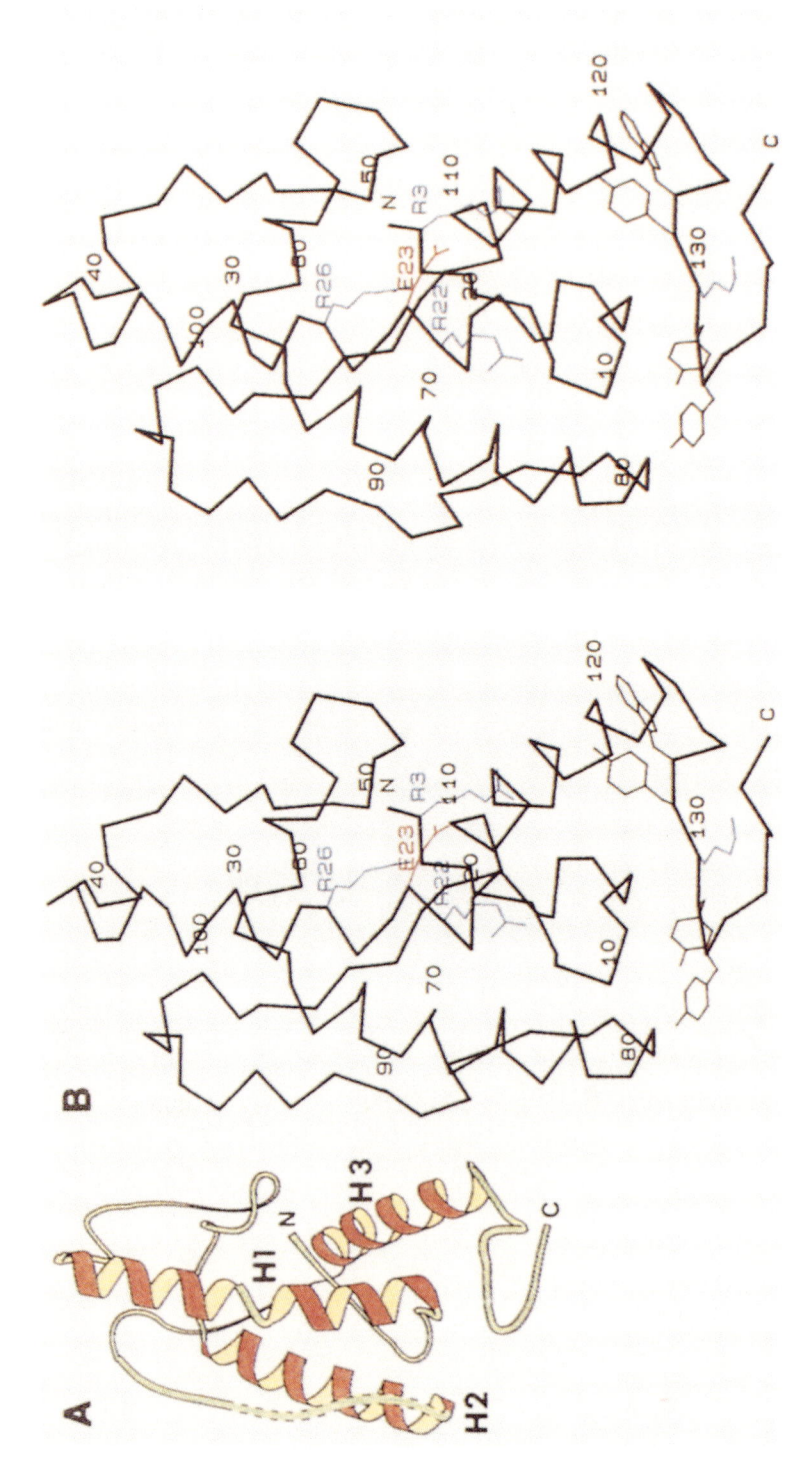

Figure 2 X-ray crystal structure of T4 endonuclease V. (Reprinted, with permission, from Morikawa et al. 1992 [copyright by the AAAS].)

structure contained a three-helix bundle with several of the most amino-terminal amino acid residues wedged between α helices 1 and 3 (Fig. 2). Specific amino acid residues previously shown to be critical for binding to "nontarget DNA," i.e., DNA regions not containing pyrimidine dimers, appeared to be clustered around the amino terminus in the structure. The carboxy-terminal 12-14 residues contain the region of the protein that has been implicated in specific binding to pyrimidine dimers (Stump and Lloyd 1988; Recinos and Lloyd 1988; Lloyd and Augustine 1989; Ishida et al. 1990). This portion of the enzyme is located approximately 15-20 Å from the putative active site in the structure (see below), which may indicate that large conformational changes take place upon substrate binding.

2. Identification of the Active Site Residue

The β-elimination mechanism of endonuclease V implies a Schiff's base intermediate, and reductive methylation of amino groups by formaldehyde and sodium cyanoborohydride inactivates the enzyme (Schrock and Lloyd 1991). Approximately one methylation event per molecule resulted in the simultaneous loss of DNA glycosylase and AP lyase activities without loss in pyrimidine-dimer-specific binding. The preferential site of methylation was shown to be the α-amino group of the amino terminus. Since the amino terminus can be covalently linked to the C1' of the 5'-pyrimidine of the dimer, this proves that the catalytic site of endonuclease V is the amino terminus (Dodson et al. 1993).

Site-directed mutagenesis at the Thr-2 residue revealed that a variety of amino acid changes could be tolerated at this site without loss of catalytic activity, although a substitution of proline or the addition of a glycine between residues 2 and 3 resulted in inactive enzymes (Schrock and Lloyd 1993). These data indicate the importance of the α-amino group position within the folded structure. These data have led to a proposed reaction mechanism that is consistent with biochemical and biophysical analyses (Fig. 3). The α-amino group of the amino terminus serves as the nucleophile attacking the deoxyribose-C1' of the 5'-pyrimidine of the dimer. The glycosyl bond is cleaved, resulting in an imino (Schiff's base) intermediate in which the protein is covalently linked to DNA. This intermediate may dissociate, giving rise to an abasic site, or, if the lifetime of the intermediate is sufficiently long, the sugar may β-eliminate, resulting in DNA strand breakage.

Glu-23 is a second catalytically important residue. The crystal structure indicates that it is the only acidic residue near the amino terminus and that it is surrounded by many basic amino acids. Site-directed muta-

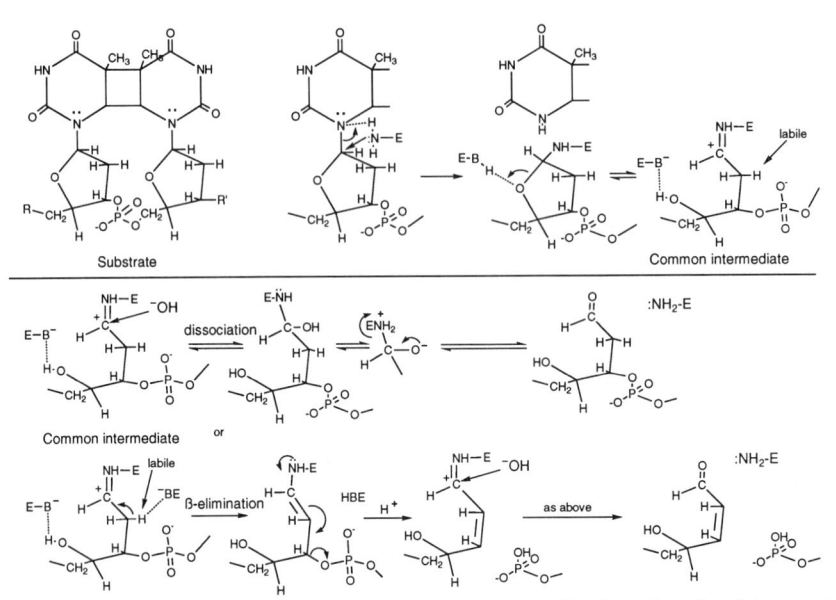

Figure 3 Reaction mechanism of T4 endonuclease V. (Reprinted, with permission, from Dodson et al. 1993 [copyright American Chemical Society].)

genesis was used to test whether this residue was important in catalysis (Morikawa et al. 1992). These and other investigations showed that the substitution of aspartic acid resulted in a loss of glycosylase activity and significant reduction in abasic (AP) lyase activity (Doi et al. 1992; Hori et al. 1992). Further studies of mutants at this site are required to assign a specific role to Glu-23 in the catalytic mechanism.

3. Amino Acids Involved in Nontarget DNA Binding

As noted above, endonuclease V locates pyrimidine dimers by a processive scanning or looping mechanism involving electrostatic interactions between the phosphate backbone of the DNA and basic amino acids on the enzyme surface. Site-directed mutagenesis of basic amino acids (Dowd and Lloyd 1989a,b) gave rise to three mutant proteins (R3N, R3K, and R3L) that were unable to complement repair-deficient *E. coli* in a cell-survival assay, but exhibited a substantial level of dimer-specific nicking and abasic lyase activity in vitro. In contrast to wild-type enzyme, which processively incises all dimers on an individual plasmid DNA molecule prior to dissociation, these mutants were found to incise DNA at pyrimidine dimer sites in a completely random distributive manner. These data suggested that multiple mutations could be tolerated at Arg-3, which resulted in substantial levels of in vitro enzymatic activity, although Arg-3 was absolutely required for the in vitro and in vivo processivity of the enzyme.

The degree of enhanced UV-resistance conferred on repair-deficient *E. coli* was hypothesized to be directly proportional to the extent of in vitro processive nicking activity by endonuclease V (Dowd and Lloyd 1990). A series of mutants at R26 and K33, whose catalytic activities in vitro were similar to that of wild-type enzyme, was used to test this hypothesis. The purified mutant enzymes had varying degrees of processivity, ranging from nearly that of wild type to completely distributive. In an in vivo plasmid DNA-repair assay (Gruskin and Lloyd 1988a,b), the kinetics of the accumulation of fully repaired form I plasmid DNA molecules was measured in cells expressing each of the mutants. The study clearly demonstrated varying degrees of processivity in vivo that correlated with the in vitro properties. These results directly relate the reduction of in vivo processivity to a change in an observed biochemical phenotype and support the proposition that the mechanisms employed by DNA-interactive proteins to locate their relevant target sites are of biological significance.

B. *E. coli* Endonuclease III

The three-dimensional crystal structure of *E. coli* endonuclease III has recently been solved (Fig. 4) (Kuo et al. 1992b). The enzyme is an elongated, bilobal protein with each lobe approximately equal in size but separated by a deep cleft or funnel with a width of one water molecule. The residues that line the cleft show very few hydrophobic interactions and no aromatic side chain packing. Instead, the major interactions between the two domains are polar side chain interactions composed of charged hydrogen bonds and salt bridges. Each domain has an extensive interior hydrophobic core and a polar charged surface, thus making it unlikely that this molecule associates as a dimer or higher multimer.

One of the domains is composed of six antiparallel α helices (αB–αG, residues 22–132) with nearest-neighbor connectivity. The six helices are arranged such that αB is centrally located within the remaining five helices. The αD helix is kinked by a 60° angle, thus dividing it into two segments of seven residues.

The second domain contains four α helices, one composed of the amino-terminal helix (residues 1–21), and the remainder composed of three carboxy-terminal α helices (residues 133–211). The 4Fe-4S cluster

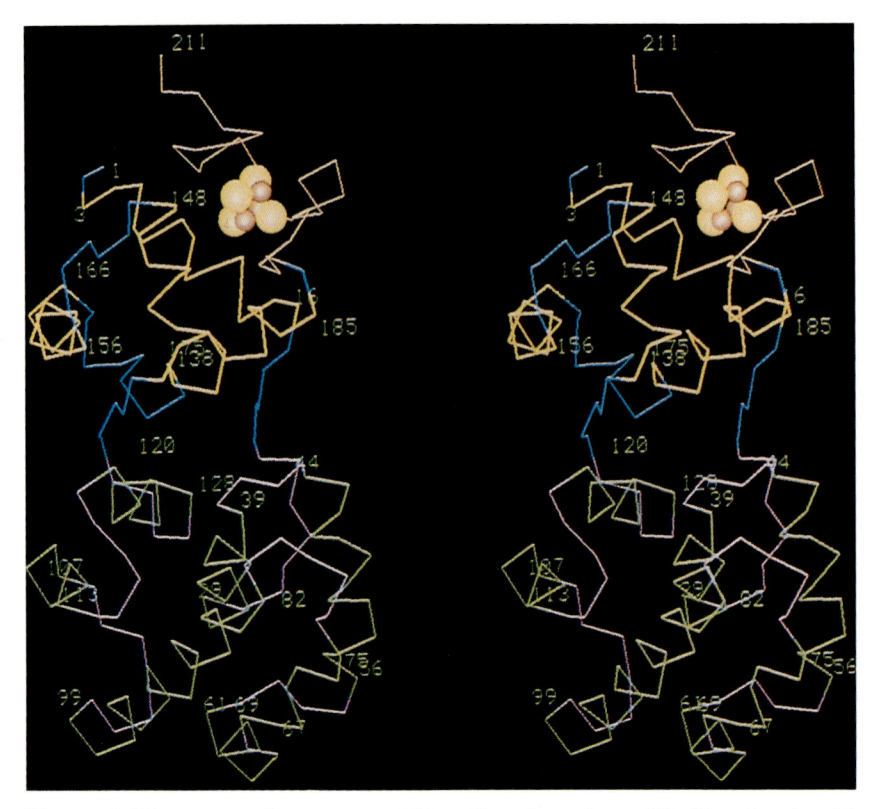

Figure 4 X-ray crystal structure of *E. coli* endonuclease III. (Reprinted, with permission, from Kuo et al. 1992b [copyright by the AAAS].)

is located completely within the carboxy-terminal α helices, with the coordination pattern being Cys-X$_6$-Cys-X$_2$-Cys-X$_5$-Cys (residues 187–203). This 17-residue segment forms a right-handed spiral around the 4Fe-4S cluster and packs next to the four-helix core.

Examination of the enzyme structure relative to DNA reveals a likely site of interaction allowing favorable electrostatic interactions to exist along the long axis of the protein. The size of this potential nucleic acid interaction site is in good agreement with footprinting studies. To help localize the enzyme's active site, crystals were soaked in a solution containing thymine glycol, a known inhibitor of the glycosylase activity. The thymine-glycol-binding site was identified as a β-hairpin structure at residues 113–119 by a difference Fourier map. The identification of the thymine-glycol-binding site will help in guiding oligonucleotide site-directed mutagenesis studies (Kuo et al. 1992b).

V. END-TRIMMING NUCLEASES

Through the action of physical or chemical DNA strand-breaking agents or enzymes such as AP endonucleases or nucleases that form 3'-phosphomonoester groups, abnormal DNA termini are formed that must be trimmed off to allow for DNA-repair synthesis and ligation. In addition, the trimming of tailed structures is requisite for a number of recombination schemes. In theory, any single-stranded DNA-specific endonuclease could serve these roles, but it appears that more specialized enzymes exist for many or all of these functions.

The most comprehensively characterized end-trimming nucleases are generally those from *E. coli*. Historically, the first observations were the discovery of DNA 3'-phosphomonoesterase associated with exonuclease III (Richardson and Kornberg 1964) and the ability of the 5'→3' exonuclease of DNA polymerase I (pol I) to remove damaged nucleotides, including pyrimidine dimers, from DNA termini (Kelly et al. 1969). This latter reaction is also presumably a part of a nick-translation DNA-repair function of DNA polymerase. For instance, DNA polymerase I appears to remove the T base from the G/T mismatch during Vsr repair (Hennecke et al. 1991). Exonuclease III also removes sugar fragments from 3' ends (e.g., phosphoglycolate) and even an unsaturated deoxyribose phosphate formed by the action of an AP β-lyase reaction. Endonuclease IV is also able to remove phosphate residues, sugar phosphate fragments, or the AP β-lyase product from DNA 3' termini (for detailed discussion of these enzymes, see Section IV). These two enzymes are generally considered the major means of "cleaning up" 3'-terminal sugar damage to facilitate repair replication. However, a recent report that exonuclease I can also remove sugar phosphate fragments (Sandigursky and Franklin 1992) may alter this notion.

An enzyme with the unique ability to remove 5'-terminal deoxyribose-5-phosphate residues, dRPase, is also known (Franklin and Lindahl 1988). The role of this enzyme seems to be to clean up 5' termini formed by class II AP endonucleases, a role that endonuclease III is unable to assume.

E. coli exonuclease VII (Chase and Richardson 1974a,b) and the RecBCD enzyme (Goldmark and Linn 1972; Karu et al. 1973; Tanaka and Sekiguchi 1975) also act with both a 3'→5' and 5'→3' polarity and can excise damaged nucleotides. Exonuclease VII is specific either for single-stranded DNA or for single-stranded termini on duplexes, whereas the RecBCD enzyme acts on both duplex and single-stranded DNAs. Both enzymes form oligonucleotides and can excise pyrimidine dimers. In addition, exonuclease VII can also act from nicks formed by the *M. luteus* UV endonuclease. Deletion mutants in the structural gene of exonuclease VII, *xseA*, are fully viable, have a normal rate of thymine dimer excision, and are only somewhat UV-sensitive. This is presumably because of back-up by other enzymes. The mutants are also "hyper-rec." However, in cells carrying defective RecBCD enzyme and a DNA polymerase I defective in its 5'→3' exonuclease, *xseA* results in reduced excision but does not completely eliminate it (Chase et al. 1979). Evidently, exonuclease VII, the RecBCD enzyme, the 5'→3' exonuclease of DNA polymerase I, and at least one other nuclease (perhaps the 5'→3' exonuclease of DNA polymerase III) can each excise pyrimidine dimers in *E. coli*.

A final *E. coli* enzyme involved in end-trimming is the RecJ protein (Lovett and Kolodner 1989). This single-stranded DNA-specific exonuclease is able to trim 5' single-stranded tails efficiently and, to a lesser extent, 3' single-stranded tails from duplex DNA. It therefore appears to be involved in some way in processing the termini of recombination intermediates, or possibly in forming gaps should it act in concert with a helicase. Its (in)ability to remove DNA damages has not been reported. The overlapping but not identical specificities of exonucleases I and VII and RecJ protein probably imply overlapping roles in DNA transactions.

In mammalian cells, DNase V is able to trim both 3' and 5' termini, and, when acting in a complex with DNA polymerase β, it catalyzes excision of damage from either terminus and simultaneous repair synthesis to fill gaps (Randahl et al. 1988). DNase IV can also excise pyrimidine dimers from DNA in a 5'→3' direction and also appears to have the properties desired for an excision exonuclease (Lindahl 1971).

Finally, a so-called "correxonuclease," which has been isolated from human placenta, acts in both the 3'→5' and 5'→3' directions to release oligonucleotides. It can release pyrimidine dimers from DNA, notably from DNA nicked by *M. luteus* UV endonuclease (Doniger and Grossman 1976). The major AP endonucleases of mammalian cells, the HAP I (HeLa) or BAP I (bovine) protein, which have sequence homology with *E. coli* exonuclease III, were discussed in Section III above. Although they lack the strong exonuclease of the bacterial enzyme, they do have the activities for removing 3'-phosphomonoesters and 3'-damaged sugar phosphate residues and presumably act in this capacity. A second AP endonuclease/3'-phosphoglycolate diesterase activity has been reported from HeLa cells (Chen et al. 1991; Demple and Levin 1991), which might also remove 5'-terminal AP sites and serve the role of dRPase for *E. coli*.

Finally, in the yeast *S. cerevisiae*, Apn1 is the major 3'-diesterase/monoesterase for processing damaged termini (Demple 1991; Demple and Levin 1991). It has sequence homology with *E. coli* endonuclease IV.

VI. NUCLEOTIDE EXCISION REPAIR

Classically, nucleotide excision repair involves recognition of and cleavage adjacent to a damaged nucleotide by an incision endonuclease, excision of the damaged region including some adjacent normal nucleotides and then filling of the resulting gap by DNA polymerase and sealing by ligase. It is now clear that there are several variations on this classic theme. On the one hand, in several systems (e.g., *E. coli* Uvr-ABC), incision and excision are one and the same event. On the other hand, in several systems noted earlier in this chapter (e.g., *E. coli* DNA polymerase I and mammalian β-polymerase/DNase V), excision and repair synthesis are coordinate. Finally, other proteins, most notably DNA helicases and DNA-binding proteins, seem also commonly to be part of the excision repair machinery.

A. UvrABC

1. Overview

This *E. coli* enzyme system, UvrABC, which has been reviewed extensively (Grossman and Yeung 1990; Selby and Sancar 1990; Van Houten 1990), recognizes a variety of bulky DNA adducts, as well as free-radical and radiation damages. It appears in vivo even to overlap in specificity with several of the DNA glycosylases. The system usually makes one break 7–8 nucleotides 5′ to the damage and another 3–4 nucleotides 3′ to the damage to give a small gap of 12–13 nucleotides. ATP is hydrolyzed during the reaction to allow a UvrA$_2$B complex to translocate DNA, acting as a helicase while searching for abnormal nucleotide material. ATP is also required for dissociation of the DNA-enzyme complex once incision is accomplished. After UvrA$_2$B is bound to a damaged region, UvrA$_2$ dissociates, and in the presence of the remaining UvrB protein bound to the DNA damage site, UvrC protein promotes the incision/excision reaction.

UvrC apparently harbors the catalytic site for the 5′ nick, whereas UvrB harbors the site for the 3′ nick (Lin et al. 1992). UvrD DNA helicase and DNA polymerase I are also required for optimal efficiency and turnover of the excision events. Details of the UvrABC reaction are given in Section VII below.

2. Functional Analyses of Domains within the Proteins That Comprise the E. coli *UvrABC Complex*

a. ATP-binding Domains. The deduced amino acid sequence of the UvrA protein reveals a series of potentially interesting motifs that may be essential for structural damage recognition and/or catalytic roles within the multisubunit complex (Doolittle et al. 1986). These include two ATP-binding sites, two zinc finger motifs, and a putative helix-turn-helix motif.

Site-directed mutagenesis of the two Walker A-type ATP-binding domains within UvrA (GLSGSGK, 31-37 and GVSGSGK, 640-646) has been carried out (Myles and Sancar 1991; Myles et al. 1991; Thiagalingam and Grossman 1991, 1993). A variety of single and double mutations were made at K37 and K646, and only one mutant, K37R, was able to complement $uvrA^-$ E. coli to levels equivalent to that of the wild-type enzyme, whereas the other mutants conferred little to no enhanced survival. These data suggest a significant role for both ATP-binding sites within the protein for proper functioning of the complex. In addition, the binding and hydrolysis of ATP appear to be tightly linked to the association and dissociation of the UvrA$_2$ complex, respectively. These data also suggest that the binding of ATP could aid in the discrimination between nontarget and target DNAs.

The UvrB protein has been shown to contain a cryptic ATP-binding site that is activated upon interaction with two UvrA molecules (Caron and Grossman 1988; Orren and Sancar 1989). Using site-directed mutagenesis, Seeley and Grossman (1989, 1990) were able to demonstrate that expression of proteins with alterations of K45 (within the ATP-binding domain) was unable to enhance UV survival in uvrB mutants, whereas K51A was indistinguishable from that of wild type. It was demonstrated that the UvrA$_2$B complex, containing UvrB K45A, was not capable of hydrolyzing ATP, whereas complexes containing wild type or K51A showed significant ATP hydrolysis. The ability to hydrolyze ATP has been demonstrated to be associated with the nontarget DNA tracking mechanism of the UvrA$_2$B complex (Seeley and Grossman 1990).

b. Zinc Finger Motifs. The role of the two zinc finger motifs in UvrA has been analyzed by constructing fragments of the protein that contain one zinc finger (Myles and Sancar 1991). Although the correct coordination of zinc was necessary and sufficient for general DNA binding, damage-specific DNA binding required the maintenance of a proper orientation between the two zinc-binding sites. Moreover, a series of carboxy-terminal deletion proteins were made in which the correct renaturation of the mutant proteins absolutely required the presence of Zn^{++} (Claassen et al. 1991).

In that study, it was also determined that the first 230 amino acids of UvrA are sufficient for interaction with UvrB, and the carboxy-terminal approximately 40 residues are essential for target/nontarget DNA discrimination.

c. Helix-Turn-Helix Motif. The role of the consensus helix-turn-helix motif (residues 494–513) within UvrA has been investigated by cassette mutagenesis (Wang and Grossman 1993). Two mutant proteins, G502D and V508D, displayed little or no enhancement in UV survival in *uvrA* mutants. Although these proteins displayed normal levels of binding for nontarget DNA, retained significant ATPase activity, and when complexed with UvrB, were able to support the generation of positively supercoiled plasmids from relaxed circular DNAs in the presence of topisomerase I and ATP, the defect in these proteins is that the UvrA$_2$B complex is unable to recognize UV-damaged DNA templates. Thus, the helix-turn-helix motif is likely to be an additional candidate for participating in damage recognition.

d. Catalytically Important Residues. UvrC is a 68-kD protein that has been strongly implicated in the catalytic incision activities of the Uvr(A)BC complex (for review, see Sancar and Sancar 1988; Snowden and Van Houten 1993). To localize the portion of UvrC required for catalytic activity, Lin and Sancar (1991) utilized linker-scanning mutagenesis and both amino-terminal and carboxy-terminal deletion mutagenesis to identify a domain of UvrC that was not only sufficient to interact with UvrB, but also sufficient to catalyze incision correctly at damaged bases and complement *uvrC*$^-$ mutations in vivo. When only segments of the amino-terminal half of the UvrC protein were produced, or when the carboxy-terminal 272 or fewer amino acid residues were expressed, no complementation of *uvrC*$^-$ mutations was observed. However, expression of the carboxy-terminal 375 (C375C) or 314 (C314C) residues gave full and partial complementation, respectively.

The UvrC protein has been demonstrated to be responsible for making the 5′ incision relative to the damaged base, whereas UvrB produces the 3′ incision (Lin and Sancar 1992; Lin et al. 1992). Lin and Sancar (1992) reasoned that the active site residues were likely to be histidine, aspartic acid, and glutamic acid, based on the reaction mechanism of other characterized nucleases in which the hydrolysis of the phosphodiester backbone occurs by acid-base catalysis via a histidine-activated or metal-activated water molecule. Of the 11 histidine, 8 aspartic acid, and 6 glutamic acid residues that were mutated in the carboxy-terminal portion of UvrC, four sites were found to be implicated in catalysis at the 5′ incision site: H538, D399, D438, and D466. When these mutated UvrC molecules were added to wild-type UvrAB, the 3′ incision site was efficiently produced. This result suggests that the 5′ incision catalyzed by UvrC is made by a metal-activated water molecule in which the metal is coordinated by aspartate and glutamate residues, similar to that observed for RNase H and the Klenow 3′→5′ exonuclease.

Site-directed mutagenesis was also performed on histidine and aspartic acid residues in UvrB that are conserved between UvrB proteins from *E. coli*, *M. luteus*, and *M. genitalium* (Lin et al. 1992). These data revealed that at least three residues, D337, D478 and D510, are associated with DNA binding and that D478 is involved in DNA bending and possibly catalysis. In addition, UvrB molecules that are missing the final 43 amino acids were found to bind and bend DNA but retain very low incision activity. Site-directed mutagenesis identified E639 as a likely candidate for a catalytic residue in making the 3′ incision.

B. The Mammalian Counterpart to UvrABC

Using mammalian cell-free extracts (Huang et al. 1992) or *Xenopus* oocytes, Svoboda et al. (1993) observed vertebrate homologs of the Uvr-ABC system. The former activity produces an oligomer of 27–29 residues, and the latter produces 27–32 residues; in both cases, one of the cleavages appears to be five nucleotides 3′ to the site of damage. Further study of these systems awaits purification of the proteins involved. One would expect more active participation of a helicase activity for removing a 30-residue fragment than for the smaller fragment formed by Uvr-ABC. Indeed, a surprisingly large number of helicases (as well as damaged DNA-binding proteins) have been found through complementation assays with DNA-repair-defective mammalian or yeast cells or cell-free extracts (Biggerstaff and Wood 1992; Friedberg 1992; Hoeijmakers 1993).

C. The RAD1/RAD10 Incision Complex

The phenotypes of RAD1 and RAD10 mutants of *S. cerevisiae* implicate these genes in recombination and nucleotide excision repair. The Rad10 protein possesses DNA-binding activity with a strong preference for single-stranded DNA (Sung et al. 1992), whereas the Rad1 protein binds to both single- and double-stranded DNAs. Neither shows preference for binding UV-damaged DNA (Sung et al. 1992, 1993). A Rad1/Rad10 complex degrades single-stranded circular DNA, and this degradation is inhibited by antibodies against either protein (Sung et al. 1993; Tomkinson et al. 1993). This activity is reminiscent of that by UvrC protein in the absence of UvrAB. The endonucleolytic activity generates 3′-hydroxyl and 5′-phosphate termini and requires Mg^{++}. Incision of double-stranded DNAs increases with negative superhelix density (Sung et al. 1993), a property that may be important in the initiation of repair in transcriptionally active genes (Ljungman and Hanawalt 1992) or in the initiation of mitotic recombination stimulated by transcriptional activity. Since RAD10 shows sequence homology with human ERCC1, a similar enzyme may also exist in mammalian cells.

D. Other Incision Endonucleases

A large number of endonucleases that prefer damaged DNA over undamaged duplexes have been reported in the literature. The large majority of these remain to be purified and characterized to discern whether indeed they are unique, whether they are classic phosphodiesterases or DNA glycosylase/AP endonucleases (lyases), or whether they are single-strand-specific endonucleases. However, there are a few noteworthy examples.

E. coli endonuclease V (Gates and Linn 1977; Demple and Linn 1982) prefers DNA damaged by OsO_4, UV light (but it does not act at pyrimidine dimers), acid treatment, or 7-bromomethylbenz-[a]anthracene. It acts most rapidly upon DNA with uracil substituted for thymine. The enzyme appears to act cooperatively and processively and makes about one double-strand break per eight single-strand nicks. Although the mechanism and specificity of the enzyme offer no simple model for its action, it is likely to serve as a back-up function for other repair systems of *E. coli*. A similar enzyme has been described in *M. luteus* (Hecht and Thielmann 1978).

Deutsch and Spiering (1982) reported a similar enzyme from *Drosophila* that acts specifically on DNA containing uracil. The activity is transient in development, appearing only in third instar larvae. The further observation that dUTPase is present only in *Drosophila* embryos has led Giroir and Deutsch (1987) to propose the existence of a coordinated scheme to incorporate uracil into DNA, creating a substrate for the uracil nuclease as part of the histolyzation process.

A novel type of nuclease activity has been observed from animal cells that cleaves a phosphodiester bond within a cyclobutane pyrimidine dimer to yield a product that can be detected as "nicked" only after altering the "bridge" formed by the pyrimidine dimer. Such structures have been observed after nucleotide excision repair in vivo (Weinfield et al. 1986). In one case, the nick can be rendered detectable with *E. coli* photolyase (Liuzzi and Paterson 1992) and in another case by T4 endonuclease V (J. Kim and S. Linn, unpubl.). The latter activity has been purified extensively and appears to be inseparable from ribosomal protein S3 (R. Fellous and S. Linn, unpubl.).

VII. TRANSCRIPTIONAL REGULATION OF REPAIR NUCLEASES DURING STRESS RESPONSES

Nuclease activities have been observed to be subject to regulation at the transcriptional, translational, and protein-processing levels. Knowledge of the regulation of DNA-repair nucleases is generally well studied only in the case of transcriptional regulation during stress responses in bacteria. Some fascinating recent observations on the active involvement of DNA-repair enzymes with transcriptional regulation, which include prokaryotes and eukaryotes, are discussed in Section IX.

A. The SOS Response

In *E. coli*, the SOS response induces UvrA and UvrB proteins from 20 to 200 and 200 to 1000 molecules per cell, respectively (Van Houten 1990). UvrC protein, present at 10 molecules per cell, is apparently not induced despite a putative "lexA box" in the *uvrC* promoter region. The induction of UvrA and UvrB is clearly to enhance nucleotide excision repair; however, the reason for the lack of stoichiometry among the three proteins is not entirely clear. It may be that damaged sites are found by A_2B, and then UvrB binds these sites to protect against exonucleases and to stall transcription (and replication). Meanwhile, A_2 becomes free and complexes another UvrB molecule to search for additional damaged sites to protect. UvrC may be limited as it acts catalytically and also so as to avoid a large number of simultaneous incisions into the genome. Parenthetically, the SOS response also results in the induction of UvrD protein but not DNA polymerase I.

recJ and *ruvC* gene expression are both enhanced during SOS induction, presumably to enhance recombinational repair (for a discussion of RecJ and RuvC nucleases in recombination, see Chapter 5).

B. The Adaptive (ada) Response

During the adaptive response to alkylating agents, 3-methyladenine–DNA glycosylase II (the *alkA* gene product) is induced (see Section III above). Whereas this induction presumably enhances base excision repair of alkylated DNA purines, no concomitant induction of nuclease or polymerase activities has been observed.

C. Responses to Oxidative Stress

Curiously, although the adaptive response to alkylating agents appears to induce only DNA glycosylase repair activity, oxidative stress results in

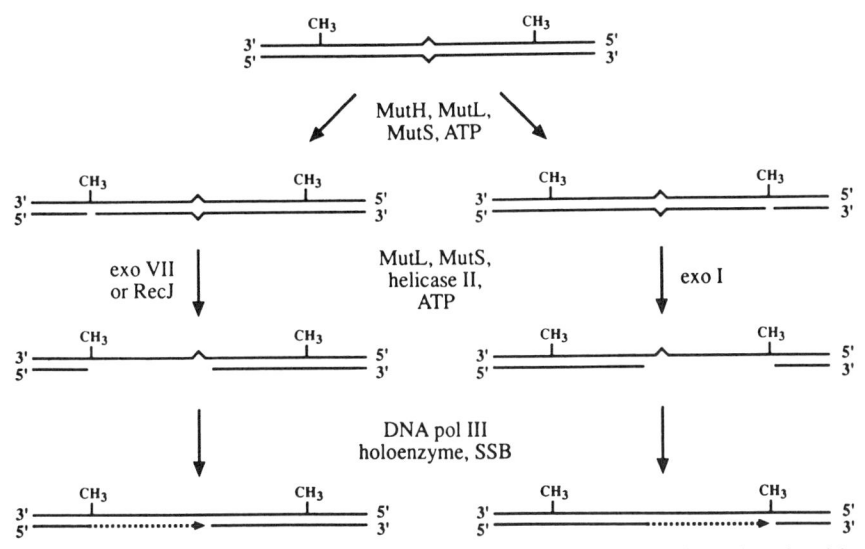

Figure 5 Basic pathway for methyl-directed mismatch repair. (Reprinted, with permission, from Grilley et al. 1993.)

the induction of repair nucleases but apparently not DNA glycosylases (Demple 1991). Hence, exonuclease III is under control of the *katF* regulon, which is mediated by the 42-kD *katF* RNA polymerase σ factor. The *katF* regulon also includes hydroperoxidase (catalase) II and responds to oxidative stress due to growth on TCA cycle intermediates (Sak et al. 1989).

The *soxRS* regulon responds to the presence of superoxide and controls the expression of endonuclease IV (as well as Mn^{++}-dependent superoxide dismutase, outer membrane porins, and NADH metabolism) (Wu and Weiss 1991). The differing regulations of exonuclease III (*xth*) and endonuclease IV (*nfo*) suggest unknown distinctions in their roles in DNA excision repair.

VIII. METHYL-DIRECTED MISMATCH REPAIR

Within *E. coli*, the correction of mismatched DNA bases that arise due to errors in DNA replication is carried out by a strand-specific pathway in which the cell utilizes the transiently hemimethylated state of the daughter strand to distinguish new from old DNA (for review, see Modrich 1991). The basic pathway is outlined in Figure 5. The strand specificity of this repair pathway is determined by the methylation status at GATC sequences that are located within several kilobase pairs of the site of the mismatched base pair. To achieve proper base pair correction, a hemimethylated GATC sequence is required such that the unmethylated strand is corrected using the methylated strand as a template. Not all mismatched bases are equally good substrates for this repair system: G/T, A/C, A/A, and G/G mispairs are good substrates, T/T, T/C, and A/G mispairs are poor intermediate substrates, and the C/C mismatch is completely resistant to correction (Dohet et al. 1986; Lu and Chang 1988a,b; Su et al. 1988). An additional DNA structure subject to methyl-directed mismatch correction is the looping out of several nucleotides that may have arisen by slip-mispairing (Dohet et al. 1986; Fishel et al. 1986; Learn and Grafstrom 1989). The complete *E. coli* repair system has been reconstituted in vitro and contains the following proteins:

- MutS (97 kD) binds to the mismatched bases in the absence of additional cofactors, but if ATP is bound to its ATP-binding site, it promotes the formation of α-shaped loop structures (Su and Modrich 1986).
- MutH (25 kD) recognizes the hemimethylated GATC sequence and catalyzes a single break at that site (discussed in more detail below).
- MutL (70 kD, but a 140-kD homodimer in solution) interacts with MutS in the presence of ATP or ATPγS at the site of a mismatched base pair and dramatically enlarges the DNase-I-resistant footprint (Grilley et al. 1989).
- UvrD (DNA helicase II) may load onto both DNA strands at the site of a MutS, MutL complex, resulting in bidirectional strand displacement until a single-strand break is encountered, presumably at the site of an incised hemimethylated GATC sequence (Längle-Rouault et al. 1987).
- Exonuclease I catalyzes the $3' \rightarrow 5'$ exonucleolytic degradation of the MutH-incised strand when the GATC sequence is located $3'$ to the mismatched base pair (Cooper et al. 1993; Grilley et al. 1993).
- Exonuclease VII or RecJ exonuclease catalyzes the $5' \rightarrow 3'$ exonucleolytic degradation of the MutH-incised strand when the GATC sequence is located $5'$ to the mismatched base pair (Cooper et al. 1993; Grilley et al. 1993).
- Single-stranded DNA-binding protein (SSB) presumably binds to the intact single-stranded DNA that is exposed by the action of exonuclease I or VII or RecJ prior to DNA polymerization.
- DNA polymerase III holoenzyme catalyzes the resynthesis of the gap formed in DNA by the action of the $5' \rightarrow 3'$ or $3' \rightarrow 5'$ exonucleases.
- DNA ligase seals the single-strand break in DNA following DNA polymerase III synthesis.

A. MutH

The substrate specificity of the MutH protein is the key feature of this repair system since it allows methyl-directed mismatch repair to discriminate between parental methylated and daughter unmethylated DNA immediately following DNA replication in a dam^+ *E. coli*. MutH is an endonuclease that binds to and incises DNA at hemimethylated or unmethylated GATC sequences, but it will not cleave at a fully methylated GATC sequence (Welsh et al. 1987). Although the site of cleavage (↓pGpApTpC) is highly specific, leaving a 3'-hydroxyl and a 5'-phosphate, the rate of incision by the MutH protein alone is extremely slow (less than one turnover per hour). This rate is unaffected by the presence or absence of mismatched bases. Prolonged incubations with completely unmethylated substrates results in some double-stranded DNA cleavage (Au et al. 1992).

Since the intrinsic nicking activity of MutH is so poor, it was postulated that this activity could be significantly enhanced through interactions with other components of the mismatch repair system. Au et al. (1992) tested this hypothesis and found that the activity of MutH can be increased more than 30-fold when MutS, MutL, and ATP are present. In addition, it was demonstrated that this enhancement in activity is maximal on supercoiled or relaxed circular DNAs compared to linear DNA substrates. Interestingly, the degree of activation of MutH is directly proportional to the relative affinities of MutS to various mismatched bases. ATP (or dATP) is also required for MutH activation, with half-maximal activity appearing at 0.3 mM ATP. GTP, CTP, UTP, dGTP, dCTP, dUTP, and ATPγS could not support the activation of MutH. The ATPγS result suggests that ATP hydrolysis is required for activation. In this regard, neither MutH nor MutL has been shown to have an ATPase activity, but MutS does possess a weak ATPase activity. The exact mechanism for this activation of MutH remains a question for future investigations.

B. Mismatch Repair in Other Prokaryotes and Eukaryotes

Although the genetics and biochemistry of methyl-directed mismatch repair have been best established in *E. coli*, it is likely that mismatch correction is common to most organisms since it enhances the overall replication fidelity 10^2–10^3-fold (for review, see Echols 1991). In addition to the Hex pathway of *S. pneumoniae* (for review, see Claverys and Lacks 1986), mismatch correction has been described in yeast (Bishop et al. 1989), *D. melanogaster* (Hare and Taylor 1985; Holmes et al. 1990; Thomas et al. 1991), *Xenopus* (Varlet et al. 1990), and humans (Holmes et al. 1990; Thomas et al. 1991; Fang and Modrich 1993).

IX. PERSPECTIVES

A good deal of progress in understanding the role of nucleases in DNA repair and their mechanisms has been made since the first edition of *Nucleases* a decade ago. The coming decade promises to be even more informative. Structural information will help us to answer a question peculiar to many DNA-repair nucleases: How do these individual enzymes specifically recognize a number of apparently dissimilar damage structures? In some cases, particularly in prokaryotes, studies of the nucleases themselves may provide us with answers. For others, we expect the discovery of a plethora of DNA-binding proteins by exploiting cell and molecular genetics techniques in fungal and vertebrate cell systems. Some of these proteins will probably be specific recognition proteins; others will probably be less specific, such as the UvrA and UvrB proteins.

A related area that should progress rapidly is the understanding of how, when, and why nucleases and DNA perturb one another's structures. We already know that UvrA$_2$B kinks DNA in its damaged region (Sancar and Hearst 1993), whereas Chi sites attenuate the nuclease activity of the RecBCD enzyme (Dixon and Kowalczykowski 1993). Such intimate and dynamic nuclease:substrate interactions are going to be plentiful and exciting to study.

Perhaps the most exciting prospect for discovery is the role of repair nucleases in nonrepair aspects of cellular regulation and metabolism. The study of the coupling of DNA repair and transcription was spurred by the observations of preferential repair of active genes and, in some cases, sense strands of those genes (Hanawalt 1991). In bacteria, the factor coupling UvrABC repair to the DNA site of transcription (including the particular strand) has been identified and characterized by Selby and Sancar (1993). That factor is the 130-kD Mfd (*m*utation *f*requency *d*ecline) protein. The protein appears to displace RNA polymerase while it recruits UvrA protein to DNA-damage sites.

The coupling of DNA repair and transcription in eukaryotes is clearly more complex, but the identification of human ERCC3 as the transcription factor BTF2 (or TFIIH) (Schaeffer et al. 1993) puts us well on the way to studying the phenomenon. ERCC3 is an apparent DNA-repair helicase that is defective in combined xeroderma pigmentosum and Cockayne's Syndrome patients who have abnormal levels of DNA repair and abnormal development. The yeast homolog of *ERCC3* is *RAD25* and the *Drosophila* homolog is *haywire*. Hence, DNA repair and developmental abnormalities are associated with defects of this gene throughout the eukaryotic spectrum.

Finally, and most unexpectedly, are the observations that repair nucleases themselves can play roles in the cell other than cleaving DNA. Binding of DNA by the Jun/Fos transcription factor (AP-1), which senses oxidative stress, is stimulated by HAP1 (human AP endonuclease 1), which itself seems to act as a redox sensor (Radler-Pohl et al. 1993). Even more intriguing, AP-1 responds to a number of other DNA-damaging stresses, and these responses might involve other DNA-repair proteins. Yet another tidbit: The HAP1 homolog in *E. coli*, exonuclease III, seems to be required to mediate the heat-shock response in *E. coli*, since *xth* mutants are defective in the expression of heat-shock proteins at high temperature (Paek and Walker 1986).

Besides evidence for coupling of DNA repair to transcription, two preliminary observations relate DNA repair to translation: Ribosomal protein S3 from human cells (J. Kim et al., unpubl.) and *Drosophila* (Wilson et al. 1993) appears to be involved in excision repair as it has AP endonuclease (β-lyase) activity and recognizes and excises a number of oxidized DNA bases (Kim and Linn 1989). Similarly, human ribosomal protein P0 is extremely homologous in sequence to *Drosophila* AP3, a putative AP endonuclease structural gene (Grabowski et al. 1992). These initial observations could indicate a coupling of DNA repair with translation or they could indicate that the ribosomal proteins present a large store of DNA-repair proteins, including nucleases, for utilization during DNA-damaging stresses. Like many other proteins, it is now clear that repair nucleases operate in concert with some disparate cellular processes and may have more than a nucleolytic function.

ACKNOWLEDGMENTS

R.S.L. thanks Katherine Latham, M.L. Dodson, David Grabowski, Ben Van Houten, and Amanda Snowden for their helpful discussions and suggestions concerning this chapter. Very special thanks also go to Heather Wyatt for her patience and tireless efforts in the preparation of this chapter. S.L. also thanks Lorraine Oller for her help in preparing the manuscript.

REFERENCES

Armel, P.R. and S.S. Wallace. 1984. DNA repair in *Saccharomyces cerevisiae*: Purification and characterization of apurinic endonucleases. *J. Bacteriol.* **160:** 895–902.

Asahara, H., P.M. Wistor, J.F. Bank, R.H. Bakerian, and R.P. Cunningham. 1989. Purification and characterization of *Escherichia coli* endonuclease III from the cloned *nth* gene. *Biochemistry* **28:** 4444–4449.

Au, K.G., K. Welsh, and P. Modrich. 1992. Initiation of methyl-directed mismatch repair. *J. Biol. Chem.* **267:** 12142–12148.

Au, K.G., M. Cabrera, J.H. Miller, and P. Modrich. 1988. *Escherichia coli mutY* gene product is required for A-G to C-G mismatch correction. *Proc. Natl. Acad. Sci.* **85**: 9163–9166.

Au, K.G., S. Clark, J.H. Miller, and P. Modrich. 1989. *Escherichia coli mutY* gene encodes an adenine glycosylase active on G-A mispairs. *Proc. Natl. Acad. Sci.* **86**: 8877–8881.

Bacchetti, S. and R. Benne. 1975. Purification and characterization of an endonuclease from calf thymus acting on irradiated DNA. *Biochim. Biophys. Acta* **390**: 285–297.

Bailly, V. and W.G. Verly. 1987. *Escherichia coli* endonuclease III is not an endonuclease but a β-elimination catalyst. *Biochem. J.* **242**: 565–572.

Bailly, V., M. Derydt, and W.G. Verly. 1989a. Delta elimination in the repair of AP (apurinic/apyrimidinic) sites in DNA. *Biochem. J.* **261**: 707–713.

Bailly, V., W.G. Verly, T.R. O'Connor, and J. Laval. 1989b. Mechanism of DNA strand nicking at apurinic/apyrimidinic sites by *Escherichia coli* [formamidopyrimidine] DNA glycosylase. *Biochem. J.* **262**: 581–589.

Berdal, K.G., M. Bjoras, S. Bjelland, and E. Seeberg. 1990. Cloning and expression in *Escherichia coli* of a gene for an alkylbase DNA glycosylase from *Saccharomyces cerevisiae;* a homolog to the bacterial *alkA* gene. *EMBO J.* **9**: 4563–4568.

Biggerstaff, M. and R.D. Wood. 1992. Requirement for *ERCC-3* gene products in DNA excision repair *in vitro. J. Biol. Chem.* **267**: 6879–6885.

Bird, A.P. 1986. CpG-rich islands and the function of DNA methylation. *Nature* **321**: 209–213.

Bishop, D.K., J. Andersen, and R.D. Kolodner. 1989. Specificity of mismatch repair following transformation of *Saccharomyces cerevisiae* with heteroduplex plasmid DNA. *Proc. Natl. Acad. Sci.* **86**: 3713–3717.

Boiteux, S., T.R. O'Connor, and J. Laval. 1987. Formamidopyrimidine-DNA glycosylase of *Escherichia coli*: Cloning and sequencing of the *fpg* structural gene and overproduction of the protein. *EMBO J.* **6**: 3177–3183.

Boiteux, S., J. Belleney, B.P. Roques, and J. Laval. 1984. Two rotameric forms of open ring 7-methylguanine are present in alkylated polynucleotides. *Nucleic Acids Res.* **12**: 5429–5439.

Boiteux, S., T.R. O'Connor, F. Lederer, A. Gougette, and J. Laval. 1990. Homogeneous *Escherichia coli* FPG protein. *J. Biol. Chem.* **265**: 3916–3922.

Boorstein, R.J., D.D. Levy, and G.W. Teebor. 1987. 5-Hydroxymethyluracil-DNA glycosylase activity may be a differentiated mammalian function. *Mutat. Res.* **183**: 257–263.

Breimer, L. and T. Lindahl. 1980. A DNA glycosylase from *Escherichia coli* that releases free urea from a polydeoxyribonucleotide containing fragments of base residues. *Nucleic Acids Res.* **8**: 6199–6210.

———. 1984. DNA glycosylase activities for thymine residues damaged by ring saturation, fragmentation, or ring contraction are function of endonuclease III in *Escherichia coli. J. Biol. Chem.* **259**: 5543–5548.

Brent, T.P. 1973. A human endonuclease activity for gamma-irradiated DNA. *Biophys. J.* **13**: 399–401.

———. 1976. Purification and characterization of human endonuclease specific for damaged DNA. Analysis of lesions induced by ultraviolet or X-radiation. *Biochim. Biophys. Acta* **454**: 172–183.

———. 1983. Properties of a human lymphoblast AP-endonuclease associated with activity for DNA damaged by ultraviolet light, gamma-rays or osmium tetroxide. *Biochemistry* **22**: 4507–4512.

Brown, T.C. and J. Jiricny. 1987. A specific mismatch repair event protects mammalian cells from loss of 5-methylcytosine. *Cell* **50:** 945-950.

Cannon, S.V., A. Cummings, and G.W. Teebor. 1988. 5-Hydroxymethylcytosine DNA glycosylase activity in mammalian tissue. *Biochem. Biophys. Res. Commun.* **151:** 1173-1179.

Cannon-Carlson, S.V., H. Gokhale, and G.W. Teebor. 1989. Purification and characterization of 5-hydroxymethyluracil-DNA glycosylase from calf thymus. *J. Biol. Chem.* **264:** 13306-13312.

Caradonna, S., D. Worrad, and R. Lirette. 1987. Isolation of a herpes simplex virus cDNA encoding the DNA repair enzyme uracil-DNA glycosylase. *J. Virol.* **61:** 3040-3047.

Caron, P.R. and L. Grossman. 1988. Involvement of cryptic ATPase activity of UvrB and its proteolysis product. UvrB in DNA repair. *Nucleic Acids Res.* **16:** 9651-9662.

Carrier, W.L. and R.B. Setlow. 1966. Excision of pyrimidine dimers from irradiated deoxyribonucleic acid *in vitro*. *Biochim. Biophys. Acta* **129:** 318-325.

Cesar, R. and W.G. Verly. 1983. The apurinic/apyrimidinic endodeoxyribonuclease of rat liver chromatin. *Eur. J. Biochem.* **129:** 509-516.

Chakravarti, D., G.C. Ibeanu, K. Taño, and S. Mitra. 1991. Cloning and expression in *Escherichia coli* of a human cDNA encoding of the DNA repair protein *N*-methylpurine DNA glycosylase. *J. Biol. Chem.* **266:** 15710-15715.

Chan, E. and B. Weiss. 1987. Endonuclease IV of *Escherichia coli* is induced by paraquat. *Proc. Natl. Acad. Sci.* **84:** 3189-3193.

Chang, C.-C., Y.W. Kow, and S.S. Wallace. 1987. Apurinic endonucleases from *Saccharomyces cerevisiae* also recognize urea residues in oxidized DNA. *J. Bacteriol.* **169:** 180-183.

Chase, J.W. and C.C. Richardson. 1974a. Exonuclease VII of *Escherichia coli*. Purification and properties. *J. Biol. Chem.* **249:** 4545-4552.

———. 1974b. Exonuclease VII of *Escherichia coli*. Mechanism of action. *J. Biol. Chem.* **249:** 4553-4561.

Chase, J.W., W.E. Masker, and J.B. Murphy. 1979. Pyrimidine dimer excision in *Escherichia coli* strains deficient in exonuclease V and VII and in the 5'→3' exonuclease of DNA polymerase I. *J. Bacteriol.* **137:** 234-242.

Chen, D.S., T. Herman, and B. Demple. 1991. Two distinct human DNA diesterases that hydrolyze 3'-blocking deoxyribose fragments from oxidized DNA. *Nucleic Acids Res.* **19:** 5907-5914.

Chen, J., B. Derfler, and L. Samson. 1990. *Saccharomyces cerevisiae* 3-methyladenine DNA glycosylase has homology to the AlkA glycosylase of *E. coli* and is induced in response to DNA alkylation damage. *EMBO J.* **9:** 4569-4675.

Chen, J., B. Derfler, A. Maskati, and L. Samson. 1989. Cloning a eukaryotic DNA glycosylase repair gene by the suppression of a DNA repair defect in *Escherichia coli*. *Proc. Natl. Acad. Sci.* **86:** 7961-7965.

Chetsanga, C.J. and T. Lindahl. 1979. Release of 7-methylguanine residues whose imidazole rings have been opened from damaged DNA by a DNA glycosylase from *Escherichia coli*. *Nucleic Acids Res.* **6:** 3673-3684.

Chung, M.H., H. Kasai, D.S. Jones, H. Inoue, H. Ishikawa, E. Ohtsuka, and S. Nishimura. 1991. An endonuclease activity of *Escherichia coli* that specifically removes 8-hydroxyguanine residues from DNA. *Mutat. Res.* **254:** 1-12.

Claassen, L.A., B. Ahn, H.-S. Koo, and L. Grossman. 1991. Construction of deletion mutants of the *Escherichia coli* UvrA protein and their purification from inclusion bodies. *J. Biol. Chem.* **266:** 11380-11387.

Clarke, N.D., M. Kvaal, and E. Seeberg. 1984. Cloning of *Eschericha coli* genes encoding 3-methyladenine DNA gycosylases I and II. *Mol. Gen. Genet.* **197:** 368–372.

Claverys, J.P. and S.A. Lacks. 1986. Heteroduplex deoxyribonucleic acid base mismatch repair bacteria. *Microbiol. Rev.* **50:**133–165.

Cooper, D.L., R.S. Lahue, and P. Modrich. 1993. Methyl-directed mismatch repair is bidirectional. *J. Biol. Chem.* **268:** 11823–11829.

Cunningham, R.P., H. Asahara, and J.F. Bank. 1989. Endonuclease III is an iron-sulfur protein. *Biochemistry* **28:** 4450–4455.

Cunningham, R.P., S.M. Saporito, S.G. Spitzer, and B. Weiss. 1986. Endonuclease IV (*nfo*) mutant of *Escherichia coli. J. Bacteriol.* **168:** 1120–1127.

Delort, A.-M., A.-M. Duplaa, D. Molko, R. Teoule, J.-P. Leblanc, and J. Laval. 1985. Excision of uracil residues in DNA: Mechanism of action of *Escherichia coli* and *Micrococcus luteus* uracil-DNA glycosylases. *Nucleic Acids Res.* **13:** 319–335.

Demple, B. 1991. Regulation of bacterial oxidative stress genes. *Annu. Rev. Genet.* **25:** 315–337.

Demple, B. and J.D. Levin. 1991. Repair systems for radical-damaged DNA. In *Oxidative stress: Oxidants and antioxidants* (ed. H. Sies), pp. 119–154. Academic Press, London.

Demple, B. and S. Linn. 1980. DNA *N*-glycosylases and UV repair. *Nature* **287:** 203–208.

———. 1982. On the recognition and cleavage mechanism of *Escherichia coli* endonuclease V, a possible DNA repair enzyme. *J. Biol. Chem.* **257:** 2848–2855.

Demple, B., T. Herman, and D.S. Chen. 1991. Cloning and expression of APE, the cDNA encoding the major human apurinic endonuclease: Definition of a family of DNA repair enzymes. *Proc. Natl. Acad. Sci.* **88:** 11450–11454.

Demple, B., A. Johnson, and D. Fund. 1986. Exonuclease III and endonuclease IV remove 3′ blocks from DNA synthesis primers in H_2O_2-damaged *Escherichia coli. Proc. Natl. Acad. Sci.* **83:** 7731–7735.

Desiraju, V., W.G. Shanabruch, and A.-L. Lu. 1993. Nucleotide sequence of the *Salmonella typhimurium nutB* gene, the homolog of *Escherichia coli mutY. J. Bacteriol.* **175:** 541–543.

Deutsch, W.A. and A.L. Spiering. 1982. A new pathway expressed during a distinct stage of *Drosophila* development for the removal of dUMP residues in DNA. *J. Biol. Chem.* **257:** 3366–3368.

Dianov, G. and T. Lindahl. 1991. Preferential recognition of I:T base-pairs in the initiation of excision-repair by hypoxanthine-DNA glycosylase. *Nucleic Acids Res.* **19:** 3829–3833.

Dixon, D.A. and S.C. Kowalczykowski. 1993. The recombination hotspot χ is a regulatory sequence that acts by attenuating the nuclease activity of the *E. coli* RecBCD enzyme. *Cell* **73:** 1–20.

Dodson, M.L., R.D. Schrock, and R.S. Lloyd. 1993. Evidence for an imino intermediate in the T4 endonuclease V reaction. *Biochemistry* **32:** 8284–8290.

Doerfler, W. 1983. DNA methylation and gene activity. *Annu. Rev. Biochem.* **52:** 93–124.

Doetsch, P.W. and R.P. Cunningham. 1990. The enzymology of apurinic/apyrimidinic endonucleases. *Mutat. Res.* **236:** 173–201.

Doetsch, P.W., D.E. Helland, and W.A. Haseltine. 1986. Mechanism of action of a mammalian DNA repair endonuclease. *Biochemistry* **25:** 2212–2220.

Dohet, C., R. Wagner, and M. Radman. 1986. Methyl-directed repair of frameshift mutations in heteroduplex DNA. *Proc. Natl. Acad. Sci.* **83:** 3395–3397.

Doi, T., A. Recktenwald, Y. Karaki, M. Kikuchi, K. Morikawa, M. Ikehara, T. Inaoka, N. Hori, and E. Ohtsuka. 1992. Role of the basic amino acid cluster and Glu-23 in pyrimidine dimer glycosylase activity of T4 endonuclease V. *Proc. Natl. Acad. Sci.* **89:** 9420–9424.

Domena, J.D. and D.W. Mosbaugh. 1985. Purification of nuclear and mitochondrial uracil-DNA glycosylase from rat liver. Identification of two distinct subcellular forms. *Biochemistry* **24:** 7320–7328.

Domena, J.D., R.T. Timmer, S.A. Dicharry, and D.W. Mosbaugh. 1988. Purification and properties of mitochondrial uracil-DNA glycosylase from rat liver. *Biochemistry* **27:** 6724–6751.

Doniger, J. and L. Grossman. 1976. Human correxonuclease. Purification and properties of DNA repair exonuclease from placentas. *J. Biol. Chem.* **251:** 4579–4587.

Doolittle, R.F., M.S. Johnson, I. Husain, B. Van Houten, D.C. Thomas, and A. Sancar. 1986. Domainal evolution of a prokaryotic DNA repair protein and its relationship to active-transport proteins. *Nature* **323:** 451–453.

Dowd, D.R. and R.S. Lloyd. 1989a. Biological consequences of a reduction in the non-target DNA scanning capacity of a DNA repair enzyme. *J. Mol. Biol.* **208:** 701–707.

———. 1989b. Site-directed mutagenesis of the T4 endonuclease V gene: The role of arginine-3 in the target search. *Biochemistry* **28:** 8699–8705.

———. 1990. Biological significance of facilitated diffusion in protein-DNA interactions. *J. Biol. Chem.* **265:** 3424–3431.

Duncan, B.K. and J.A. Chambers. 1984. The cloning and overproduction of *Escherichia coli* uracil-DNA glycosylase. *Gene* **28:** 211–219.

Eberle, G., A. Barbin, R.J. Laib, F. Ciroussel, J. Thomale, H. Bartsch, and M.F. Rajewsky. 1989. 1,N^6-Etheno-2'-deoxyadenosine and 3,N^4-etheno-2'-deoxycytidine detected by monoclonal antibodies in lung and liver DNA of rats exposed to vinyl chloride. *Carcinogenesis* **10:** 209–212.

Echols, H. 1991. Fidelity mechanisms in DNA replication. *Annu. Rev. Biochem.* **60:** 477–511.

Engelward, B. P., M.S. Boosalis, B.J. Chen, D. Zuoming, M.J. Siciliano, and L.D. Samson. 1993. Cloning and characterization of a mouse 3-methyladenine/7-methylguanine/3-methylguanine DNA glycosylase cDNA whose gene maps to chromosome 11. *Carcinogenesis* **14:** 175–181.

Fang, W.-H. and P. Modrich. 1993. Human strand-specific mismatch repair occurs by a bidirectional mechanism similar to that of the bacterial reaction. *J. Biol. Chem.* **268:** 11838–11844.

Fishel, R.A., E.C. Siegel, and R. Kolodner. 1986. Gene conversion in *Escherichia coli*. Resolution of heteroallelic mismatched nucleotides by co-repair. *J. Mol. Biol.* **188:** 147–157.

Foote, R.S., S. Mitra, and B.C. Pal. 1980. Demethylation of O^6-methylguanine in a synthetic DNA polymer by an inducible activity in *Escherichia coli*. **97:** 654–659.

Foster, P.L. and E.F. Davis. 1987. Loss of apurinic/apyrimidinic site endonuclease increases the mutagenicity of N-methyl-N'-nitro-N-nitrosoguanidine to *Escherichia coli*. *Proc. Natl. Acad. Sci.* **84:** 2891–2895.

Franklin, W.A. and T. Lindahl. 1988. DNA deoxyribophosphodiesterase. *EMBO J.* **7:** 3617–3622.

Friedberg, E.C. 1992. Xeroderma pigmentosum, Cockayne's syndrome, helicases, and DNA repair: What's the relationship? *Cell* **71:** 887–889.

Fu, W., S. O'Handley, R.P. Cunningham, and M.K. Johnson. 1992. The role of the iron-sulfur cluster in *Escherichia coli* endonuclease III—A resonance Raman study. *J. Biol.*

Chem. **267:** 16135-16137.

Ganesan, A. K., P. Seawell, R.J. Lewis, and P.C. Hanawalt. 1986. Processivity of T4 endonuclease V is sensitive to NaCl concentration. *Biochemistry* **25:** 5751-5755.

Gates, F.T., III and S. Linn. 1977. Endonuclease from *Escherichia coli* that acts specifically upon duplex DNA damaged by ultraviolet light, osmium tetroxide, acid, or X-rays. *J. Biol. Chem.* **252:** 2802-2807.

Giroir, L.E. and W.A. Deutsch. 1987. *Drosophila* deoxyuridine triphosphatase. *J. Biol. Chem.* **262:** 130-134.

Goldmark, P.J. and S. Linn. 1972. Purification and properties of the *recBC* DNase of *E. coli* K-12. *J. Biol. Chem.* **247:** 1849-1860.

Gordon, L.K. and W.A. Haseltine. 1980. Comparison of the cleavage of pyrimidine dimers by the bacteriophage T4 and *Micrococcus luteus* UV-specific endonucleases. *J. Biol. Chem.* **255:** 12047-12050.

Gossett, J., K. Lee, R.P. Cunningham, and P.W. Doetsch. 1988. Yeast redoxyendonuclease, a DNA repair enzyme similar to *Escherichia coli* endonuclease III. *Biochemistry* **27:** 2629-2634.

Grabowski, D.T., W.A. Deutsch, D. Derda, and M.R. Kelley. 1991. *Drosophila* AP3, a presumptive DNA repair protein, is homologous to human ribosomal associated protein PO. *Nucleic Acids Res.* **19:** 4297.

Grabowski, D.T., R.O. Pieper, B.W. Futscher, W.A. Deutsch, L.C. Erickson, and M.R. Kelley. 1992. Expression of ribosomal phosphoprotein PO is induced by antitumor agents and increased in Mer-human tumor cell lines. *Carcinogenesis* **13:** 259-263.

Grafstrom, R.H., N.L. Shaper, and L. Grossman. 1982. Human placental apurinic/apyrimidinic endonuclease. *J. Biol. Chem.* **257:** 13459-13464.

Graves, R., J. Laval, and A.E. Pegg. 1992a. Sequence specificity of DNA repair by *Escherichia coli* Fpg protein. *Carcinogenesis* **13:** 1455-1459.

Graves, R.J., I. Felzenszwalb, J. Laval, and T.R. O'Connor. 1992b. Excision of 5'-terminal deoxyribose phosphate from damaged DNA is catalyzed by the Fpg protein of *Escherichia coli*. *J. Biol. Chem.* **267:** 14429-14435.

Greenberg, J.T., P. Monach, J.H. Chou, P.D. Josephy, and B. Demple. 1990. Positive control of a global antioxidant defense regulon activated by superoxide-generating agents in *Escherichia coli*. *Proc. Natl. Acad. Sci.* **87:** 6181-6185.

Grilley, M., J. Griffith, and P. Modrich. 1993. Bidirectional excision in methyl-directed mismatch repair. *J. Biol. Chem.* **268:** 11830-11837.

Grilley, M., K.M. Welsh, S. Su, and P. Modrich. 1989. Isolation and characterization of the *Escherichia coli mutL* gene product. *J. Biol. Chem.* **264:** 1000-1004.

Grossman, L.I. and A.T. Yeung. 1990. The uvrABC endonuclease system—A view from Baltimore. *Mutat. Res.* **236:** 213-221.

Gruskin, E.A. and R.S. Lloyd. 1986. The DNA scanning mechanism of T4 endonuclease V. *J. Biol. Chem.* **261:** 9607-9613.

———. 1988a. Molecular analysis of plasmid DNA repair within ultraviolet-irradiated *Escherichia coli*. *J. Biol. Chem.* **263:** 12728-12737.

———. 1988b. UvrABC-initiated excision repair and photolyase-catalysed dimer monomerization. *J. Biol. Chem.* **263:** 12738-12743.

Hamilton, R.W. and R.S. Lloyd. 1989. Modulation of the DNA scanning activity of the *Micrococcus luteus* UV endonuclease. *J. Biol. Chem.* **264:** 17422-17427.

Hamilton, K.K., P.M.J. Kim, and P.W. Doetsch. 1992. A eukaryotic DNA glycosylase/lyase recognizing ultraviolet light-induced pyrimidine dimers. *Nature* **356:** 725-728.

Hanawalt, P.C. 1991. Heterogeneity of DNA repair at the gene level. *Mutat. Res.* **247:**

203-211.
Hare, J.T. and J.H. Taylor. 1985. One role for DNA methylation in vertebrate cells is strand discrimination in mismatch repair. *Proc. Natl. Acad. Sci.* **82:** 7350-7354.
Harosh, I. and J. Sperling. 1988. Hypoxanthine-DNA glycosylase from *Escherichia coli. J. Biol. Chem.* **263:** 3328-3334.
Harrison, L., G. Ascione, J.C. Menninger, D.C. Ward, and B. Demple. 1992. Human apurinic endonuclease gene (*SPE*): Structure and genomic mapping (chromosome 14q11.2-12). *Hum. Mol. Genet.* **1:** 677-680.
Haseltine, W.A., L.K. Gordon, C.P. Lindan, R.H. Grafstrom, N.L. Shaper, and L. Grossman. 1980. Cleavage of pyrimidine dimers in specific DNA sequences by a pyrimidine dimer DNA-glycosylase of *M. luteus. Nature* **285:** 634-640.
Hecht, R. and H.W. Thielmann. 1978. Purification and characterization of an endonuclease from *Micrococcus luteus* that acts on depurinated and carcinogen-modified DNA. *Eur. J. Biochem.* **89:** 607-618.
Helland, D.E., P.W. Doetsch, and W.A. Haseltine. 1986. Substrate specificity of a mammalian DNA repair endonuclease that recognizes oxidative base damage. *Mol. Cell. Biol.* **6:** 1983-1990.
Helland, D.E., A.J. Raae, P. Fadnes, and K. Kleppe. 1985. Properties of a DNA repair endonuclease from mouse plasmacytoma cells. *Eur. J. Biochem.* **148:** 471-477.
Hennecke, F., H. Kolmar, K. Bründl, and H.-J. Fritz. 1991. The *vsr* gene product of *E. coli* K-12 is a strand- and sequence- specific DNA mismatch endonuclease. *Nature* **353:** 776-778.
Henner, W.D., S.M. Grunberg, and W.A. Haseltine. 1983. Enzyme action at the 3' termini of ionizing radiation-induced DNA-strand breaks. *J. Biol. Chem.* **258:** 15198-15205.
Henner, W.D., N.P. Kiker, T.J. Jorgensen, and J.-N. Munck. 1987. Purification and amino acid terminal sequence of an apurinic/apyrimidinic endonuclease from calf-thymus. *Nucleic Acids Res.* **15:** 5529-5544.
Higley, M. and R.S. Lloyd. 1993. Processivity of uracil DNA glycosylase. *Mutat. Res.* **294:** 101-108.
Hoeijmakers, J.H.J. 1993. Nucleotide excision repair II: From yeast to mammals. *Trends Genet.* **9:** 211-217.
Hollstein, M.C., P. Brooks, S. Linn. and B.N. Ames. 1984. Hydroxymethyluracil DNA glycosylase in mammalian cells. *Proc. Natl. Acad. Sci.* **81:** 4003-4007.
Holmes, J.J., S. Clark, and P. Modrich. 1990. Strand-specific mismatch correction in nuclear extracts of human and *Drosophila melanogaster* cell lines. *Proc. Natl. Acad. Sci.* **87:** 5837-5841.
Hori, N., T. Doi, Y. Karaki, M. Kikuchi, M. Ikehara, and E. Ohtsuka. 1992. Participation of glutamic acid 23 and T4 endonuclease V in the β-elimination reaction of an abasic site in a synthetic duplex DNA. *Nucleic Acids Res.* **20:** 4761-4764.
Huang, J.-C., D.L. Svoboda, J.T. Reardon, and A. Sancar. 1992. Human nucleotide excision nuclease removes thymine dimers from DNA by incising the 22nd phosphodiester bond 5' and the 6th phosphodiester bond 3' to the photodimer. *Proc. Natl. Acad. Sci.* **89:** 3664-3668.
Ishida, M., Y. Kanamori, N. Hori, T. Inaoka, and E. Ohtsuka. 1990. *In vitro* and *in vivo* activities of T4 endonuclease V mutants altered in the C-terminal aromatic region. *Biochemistry* **29:** 3817-3821.
Johnson, A.W. and B. Demple. 1988a. Yeast DNA diesterase for 3'-fragments of deoxyribose: Purification and physical properties of a repair enzyme for oxidative DNA damage. *J. Biol. Chem.* **263:** 18009-18016.

———. 1988b. Yeast 3'-repair diesterase is the major cellular/apyrimidinic endonuclease: Substrate specificity and kinetics. *J. Biol. Chem.* **263**: 18017-18022.

Jones, M., R. Wagner, and M. Radman. 1987. Mismatch repair of deaminated 5-methylcytosine. *J. Mol. Biol.* **194**: 155-159.

Jorgensen, T.J., Y.W. Kow, S.S. Wallace, and W.D. Henner. 1987. Mechanism of action of *Micrococcus luteus* gamma-endonuclease. *Biochemistry* **26**: 6436-6443.

Kane, C.M. and S. Linn. 1981. Purification and characterization of an apurinic/apyrimidinic endonuclease from HeLa cells. *J. Biol. Chem.* **256**: 2405-3414.

Karran, P. and T. Lindahl. 1978. Enzymatic excision of free hypoxanthine from polydeoxynucleotides and DNA containing deoxyinosine monophosphate residues. *J. Biol. Chem.* **253**: 5877-5879.

———. 1980. Hypoxanthine in deoxyribonucleic acid: Generation by the heat-induced hydrolysis of adenine residues and release in free form by a deoxyribonucleic acid glycosylase from calf thymus. *Biochemistry* **19**: 6005-6011.

Karran, P., T. Hjelmgren, and T. Lindahl. 1982. Induction of a DNA glycosylase for N-methylated purines is part of the adaptive response to alkylating agents. *Nature* **296**: 770-773.

Karran, P., T. Lindahl, I. Ofsteng, G.B. Evensen, and E. Seeberg. 1980. *Escherichia coli* mutants deficient in 3-methyladenine-DNA glycosylase. *J. Mol. Biol.* **140**: 101-127.

Karu, A., V. MacKay, P. Goldmark, and S. Linn. 1973. The *recBC* deoxyribonuclease of *Escherichia coli* K-12: Substrate specificity and reaction intermediates. *J. Biol. Chem.* **248**: 4874-4884.

Katcher, H.L. and S.S. Wallace. 1983. Characterization of the *Escherichia coli* X-ray endonuclease, endonuclease III. *Biochemistry* **22**: 4071-4081.

Keller, W. and R. Crouch. 1972. Degradation of DNA RNA hybrids by ribonuclease H and DNA polymerases of cellular and viral origin. *Proc. Natl. Acad. Sci.* **69**: 3360-3364.

Kelley, M.R., S. Venugopal, J. Harless, and W.A. Deutsch. 1989. Antibody to a human DNA repair protein allows for cloning of a *Drosophila* cDNA that encodes an apurinic endonuclease. *Mol. Cell. Biol.* **9**: 965-973.

Kelly, R.B., M.R. Atkinson, J.A. Huberman, and A. Kornberg. 1969. Excision of thymine dimers and other mismatched sequences by DNA polymerase of *Escherichia coli*. *Nature* **224**: 495-501.

Kim, J. and S. Linn. 1989. Purification and characterization of UV endonucleases I and II from murine plasmacytoma cells. *J. Biol. Chem.* **264**: 2739-2745.

Kow, Y.W. 1989. Mechanism of action of *Escherichia coli* exonuclease III. *Biochemistry* **28**: 3280-3287.

Kow, Y.W. and S.S. Wallace. 1987. Exonuclease III recognizes urea residues in oxidized DNA. *Proc. Natl. Acad. Sci.* **82**: 8354-8358.

Kuhnlein, U. 1985. Comparison of apurinic DNA binding protein from an ataxia telangiectasia and a HeLa cell line. *J. Biol. Chem.* **260**: 14918-14929.

Kuo, C.-F., D.E. McRee, R.P. Cunningham, and J.A. Tainer. 1992a. Crystallization and crystallographic characterization of the iron-sulfur-containing DNA-repair enzyme endonuclease III from *Escherichia coli*. *J. Mol. Biol.* **227**: 347-351.

Kuo, C.-F., D.E. McRee, C.L. Fisher, S.F. O'Handley, R.P. Cunningham, and J.A. Tainer. 1992b. Atomic structure of the DNA repair [4Fe-4S] enzyme endonuclease III. *Science* **258**: 434-440.

Laib, R.J., L.M. Gwinner, and H.M. Bolt. 1981. DNA alkylation by vinyl chloride metabolites: Etheno derivatives or 7-alkylation of guanine? *Chem. Biol. Interact.* **37**: 219-231.

Längle-Rouault, F., G. Maenhaut-Michel, and M. Radman. 1987. GATC sequences, DNA nicks and the MutH function in *Escherichia coli* mismatch repair. *EMBO J.* **6:** 1121–1127.

Laval, J. 1977. Two enzymes are required from strand incision in repair of alkylated DNA. *Nature* **269:** 829–832.

Learn, B.A. and R.H. Grafstrom. 1989. Methyl-directed repair of frameshift heteroduplexes in cell extracts from *Escherichia coli. J. Bacteriol.* **171:** 6473–6481.

Leithauser, M.T., A. Liem, B.C. Stewart, E.C. Miller, and J.A. Miller. 1990. 1,N^6-Ethenoadenosine formation, mutagenicity and murine tumor induction as indicators of the generation of an electrophilic epoxide metabolite of the closely related carcinogens ethyl carbamate (urethane) and vinyl carbamate. *Carcinogenesis* **11:** 463–473.

Levin, J.D., A.W. Johnson, and B. Demple. 1988. Homogenous *Escherichia coli* endonuclease IV. *J. Biol. Chem.* **263:** 8066–8071.

Levin, J.D., R. Shapiro, and B. Demple. 1991. Metalloenzymes in DNA repair. *J. Biol. Chem.* **266:** 22893–22898.

Lieb, M. 1983. Specific mismatch correction in bacteriophage lambda crosses by very short patch repair. *Mol. Gen. Genet.* **191:** 118–125.

———. 1985. Recombination in the lambda repressor gene: Evidence that very short patch (VSP) mismatch correction restores a specific sequence. *Mol. Gen. Genet.* **199:** 465–470.

Lin, J.-J. and A. Sancar. 1991. The carboxy terminal half of UvrC protein is sufficient to reconstitute (A)BC excinuclease. *Proc. Natl. Acad. Sci.* **88:** 6824–6828.

———. 1992. Active site of (A)BC excinuclease. *J. Biol. Chem.* **267:** 17688–17692.

Lin, J.-J., A.M. Phillips, J.E. Hearst, and A. Sancar. 1992. Active site of (A)BC excinuclease. I. Binding, bending and catalysis mutants of UvrB reveal a direct role in 3' and an indirect role in 5' incision. *J. Biol. Chem.* **267:** 17693–17700.

Lindahl, T. 1971. Excision of pyrimidine dimers from ultraviolet-irradiated DNA by exonucleases from mammalian cells. *Eur. J. Biochem.* **18:** 407–414.

———. 1974. An N-glycosidase from *Escherichia coli* that releases free uracil from DNA containing deaminated cytosine residues. *Proc. Natl. Acad. Sci.* **71:** 3649–3653.

———. 1979. DNA glycosylases, endonucleases for apurinic/apyrimidinic sites and base excision repair. *Prog. Nucleic Acid Res.* **22:** 135–192.

———. 1982. DNA repair enzymes. *Annu. Rev. Biochem.* **51:** 61–87.

Lindahl, T. and B. Nyberg. 1974. Heat-induced deamination of cytosine residues in deoxyribonucleic acid. *Biochemistry* **13:** 3405–3410.

Lindahl, T., S. Ljungquist, W. Siegert, B. Nyberg, and B. Sperens. 1977. DNA N-glycosidases: Properties of uracil-DNA glycosidase from *Escherichia coli. J. Biol. Chem.* **252:** 3286–3294.

Linsley, W.S., E.E. Penhoet, and S. Linn. 1977. Human endonuclease specific for apurinic/apyrimidinic sites in DNA. *J. Biol. Chem.* **252:** 1235–1242.

Liuzzi, M. and M.C. Paterson. 1992. Enzymatic analysis of oligonucleotides containing cyclobutane pyrimidine photodimers with a cleaved intradimer phosphodiester linkage. *J. Biol. Chem.* **267:** 22421–22427.

Ljungman, S. and P.C. Hanawalt. 1992. Localized torsional tension in the DNA of human cells. *Proc. Natl. Acad. Sci.* **89:** 6055–6059.

Ljungquist, S. 1977. A new endonuclease from *Escherichia coli* acting at apurinic sites in DNA. *J. Biol. Chem.* **252:** 2808–2814.

Ljungquist, S. and T. Lindahl. 1977. Relation between *Escherichia coli* endonuclease specific for apurinic sites in DNA and exonuclease III. *Nucleic Acids Res.* **4:** 2871–2879.

Ljungquist, S., T. Lindahl, and P. Howard-Flanders. 1976. Methyl methane sulfonate-sensitive mutant of *Escherichia coli* deficient in an endonuclease specific for apurinic sites in deoxyribonucleic acid. *J. Bacteriol.* **126:** 646-653.

Lloyd, R. S. and M.L. Augustine. 1989. Site-directed mutagenesis of the T4 endonuclease V gene: Mutations which enhance enzyme specific activity at low salt concentrations. *Proteins* **6:** 128-138.

Lloyd, R.S., P.C. Hanawalt, and M.L. Dodson. 1980. Processive action of T4 endonuclease V on ultraviolet-irradiated DNA. *Nucleic Acids Res.* **8:** 5113-5127.

Lloyd, R.S., M.L. Dodson, E.A. Gruskin, and D.L. Robberson. 1987. T4 endonuclease V promotes the formation of multimeric DNA structures. *Mutat. Res.* **183:** 109-115.

Lovett, S.T. and R.D. Kolodner. 1989. Identification and purification of a single-stranded-DNA-specific exonuclease encoded by the *recJ* gene of *Escherichia coli*. *Proc. Natl. Acad. Sci.* **86:** 2627-2631.

Lu, A.L. and D.-Y. Chang. 1988a. Repair of single base-pair transversion mismatches of *Escherichia coli in vitro*: Correction of certain A/G mismatches is independent of *dam* methylation and host *mutHLS* gene function. *Genetics* **118:** 593-600.

———. 1988b. A novel nucleotide excision repair for the conversion of an A/G mismatch to C/G base pair in *E. coli*. *Cell* **54:** 805-812.

Lu, A.-L., J.J. Cuipa, M.S. Ip, and W.G. Shanabruch. 1990. Specific A/G-to-C·G mismatch repair in *Salmonella typhimurium* LT2 requires the *mutB* gene product. *J. Bacteriol.* **172:** 1232-1240.

Male, R., D.E. Helland, and K. Kleppe. 1985. Purification and characterization of 3-methyladenine-DNA glycosylase from calf thymus. *J. Biol. Chem.* **260:** 1623-1629.

Male, R., B.I. Haukanes, D.E. Helland, and K. Kleppe. 1987. Substrate specificity of 3-methyladenine-DNA glycosylase from calf thymus. *Eur. J. Biochem.* **165:** 13-19.

Margulies, L. and S.S. Wallace. 1984. Apurinic endonuclease activity remains constant during early *Drosophila* development. *Cell Biol. Toxicol.* **1:** 127-143.

Mazumder, A., J.A. Gerlt, M.J. Absalon, J. Stubbe, R.P. Cunningham, J. Withka, and P.H. Bolton. 1991. Stereochemical studies of the β-elimination reactions at aldehydic abasic sites in DNA: Endonuclease III from *Escherichia coli*, sodium hydroxide, and Lys-Trp-Lys. *Biochemistry* **39:** 1119-1126.

Méjean, V., I. Rives, and J.P. Claverys. 1990. Nucleotide sequence of the *Streptococcus pneumoniae ung* gene encoding uracil-DNA glycosylase. *Nucleic Acids Res.* **18:** 6693.

Meyer-Siegler, K., D.J. Mauro, G. Seal, J. Wurzer, J.K. DeRiel, and M.A. Sirover. 1991. A human nuclear uracil DNA glycosylase is the 37-kDa subunit of glyceraldehyde-3-phosphate dehydrogenase. *Proc. Natl. Acad. Sci.* **88:** 8460-8464.

Michaels, M.L. and J.H. Miller. 1992. The GO system protects organisms from the mutagenic effect of the spontaneous lesion 8-hydroxyguanine (7,8-dihydro-8-oxoguanine). *J. Bacteriol.* **174:** 6321-6325.

Michaels, M.L., C. Cruz, A.P. Grollman, and J.H. Miller. 1992. Evidence that MutY and MutM combine to prevent mutation by an oxidatively damaged form of guanine in DNA. *Proc. Natl. Acad. Sci.* **89:** 7022-7025.

Michaels, M.L., Y. Pham, C. Cruz, and J.H. Miller. 1991. MutM, a protein that prevents G·C→T·A transversions, is formamidopyrimidine-DNA glycosylase. *Nucleic Acids Res.* **19:** 3629-3632.

Michaels, M.L., L. Pham, Y. Ngheim, C. Cruz, and J.H. Miller. 1990. MutY, and adenine glycosylase active on G-A mispairs, has homology to endonuclease III. *Nucleic Acids Res.* **18:** 3841-3845.

Mitra, S. and B. Kaina. 1993. Regulation of repair of alkylation damage in mammalian genomes. *Prog. Nucleic Acid Res.* **44:** 109-141.

Modrich, P. 1991. Mechanisms and biological effects of mismatch repair. *Annu. Rev. Genet.* **25:** 229–253.

Morikawa, K., O. Matsumoto, M. Tsujimoto, K. Katayanagi, M. Ariyoshi, T. Doi, M. Ikehara, T. Inaoka, and R. Ohtsuka. 1992. X-ray structure of T4 endonuclease V: An excision repair enzyme specific for a pyrimidine dimer. *Science* **256:** 523–526.

Moriya, M., C. Ou, V. Bodepudi, F. Johnson, M. Takeshita, and A. Grollman. 1991. Site-specific mutagenesis using a gapped duplex vector: A study of translesion synthesis past 8-oxodeoxyguanosine in *E. coli. Mutat. Res.* **254:** 281–288.

Mosbaugh, D. and S. Linn. 1980. Further characterization of human fibroblast apurinic/apyrimidinic DNA endonucleases. *J. Biol. Chem.* **25:** 11743–11752.

Myles, G.M. and A. Sancar. 1989. DNA repair. *Chem. Res. Toxicol.* **2:** 197–226.

——— . 1991. Isolation and characterization of functional domains of UvrA. *Biochemistry* **30:** 3834–3840.

Myles, G.M., J.E. Hearst, and A. Sancar. 1991. Site-specific mutagenesis of conserved residues within Walker A and B sequences of *Escherichia coli* UvrA protein. *Biochemistry* **30:** 3824–3834.

Myrnes, B., P.H. Guddal, and H. Krohan. 1982. Metabolism of dITP in HeLa cell extracts, incorporation into DNA by isolated nuclei and release of hypoxanthine from DNA by a hypoxanthine-DNA glycosylase activity. *Nucleic Acids Res.* **10:** 3693–3701.

Nakabeppu, Y., T. Miyata, H. Kondo, S. Iwanaga, and M. Sekiguchi. 1984. Structure and expression of the *alkA* gene of *Escherichia coli* involved in adaptive response to alkylating agents. *J. Biol. Chem.* **259:** 13730–13736.

Nakayama, H., S. Okubo, M. Sekiguchi, and Y. Takagi. 1967. A deoxyribonuclease activity specific for ultraviolet-irradiated DNA: A chromatographic analysis. *Biochem. Biophys. Res. Commun.* **27:** 217–223.

Nes, I.F. 1980. Purification and properties of a mouse-cell DNA-repair endonuclease, which recognizes lesions in DNA induced by ultraviolet light, depurination, γ-rays, and OsO_4 treatment. *Eur. J. Biochem.* **112:** 161–168.

Nes, I.F. and J. Nissen-Mayer. 1978. Endonuclease activities from a permanently established mouse cell line that act upon DNA damaged by ultraviolet light, acid and osmium tetroxide. *Biochim. Biophys. Acta* **520:** 111–121.

Niwa, O. and R.E. Moses. 1981. Synthesis by DNA polymerase I on bleomycin-treated deoxyribonucleic acid; a requirement of exonuclease III. *Biochemistry* **20:** 238–244.

O'Connor, T.R. and F. Laval. 1990. Isolation and structure of a cDNA expressing a mammalian 3-methyl-adenine-DNA glycosylase. *EMBO J.* **9:** 3337–3342.

——— . 1991. Human cDNA expressing a functional DNA glycosylase excising 3-methyladenine and 7-methylguanine. *Biochem. Biophys. Res. Commun.* **176:** 1170–1177.

O'Connor, T.R., S. Boiteux, and J. Laval. 1988. Ring-opened 7-methylguanine residues in DNA are a block to *in vitro* DNA synthesis. *Nucleic Acids Res.* **16:** 5879–5894.

O'Connor, T.R., R.J. Graves, G. de Murcia, B. Castaing, and J. Laval. 1993. Fpg protein of *Escherichia coli* is a zinc finger protein whose cysteine residues have a structural and/or functional role. *J. Biol. Chem.* **268:** 9063–9070.

Olsen, L.C., R. Aasland, C.U. Wittwer, H.E. Krokan, and D.E. Helland. 1989. Molecular cloning of human uracil-DNA glycosylase, a highly conserved DNA repair enzyme. *EMBO J.* **8:** 3121–3125.

Olsson, M. and T. Lindahl. 1980. Repair of alkylated DNA in *E. coli:* Methyl group transfer from O^6-methyl-guanine to a protein cysteine residue. *J. Biol. Chem.* **255:** 10569–10571.

Orren, D.K. and A. Sancar. 1989. The (A)BC excinuclease of *Escherichia coli* has only

the UvrB and UvrC subunits in the incision complex. *Proc. Natl. Acad. Sci.* **86:** 5237–5241.
Paek, K.-H. and G.C. Walker. 1986. Defect in expression of heat-shock proteins at high temperature in *xthA* mutants. *J. Bacteriol.* **165:** 763–770.
Percival, K.J., M.B. Klein, and M.J. Burgers. 1989. Molecular cloning and primary structure of the uracil-DNA-glycosylase gene from *Saccharomyces cerevisiae*. *J. Biol. Chem.* **264:** 2593–2598.
Pierre, J. and J. Laval. 1986. Cloning of *Micrococcus luteus* 3-methyladenine-DNA glycosylase genes in *Escherichia coli*. *Gene* **43:** 139–146.
Popoff, S.C., A.I Spira, A.W. Johnson, and B. Demple. 1990. Yeast structural gene (*APN1*) for the major apurinic endonuclease: Homology to *Escherichia coli* endonuclease IV. *Proc. Natl. Acad. Sci.* **87:** 4193–4197.
Radany, E.H. and E.C. Friedberg. 1980. A pyrimidine dimer-DNA glycosylase activity associated with the *v* gene product of bacteriophage T4. *Nature* **286:** 182–185.
Radicella, J.P., E.A. Clark, and M.S. Fox. 1988. Some novel mismatch repair activities in *Escherichia coli*. *Proc. Natl. Acad. Sci.* **85:** 9674–9678.
Radler-Pohl, A., C. Sachsenmaier, S. Gebel, H.-P. Auer, J.T. Bruder, U. Rapp, P. Angel, H.J. Rahmsdorf, and P. Herrlich. 1993. UV-induced activation of AP-1 involves obligatory extranuclear steps including Raf-1 kinase. *EMBO J.* **12:** 1005–1012.
Ramotar, D., S.C. Popoff, and B. Demple. 1991a. Complementation of DNA repair-deficient *Escherichia coli* by the yeast Apn1 apurinic/apyrimidinic endonuclease gene. *Mol. Microbiol.* **5:** 149–155.
Ramotar, D., S.C. Popoff, E.B. Gralla, and B. Demple. 1991b. Cellular role of yeast Apn1 apurinic endonuclease/3'-diesterase: Repair of oxidative and alkylation DNA damage and control of spontaneous mutation. *Mol. Cell. Biol.* **11:** 4537–4544.
Randahl, H., G.C. Elliott, and S. Linn. 1988. DNA repair reactions by purified HeLa DNA polymerase α and exonucleases. *J. Biol. Chem.* **263:** 12228–12234.
Recinos, A. and R.S. Lloyd. 1988. Site-directed mutagenesis of the T4 endonuclease V gene: Role of lysine-130. *Biochem.* **27:** 1832–1838.
Riazuddin, S. and L. Grossman. 1977a. *Micrococcus luteus* correndonucleases. I. Resolution and purification of two endonucleases specific for DNA containing pyrimidine dimers. *J. Biol. Chem.* **252:** 6280–6286.
———. 1977b. *Micrococcus luteus* correndonucleases. II. Mechanism of action of two endonucleases specific for DNA containing pyrimidine dimers. *J. Biol. Chem.* **252:** 6287–6293.
———. 1977c. *Micrococcus luteus* correndonucleases. III. Evidence for involvement in repair *in vivo* of two endonucleases specific for DNA containing pyrimidine dimers. *J. Biol. Chem.* **252:** 6294–6298.
Riazuddin, S. and T. Lindahl. 1978. Properties of 3-methyladenine-DNA glycosylase from *Escherichia coli*. *Biochemistry* **17:** 2110–2118.
Rich, B.E. and J.A. Steitz. 1987. Human acidic ribosomal phosphoproteins P0, P1, and P2: Analysis of cDNA clones, *in vitro* synthesis, and assembly. *Mol. Cell. Biol.* **7:** 4065–4074.
Richardson, C.C. and A. Kornberg. 1964. A deoxyribonucleic acid phosphatase-exonuclease from *Escherichia coli*. I. Purification of the enzyme and characterization of the phosphatase activity. *J. Biol. Chem.* **239:** 242–250.
Riggs, A.D. and P.A. Jones. 1983. 5-Methylcytosine, gene regulation and cancer. *Adv. Cancer Res.* **40:** 1–30.
Robson, C.N. and I.D. Hickson. 1991. Isolation of cDNA clones encoding a human apurinic/apyrimidinic endonuclease that corrects DNA repair and mutagenesis defects

in *E. coli xth* (exonuclease III) mutants. *Nucleic Acids Res.* **19**: 5519–5523.

Robson, C.N., A.M. Milne, D.J. Pappin, and I.D. Hickson. 1991. Isolation of cDNA clones encoding an enzyme from bovine cells that repairs oxidative DNA damage in vitro: Homology with bacterial repair enzymes. *Nucleic Acids Res.* **19**: 1087–1092.

Robson, C.N., D. Hochhauser, R. Craig, K. Rack, V.J. Buckle, and I.D. Hickson. 1992. Structure of the human DNA repair gene *HAP1* and its localization to chromosome 14q 11.2-12. *Nucleic Acids Res.* **20**: 4417–4421.

Rydberg, B., M.K. Dosanjh, and B. Singer. 1991. Human cells contain protein specifically binding to a single 1, N^6-ethenoadenine in a DNA fragment. *Proc. Natl. Acad. Sci.* **88**: 6839–6842.

Rydberg, B., Z.-H. Qiu, M.K. Dosanjh, and B. Singer. 1992. Partial purification of a human DNA glycosylase acting on the cyclic carcinogen adduct 1,N^6-ethenodeoxyadenosine. *Cancer Res.* **52**: 1377–1379.

Sak, B.D., A. Eisenstark, and D. Touati. 1989. Exonuclease III and the catalase hydroperoxidase II in *Escherichia coli* are both regulated by the KatF gene product. *Proc. Natl. Acad. Sci.* **86**: 3271–3275.

Sakumi, K. 1986. Purification and structure of 3-methyladenine-DNA glycosylase I of *Escherichia coli*. *J. Biol. Chem.* **261**: 15761–15766.

Samson, L.D. and J. Cairns. 1977. A new pathway for DNA repair in *Escherichia coli*. *Nature* **267**: 281–282.

Samson, L., B. Derfler, M. Boosalis, and K. Call. 1991. Cloning and characterization of a 3-methyladenine DNA glycosylase cDNA from human cells whose gene maps to chromosome 16. *Proc. Natl. Acad. Sci.* **88**: 9127–9131.

Sancar, A. and J.E. Hearst. 1993. Molecular matchmakers. *Science* **259**: 1415–1420.

Sancar, A. and G.B. Sancar. 1988. DNA repair enzymes. *Annu. Rev. Biochem.* **57**: 29–67.

Sander, M., K. Lowenhaupt, and A. Rich. 1991. *Drosophila* Rrp1 protein: An apurinic endonuclease with homologous recombination activities. *Proc. Natl. Acad. Sci.* **88**: 6780–6784.

Sanderson, B.J.S., C.-N. Chang, A.P. Grollman, and W.D. Henner. 1989. Mechanism of DNA cleavage and substrate recognition by a bovine apurinic endonuclease. *Biochemistry* **28**: 3894–3901.

Sandigursky, M. and W.A. Franklin. 1992. DNA deoxyribophosphodiesterase of *Escherichia coli* is associated with exonuclease I. *Nucleic Acids Res.* **20**: 4699–4703.

Saporito, S.M. and R.P. Cunningham. 1988. Nucleotide sequence of the *nfo* gene of *Escherichia coli* K-12. *J. Bacteriol.* **170**: 5141–5145.

Saporito, S.M., B.J. Smith-White, and R.P. Cunningham. 1988. Nucleotide sequence of the *xth* gene of *Escherichia coli* K-12. *J. Bacteriol.* **170**: 4542–4547.

Schaeffer, L., R. Roy, S. Humbert, V. Moncollin, W. Vermeulen, J.H.J. Hoeijmakers, P. Chambon, and J.-M. Egly. 1993. DNA repair helicase: A component of BTF2 (TFIIH) basic transcription factor. *Science* **260**: 58–63.

Schrock, R.D., III and R.S. Lloyd. 1991. Reductive methylation of the N-terminus of endonuclease V eradicates catalytic activities—Evidence for an essential role of the N-terminus in the chemical mechanisms of catalysis. *J. Biol. Chem.* **266**: 17631–17639.

———. 1993. Site directed mutagenesis of the N-terminus of T4 endonuclease V: The position of the αNH_2 moiety affects catalytic activity. *J. Biol. Chem.* **268**: 880–886.

Seawell, P.C., C.A. Smith, and A.K. Ganesan. 1980. *denV* gene of bacteriophage T4 determines a DNA glycosylase specific for pyrimidine dimers in DNA. *J. Virol.* **35**: 790–797.

Seeley, T.W. and L. Grossman. 1989. Mutations in the *Escherichia coli* UvrB ATPase motif compromise excision repair capacity. *Proc. Natl. Acad. Sci.* **86**: 6577–6581.

———. 1990. The role of *Escherichia coli* UvrB in nucleotide excision repair. *J. Biol. Chem.* **265:** 7158-7165.

Selby, C.P. and A. Sancar. 1990. Structure and function of the (A)BC excinuclease of *Escherichia coli*. *Mutat. Res.* **236:** 203-211.

———. 1993. Molecular mechanism of transcription-repair coupling. *Science* **260:** 53-58.

Setlow, R.B., J.K. Setlow, and W.L. Carrier. 1970. Endonuclease from *Micrococcus luteus* which has activity toward ultraviolet-irradiated deoxyribonucleic acid: Its action on transforming deoxyribonucleic acid. *J. Bacteriol.* **102:** 187-192.

Shackleton, J., W. Warren, and J.J. Roberts. 1979. The excision of *N*-methyl-*N*-nitrosourea-induced lesions from the DNA of Chinese hamster cells is measured by the loss of sites sensitive to an enzyme extract that excises 3-methylpurines but not O^6-methylguanine. *Eur. J. Biochem.* **97:** 425-433.

Shaper, N.L., R.H. Grafstrom, and L. Grossman. 1982. Human placental apurinic/apyrimidinic endonuclease: Its isolation and purification. *J. Biol. Chem.* **257:** 13455-13458.

Shibutani, S., M. Takeshita, and A.P. Grollman. 1991. Insertion of specific bases during DNA synthesis past the oxidation-damaged base 8-oxodG. *Nature* **349:** 431-434.

Shimada, K., H. Nakayama, S. Okubo, M. Sekiguchi, and Y. Takagi. 1967. An endonucleolytic activity specific for ultraviolet-irradiated DNA in wild type and mutant strains of *Micrococcus lysodeikticus*. *Biochem. Biophys. Res. Commun.* **27:** 539-545.

Singer, B. and J.T. Kusmierek. 1982. Chemical reactions of metabolically activated aromatic mutagens. *Annu. Rev. Biochem.* **52:** 655-693.

Snowden, A. and B. Van Houten. 1991. Initiation of the UvrABC nuclease cleavage reaction efficiency of incision is not correlated with UvrA binding affinity. *J. Mol. Biol.* **220:** 19-33.

———. 1993. Mechanism of action of the *Escherichia coli* UvrABC nuclease: Clues to the damage recognition problem. *BioEssays* **15:** 51-59.

Snowden, A., Y.W. Kow, and B. Van Houten. 1990. Damage repertoire of the *Escherichia coli* UvrABC nuclease complex includes abasic sites, base-damage analogues, and lesions containing adjacent 5' or 3' nicks. *Biochemistry* **29:** 7251-7259.

Spiering, A.L. and W.A. Deutsch. 1981. Apurinic DNA endonucleases from *Drosophila melanogaster* embryos. *Mol. Gen. Genet.* **183:** 171-174.

———. 1986. *Drosophila* apurinic/apyrimidinic DNA endonucleases. *J. Biol. Chem.* **261:** 3222-3228.

Strauss, B.S. 1962. Differential destruction of the transforming activity of damaged deoxyribonucleic acid by a bacterial enzyme. *Proc. Natl. Acad. Sci.* **48:** 1670-1675.

Strauss, B., T. Searashi, and M. Robbins. 1966. Repair of DNA studied with a nuclease specific for UV-induced lesions. *Proc. Natl. Acad. Sci.* **56:** 932-939.

Stump, D.G. and R.S. Lloyd. 1988. Site-directed mutagenesis of the T4 endonuclease V gene: Role of tyrosine-129 and -131 in pyrimidine dimer-specific binding. *Biochemistry* **27:** 1839-1843.

Su, S.-S. and P. Modrich. 1986. *Escherichia coli mutS*-encoded protein binds to mismatched DNA base pairs. *Proc. Natl. Acad. Sci.* **83:** 5057-5061.

Su, S.-S., R.S. Lahue, K.G. Au, and P. Modrich. 1988. Mispair specificity of methyl-directed DNA mismatch correction *in vitro*. *J. Biol. Chem.* **263:** 5057-5061.

Sung, P., L. Prakash, and S. Prakash. 1992. Renaturation of DNA catalysed by yeast DNA repair and recombination protein RAD10. *Nature* **355:** 743-745.

Sung, P., P. Reynolds, L. Prakash, and S. Prakash. 1993. Purification and characterization of the *Saccharomyces cerevisiae* RAD1/RAD10 endonuclease. *J. Biol. Chem.* **268:** 26391-26399.

Svoboda, D.L., J.-S. Taylor, J.E. Hearst, and A. Sancar. 1993. DNA repair by eukaryotic nucleotide excision nuclease. *J. Biol. Chem.* **268:** 1931-1936.

Tanaka, J.-I. and M. Sekiguchi. 1975. Action of exonuclease V (the *rec*BC enzyme) on ultraviolet-irradiated DNA. *Biochim. Biophys. Acta* **383:** 178-187.

Tchou, J. and A.P. Grollman. 1993. Repair of DNA containing the oxidatively-damaged base, 8-oxoguanine. *Mutat. Res.* **299:** 277-287.

Tchou, J., H. Kasai, S. Shibutani, M.-H. Chung, J. Laval, A.P. Grollman, and S.Nishimura. 1991. 8-Oxoguanine (8-hydroxyguanine) DNA glycosylase and its substrate specificity. *Proc. Natl. Acad. Sci.* **88:** 4690-4694.

Thiagalingam, S. and L. Grossman. 1991. Both ATPase sites of *Escherichia coli* UvrA have functional roles in nucleotide excision repair. *J. Biol. Chem.* **266:** 11395-11403.

———. 1993. The multiple roles for ATP in the *Escherichia coli* UvrABC endonuclease catalysed incision reaction. *J. Biol. Chem.* **268:** 18382-18389.

Thomas, D.C., J.D. Roberts, and T.A. Kunkel. 1991. Heteroduplex repair in extracts of human HeLa cells. *J. Biol. Chem.* **266:** 3744-3751.

Thomas, L., C.H. Yuang, and D.A. Goldthwait. 1982. Two DNA glycosylases in *Escherichia coli* which release primarily 3-methyladenine. *Biochemistry* **21:** 1162-1169.

Tomilin, N.V. and O.N. Aprelikova. 1989. Uracil-DNA glycosylases and DNA uracil repair. *Int. Rev. Cytol.* **114:** 125-179.

Tomkinson, A.E., A.J. Bardwell, L. Bardwell, N.J. Rappe, and E.C. Friedberg. 1993. Yeast DNA repair and recombination proteins Rad1 and Rad10 are subunits of a single-stranded DNA endonuclease. *Nature* **362:** 860-862.

Tomkinson, A.E., R.T. Bonk, J. Kim, N. Bartfeld, and S. Linn. 1990. Mammalian mitochondrial endonuclease activities specific for ultraviolet-irradiated DNA. *Nucleic Acids Res.* **18:** 929-935.

Tsai-Wu, J.-J., H.-F. Liu, and A.-L. Lu. 1992. *Escherichia coli* MutY protein has both N-glycosylase and apurinic/apyrimidinic endonuclease activities on A•C and A•G mispairs. *Proc. Natl. Acad. Sci.* **89:** 8779-8783.

Tsai-Wu, J.-J., J.P. Radicella, and A.-L. Lu. 1991. Nucleotide sequence of the *Escherichia coli micA* gene required for A/G-specific mismatch repair: Identity of MicA and MutY. *J. Bacteriol.* **173:** 1902-1910.

Upton, C., D.T. Stuart, and G. McFadden. 1993. Identification of a poxvirus gene encoding a uracil DNA glycosylase. *Proc. Natl. Acad. Sci.* **90:** 4518-4522.

Van Houten, B. 1990. Nucleotide excision repair in *Escherichia coli*. *Microbiol. Rev.* **54:** 18-51.

Varlet, I., M. Radman, and P. Brooks. 1990. DNA mismatch repair in *Xenopus* egg extracts: Repair efficiency and DNA repair synthesis for all single base-pair mismatches. *Proc. Natl. Acad. Sci.* **87:** 7883-7887.

Varshney, U., T. Hutcheon, and J.H. van de Sande. 1988. Sequence analysis, expression, and conservation of *Escherichia coli* uracil DNA glycosylase and its gene (*ung*). *J. Biol. Chem.* **263:** 7776-7784.

Venugopal, S., S.N. Guzder, and W.A. Deutsch. 1990. Apurinic endonuclease activity from wild-type and repair-deficient *mei-9 Drosophila* ovaries. *Mol. Gen. Genet.* **221:** 421-426.

Vickers, M.A., P. Vyas, P.C. Harris, D.L. Simmons, and D.R. Higgs. 1993. Structure of the human 3-methyladenine DNA glycosylase gene and localization close to the 16p telomere. *Proc. Natl. Acad. Sci.* **90:** 3437-3441.

Voigt, J.M., B. Van Houten, A. Sancar, and M.D. Topal. 1989. Repair of O^6-methylguanine by ABC excinuclease of *Escherichia coli in vitro. J. Biol. Chem.* **264:** 5172–5176.

Walker, G.C. 1984. Mutagenesis and inducible responses to deoxyribonucleic acid damage in *Escherichia coli. Microbiol. Rev.* **48:** 60–93.

Wang, J. and L. Grossman. 1993. Mutations in the helix-turn-helix motif of the *Escherichia coli* UvrA protein eliminate its specificity for UV-damaged DNA. *J. Biol. Chem.* **268:** 5323–5331.

Warner, H.R., B.F. Demple, W.A. Deutsch, C.M. Kane, and S. Linn. 1980. Apurinic/apyrimidinic endonucleases in repair of pyrimidine dimers and other lesions in DNA. *Proc. Natl. Acad. Sci.* **77:** 4602–4606.

Weinfield, M., N. Gentner, L. Johnson, and M. Paterson. 1986. Photoreversal-dependent release of thymidine and thymidine monophosphate from pyrimidine dimer-containing DNA excision fragments isolated from ultraviolet-damaged human fibroblasts. *Biochemistry* **25:** 2656–2664.

Welsh, K.M., A.L. Lu, S. Clark, and P. Modrich. 1987. Isolation and characterization of the *Escherichia coli mutH* gene product. *J. Biol. Chem.* **262:** 15624–15629.

Wiebauer, K. and J. Jiricny. 1989. *In vitro* correction of G/T mispairs to G/C pairs in nuclear extracts from human cells. *Nature* **339:** 234–236.

———. 1990. Mismatch-specific thymine DNA glycosylase and DNA polymerase β mediate the correction of G/T mispairs in nuclear extracts from human cells. *Proc. Natl. Acad. Sci.* **87:** 5842–5845.

Wilson, D.M., III, W.A. Deutsch and M.R. Kelley. 1993. Cloning of the *Drosophila* ribosomal protein S3: Another multifunctional ribosomal protein with AP endonuclease DNA repair activity. *Nucleic Acids Res.* **21:** 2516.

Wittwer, C.U. and H. Krokan. 1985. Uracil-DNA glycosylase in HeLa S3 cells: Interconvertibility of 50 and 20 kDa forms and similarity of the nuclear and mitochondrial form of the enzyme. *Biochim. Biophys. Acta* **832:** 308–318.

Wood, M.L., M. Dizdaroglu, E. Gajewksi, and J.M. Essigmann. 1990. Mechanistic studies of ionizing radiation and oxidative mutgenesis: Genetic effects of a single 8-hydroxyguanine (7-hydro-8-oxoguanine) residue inserted at a unique site in a viral genome. *Biochemistry* **29:** 7024–7032.

Worrad, D.M. and S. Caradonna. 1988. Identification of the coding sequence for herpes simplex virus uracil-DNA glycosylase. *J. Virol.* **62:** 4774–4777.

Wu, J. and B. Weiss. 1991. Two divergently transcribed genes, *soxR* and *soxS*, control a superoxide response regulon of *Escherichia coli. J. Bacteriol.* **173:** 2864–2871.

Zell, R. and H.J. Fritz. 1987. DNA mismatch-repair in *Escherichia coli* counteracting the hydrolytic deamination of 5-methyl-cytosine residues. *EMBO J.* **6:** 1809–1815.

10
Artificial Nucleases

Dehua Pei* and Peter G. Schultz
Department of Chemistry
University of California
Berkeley, California 94720

I. Introduction
II. Cleavage Chemistry
III. Sequence-specific Cleavage
   A. Small Molecules
   B. Oligonucleotides as Targeting Ligands
      1. Recognition of Single-stranded DNA (RNA)
      2. Recognition of Duplex DNA via Triple-helix Formation
      3. Oligonucleotides Modified with Nonhydrolytic Nucleolytic Agents
      4. Oligonucleotide-Staphylococcal Nuclease Conjugates
      5. Recognition of Duplex DNA via D-loop Formation
      6. Oligonucleotide-directed Restriction Endonuclease Cleavage
      7. DNA Isolation by Amplification
   C. DNA-binding Proteins and Peptides as Targeting Ligands
   D. Other Applications
IV. Conclusions

I. INTRODUCTION

Nucleases, particularly the sequence-specific restriction endonucleases, have become indispensable tools in modern biochemistry and molecular biology. They have made possible gene isolation, DNA sequencing, and recombinant DNA technology. However, their usefulness is, in many instances, limited by their small recognition site sizes (4–8 bp) and limited availability. For example, the restriction enzymes cleave too frequently to allow analysis of large chromosomal DNAs. Most restriction enzymes have recognition sites of 4–6 bp, corresponding to cleavage frequencies of once in every 136 and 2080 bp, respectively. Because a single mammalian chromosome can contain more than 100 million base pairs, these enzymes generate far too many fragments to be analyzed by current gel electrophoresis methods. Second, although 2400 restriction endonucleases have been reported, and even more will likely be discovered over time, the number of different sequences they recognize has not exceeded 200 (see Chapter 2; Appendix A). More specificities will surely be added

*Present address: Department of Biological Chemistry and Molecular Pharmacology, Harvard Medical School, 240 Longwood Avenue, Boston, Massachusetts 02115.

Nucleases, 2nd Edition
© 1993 Cold Spring Harbor Laboratory Press 0-87969-426-2/93 $5 + .00

to the existing pool in the future, but there will remain many sequences for which no restriction endonuclease is available. A final limitation is that restriction enzymes do not in general cleave single-stranded DNA or RNA.

Currently, both chemists and biologists are focusing considerable effort on generating "artificial nucleases" capable of cleaving DNA and RNA with high selectivity at any desired site. Work has focused on both the design and synthesis of efficient DNA-cleaving agents, as well as the development of strategies for selectively recognizing nucleic acid sequences of any defined sequence and length. Oxidative and alkylating agents, as well as nonspecific nucleases, have been attached to many selective DNA-binding molecules, including small organic molecules, oligonucleotides, proteins, and peptides. In addition, methods have been developed for enhancing the specificities of existing enzymes. In this chapter, we review a number of approaches currently being pursued toward the development of artificial nucleases.

II. CLEAVAGE CHEMISTRY

The backbone of DNA is rather resistant to hydrolytic cleavage due to the negative charge of the phosphodiester group. Consequently, many of the naturally occurring antibiotic, antiviral, and anticancer agents that cleave DNA do so by oxidation of the ribose ring or electrophilic alkylation of the base. For example, bleomycin (Fig. 1), a widely used anticancer drug, forms a complex with Fe(II) that binds and cleaves DNA in the presence of O_2 and a reductant (Umezawa et al. 1984; Stubbe and Kozarich 1987). DNA cleavage is initiated by abstraction of a hydrogen atom from the C4' position of the deoxyribose ring, leading to a series of elimination and rearrangement reactions that ultimately result in strand scission (Stubbe and Kozarich 1987). Recent studies have shown that metallobleomycin is also capable of cleaving RNA substrates (Holms et al. 1993).

Another family of naturally occurring oxidative DNA-cleaving agents are the enediynes, represented by calicheamicin γ_1^I (Zein et al. 1988; Lee et al. 1991), dynemicin A (Konishi et al. 1989), esperamicin A1 (Golik et al. 1987a,b), and the neocarzinostatin chromophore (Goldberg 1991). With the exception of the neocarzinostatin chromophore, all of these molecules contain a 1,5-diyn-3-ene embedded within a strained ten-member ring. Upon chemical activation, diyn-3-ene undergoes a cyclo-aromatization reaction, known as the Bergman cyclization (Bergman 1973), to generate a 1,4-benzenoid diradical. The benzenoid diradical subsequently abstracts hydrogen atoms from various positions on the

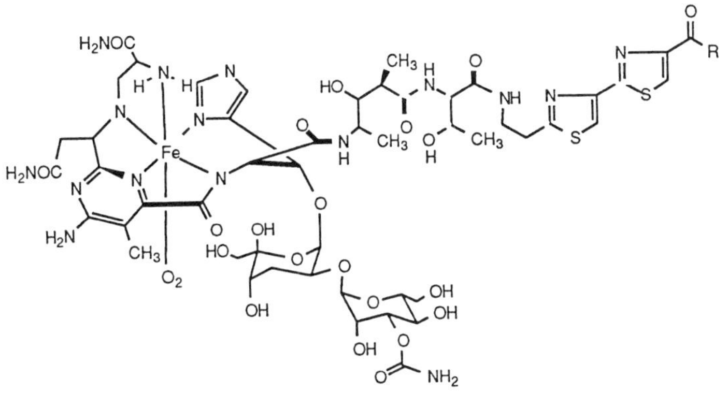

Figure 1 The bleomycin•Fe(II) complex.

deoxyribose ring of DNA, leading to strand cleavage. This mechanism has been supported by the studies of Nicolau and co-workers, who have synthesized a variety of enediyne model compounds that undergo Bergman cyclization under appropriate conditions and cleave φX174 DNA (Nicolaou and Smith 1992).

These naturally occurring chemical nucleases have inspired the design of a number of synthetic nucleolytic agents that also degrade DNA or RNA through radical chemistry. Ferrous-EDTA (Fe[II]•EDTA) (Fig. 2a) (Hertzberg and Dervan 1982, 1984; Tullius and Dombroski 1986), 1,10-phenanthroline-cuprous ion (OP-Cu) (Fig. 2b) (Sigman 1986), GlyGlyHis-cupric ion (GGH•Cu[(II)]) (Fig. 2d) (Chiou 1983; Mack et al. 1988; Mack and Dervan 1990), and metalloporphyrin derivatives (Fig. 2f) (Le Doan et al. 1986; Ward et al. 1986) can be activated in the presence of oxygen and/or reducing agents (e.g., DTT and ascorbate) to generate an oxidizing species capable of cleaving DNA. The mechanism of DNA cleavage is again believed to involve initial hydrogen abstraction from the deoxyribose ring. The sequence independence of DNA cleavage by EDTA•Fe(II) (which likely involves diffusible hydroxyl radicals) makes this reagent a useful tool for both DNA footprinting and the study of conformation variability in DNA (Tullius 1988).

In contrast to the oxidative DNA-cleaving agents described above, a second class of small molecules function by alkylation of the base moiety of nucleic acids. Examples include mitomycin and the nitrogen mustards, both of which have been used in clinical cancer chemotherapy. Enzymatic or chemical reduction of mitomycin triggers a sequence of reactions that lead to the alkylation (and cross-linking) of N2 of guanosine (Moore 1977; Tomasz et al. 1986, 1988). The nitrogen mustards primarily alkylate N7 of guanine. In many instances, alkylation of DNA

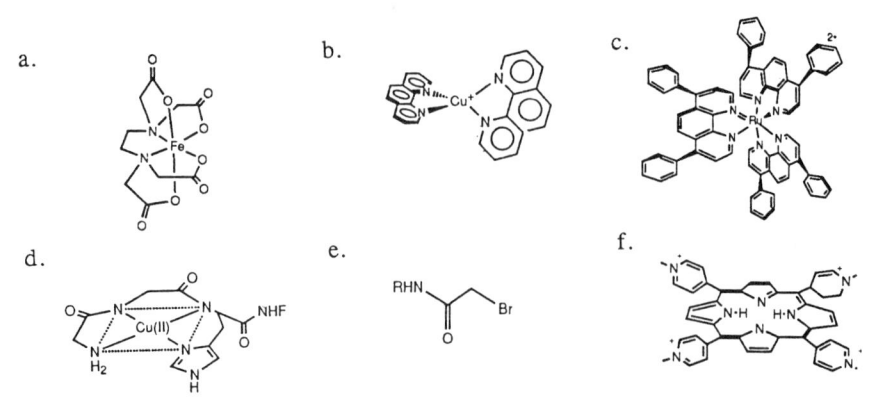

Figure 2 Structures of synthetic nucleolytic agents. (*a*) Fe(II)·EDTA; (*b*) 1,10-phenanthroline-cuprous ion; (*c*) tris(4,7-diphenyl-1,10-phenanthroline) ruthenium(II); (*d*) GGH·Cu(II); (*e*) *N*-bromoacetyl; (*f*) metalloporphyrin.

leads to strand scission in much the same way as dimethylsulfate produces strand scission in the Maxam-Gilbert G reaction (Maxam and Gilbert 1977).

Again, these agents have inspired chemists to synthesize a variety of DNA alkylating agents that can be selectively delivered to target sites in DNA. These include the bromoacetyl group (Fig. 2e) (Baker and Dervan 1989), the *p*-(N-2-chloroethyl-N-methylamino)benzyl group (Vlassov et al. 1988), and 5-[3-{(3-methylthio)propionyl] amino}-*trans*-1-propenyl] 2'-deoxyuridine (MT-dU) (Iverson and Dervan 1988). The bromoacetyl group, when appropriately positioned in the minor groove of double-helical DNA, selectively alkylates the N3 of adenine, resulting in the depurination of the nucleotide (Baker and Dervan 1989). The other two alkylating agents (alkylation by MT-dU requires pre-activation with cyanogen bromide) have been shown to attack predominantly at the N7 position of guanosine residues. Treatment of the alkylated DNA with base (e.g., piperidine) produces DNA strand scission.

A third class of nucleolytic agents are the photosensitizers. Examples include porphyrins (Ward et al. 1986), octahedral tris(phenanthroline)cobalt(III) (Barton and Raphael 1985) or ruthenium(II) (Fig. 2c) (Barton 1986) derivatives, azidophenacyl derivatives (Praseuth et al. 1988), and ellipticine derivatives (Perrouault et al. 1990) that can be photoactivated by irradiation with UV light to produce strand scission and/or cross-linking of DNA.

Finally, Breslow and colleagues have designed enzyme mimics for ribonuclease-A-catalyzed RNA hydrolysis (Breslow 1991). These investigators derivatized β-cyclodextrin with two imidazole groups and found that the adducts were capable of hydrolyzing RNA analogs with increased rates. Chin et al. (1989) have developed a number of metal complexes that hydrolyze model phosphodiester substrates quite efficiently.

III. SEQUENCE-SPECIFIC CLEAVAGE

A. Small Molecules

A number of naturally occurring small molecules interact with double-stranded DNA in a sequence-specific manner, although the recognition sites are typically small and specificity is not absolute. These small DNA-binding drugs have been used by chemists to target synthetic DNA-cleaving reagents to the DNA backbone. Delivery of the cleaving reagent to the DNA backbone is expected to increase significantly the efficiency of strand scission based on proximity effects. One of the first examples of this approach was work by Dervan and co-workers, who tethered the DNA-cleaving agent, Fe(II)·EDTA, to a DNA intercalator, methidium (Hertzberg and Dervan 1982, 1984). The resulting methidiumpropyl-EDTA·Fe(II) (MPE·Fe[II]) complex efficiently cleaved one strand of duplex DNA at submicromolar concentrations in the presence of O_2 and reducing agents (e.g., dithiothreitol). The low binding specificity of methidium is reflected in relatively nonspecific cleavage of DNA by MPE·Fe(II). These properties have made MPE·Fe(II) a popular footprinting agent, useful in defining the location of small ligands that bind to DNA (Van Dyke and Dervan 1984). This reagent, because of its small size and lack of specificity, allows a more accurate resolution of small molecule-binding sites than DNase I.

To increase cleavage specificity, EDTA·Fe(II) was appended to either the amino or carboxyl terminus of distamycin (Schultz et al. 1982; Taylor et al. 1984). Distamycin (Fig. 3, left) is a tripeptide antibiotic containing three N-methylpyrrole carboxamide groups that binds in the minor groove of double-helical DNA with a strong preference for AT-rich regions (Reinert 1972; Luck et al. 1977; Nosikov and Jain 1977; Fagan and Wemmer 1992). Upon addition of dithiothreitol, the distamycin-EDTA·Fe(II) conjugate (DE·Fe[II]) (Fig. 3, middle) nicks DNA at multiple sites adjacent to AT-rich regions. More recently, Baker and Dervan (1989) have derivatized distamycin with an electrophilic bromoacetyl group at its amino terminus. The N-bromoacetyldistamycin selectively alkylates an adenosine residue within the AT-rich distamycin-binding site, and following piperidine treatment, it cleaves DNA predominantly at one nucleotide. The fluorescent dye Hoechst 33258 has been conjugated to 5-iodoacetyl-1,10-phenanthroline. The resulting product is an efficient DNA scission reagent that also cleaves DNA adjacent to AT-

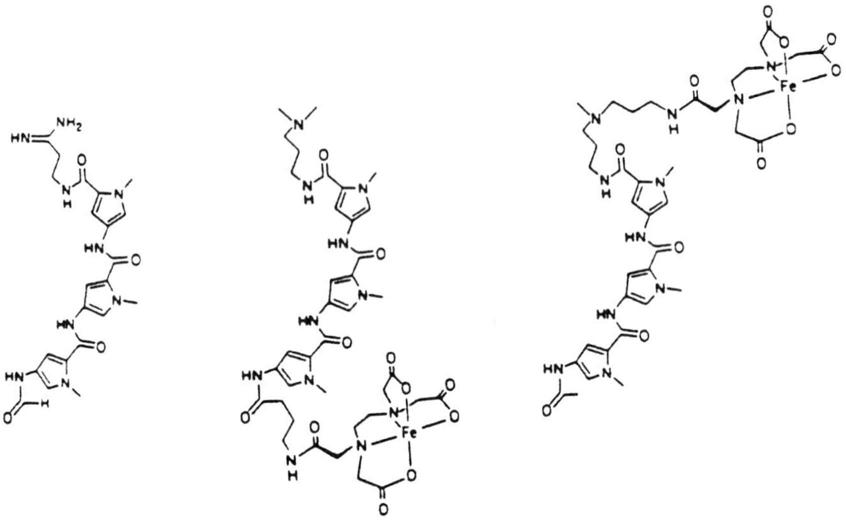

Figure 3 Structures of distamycin and distamycin/EDTA·Fe(II) conjugates.

rich regions, as expected from the binding specificity of Hoechst (Sigman and Chen 1990).

Because of the small sizes of these DNA-binding drugs, these chemical nucleases have relatively low sequence specificities. Moreover, they only effect single-strand cleavage of DNA. Dervan and co-workers have synthesized longer analogs of distamycin that bind to larger recognition sites. These analogs include tetra-, penta-, and hexa-(N-methylpyrrolecarboxamide)-EDTA·Fe(II), which bind to recognition sites of 6, 7, and 8 bp, respectively (Youngquist and Dervan 1985a; Dervan 1986). Other analogs, such as bis(EDTA-distamycin·Fe[II]) (Schultz and Dervan 1983), EDTA-bis(distamycin)·Fe(II) (Schultz and Dervan 1983), bis(EDTA-distamycin) fumaramide (Youngquist and Dervan 1985b), and bis(Fe[II]·EDTA-distamycin)phenoxazone (Dervan 1986) also showed enhanced sequence specificity, recognizing AT-rich duplex DNA sequences up to 10 bp in length. Moreover, a number of these reagents were capable of cleaving duplex DNA on both strands, and in a catalytic fashion, to give fragments of discrete size. These results demonstrate that sequence-specific DNA-cleaving agents can be constructed by tethering a DNA-binding molecule (natural or synthetic) with a nonspecific DNA-cleaving agent. However, in general, they cleave DNA with lower selectivity than the restriction enzymes.

Some synthetic organometallic complexes also show remarkable sequence-specific DNA-binding and -cleaving properties. Incubation of metalloporphyrins with DNA in the presence of ascorbate, superoxide ion, or iodosobenzene results in DNA breakage at d(A·T)$_3$ sites (Ward et al. 1986). Interestingly, the porphyrin-induced DNA scission appears to be catalytic, provided excess activating agents are present. Barton and co-workers have demonstrated that tri(phenanthroline) complexes of ruthenium(II) display enantiomeric selectivity in binding to DNA; the Λ isomer binds avidly to left-handed Z-DNA but not to right-handed B-DNA, and vice versa for the Δ isomer (Barton et al. 1984). When these workers subsequently replaced the ruthenium(II) ion with the redox-active metal ion, cobalt(III), the resulting tris(4,7-diphenyl-1,10-phenanthroline) cobalt(III) complexes cleaved DNA stereospecifically upon photoactivation (Barton and Raphael 1984, 1985). These compounds provide a sensitive probe of DNA conformation.

B. Oligonucleotides as Targeting Ligands

Oligonucleotides are particularly useful for recognizing defined sequences of single-stranded DNA and RNA and duplex DNA. The sugar-phosphate backbone provides a natural framework for building specific hydrogen-bonding interactions using the functional groups of the bases. In addition, oligodeoxyribonucleotides (Gait 1984) are readily available using solid-phase phosphoramidite chemistry.

1. Recognition of Single-stranded DNA (RNA)

For single-stranded DNA (RNA) substrates, an oligodeoxyribonucleotide can bind the target sequence via the formation of highly specific Watson-Crick base pairs (A·T [U] and G·C) (Watson and Crick 1953). A 16-nucleotide oligonucleotide should be capable of recognizing a unique target sequence in a DNA approximately the size of the human genome. A concern is that oligonucleotides of this length may also bind other targets. However, it has been shown that under appropriate hybridization conditions, the formation of mismatched duplexes can be virtually eliminated (Dodgso and Wells 1977; Wallace et al. 1979). Kool (1991) has demonstrated that circular oligonucleotides bind to single-stranded DNAs with higher sequence selectivity and higher affinity than a linear DNA oligomer.

2. Recognition of Duplex DNA via Triple-helix Formation

Currently, a general solution to the recognition of defined target sequences in duplex DNA does not exist. One promising approach to this problem involves the formation of triple-stranded DNA. It has been known for more than three decades that homopyrimidine and homopurine oligonucleotides can associate in aqueous solution to form a stable 2:1 complex (Felsenfeld et al. 1957; Felsenfeld and Miles 1967). The third pyrimidine strand binds to the purine sequences in the major groove of duplex DNA to form a triple-helical structure, in which the third strand is parallel to the purine strand (Arnott and Selsing 1974; Moser and Dervan 1987; de los Santos et al. 1989; Rajagopal and Feigon 1989). Specificity is determined by the formation of T·AT and C$^+$·GC base triplets (C$^+$ represents a protonated C) through Hoogsteen hydrogen bonding (Hoogsteen 1959) (Fig. 4). It is also known that polypurine oligonucleotides bind to duplex DNA in the major groove, parallel to the pyrimidine strand, through the formation of G·GC and A·AT triplets (Lipsett 1964; Broitman et al. 1987; Kohwi and Kohwi-Shigematsu 1988; Beal and Dervan 1991).

Slightly acidic conditions favor the formation of a stable triple helix of mixed sequence, since protonation at the N3 position of cytosine is required for stable Hoogsteen base pairing in the major groove. Multivalent cations such as Mg^{++}, Co(NH$_3$)$_6^{+++}$, and spermine also stabilize the triple-stranded structure, presumably by reducing the electrostatic repulsion between the Watson-Crick duplex and the negatively charged phosphodiester backbone of the third pyrimidine strand (Felsenfeld and Miles 1967; Glaser and Gabbay 1968; Moser and Dervan 1987). The presence of organic solvents (e.g., ethylene glycol, methanol, ethanol, dioxane, and DMF) also contributes significantly to the stability of a triple helix (Moser and Dervan 1987). To further stabilize the triple helix (and duplex as well) formed between the oligonucleotide probes and the target DNAs (RNAs), Hèléne and co-workers have linked the DNA intercalating agent acridine to either the 5′ or 3′ termini of an oligonucleotide probe (Asseline et al. 1986; Thuong and Chassignol 1988). In one experiment, an 11-nucleotide pyrimidine oligonucleotide with acridine attached to its 5′ terminus formed a triple helix with the complementary duplex DNA with a melting temperature 13–18°C higher than the underivatized oligonucleotide analog (Sun et al. 1989). These oligonucleotide-acridine conjugates were also more potent inhibitors of viral gene expression than the underivatized oligonucleotides (Zerial et al. 1987).

The presence of any mismatches significantly destabilizes the triple-helical structure (Mirkin et al. 1987; Griffin and Dervan 1989; Belotserkovskii et al. 1990). This makes triple-helix formation highly specific and very promising for recognition of target sequences in very large duplex DNAs (Strobel and Dervan 1990, 1991). The main drawback is that this method is largely limited to the recognition of homopurine-

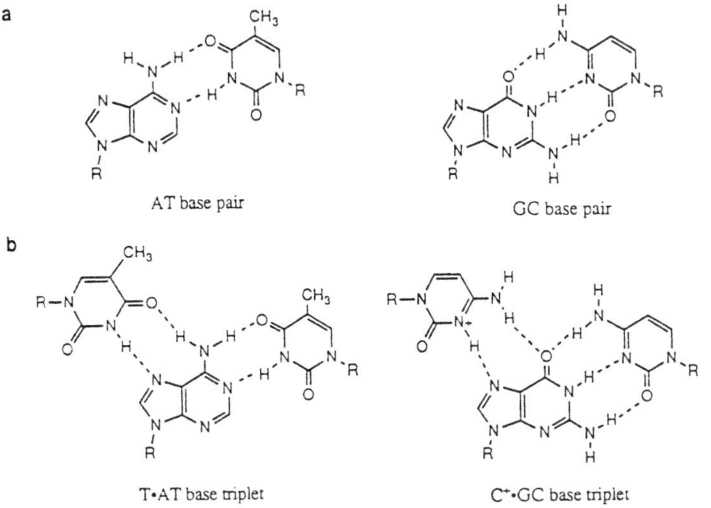

Figure 4 (*A*) Watson-Crick A·T and G·C base pairs; (*B*) isomorphous base triplets of T·AT and C⁺·GC.

homopyrimidine sequences. However, Dervan and others have shown that this sequence limitation can be relaxed to some degree through alternative strand triple-helix formation (Horne and Dervan 1990), the use of the G·TA triplet (Griffin and Dervan 1989), and the design of unnatural bases for completion of the triplet code (Griffin et al. 1992). Screening of large RNA libraries is also being carried out in order to identify new triple-helix motifs (Pei et al. 1991 and unpubl.).

3. Oligonucleotides Modified with Nonhydrolytic Nucleolytic Agents

Both oligonucleotide and oligonucleotide-acridine conjugates have been used to deliver a variety of DNA-cleaving agents to target sequences within single-stranded DNA (RNA) and double-stranded DNA substrates. One of the most widely used cleavage reagents has been Fe(II)·EDTA (Chu and Orgel 1985; Dreyer and Dervan 1985). The resulting oligonucleotide-Fe(II)·EDTA adducts bind defined target sequences in single-stranded and double-stranded DNAs via Watson-Crick and Hoogsteen hydrogen bonds, respectively (see Fig. 5). Upon addition of dithiothreitol, cleavage of single-stranded DNA (Chu and Orgel 1985; Dreyer and Dervan 1985) and both strands of duplex DNA (Moser and Dervan 1987) occurs over several nucleotides near or within the oligonucleotide-binding site. λ DNA (48.5 kb) (Strobel et al. 1988) and yeast chromosome III (340 kb) (Strobel and Dervan 1990) have been

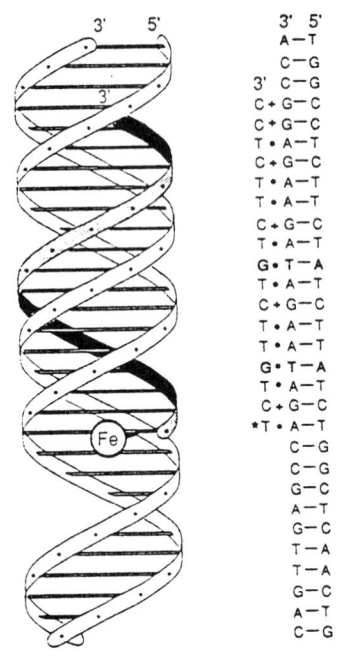

```
            3'  5'
            A—T
            C—G
         3' C—G
            C+G—C
            C+G—C
            T•A—T
            C+G—C
            T•A—T
            T•A—T
            C+G—C
            T•A—T
            G•T—A
            T•A—T
            C+G—C
            T•A—T
            T•A—T
            G•T—A
            T•A—T
            C+G—C
           *T•A—T
            C—G
            C—G
            G—C
            A—T
            G—C
            T—A
            T—A
            G—C
            A—T
            C—G
```

Figure 5 Triple helix formation: T* is a uridine derivatized with EDTA·Fe(II) (Strobel et al. 1988).

cleaved at single predetermined sites using this approach. These oligonucleotide-EDTA·Fe(II) adducts might provide a powerful tool for chromosomal mapping.

Oligonucleotides have also been derivatized with 1,10-phenanthroline. In the presence of cupric ion, O_2, and reducing agents, these adducts have been shown to cleave site-specifically single-stranded DNA (Chen and Sigman 1986; Francois et al. 1989a), double-stranded DNA (Francois et al. 1989a,b), and RNA (Chen and Sigman 1988; Murakawa et al. 1989). Hèléne and co-workers have synthesized a site-specific artificial photoendonuclease by tethering a pyrimidine oligonucleotide at its 3' terminus with an ellipticine derivative (Perrouault et al. 1990). The photonuclease binds to the homopyrimidine-homopurine region of a duplex DNA and cleaves the duplex DNA at the target site upon irradiation at wavelengths greater than 300 nm. Because of the ease of activation, this type of photoendonuclease should be particularly useful for in vivo studies (e.g., regulation of gene transcription and translation) where activation by other means is not possible. Besides the DNA-cleaving agents discussed above, an azidophenacyl derivative (Praseuth et al. 1988; Birg et al. 1990) and *p*-(N-2-chloroethyl-N-methyl-amino)benzyl group (Vlassov et al. 1988) have also been appended to oligonucleotides. The resulting chemical nucleases cleave single-stranded and duplex DNAs in a sequence-specific manner.

All of the oxidative strategies described so far for cleaving DNA (RNA) result in abnormal 5′ and/or 3′ termini that must be removed to create substrates for many of the enzymes used to manipulate nucleic acids (e.g., ligases and kinases). In addition, for a number of these synthetic chemical nucleases, the efficiency for either single-stranded or double-stranded DNA cleavage generally does not exceed 30%, even with a large excess of the nuclease. The factors responsible for incomplete reaction are not clear, but they could include self-destruction of the cleavage reagents. It is also unknown whether these chemical nucleases will be useful for cleaving relatively large quantities of DNA. These drawbacks have no doubt slowed the widespread use of these reagents as tools in molecular biology.

4. Oligonucleotide–Staphylococcal Nuclease Conjugates

To improve the properties of artificial nucleases, Schultz and co-workers have generated semisynthetic nucleases by attaching the enzyme, staphylococcal nuclease, to oligonucleotides (for review, see Corey et al. 1991). Staphylococcal nuclease is a stable, relatively nonspecific phosphodiesterase that hydrolyzes either single-stranded DNA, double-stranded DNA, or RNA to generate products with 3′-phosphate and 5′-hydroxyl termini (Tucker et al. 1978, 1979a,b,c). The resulting semisynthetic nucleases offer several advantages over the chemical nucleases described above; most notably, they produce undamaged termini that are suitable for subsequent enzymatic manipulations. Also important is the fact that staphylococcal nuclease is completely dependent on Ca^{++} for activity, providing a mechanism for modulating the enzyme activity; the enzyme can be reversibly activated and inactivated by the addition of Ca^{++} and Ca^{++} chelators, respectively.

A K116C mutant staphylococcal nuclease was coupled to the 3′ terminus of a 3′-thiol-containing 15-nucleotide oligonucleotide via a disulfide bond (Fig. 6). The resulting oligonucleotide-nuclease adduct hydrolyzed a synthetic DNA substrate over four phosphodiester bonds adjacent to the complementary target site (Corey and Schultz 1987). The use of a Y113A mutant nuclease (which has reduced affinity for DNA) significantly improved the selectivity of the semisynthetic nuclease (Corey et al. 1989a). This semisynthetic nuclease selectively hydrolyzed single-stranded M13 DNA at a single site within a few seconds. In addition, the hydrolysis of DNA substrates was catalytic when the cleavage

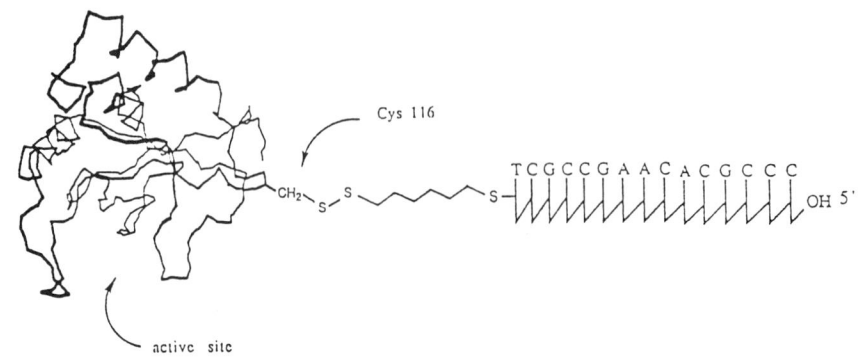

Figure 6 Structure of a semisynthetic nuclease that contains the K116C mutant staphylococcal nuclease and a 15-nucleotide oligonucleotide-binding domain (Zuckermann et al. 1988).

reaction was carried out near the melting temperature of the duplex formed by the oligonucleotide-binding domain and the substrate (Corey et al. 1989a). These semisynthetic nucleases have also been shown to cleave selectively both oligoribonucleotides (Zuckermann et al. 1988) and structured RNAs (tRNA and M1 RNA) (Zuckermann and Schultz 1989).

By coupling staphylococcal nuclease to the 5′ terminus of a homopyrimidine oligonucleotide (Pei et al. 1990), semisynthetic nucleases can be generated that selectively cleave both DNA strands of duplex DNA adjacent to a homopurine-homopyrimidine target sequence via triple-helix formation. Microgram quantities of plasmid DNA (4.4 kb) were converted into desired products with greater than 75% yield using less than five molar equivalents of the nuclease. To demonstrate the general utility of this reagent, the gene encoding β-lactamase from the 2.7-kb plasmid pUC19 was isolated using two such semisynthetic nucleases (Pei and Schultz 1991), and the resulting DNA fragment was cloned into several expression vectors. Earlier, Zuckermann and Schultz (1989) had shown that fragments from a tRNA hydrolyzed by a semisynthetic nuclease could be ligated to regenerate the original tRNA molecule. These results demonstrate the utility of these semisynthetic nucleases for molecular cloning and other manipulations of DNA or RNA.

5. Recognition of Duplex DNA via D-loop Formation

Because of the sequence limitations involved in triple-helix formation, efforts are being made to find alternative ways to recognize targets in duplex DNA. One possible alternative is the formation of a displacement loop (D-loop), in which one strand of a duplex DNA target sequence is displaced by an oligonucleotide via Watson-Crick hydrogen-bonding interactions. Negative superhelicity in duplex DNA stabilizes a D-loop structure, and supercoiled DNA can spontaneously incorporate complementary single strands of DNA (Wiegand et al. 1977). Prior denaturation of a supercoiled duplex DNA with alkali also facilitates the incorporation of single-stranded fragments. In addition, the recombination protein, RecA, is known to promote this process (Cox and Lehman 1987).

D-loop formation has been used to deliver oligonucleotides to complementary sequences within duplex DNA for the sequencing of duplex DNA (Tonequzzo et al. 1988) and as hybridization probes (Landgren et al. 1988). Vlassov et al. (1988) have used alkali-mediated D-loop formation to deliver selectively an oligonucleotide-linked alkylating agent into supercoiled duplex DNA. Hybridization of the derivatized oligonucleotide induced alkylation of the duplex DNA at the target sequence. Subsequent treatment of the alkylated duplex DNA with piperidine resulted in selective cleavage of one strand of the duplex substrate. This protocol was also adopted by Corey et al. (1989b) to cleave selectively a supercoiled plasmid DNA with a 19-nucleotide oligonucleotide–staphylococcal nuclease adduct.

Recently, several laboratories have synthesized peptide nucleic acid (PNA) analogs (Nielsen et al. 1991; Egholm et al. 1992) that form D-loops with relatively high efficiencies (Hanvey et al. 1992). Addition of nuclease S1 to the D-loop structures resulted in selective double-strand cleavage at the D-loop region (Demidov et al. 1993). RNA transcripts modified with 1,10-phenanthroline-copper(I) can also selectively cleave double-stranded DNA through the formation of R-loops (Chen et al. 1993).

6. Oligonucleotide-directed Restriction Endonuclease Cleavage

Oligonucleotides have been used to direct type II restriction endonucleases to specific sites in single-stranded DNAs (Corey et al. 1989a; Milavetz 1989). A synthetic oligonucleotide complementary to the restriction site of interest is hybridized to the single-stranded DNA to form a duplex substrate which is then cleaved with the appropriate endonuclease. This method is straightforward to perform and can be used to cleave virtually any restriction site in a cloned single-stranded DNA efficiently. However, the method does not allow cleavage of DNA at sites that are not recognized by restriction endonucleases. Szybalski and co-workers have developed another approach to the cleavage of single-stranded DNA which makes use of type IIS restriction endonucleases in conjunction with an oligonucleotide adapter (Podhajska and Szybalski 1985; Szybalski 1985; Kim et al. 1988) (type IIS restriction endonuclease [e.g., *Fok*I] cleaves double-stranded DNA at sites adjacent to their recognition sequences [see Chapter 2]). A properly designed oligonucleotide adapter, which folds into a partially double-strand hairpin loop containing the recognition sequence of *Fok*I, is hybridized to the target sequence. The enzyme binds to its cognate sequence within the hairpin loop but cleaves individual strands of the duplex DNA formed between the adapter and the target sequence at sites 9 and 13 nucleotides from the recognition sequence. This approach should allow one to cleave single-stranded DNA selectively at any desired sequence.

Triple-helix formation between oligonucleotides and duplex DNA has also been used to protect specific DNA target sites from cleavage by restriction enzymes, from methylation, and from binding by transcription factors (Maher et al. 1989). By using a modified protocol of Koob et al. (1988) termed "Achilles heel" (see below), Strobel and Dervan (1991) cleaved yeast chromosome III (340 kb) with the restriction endonuclease *Eco*RI at a single site, which had been protected from methylation by the formation of a triple helix with a pyrimidine oligonucleotide.

7. DNA Isolation by Amplification

Recently, the polymerase chain reaction (PCR) (Saiki et al. 1988) has also been used to isolate DNA fragments of interest. Primers can be designed with restriction endonuclease recognition sites added at the 5' termini. After the amplification reaction, the product contains new restriction sites flanking the DNA fragment of interest. When treated with the appropriate restriction endonucleases, the DNA fragment can be cloned into desired vectors. This method can "cleave" DNA at virtually any predetermined site and has become increasingly popular in molecular cloning experiments. However, this method does not provide a means to map large DNAs. Moreover, PCR may introduce undesired mutations into a DNA fragment.

C. DNA-binding Proteins and Peptides as Targeting Ligands

A large number of DNA-binding proteins such as repressors and transcription factors have been identified, many of which are involved in different aspects of the regulation of prokaryotic and eukaryotic gene expression. These proteins can be grouped into a number of families including the helix-turn-helix motif-containing proteins (Pabo and Sauer

1984; Harrison and Aggarwal 1990), the homeodomain proteins (Levine and Hoey 1988; Mitchell and Tjian 1989; Scott et al. 1989), the zinc finger proteins (Evans and Hollenberg 1988), and the leucine zipper proteins (Johnson et al. 1987; Landschulz et al. 1988). These proteins typically recognize specific DNA sequences longer than 10 bp in length and have extremely high sequence specificities and binding affinities. The specificity is largely determined by hydrogen-bonding and hydrophobic interactions between the amino acid side chains in an α-helical region of the protein and the DNA base pairs.

The first DNA-binding protein to be chemically converted into a site-specific nuclease was the *Escherichia coli trp* repressor (Chen and Sigman 1987). Derivatization of the protein involved covalent attachment of 1,10-phenanthroline to the lysine residues (and probably the amino terminus) via an iminothiolane linker. The modified repressor retained both its high affinity and specificity for the cognate operator sequence. Upon the addition of cupric ion and thiol, the phenanthroline-containing repressor selectively cleaved duplex DNA within the repressor-binding site. Similarly, Bruice et al. (1991) have linked 1,10-phenanthroline to the carboxyl terminus of an A66C mutant Cro repressor. The resulting nuclease again afforded specific cleavage within the recognition sequence.

Ebright et al. (1990) chemically converted *E. coli* catabolite-activating protein (CAP) into a site-specific DNA cleavage agent by incorporation of 1,10-phenanthroline at a unique cysteine residue (position 178). Although the affinity of the modified CAP repressor for a 22-bp DNA recognition sequence was reduced by a factor of 400, specific DNA cleavage occurred within the DNA recognition site. Pei and Schultz (1990) coupled staphylococcal nuclease to a genetically engineered amino-terminal domain of the λ repressor through a flexible polyamine linker. The chimeric protein retained its specificity for λ operator sequences. Upon activation with Ca^{++}, the chimera selectively hydrolyzed a 2704-bp plasmid DNA containing a λ operator sequence with approximately 50% efficiency. In addition, when the Y113A mutant staphylococcal nuclease was used instead of the wild-type enzyme, the resulting chimera specifically cleaved the bacteriophage λ chromosome (48.5 kb) at the two operator sites, O_R1 and O_L1 (Pei 1991).

Peptides may also be used to target DNA cleavage agents. Dervan and colleagues chemically synthesized a 52-residue peptide derived from the sequence-specific DNA-binding domain of *Hin* recombinase and attached an EDTA·Fe(II) moiety to its amino terminus (Sluka et al. 1987). This modified peptide cleaved both strands of a plasmid DNA containing the recognition site of *Hin* recombinase. These investigators also ex-

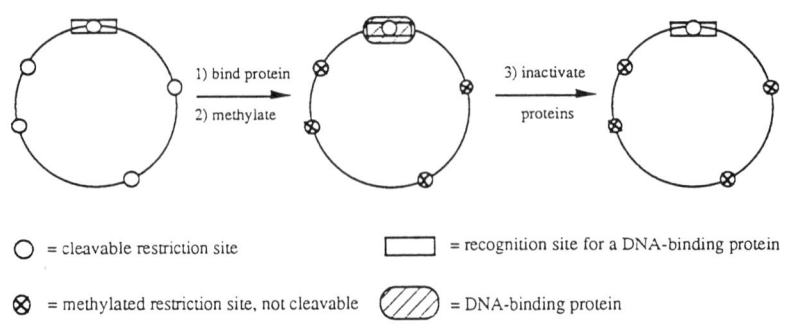

Figure 7 Increasing the site specificity of restriction endonucleases using DNA-binding proteins and oligonucleotides.

tended the 52-residue peptide at the amino terminus with the tripeptide, NH_2-GlyGlyHis-OH (GGH) (Mack et al. 1988; Mack and Dervan 1990). Since this tripeptide has high affinity for cupric ion, the peptide delivered the GGH·Cu(II) complex to the recognition site of *Hin* recombinase, resulting in sequence-specific cleavage of DNA.

Szybalski and co-workers have designed a restriction endonuclease/methylase/repressor system, designated as Achilles heel, to extend the effective recognition sequence of a restriction endonuclease (Fig. 7) (Koob et al. 1988). The DNA to be cleaved is treated with a DNA-binding protein such as the λ repressor and then treated with a methylase that has the same recognition sequence as the restriction endonuclease to be used for cleavage. All restriction sites are methylated except the one that overlaps the repressor-binding site. Removal of the repressor and subsequent addition of the appropriate restriction endonuclease result in cleavage at the protected site. Since repressor recognition sequences are as long as 20 bp, this method allows unique cleavage of large DNAs, providing another powerful tool for genomic manipulation. A major drawback of the method is that a restriction site must overlap the repressor-binding sequence, and for most repressors, this overlap does not exist. Again, large libraries of DNA-binding protein mutants are being generated and screened in order to find new proteins with novel specificities (L. Huang and P. Schultz, pers. comm.).

D. Other Applications

Small molecules, oligonucleotides, proteins, and peptides modified with DNA-cleaving agents also provide insight into the nature of interactions of ligands with nucleic acids. Affinity cleavage experiments have provided information on the binding specificities and modes of binding of small molecules, oligonucleotides, and proteins to duplex DNA (Taylor et al. 1984). For example, the parallel relationship between the third pyrimidine strand and the purine strand of B-DNA (Moser and Dervan 1987) was determined by affinity cleavage. Affinity cleavage experiments with synthetic GCN4 protein containing Fe(II)•EDTA at its amino terminus confirmed that the amino-terminal basic regions of leucine zipper proteins lie in the major groove of DNA (Oakley and Dervan 1990).

Finally, these artificial nucleases may find potential use as therapeutic agents. A viral capsid staphylococcal nuclease fusion protein has been shown to degrade viral genomic RNA, leading to viral inactivation (Natsoulis and Boeke 1991). This may offer a promising strategy for interfering with the replication of retroviruses and other viruses. D-loop or triple-helix formation may be useful for the delivery of reactive agents to specific sites in viral genomes.

IV. CONCLUSIONS

Sequence-specific artificial nucleases have been generated by coupling various nucleic acid cleaving agents to a variety of nucleic-acid-binding molecules. These synthetic nucleases are capable of selectively cleaving RNAs, single-stranded DNAs, and double-stranded DNAs. Some function catalytically, and in a number of cases, these artificial nucleases have been used to cleave large DNAs at a few specific sites. These reagents are likely to prove to be useful tools for genomic mapping and structural studies of nucleic acids and may have therapeutic potential.

ACKNOWLEDGMENT

We are grateful for financial support for this work from the National Institutes of Health (grant R01-GM-41679).

REFERENCES

Arnott, S. and E. Selsing. 1974. Structures for the polynucleotide complexes poly(dA)•poly(dT) and poly(dT)•poly(dA)•poly(dT). *J. Mol. Biol.* **88:** 509–521.

Asseline, U., N.T. Thuong, and C. Helene. 1986. Oligothymidylates substituted at 3′ end by an acridine derivative. *Nucleosides Nucleotides.* **5:** 45–63.

Baker, B.F. and P.B. Dervan. 1989. Sequence-specific cleavage of DNA by *N*-bromoacetyldistamycin. Product and kinetic analysis. *J. Am. Chem. Soc.* **111:** 2700–2712.

Barton, J.B. and A.L. Raphael. 1984. Photoactivated stereospecific cleavage of doublehelical DNA by cobalt (III) complexes. *J. Am. Chem. Soc.* **106:** 2466–2468.

———. 1985. Site-specific cleavage of left-handed DNA in pBR322 by Λ-tris (diphenylphenanthroline)cobalt(III). *Proc. Natl. Acad. Sci.* **82:** 6460–6464.

Barton, J.K. 1986. Metals and DNA: Molecular and left-handed complements. *Science* **233:** 727–734.
Barton, J.K., L.A. Basile, A. Danishefsky, and A. Alexandrescu. 1984. Chiral probes for the handedness of DNA helices: Enantiomers of tris (4,7-diphenylphenanthroline)ruthenium(II). *Proc. Natl. Acad. Sci.* **81:** 1961–1965.
Beal, P.A. and P.B. Dervan. 1991. Second structural motif for recognition of DNA by oligonucleotide-directed triple helix formation. *Science* **251:** 1360–1363.
Belotserkovskii, B.P., A.G. Veselkov, S.A. Filippov, V.N. Dobrynin, S.M. Mirkin, and M.D. Frank-Kamenetskii. 1990. Formation of intramolecular triplex in homopurine-homopyrimidine mirror repeats with point substitutions. *Nucleic Acids Res.* **18:** 6621–6624.
Bergman, R.G. 1973. Reactive 1,4-dehydroaromatics. *Accts. Chem. Res.* **6:** 25–31.
Birg, F., D. Praseuth, A. Zerial, N.T. Thuong, U. Asseline, T. Le Doan, and C. Helene. 1990. Inhibition of simian virus 40 DNA replication in CV-1 cells by an oligodeoxynucleotide covalently linked to an intercalating agent. *Nucleic Acids Res.* **18:** 2901–2907.
Breslow, R. 1991. How do imidazole groups catalyze the cleavage of RNA in enzyme models and in enzymes? Evidence from "negative catalysis". *Accts. Chem. Res.* **24:** 317–324.
Broitman, S.L., D.D. Im, and J.R. Fresco. 1987. Formation of the triple-stranded polynucleotide helix, poly (A·A·U)*. *Proc. Natl. Acad. Sci.* **84:** 5120–5124.
Bruice, T.W., J.G. Wise, D.S.E. Rosser, and D.S. Sigman. 1991. Conversion of λ phage *cro* into an operator-specific nuclease. *J. Am. Chem. Soc.* **113:** 5446–5447.
Chen, C.-H.B. and D.S. Sigman. 1986. Nuclease activity of 1,10-phenanthroline-copper: Sequence-specific targeting. *Proc. Natl. Acad. Sci.* **83:** 7147–7151.
———. 1987. Chemical conversion of a DNA-binding protein into site-specific nuclease. *Science* **237:** 1197–1201.
———. 1988. Sequence-specific scisson of RNA by 1,10-phenanthroline-copper linked to deoxyoligonucleotides. *J. Am. Chem. Soc.* **110:** 6570–6572.
Chen, C.-C.B., M.B. Gorin, and D.S. Sigman. 1993. Sequence-specific scission of DNA by the chemical nuclease activity of 1,10-phenanthroline-copper(I) targeted by RNA. *Proc. Natl. Acad. Sci.* **90:** 4206–4210.
Chin, J., M. Banaszczyk, V. Jubian, and X. Zou. 1989. Co(III) complex promoted hydrolyis of phosphate diesters: Comparison in reactivity of rigid *cis*-diaqaotetra-azacobalt (III) complexes. *J. Am. Chem. Soc.* **111:** 186–190.
Chiou, S.-H. 1983. DNA and protein-scission activities of ascorbate in the presence of copper ion and a copper-peptide complex. *J. Biochem.* **94:** 1259–1267.
Chu, B.C.F. and L.E. Orgel. 1985. Nonenzymatic sequence-specific cleavage of single-stranded DNA. *Proc. Natl. Acad. Sci.* **82:** 963–967.
Corey, D.R. and P.G. Schultz. 1987. Generation of a hybrid sequence-specific single-stranded deoxyribonuclease. *Science* **238:** 1401–1403.
Corey, D.R., D. Pei, and P.G. Schultz. 1989a. Generation of a catalytic sequence-specific hybrid DNase. *Biochemistry* **28:** 8277–8286.
———. 1989b. Sequence-selective hydrolysis of duplex DNA by an oligonucleotide-directed nuclease. *J. Am. Chem. Soc.* **111:** 8523–8525.
Corey, D.R., R.N. Zuckermann, and P.G. Schultz. 1991. Hybrid enzymes and the sequence-specific cleavage of nucleic acids. In *Frontiers of bioorganic chemistry* (ed. H. Dugas), pp. 3–31. Springer-Verlag, Berlin.
Cox, M.M. and I.R. Lehman. 1987. Enzymes of general recombination. *Annu. Rev. Biochem.* **56:** 229–262.

de los Santos, C., M. Rosen, and D. Patel. 1989. NMR studies of DNA $(R+)_n \cdot (Y-)_n \cdot (Y+)_n$ Triple helices in solution. Imino and amino proton markers of T-A-T and C-G-C base-triple formation. *Biochemistry* **28**: 7282-7289.

Demidov, V., M.D. Frank-Kamenetskii, M. Egholm, O. Buchardt, and P.E. Nielsen. 1993. Sequence selective double strand DNA cleavage by peptide nucleic acid (PNA) targeting using nuclease S1. *Nucleic Acids Res.* **21**: 2103-2107.

Dervan, P.B. 1986. Design of sequence-specific DNA-binding molecules. *Science* **232**: 464-471.

Dodgso, J.B. and R.D. Wells. 1977. Synthesis and thermal melting behavior of oligomer·polymer complexes containing defined lengths of mismatched dA·dG and dG·dG nucleotides. *Biochemistry* **16**: 2367-2374.

Dreyer, G.B. and P.B. Dervan. 1985. Sequence-specific cleavage of single-stranded DNA: Oligodeoxynucleotide-EDTA·Fe(II). *Proc. Natl. Acad. Sci.* **82**: 968-972.

Ebright, R.H., Y.W. Ebright, P.S. Pendergrast, and A. Gunasekera. 1990. Conversion of a helix-turn-helix motif sequence-specific DNA binding protein into a site-specific DNA cleaving agent. *Proc. Natl. Acad. Sci.* **87**: 2882-2886.

Egholm, M., O. Buchardt, P.E. Nielsen, and R.H. Berg. 1992. Peptide nucleic acids (PNA). Oligonucleotide analogues with an achiral peptide backbone. *J. Am. Chem. Soc.* **114**: 1895-1897.

Evans, R.M. and S.M. Hollenberg. 1988. Zinc fingers: Gilt by association. *Cell* **52**: 1-3.

Fagan, P. and D.E. Wemmer. 1992. Cooperative binding of distamycin-A to DNA in the 2:1 mode. *J. Am. Chem. Soc.* **114**: 1080-1081.

Felsenfeld, G. and H.T. Miles. 1967. The physical and chemical properties of nucleic acid. *Annu. Rev. Biochem.* **36**: 407-448.

Felsenfeld, G., D.R. Davis, and A. Rich. 1957. Formation of a three-stranded polynucleotide molecule. *J. Am. Chem. Soc.* **79**: 2023-2024.

Francois, J.-C., T. Saison-Behmoaras, M. Chassignol, N.T. Thuong, and C. Helene. 1989a. Sequence-targeted cleavage of single- and double-stranded DNA by oligothymidylates covalently linked to 1,10-phenanthroline. *J. Biol. Chem.* **264**: 5891-5898.

Francois, J.-C., T. Saison-Behmoaras, C. Barbier, M. Chassignol, N.T. Thuong, and C. Helene. 1989b. Sequence-specific recognition and cleavage of duplex DNA via triple helix formation by oligonucleotides convalently linked to a phenanthroline-copper chelate. *Proc. Natl. Acad. Sci.* **86**: 9702-9706.

Gait, M. 1984. *Oligonucleotide synthesis: A practical approach.* IRL Press, Washington, D.C.

Glaser, R. and E.J. Gabbay. 1968. Topography of nucleic acid helices in solutions. III. Interactions of spermine and spermidine derivatives with polyadenylic-polyuridylic and polyinosinic-polycytidylic acid helices. *Biopolymers* **6**: 243-254.

Goldberg, I.H. 1991. Mechanism of neocarzinostatin action: Role of DNA microstructure in determination of chemistry of bistranded oxidative damage. *Accts. Chem. Res.* **24**: 191-198.

Golik, J., J. Clardy, G. Dubay, G. Groeneword, H. Kawaguchi, M. Konishi, B. Krishman, H. Ohkuma, K. Saitoh, and T.W. Doyle. 1987a. Esperamicins, a novel class of potent antitumor antibiotics. 2. Structure of esperamicin X. *J. Am. Chem. Soc.* **109**: 3461-3462.

———. 1987b. Esperamicins, a novel class of potent antitumor antibiotics. 3. Structures of esperamicins A_1, A_2 and A_{1b}. *J. Am. Chem. Soc.* **109**: 3462-3464.

Griffin, L.C. and P.B. Dervan. 1989. Recognition of thymine·adenine base pairs by guanine in a pyrimidine triple helix motif. *Science* **24**: 967-971.

Griffin, L.C., L.L. Kiessling, P.A. Beal, P. Gillespie, and P.B. Dervan. 1992. Recognition of all four base pairs of double-helical DNA by triple-helix formation: Design of nonnatural deoxyribonucleotides for pyrimidine–purine base pair binding. *J. Am. Chem. Soc.* **114:** 7976–7982.

Hanvey, J.C., N.J. Peffer, J.E. Bisi, S.A. Thomson, R. Cadilla, J.A. Josey, D.J. Ricca, C.F. Hassman, M.A. Bonham, K.G. Au, S.G. Carter, D.A. Btuckenstein, A.L. Boyd, S.A. Noble, and L.E. Babiss. 1992. Antisense and antigene properties of peptide nucleic acids. *Science* **258:** 1481–1485.

Harrison, S.C. and A.K. Aggarwal. 1990. DNA recognition by proteins with the helix-motif. *Annu. Rev. Biochem.* **59:** 933–969.

Hertzberg, R.P. and P.B. Dervan. 1982. Cleavage of double helical DNA by (methidiumpropyl-EDTA)iron (II). *J. Am. Chem. Soc.* **104:** 313–315.

———. 1984. Cleavage of DNA with methidiumpropyl-EDTA-iron (II): Reaction conditions and product analysis. *Biochemistry* **23:** 3934–3945.

Holms, C.E., B.J. Carter, and S.M. Hecht. 1993. Characterization of iron (II)·bleomycin-mediated RNA strand scission. *Biochemistry* **32:** 4293–4307.

Hoogsteen, K. 1959. The structure of crystals containing a hydrogen-bonded complex of 1-methylthymine and 9-methyladenine. *Acta Crystallogr.* **12:** 822–823.

Horne, D.A. and P.B. Dervan. 1990. Recognition of mixed sequence duplex DNA by alternate strand triple-helix formation. *J. Am. Chem. Soc.* **112:** 2435–2437.

Iverson, B.L. and P.B. Dervan. 1988. Nonenzymatic sequence-specific methyl transfer to single-stranded DNA. *Proc. Natl. Acad. Sci.* **85:** 4615–4619.

Johnson, P.F., W.H. Landschulz, B.J. Graves, and S.L. McKnight. 1987. Identification of a rat liver nuclear protein that binds to the enhancer core element of three animal viruses. *Genes Dev.* **1:** 133–146.

Kim, S.C., A.J. Podhajska, and W. Szybalski. 1988. Cleaving DNA at any predetermined site with adaptor-primers and class IIs restriction enzymes. *Science* **240:** 504–506.

Kohwi, Y. and T. Kohwi-Shigematsu. 1988. Magnesium ion-dependent triple-helix structure formed by homopurine-homopyrimidine sequences in supercoiled plasmid DNA. *Proc. Natl. Acad. Sci.* **85:** 3781–3785.

Konishi, M., H. Ohkuma, K. Matsumoto, T. Tsuno, H. Kame, T. Miyaki, T. Oki, H. Kawaguchi, G. VanDuyne, and J. Clardy. 1989. Dynemicin A, a novel antibiotic with the anthraquinone and 1,5-diyn-3-ene subunit. *J. Antibiot.* **42:** 1449–1452.

Koob, M., E. Grimes, and W. Szybalski. 1988. Conferring operator specificity on restriction endonucleases. *Science* **241:** 1084–1086.

Kool, E.T. 1991. Molecular recognition by circular oligonucleotides: Increasing the selectivity of DNA binding. *J. Am. Chem. Soc.* **113:** 6265–6266.

Landgren, V., R. Kaiser, J. Sanders, and L. Hood. 1988. A ligase-mediated gene detection technique. *Science* **241:** 1077–1080.

Landschulz, W.H., P.F. Johnson, and S.L. McKnight. 1988. The leucine zipper: A hypothetical structure common to a new class of DNA binding proteins. *Science* **240:** 1759–1764.

Le Doan, T., L. Perrouault, C. Helene, M. Chassignol, and N.T. Thuong. 1986. Targeted cleavage of polynucleotides by complementary oligonucleotides covalently linked to iron-porphyrins. *Biochemistry* **25:** 6736–6739.

Lee, M.D., G.A. Ellestad, and D.B. Borders. 1991. Calicheamicins: Discovery, structure, chemistry and interaction with DNA. *Accts. Chem. Res.* **24:** 235–243.

Levine, M. and T. Hoey. 1988. Homeobox proteins as sequence-specific transcription factors. *Cell* **55:** 537–540.

Lipsett, M.N. 1964. Complex formation between polycytidylic acid and guanine oligonucleotides. *J. Biol. Chem.* **239:** 1256–1260.

Luck, G., C. Zimmer, K.E. Reinert, and F. Arcamone. 1977. Specific interactions of distamycin A and its analogs with (A•T)rich and (G•C) rich duplex regions of DNA and deoxypolynucleotides. *Nucleic Acids Res.* **4:** 2655–2670.

Mack, D.P. and P.B. Dervan. 1990. Nickel-mediated sequence specific oxidative cleavage of DNA by a designed metalloprotein. *J. Am. Chem. Soc.* **112:** 4604–4606.

Mack, D.P., B.L. Iverson, and P.B. Dervan. 1988. Designed and chemical synthesis of sequence-specific DNA-cleaving proteins. *J. Am. Chem. Soc.* **110:** 7572–7574.

Maher, L.T., III, B. Wold, and P.B. Dervan. 1989. Inhibition of DNA binding proteins by oligonucleotide-directed triple helix formation. *Science* **245:** 725–730.

Maxam, A.M. and W. Gilbert. 1977. A new method for sequencing DNA. *Proc. Natl. Acad. Sci.* **74:** 560–564.

Milavetz, B. 1989. Oligonucleotide-directed restriction endonuclease digestion. *Nucleic Acids Res.* **17:** 3322–3323.

Mirkin, S.M., V.I. Lyamichev, K.N. Drushlyak, V.N. Dobrynin, S.A. Filippov, and M.D. Frank-Kamenetskii. 1987. DNA H form requires a homopurine-homopyrimidine mirror repeat. *Nature* **330:** 495–497.

Mitchell, P.J. and R. Tjian. 1989. Transcriptional regulation in mammalian cells by sequence-specific DNA binding proteins. *Science* **245:** 371–378.

Moore, H.W. 1977. Bioactivation as a model for drug design bioreductive alkylation. *Science* **197:** 527–532.

Moser, H.E. and P.B. Dervan. 1987. Sequence-specific cleavage of double helical DNA by triple helix formation. *Science* **238:** 645–659.

Murakawa, G.J., C.-H.B. Chen, M.D. Kuwabara, D.P. Nierlich, and D.S. Sigman. 1989. Scission of RNA by the chemical nuclease of 1,10-phenanthroline-copper ion: Preference for single-stranded loops. *Nucleic Acids Res.* **17:** 5361–5375.

Natsoulis, G. and J.D. Boeke. 1991. New antiviral strategy using capsid-nuclease fusion proteins. *Nature* **352:** 632–635.

Nicolaou, K.C. and A.L. Smith. 1992. Molecular design, chemical synthesis, and biological action of enediynes. *Accts. Chem. Res.* **25:** 497–503.

Nielsen, P.E., M. Egholm, R.H. Berg, and O. Buchardt. 1991. Sequence-selective recognition of DNA by strand displacement with a thymine-substituted polyamide. *Science* **254:** 1497–1500.

Nosikov, B. and B. Jain. 1977. Protection of particular endonuclease R. *Hin*dIII cleavage sites by distamycin A, propyl-distamycin and netropsin. *Nucleic Acids Res.* **4:** 2263–2273.

Oakley, M.G. and P.B. Dervan. 1990. Structural motif of the GCN4 DNA binding domain characterized by affinity cleaving. *Science* **248:** 847–850.

Pabo, C.O. and R.T. Sauer. 1984. Protein-DNA recognition. *Annu. Rev. Biochem.* **53:** 293–321.

Pei, D. 1991. "Sequence-specific DNA recognition and cleavage." Ph.D. thesis, University of California, Berkeley.

Pei, D. and P.G. Schultz. 1990. Site-specific cleavage of duplex DNA with a λ repressor-staphylococcal nuclease hybrid. *J. Am. Chem. Soc.* **112:** 4579–4580.

———. 1991. Engineering protein specificity: Gene manipulation with semisynthetic nucleases. *J. Am. Chem. Soc.* **113:** 9391–9392.

Pei, D., D.R. Corey, and P.G. Schultz. 1990. Site-selective cleavage of duplex DNA by a semisynthetic nuclease via triple-helix formation. *Proc. Natl. Acad. Sci.* **87:** 9858–9862.

Pei, D., H. Ulrich, and P.G. Schultz. 1991. A combinatorial approach toward DNA recognition. *Science.* **253:** 1048–1411.

Perrouault, L., U. Asseline, C. Rivalle, N.T. Thuong, E. Bisagui, C. Giovannangeli, T. Le Doan, and C. Helene. 1990. Sequence-specific artificial photoinduced endonucleases based on triple-helix-forming oligonucleotides. *Nature* **344:** 358–360.

Podhajska, A.J. and W. Szybalski. 1985. Conversion of the *Fok* I endonuclease to a universal restriction enzyme: Cleavage of phage M13mp7 DNA at predetermined sites. *Gene* **40:** 175–182.

Praseuth, D., L. Perrouault, T. Le Doan, M. Chassignol, N. Thuong, and C. Helene. 1988. Sequence-specific binding and photocrosslinking of α and β oligodeoxynucleotides to the major groove of DNA via triple-helix formation. *Proc. Natl. Acad. Sci.* **85:** 1349–1353.

Rajagopal, P. and J. Feigon. 1989. NMR studies of triple-strand formation from the homopurine-homopyrimidine deoxyribonucleotides $d(GA)_4$ and $d(TC)_4$. *Biochemistry* **28:** 7859–7870.

Reinert, K.E. 1972. Adenosine•thymidine cluster-specific elongation and stiffening of DNA induced by the oligopeptide antibiotic netropsin. *J. Mol. Biol.* **72:** 593–607.

Saiki, R.K., D.H. Gelfand, S. Stoffel, S.J. Scharf, R. Higuchi, G.T. Horn, K.B. Mullis, and H.A. Erlich. 1988. Primer-directed enzymatic amplification of DNA with a thermostable DNA polymerase. *Science* **239:** 487–491.

Schultz, P.G. and P.B. Dervan. 1983. Sequence-specific double-strand cleavage of DNA by bis (EDTA-distamycin•Fe^{II}) and EDTA-bis(distamycin)•Fe^{II}. *J. Am. Chem. Soc.* **105:** 7748–7750.

Schultz, P.G., J.S. Taylor, and P.B. Dervan. 1982. Design and synthesis of a sequence-specific DNA cleaving molecule (distamycin-EDTA) iron (II). *J. Am. Chem. Soc.* **104:** 6861–6863.

Scott, M.P., J.W. Tamkun, and G.W.I. Hartzell. 1989. The structure and function of the homeodomain. *Biochim. Biophys. Acta* **989:** 25–48.

Sigman, D.S. 1986. Nuclease activity of 1,10-phenanthroline-copper ion. *Accts. Chem. Res.* **19:** 180–186.

Sigman, D.S. and C.-H.B. Chen. 1990. Chemical nucleases: New reagents in molecular biology. *Annu. Rev. Biochem.* **59:** 207–236.

Sluka, J.P., S.J. Horvath, M.F. Bruist, M.I. Simon, and P.B. Dervan. 1987. Synthesis of a sequence-specific DNA-cleaving peptide. *Science* **238:** 1129–1132.

Strobel, S.A. and P.B. Dervan. 1990. Site-specific cleavage of a yeast chromosome by oligonucleotide-directed triple helix formation. *Science* **249:** 73–75.

———. 1991. Single-site enzymatic cleavage of yeast genomic DNA by triple helix formation. *Nature* **350:** 172–174.

Strobel, S.A., H.E. Moser, and P.B. Dervan. 1988. Double-strand cleavage of genomic DNA at a single site by triple-helix formation. *J. Am. Chem. Soc.* **110:** 7927–7929.

Stubbe, J. and J.W. Kozarich. 1987. Mechanisms of bleomycin-induced DNA degradation. *Chem. Rev.* **87:** 1107–1136.

Sun, J.-S., J.-C. Francois, T. Montenay-Garestier, T. Saison-Behmoaras, V. Roig, N.T. Thuong, and C. Helene. 1989. Sequence-specific intercalating agents: Intercalation at specific sequences on duplex DNA via major groove recognition by oligonucleotide-intercalator conjugates. *Proc. Natl. Acad. Sci.* **86:** 9198–9202.

Szybalski, W. 1985. Universal restriction endonucleases: Designing novel cleavage specificities by combining adapter oligodeoxynucleotide and enzyme moieties. *Gene* **40:** 169–173.

Taylor, J.S., P.G. Schultz, and P.B. Dervan. 1984. DNA affinity cleaving. *Tetrahedron* **40:** 457–465.

Thuong, N.T. and M. Chassignol. 1988. Solid phase synthesis of oligo-2- and oligo-β-

deoxynucleotides. *Tetrahedron Lett.* **29**: 5905-5908.
Tomasz, M., A.K. Chawla, and R. Lipman. 1988. Mechanism of monofunctional and bifunctional alkylation of DNA by mitomycin C. *Biochemistry* **27**: 3182-3187.
Tomasz, M., R. Lipman, G.L. Verdine, and K. Nakanishi. 1986. Reassignment of the guanine-binding mode of reduced mitomycin C. *Biochemistry* **25**: 4337-4344.
Tonequzzo, F., S. Glynn, E. Levi, S. Mjolsness, and A. Hayday. 1988. Use of a chemically modified T7 DNA polymerase for manual and automated sequencing of supercoiled DNA. *BioTechniques* **6**: 460-469.
Tucker, P.W., E.E. Hazen, and F.A. Cotton. 1978. Staphylococcal nuclease reviewed: A prototypic study in contemporary enzymology. I. Isolation, physical and enzyme properties. *Mol. Cell. Biochem.* **22**: 67-77.
―――. 1979a. Staphylococcal nuclease reviewed: A prototypic study in contemporary enzymology. II. Solution studies of the nucleotide binding site and the effects of nucleotide binding. *Mol. Cell. Biochem.* **23**: 3-16.
―――. 1979b. Staphylococcal nuclease reviewed: A prototypic study in contemporary enzymology. III. Correlation of the three-dimensional structure with the mechanisms of enzymatic action. *Mol. Cell. Biochem.* **23**: 67-86.
―――. 1979c. Staphylococcal nuclease reviewed: A prototypic study in contemporary enzymology. IV. The nuclease as a model for protein folding. *Mol. Cell. Biochem.* **23**: 131-141.
Tullius, T.D. 1988. DNA footprinting with hydroxy radical. *Nature* **332**: 663-664.
Tullius, T.D. and B.A. Dombroski. 1986. Hydroxyl radical "footprinting:" High-resolution information about DNA-protein contacts and application to λ repressor and Cro protein. *Proc. Natl. Acad. Sci.* **83**: 5469-5473.
Umezawa, H., T. Takita, Y. Sugiura, M. Otsuka, S. Kobayashi, and M. Ohno. 1984. DNA-bleomycin interaction: Nucleotide sequence-specific binding and cleavage of DNA by bleomycin. *Tetrahedron.* **40**: 501-509.
Van Dyke, M.W. and P.B. Dervan. 1984. Echinomycin binding sites on DNA. *Science* **225**: 1122-1126.
Vlassov, V.V., S.A. Gaidamakov, V.F. Zarytova, D.G. Knorre, A.S. Levina, A.A. Nikonova, L.M. Podust, and O.S. Fedorova. 1988. Sequence-specific chemical modification of double-stranded DNA with alkylating oligodeoxyribonucleotide derivatives. *Gene* **72**: 313-322.
Wallace, P.B., J. Shaffer, R.F. Murphy, J. Bonner, T. Hirose, and K. Itakura. 1979. Hybridization of synthetic oligodeoxyribonucleotides to ϕX174 DNA: The effect of single base mismatch. *Nucleic Acids Res.* **6**: 3543-3557.
Ward, B., A. Skorobogaty, and J.C. Dabrowiak. 1986. DNA cleavage specificity of a group of cationic metalloporphyrins. *Biochemistry* **25**: 6875-6883.
Watson, J.D. and F.H.C. Crick. 1953. Molecular structure of nucleic acids, a structure for deoxyribose nucleic acid. *Nature* **171**: 737-738.
Wiegand, R.C., K.L. Beattie, W.K. Holloman, and C.M. Radding. 1977. Uptake of homologous single-stranded fragments by superhelical DNA. III. The product and its enzymatic conversion to a recombinant molecule. *J. Mol. Biol.* **116**: 805-824.
Youngquist, R.S. and P.B. Dervan. 1985a. Sequence-specific recognition of B DNA by bis (EDTA-distamycin) fumaramide. *J. Am. Chem. Soc.* **107**: 5528-5529.
―――. 1985b. Sequence-specific recognition of B-DNA by oligo(*N*-methylpyrrolecarboxamide)S. *Proc. Natl. Acad. Sci.* **82**: 2565-2569.
Zein, N., A.M. Sinha, W.J. McGahren, and G.A. Ellestad. 1988. Calicheamicin γ_1^I: An antitumor antibiotic that cleaves double-stranded DNA site specificity. *Science* **240**: 1198-1201.

Zerial, A., N.T. Thuong, and C. Helene. 1987. Selective inhibition of the cytopathic effect of type A influenza viruses by oligodeoxynucleotides covalently linked to an intercalating agent. *Nucleic Acids Res.* **15:** 9909–9919.

Zuckermann, R.N. and P.G. Schultz. 1989. Site-selective cleavage of structured RNA by a staphylococcal nuclease-DNA ligand. *Proc. Natl. Acad. Sci.* **86:** 1766–1770.

Zuckerman, R.N., D.R. Corey, and P.G. Schultz. 1988. Site-selective cleavage of RNA by a hybrid. *J. Am. Chem. Soc.* **110:** 1614–1615.

11
Ribonucleases H

**Zdenek Hostomsky, Zuzana Hostomska,
and David A. Matthews**
Agouron Pharmaceuticals, Inc.
San Diego, California 92121

I. Introduction
II. Bacterial RNases H
   A. Biological Roles in Replication
   B. Crystal Structures of *Escherichia coli* RNase HI
   C. Structure of *Thermus thermophilus* RNase H
III. Retroviral RNase H
   A. Roles of RT-associated RNase H in the Retroviral Life Cycle
   B. RNase H as an Antiretroviral Target
   C. Structure of the RNase H Domain of HIV-1 RT
   D. Structure of the RNase H Domain of HIV-2 RT
   E. Structural Role of the RNase H Domain in the RT Heterodimer
   F. Maturation of Heterodimeric RT
   G. Origin and Evolution of Retroviral RNase H as an RT-associated Activity
IV. Eukaryotic RNases H: Human RNase H
V. Detection of RNase H Activity
   A. Activity of the Isolated RNase H Domain of RT
   B. Altered Substrate Specificities
VI. Mechanism of RNase H Action
   A. Endonucleolytic Character of RNase H Cleavage
   B. Catalytic Role of Metals
   C. Recogniton of RNA in the Heteroduplex
VII. Inhibitors of Retroviral RNase H
VIII. Antisense Effect
IX. Conclusions and Perspectives

I. INTRODUCTION

Ribonuclease H (RNase H) (Ribonucleate [in RNA:DNA hybrids] 5′-oligonucleotidohydrolase, E.C.3.1.26.4) specifically cleaves the RNA strand in an RNA:DNA hybrid duplex. The designation H (for hybrid) was introduced by Hausen and Stein (1970) to describe a new ribonuclease activity discovered in extracts from calf thymus (Stein and Hausen 1969). All known nucleases with RNase H activity require divalent metal cations for hydrolysis of the RNA phosphodiester bond producing 5′-phosphate and 3′-hydroxyl termini. The presence of the free 3'-hydroxyl terminus makes the resulting RNA fragment suitable as a primer for DNA polymerase.

RNases H comprise a family of ubiquitous enzymes, detected in multiple forms in all organisms from bacteria to mammals. However, with the exception of their well-understood function in the life cycle of retroviruses, the biological roles of other cellular RNases H are much less clear. In addition to their presumed involvement in DNA replication (and the practical usefulness of the isolated enzyme as a reagent for in vitro manipulations with nucleic acids), two other factors with possible therapeutic implications have contributed to the recent increased interest in RNases H: (1) the presumed role of endogenous RNase H activity in the mechanism of action of antisense oligonucleotides and (2) the potential importance of the enzyme as an antiviral target, especially in connection with human immunodeficiency virus (HIV) infection and acquired immunodeficiency syndrome (AIDS).

Since the first comprehensive overview of RNases H by Crouch and Dirksen (1982), two more recent reviews have appeared (Crouch 1990; Wintersberger 1990) that deal with these enzymes from a historical perspective and in an evolutionary context. In recent years, research on RNases H has been strongly influenced by the increasing availability of three-dimensional structural information. Building on these developments, this chapter focuses primarily on bacterial and retroviral RNases H, where a clearer picture of structure-function relationship for these fascinating enzymes is beginning to emerge.

II. BACTERIAL RNASES H

Two enzymes with RNase H activity have been identified in *Escherichia coli*. RNase HI, the major and best-characterized bacterial RNase H, is an 18-kD protein encoded by the *rnhA* gene (Kanaya and Crouch 1983). RNase HII is a much less studied 23-kD protein encoded by the *rnhB* gene (Itaya 1990). The amino acid sequences of these two enzymes are quite different.

A. Biological Roles in Replication

Various functions for RNase HI in bacterial cells have been proposed based on genetic studies with *E. coli* mutants having alterations in the *rnhA* gene (for review, see Kogoma 1986). These functions include (1) participation in removal of the RNA primers from Okazaki fragments in discontinuous synthesis of DNA (Ogawa and Okazaki 1984), (2) processing of RNA transcripts to be used as primers for DNA polymerase I in initiation of replication of ColE1-type plasmids (Itoh and Tomizawa

1980), and (3) removal of R-loops, consisting of RNA:DNA hybrids and the displaced single-stranded DNA, to ensure proper initiation of replication from *OriC* (von Meyenburg et al. 1987). In the absence of functional RNase HI, the persisting R-loops can become sites for opportunistic, unregulated initiation of replication of the *E. coli* chromosome.

Despite the involvement of RNase HI in fundamental processes of DNA replication, its activity appears to be nonessential and can be substituted by other enzymes. For example, it has been proposed that in the absence of RNase HI activity, R-loops can be removed by two alternative routes of DNA repair: one involving the RecBCD enzyme, exonuclease I, and DNA polymerase I and the other involving both the RecBCD and the RecF pathways of homologous recombination (Kogoma et al. 1993). *rnhA* mutants are therefore lethal only in certain genetic backgrounds, where alternative pathways are also inactivated, or under certain growth conditions when tight regulation of all steps in DNA replication becomes critical.

The combination of an insertionally inactivated *rnhA* gene with a temperature-sensitive mutation in one of the genes for the multisubunit RecBCD protein led to the creation of an *E. coli* strain that is temperature sensitive for growth (Itaya and Crouch 1991). This strain (*rnhA-339:cat, recB270*) can be used for cloning of RNase H genes from other organisms that can functionally complement the conditionally lethal phenotype. Clones expressing active forms of RNase H from *Salmonella typhimurium* and *Saccharomyces cerevisiae* (Itaya et al. 1991) have been obtained using this approach.

B. Crystal Structures of *Escherichia coli* RNase HI

X-ray studies of an RNase H were first carried out on crystals of the enzyme from *E. coli*. A high-resolution structure of this protein was determined independently in two laboratories (Katayanagi et al. 1990; Yang et al. 1990). In a subsequent publication by Katayanagi et al. (1992), resolution was extended to 1.48 Å. The RNase H polypeptide chain folds into a five-stranded mixed β-sheet surrounded by an asymmetric distribution of five α helices, four of which cluster proximate to one face of the sheet (Fig. 1). The fifth helix is located on the opposite side of the central β-sheet near the carboxyl terminus of the polypeptide chain, where it runs approximately diagonally across strands β1, β2, and β3. A marked splaying apart of two inner strands, β3 and β4, occurs as they propagate toward one edge of the sheet. This strand separation provides a structural framework for the placement and positioning of several amino acids known to be crucial for enzyme catalysis.

HIV-1 RNase H　　　　E.coli RNase H

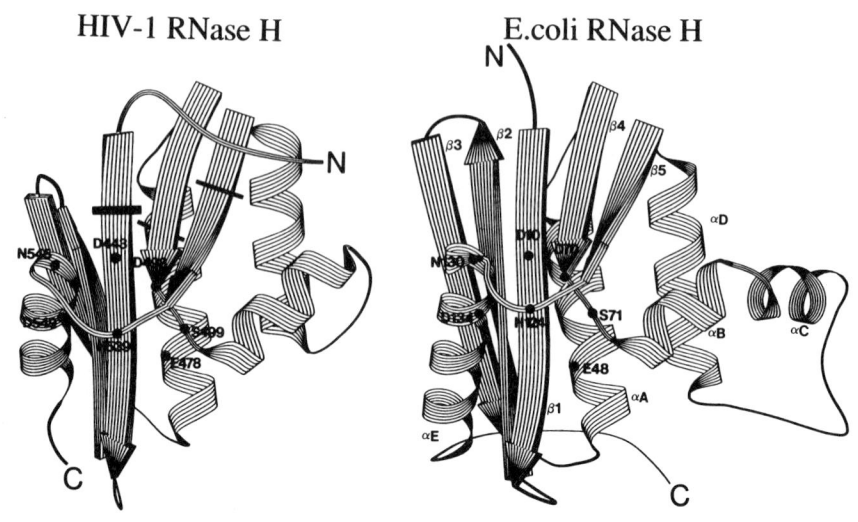

Figure 1 Comparison of the overall folding of RNase HI from *E. coli* and of the isolated RNase H domain of HIV-1 RT. Black dots in the ribbon indicate the positions of the seven invariant residues that constitute the catalytic site. The perpendicular bars indicate sites of cleavage by HIV-1 protease occurring during processing of HIV-1 RT. The thicker of these bars indicates the cleavage site between Phe-440 and Tyr-441 that generates the carboxyl terminus of the p51 subunit (compare paragraph on maturation of the heterodimeric RT and Fig. 5). α helices and β strands in *E. coli* RNase H are labeled to conform to the nomenclature used for HIV-1 RNase H. Amino (N) and carboxyl (C) termini are indicated.

At one edge of the sheet separated from the central core of the protein is a region of approximately 18 contiguous amino acids variously described as a "handle region" (Katayanagi et al. 1990), "minor domain" (Yang et al. 1990), or "basic protrusion" (Katayanagi et al. 1992). Helix αC, which forms the amino-terminal portion of the handle region, is tightly coupled to another helix, αB, immediately preceding it in amino acid sequence, so as to produce an abrupt 35° bend in their respective helix axes. Helix αC and the following loop connecting it with αD are important for binding the RNA:DNA heteroduplex and are discussed in more detail below.

Primary structure comparisons for retroviral and bacterial RNases H reveal that seven amino acids are identically conserved in all of these sequences (Doolittle et al. 1989). Kanaya et al. (1990), using site-directed mutagenesis, have shown that amidation of any of three conserved acidic residues, Asp-10, Glu-48, or Asp-70, eliminates catalytic activity. These findings were placed in a structural context by the X-ray diffraction studies of *E. coli* RNase HI which indicated that all three carboxylate side chains are clustered on one edge of a broad concave surface that most probably represents a binding cleft for the RNA:DNA substrate. Moreover, the other four identically conserved residues are located nearby. Katayanagi et al. (1990) went on to show that a single divalent metal ion binds in the vicinity of this essential carboxylate triad. These observations have led to two rather different proposals concerning how RNase H may cleave a hybrid duplex substrate. Possible mechanisms of action for this enzyme are discussed in more detail below.

The handle region of *E. coli* RNase HI, consisting of αC and the following loop, is rich in basic amino acids. Alanine-scanning mutagenesis has indicated that a cluster of six lysine and arginine residues in a stretch of 13 amino acids beginning with Lys-87 is important for substrate binding (Kanaya et al. 1991). Single amino acid replacements were found to elevate the K_m for RNA:DNA hybrids three- to fivefold compared to wild-type values. The effects of mutating several basic residues simultaneously were cumulative. Additional information concerning substrate binding has come from analysis of changes in ^{1}H-^{15}N heteronuclear two-dimensional nuclear magnetic resonance (NMR) spectra for *E. coli* RNase HI upon addition of a synthetic nonanucleotide RNA:DNA hybrid (Nakamura et al. 1991). Chemical shifts for backbone protons and ^{15}N had been assigned recently (Yamazaki et al. 1991), so by observing shifts in cross peaks during oligonucleotide titration, residues responsible for hybrid recognition could be mapped. A model for *E. coli* RNase HI with an RNA:DNA hybrid was proposed on the basis of the X-ray, mutagenesis, and NMR results (Fig. 2) that is consistent in its gross features with ideas originally suggested by Yang et al. (1990).

D. Structure of *Thermus thermophilus* RNase H

The crystal structure of a second bacterial RNase H, that from *Thermus thermophilus*, has recently been reported at 2.8 Å resolution (Ishikawa et al. 1993b). The overall sequence identity between the *T. thermophilus* and *E. coli* enzymes is 52%. Compared to *E. coli* RNase HI, the *T. thermophilus* enzyme has four and six additional residues at its amino and carboxyl termini, respectively, plus a single glycine insertion in the sharp transition region between αB and αC (Fig. 3). Structural comparison shows that these two bacterial enzymes are very similar, with an overall root-mean-displacement of only 0.95 Å between equivalent α-carbon atoms from all elements of secondary structure. Of particular interest is the observation that the loop connecting β5 to αE, which contains conserved His-124 (*E. coli* numbering), is characterized by very

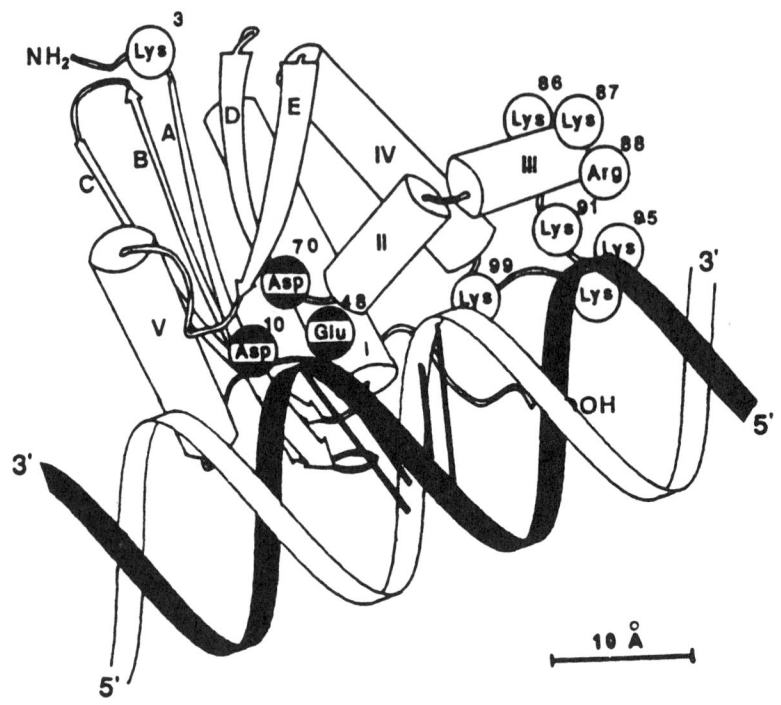

Figure 2 Model for the interaction of *E. coli* RNase HI with an RNA:DNA hybrid. (*Open circles*) Basic residues of the "handle region" (residues 80–99); (*closed circles*) three essential acidic residues of the catalytic center. The RNA strand of the A-form RNA:DNA hybrid is black, and the DNA strand is white. (Reprinted, with permission, from Kanaya et al. 1991.)

weak electron density, suggesting considerable thermal mobility and/or multiple conformational states. There is considerable interest in attempting to understand the thermal stability of the *T. thermophilus* RNase H, which has a 34°C higher thermal unfolding temperature than the *E. coli* enzyme. Systematic replacement of residues in *E. coli* RNase HI with corresponding residues in *T. thermophilus* RNase H combined with thermal stability measurements on the mutant proteins has already begun to provide useful information on structure-stability relationships (Ishikawa et al. 1993a).

III. RETROVIRAL RNASE H

Specific biological functions of RNase H in the life cycle of retroviruses are extensively reviewed in Skalka and Goff (1993). In addition to a general overview of the roles of RNase H in the retroviral life cycle (Cham-

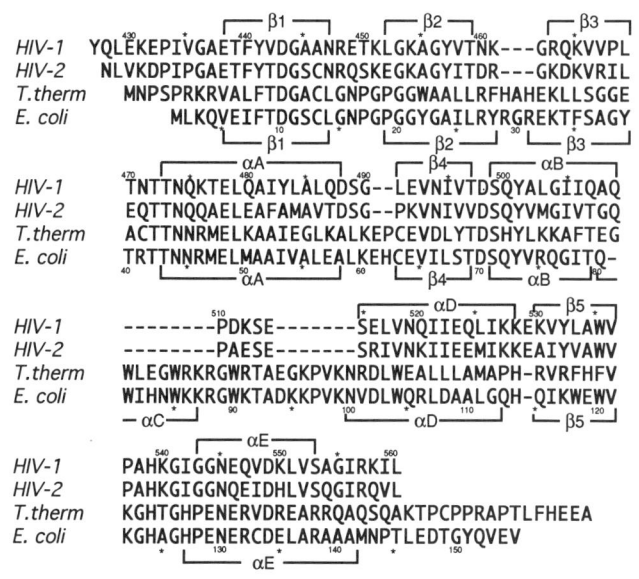

```
            ┌──── β1 ────┐      ┌──── β2 ────┐        ┌─ β3 ─┐
HIV-1    YQLEKEPIVGAETFYVDGAANRETKLGKAGYVTNK---GRQKVVPL
HIV-2    NLVKDPIPGAETFYTDGSCNRQSKEGKAGYITDR---GKDKVRIL
T.therm  MNPSPRKRVALFTDGACLGNPGPGGWAALLRFHAHEKLLSGGE
E. coli  MLKQVEIFTDGSCLGNPGPGGYGAILRYRGREKTFSAGY
            └── β1 ──┘       └── β2 ──┘        └─ β3 ─┘
                   ┌──── αA ────┐        ┌─ β4 ─┐  ┌──── αB ────┐
HIV-1    TNTTNQKTELQAIYLALQDSG--LEVNIVTDSQYALGIIQAQ
HIV-2    EQTTNQQAELEAFAMAVTDSG--PKVNIVVDSQYVMGIVTGQ
T.therm  ACTTNNRMELKAAIEGLKALKEPCEVDLYTDSHYLKKAFTEG
E. coli  TRTTNNRMELMAAIVALEALKEHCEVILSTDSQYVRQGITQ-
              └──── αA ────┘       └─ β4 ─┘  └── αB ──┘
                                      ┌──── αD ────┐  ┌─ β5 ─┐
HIV-1    --------PDKSE-------SELVNQIIEQLIKKEKVYLAWV
HIV-2    --------PAESE-------SRIVNKIIEEMIKKEAIYVAWV
T.therm  WLEGWRKRGWRTAEGKPVKNRDLWEALLLAMAPH-RVRFHFV
E. coli  WIHNWKKRGWKTADKKPVKNVDLWQRLDAALGQH-QIKWEWV
           └─ αC ─┘              └──── αD ────┘  └─ β5 ─┘
                 ┌──── αE ────┐
HIV-1    PAHKGIGGNEQVDKLVSAGIRKIL
HIV-2    PAHKGIGGNQEIDHLVSQGIRQVL
T.therm  KGHTGHPENERVDREARRQAQSQAKTPCPPRAPTLFHEEA
E. coli  KGHAGHPENERCDELARAAAMNPTLEDTGYQVEV
           └──── αE ────┘
```

Figure 3 Geometrical alignment of secondary structure elements in the four RNases H whose three-dimensional structures are known. Numbering of HIV-1 RT is used for the HIV-1 and HIV-2 RNase H domains. Numbering of *E. coli* RNase HI is applied also for RNase H from *T. thermophilus*, with the extra glycine between αB and αC in the *T. thermophilus* structure designated as Gly-80b. Every fifth residue is marked with an asterisk (*) or the appropriate residue number.

poux 1993), it contains chapters discussing individual steps of reverse transcription in detail. Another recent review of interest focusing on retroviral reverse transcription is that by Whitcomb and Hughes (1992). We briefly summarize here current understanding of the functions of retroviral RT-associated RNase H in reverse transcription.

A. Roles of RT-associated RNase H in the Retroviral Life Cycle

In the life cycle of retroviruses, viral genomic RNA is converted to double-stranded DNA of the provirus in a complex process of reverse transcription, catalyzed by a virus-encoded multifunctional enzyme, reverse transcriptase (RT). RNase H activity of RT is required at several stages of reverse transcription (Fig. 4). It has well-defined roles in (1) degradation of the original RNA template, (2) generation of the plus-strand primer, and (3) removal of both plus- and minus-strand primers.

The synthesis of minus-strand DNA is initiated from a transfer RNA primer and proceeds to the 5' end of the genomic RNA. The RNA in this

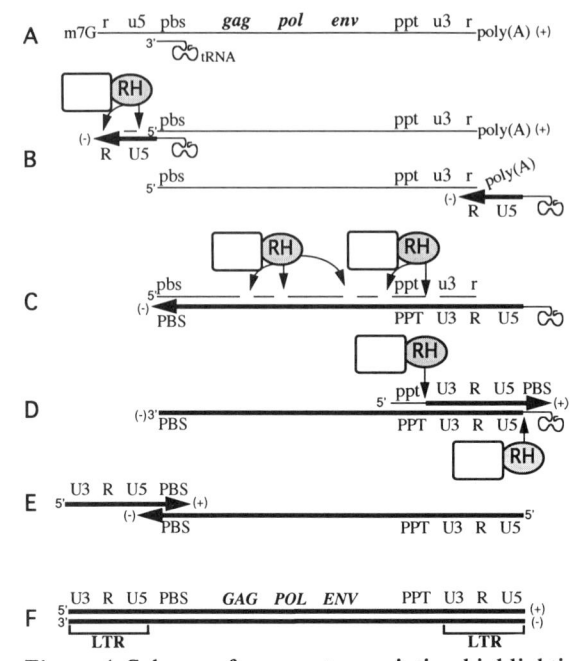

Figure 4 Scheme of reverse transcription highlighting the role of RT-associated RNase H. (*Thin lines*) RNA sequences; (*thick lines*) DNA; (*horizontal arrows*) direction of DNA synthesis. Polarity of the strands, (+) or (–), is indicated, as are the 3' and 5' termini. Regions in RNA are indicated by lowercase letters, and capital letters are used for the corresponding regions in DNA. The retroviral genes are in italics (not to scale). (*A*) Genomic RNA of a retrovirus. The genome has short direct repeats (r) at each end, as well as unique 5'-terminal (u5) and 3'-terminal (u3) sequences. The 5' end is capped (m7G), and the 3' end is polyadenylated. Synthesis of the minus-strand DNA is initiated from a host tRNA primer annealed to the primer-binding site (pbs). (*B*) RNase-H-catalyzed degradation of the 5'-terminal region of the genomic RNA facilitates the first-strand transfer, mediated by base pairing between the complementary r and R segments. The transfer can be either intramolecular—to the 3'-terminal r region of the same RNA molecule—or intermolecular—to the 3'-terminal r region of the second copy of genomic RNA packaged in the virion. The m7G-cap and poly(A) tail are not reverse-transcribed. (*C*) RNase H degrades most of the template RNA. Specific cut at the polypurine tract (ppt) region generates a primer for the plus-strand DNA synthesis. (*D*) Initiation of the plus-strand DNA synthesis, followed by removal of the RNA primers. Note that specific cuts by RNase H at the ppt and tRNA primers define the future boundaries of the linear proviral DNA. (*E*) Intramolecular second-strand transfer is mediated by base pairing between the complementary sequences of the PBS region. (*F*) Completed preintegrative DNA of the provirus. Duplication of long terminal repeats (LTR), composed of U3-R-U5 sequences, is a result of second-strand transfer, followed by completion of DNA synthesis involving strand displacement.

terminal region must be removed by RNase H to enable the efficient first-strand-transfer reaction (first jump) (Telesnitsky and Goff 1993). Although the same molecule of RT is capable, in principle, of performing DNA synthesis and RNA degradation simultaneously, it appears that during the minus-strand synthesis, the RT-associated RNase H cleaves the template RNA only infrequently (DeStefano et al. 1991b). Most of the resulting RNA fragments remain associated with the nascent DNA and may be further degraded by the RNase H activity of other RT molecules not participating in DNA synthesis. A specific product of the RNase-H-catalyzed RNA degradation, generated in the polypurine track region of the retroviral genome, is recognized as primer for plus-strand synthesis. Removal of the tRNA primer by a single endonucleolytic cut by RNase H at or close to the RNA-DNA junction facilitates the second-strand-transfer reaction and continuation of the minus-strand synthesis. The polypurine track primer also appears to be removed intact, suggesting a sequence-specific recognition by RT/RNase H (Champoux 1993). It has been proposed that the high rate of retrovirus variation is a direct consequence of the requirement for the two strand transfers during retroviral DNA synthesis (Temin 1993).

B. RNase H as an Antiretroviral Target

Since RNase H activity has several functions required for ordered progression of reverse transcription, selective inhibition of any of these functions would be expected to block virus replication. Point mutations inactivating RNase H in Moloney murine leukemia virus (Mo-MLV) (Repaske et al. 1989) and HIV-1 (Schatz et al. 1990b; Tisdale et al. 1991) led to production of noninfectious virions. These experiments prove that virus-encoded RNase H is essential for retrovirus replication and that this activity cannot be substituted by cellular enzymes. Interestingly, mutations of His-539 in the RNase H domain of HIV-1 RT, which displayed reduced but still detectable RNase H activity in enzymatic assays in vitro, rendered the virus noninfectious, similar to mutations of Glu-478 which completely inactivated RNase H in vitro. These data suggest that even a partial inhibition of RNase H function is sufficient to impair virus replication severely. It thus appears that RNase H represents a valid target for therapy of infections by retroviruses and possibly by other pathogenic viruses that employ RT for their replication, such as hepatitis B virus (Radziwill et al. 1990).

C. Structure of the RNase H Domain of HIV-1 RT

Unlike Mo-MLV RT, which is monomeric, HIV-1 RT is a heterodimer composed of two subunits, p66 and p51, which have identical amino

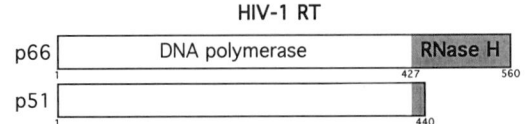

```
              HIV-1 RT
p66 |    DNA polymerase    | RNase H |
    1                     427        560
p51 |                                |
    1                              440
```

Figure 5 Subunit composition of the mature p66/p51 heterodimer of HIV-1 RT. The p51 subunit arises by protease cleavage within the RNase H domain of one of the p66 subunits in the p66/p66 homodimeric RT precursor, between Phe-440 and Tyr-441 (compare paragraph on maturation of the heterodimeric RT and Fig. 1). The boundaries of the individual subunits and the RNase H domain *(shaded area)* are indicated by residue numbers. Despite the identical amino acid sequences of the p51 subunit and the DNA polymerase portion of p66, the two subunits have a markedly different three-dimensional arrangement. Although DNA polymerase and RNase H catalytic sites are both in p66, the p51 subunit appears to have a supporting structural role in the heterodimer.

termini (Fig. 5). The DNA polymerase and RNase H catalytic sites are both located on the p66 subunit of the p66/p51 heterodimer (Le Grice et al. 1991; Hostomsky et al. 1992a).

An isolated carboxy-terminal portion of the p66 subunit of HIV-1 RT, representing the RNase H domain, has been crystallized (Hostomska et al. 1991) and its structure determined at 2.4 Å resolution by X-ray diffraction (Davies et al. 1991). This protein comprises residues 427 to 560 of RT. Overall folding of the HIV-1 RNase H domain is very similar to that of *E. coli* RNase HI (see Fig. 1), even though only 32 amino acids are identical when the two sequences are aligned according to geometrical equivalence. As shown in Figure 3, there are four loop regions in which deletions occur in one of the sequences relative to the other. The most significant of these occurs in the connecting region between αB and αD. In *E. coli* RNase HI, this connection is made by two turns of α helix followed by a 12-residue loop that we have previously referred to as the handle region or basic protrusion. In HIV-1 RNase H, this feature is absent and the connection between αB and αD is made by a simple five-residue loop in extended conformation. Davies et al. (1991) originally pointed out that the absence of this basic region in the HIV-1 protein probably contributes to the inability of this isolated domain to bind the RNA:DNA substrate. Recent structural data on heterodimeric HIV-1 RT suggest that protein binding to oligonucleotide duplex occurs over a range of about 20 base pairs separating the polymerase and RNase catalytic sites (Fig. 6) (Arnold et al. 1992; Kohlstaedt et al. 1992). Apparently, substrate binding determinants in regions adjacent to the catalytic site of bacterial RNases H have been supplanted in the case of the HIV-1 enzyme by the polymerase domain of p66.

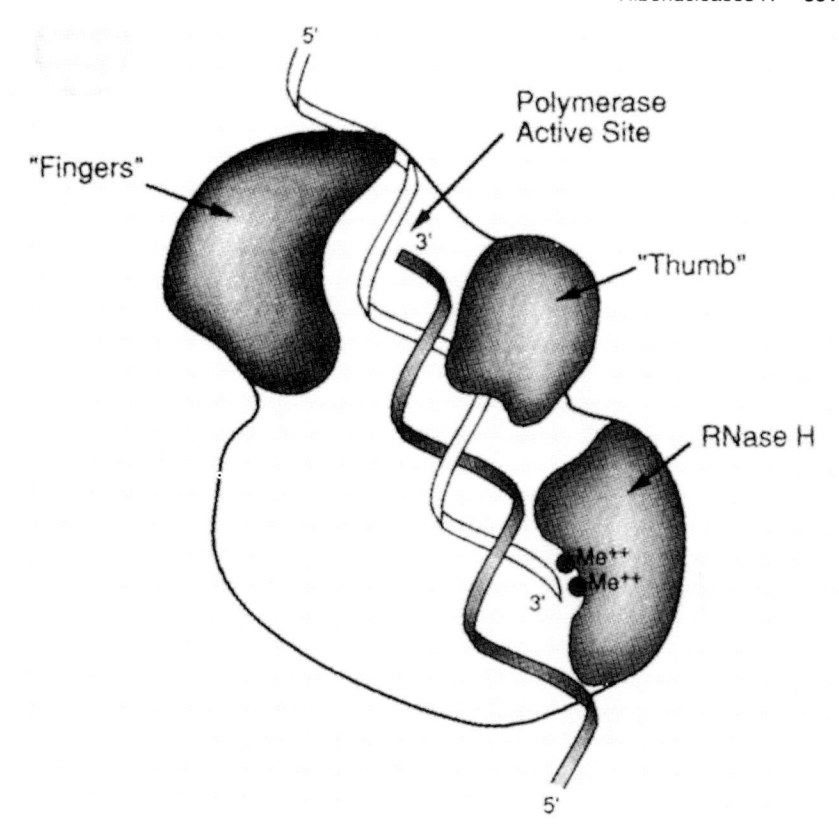

Figure 6 Schematic representation of the heterodimeric HIV-1 RT showing domains of the p66 subunit (*shaded areas*) interacting with an A-form RNA:DNA hybrid. The RNA template strand is white, and the DNA strand is black. Two divalent metal ions (Me^{++}) are in the RNase H catalytic site. (Reprinted, with permission, from Kohlstaedt et al. 1992.)

Residues connecting β5 to αE are disordered in the isolated HIV-1 RNase H domain, in contrast to the corresponding segment in the *E. coli* protein, which is clearly defined in the high-resolution electron density maps. This loop orients a conserved histidine residue proximate to the cluster of carboxylate side chains that have been shown by directed mutagenesis to be crucial for catalysis. Recent NMR studies of an HIV-1 RT RNase H domain differing in sequence by only five amino acids from that characterized crystallographically indicated disorder for all residues beyond β5, including not only the "His loop," but the succeeding helix αE as well (Powers et al. 1992). We have observed similar disordering of residues carboxy-terminal to β5 in the isolated RNase H domain from HIV-2 RT (see below). In the recently reported structure of hetero-

Figure 7 Stereoview of the α-carbon backbone of the RNase H domain (*orange*) interacting with the surrounding regions of the p66 and p51 subunits of HIV-1 RT. The loop with His-539 (*pink*), which is disordered in the structure of the isolated domain, becomes ordered in the heterodimer as a result of direct interaction with the "thumb" domain (*green*) of p51. The "thumb" domain of p66, which is not in contact with the RNase H domain, is at the top. White dots indicate the positions of the two divalent metal ions. A beaded white line represents the "handle region" of *E. coli* RNase HI, which is absent in the HIV-1 RNase H domain. (Reprinted, with permission, from Kohlstaedt et al. 1992.)

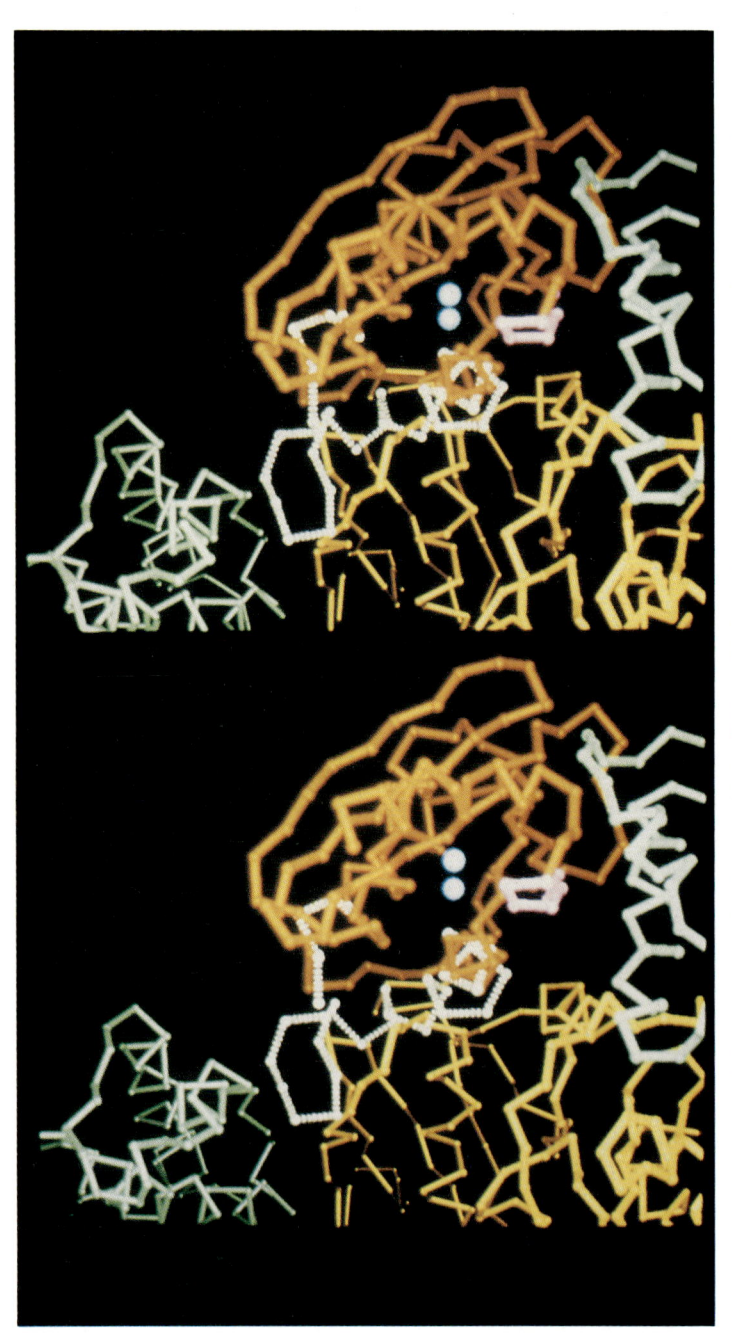

dimeric HIV-1 RT, the carboxyl terminus of the RNase H domain, including the "His loop," is well-ordered as a result of direct interactions with the "thumb" domain of p51 (Fig. 7). This may indicate that p51-mediated positioning of the "His loop" is important for substrate binding and catalysis. The absence of such stabilizing interactions could be a contributing factor to inactivity of the isolated HIV-1 RNase H domain and is consistent with the observed requirement of p51 for RNase H activity in reconstitution experiments (Hostomsky et al. 1991). However, this suggestion must be reconciled with the observation that the loop in question is at least partially disordered in the recently reported structure of a highly active *T. thermophilus* RNase H (Ishikawa et al. 1993b).

D. Structure of the RNase H Domain of HIV-2 RT

Interestingly, activity of the isolated HIV-2 RNase H domain can be reconstituted in the presence of the HIV-1 p51 subunit (Z. Hostomska et al., unpubl.), suggesting that sites involved in specific interdomain interactions in the RT heterodimer are conserved between HIV-1 and HIV-2. The RNase H portion of the HIV-2 RT primary sequence contains no insertions or deletions relative to the HIV-1 sequence and is 61% identical to that of the HIV-1 portion (cf. Fig. 3). The X-ray structure of the isolated RNase H domain of HIV-2 RT has recently been determined at 2.7 Å resolution (D.A. Matthews and D. Knighton, unpubl.). Its structure in general parallels that of the HIV-1 domain, including disorder of the conserved His-539-containing loop. Superposition of the α-carbon atoms of residues 431–498 and 515–537 in the HIV-1 and HIV-2 RNase H domains, respectively, results in a root mean square position difference of only 0.6 Å. Some minor backbone structural differences between the two domains do occur, probably as a result of different intermolecular contacts in the respective crystal unit cells. In the HIV-2 RNase H crystal form, the carboxy-terminal sequence corresponding to αE and beyond (following Pro-537) is not visible and would interfere with a neighboring molecule in the crystal unit cell were it to follow the HIV-1 conformation. Helix αB is only slightly shifted relative to the corresponding helix in HIV-1 RNase H, but its connection to αD cannot adopt the HIV-1 conformation because of crystal packing interaction and therefore differs somewhat from the corresponding HIV-1 connection.

E. Structural Role of the RNase H Domain in the RT Heterodimer

The RNase H domain of HIV-1 RT, in addition to its catalytic function, has acquired a structural role in stabilizing the active configuration of the p66/p51 heterodimer. These two roles appear to be separable. The p51 subunit, corresponding to the amino-terminal fragment of p66 without the RNase H domain, is primarily monomeric in solution (Hostomsky et al. 1991), with reduced polymerase activity attributable to the presence of low concentrations of weakly associated p51/p51 homodimers (Restle et al. 1990). On the other hand, heterodimeric RTs, which have point mutations at the RNase H catalytic site that eliminate nuclease activity, are fully functional as polymerases because the inactive RNase H domain retains its crucial structural role in stabilizing the RT heterodimer (Mizrahi et al. 1990; Hostomsky et al. 1992a).

F. Maturation of Heterodimeric RT

Mature heterodimeric RT results from asymmetric processing of a single chain in the p66/p66 homodimeric RT precursor. It has been shown by carboxy-terminal sequence analysis of the resulting p51 subunit that this proteolytic cut occurs between Phe-440 and Tyr-441 (Le Grice et al. 1989; Mizrahi et al. 1989b; Chattopadhyay et al. 1992). It was anticipated that this cleavage site would be located in a solvent-accessible interdomain region separating DNA polymerase and RNase H. However, the X-ray structure of the isolated HIV-1 RNase H domain revealed that the peptide bond in question is, in fact, part of a structurally important central five-stranded β-sheet that is inaccessible to solvent in the properly folded RNase H structure (Davies et al. 1991). This is consistent with the remarkable stability of the isolated RNase H domain in the presence of various proteolytic enzymes, including HIV-1 protease (Hostomska et al. 1991; Hostomsky et al. 1991). Since homodimeric RT has RNase H activity comparable to that of the mature enzyme (Schatz et al. 1989), at least one of the two RNase H domains in the homodimer must be correctly folded. On the basis of experiments involving mutagenesis of the protease cleavage site and in vitro proteolysis of the isolated RNase H domain, as well as of interpretation of the crystal structure of this domain, a model for asymmetric processing of a homodimeric RT precursor was proposed (Davies et al. 1991; Hostomska et al. 1991). This model suggests an asymmetric structure for p66/p66 in which one RNase H domain is at least partially unfolded so as to expose the Phe-440–Tyr-441 peptide bond to proteolytic processing. Whether or not this putative asymmetry of the homodimeric RT is as extensive as the remarkable asymmetry recently revealed in the crystal structure of the heterodimeric RT (Kohlstaedt et al. 1992) remains to be demonstrated.

The carboxy-terminal fragment of 120 residues, representing an incomplete RNase H domain, should be the other product of this unusual processing strategy. It seems probable that this fragment can be further degraded by various proteases, since as an isolated polypeptide chain, it almost certainly cannot assume a native-like fold, having lost the first three residues needed to form the middle β strand of the central β-sheet. This prediction is supported by analysis of cleavage products resulting from in vitro processing of the recombinant p66 form of HIV-1 RT by HIV-1 protease. The analysis of the digest revealed two additional cleavage sites within the RNase H domain (Tyr-483–Leu-484 and Tyr-532–Leu-533) besides the site Phe-440–Tyr-441 that is cleaved to produce p51 (cf. Fig. 1) (Chattopadhyay et al. 1992). All three of these HIV-1 protease cleavage sites are refractory to hydrolysis in the properly folded RNase H domain of the p66 subunit.

The in vitro experiments may also explain unsuccessful attempts by several groups to detect a carboxy-terminal fragment released during processing of HIV-1 RT in bacterial extracts (Lowe et al. 1988; Ferris et al. 1990). On the other hand, Hansen et al. (1988), Schulze et al. (1991), and Hafkemeyer et al. (1991a) report enzymatically active forms of what appear to be proteolytically released proteins derived from the carboxyl terminus of HIV-1 RT. A stable carboxy-terminal domain could presumably arise from a proteolytic cleavage upstream of the Phe-440–Tyr-441 cleavage site. Alternatively, it could be a product of an internal initiation of translation within the polymerase gene, neither of which has as yet been confirmed.

G. Origin and Evolution of Retroviral RNase H as an RT-associated Activity

Several lines of reasoning point to an ancient evolutionary origin of RTs. This activity is suggested to have played a major role in mediating transition from the RNA-based world which is widely believed to have preceded the modern world of DNA-dominated genetic systems (see, e.g., Wintersberger and Wintersberger 1987; Lazcano et al. 1988). The evolutionary history of RTs deduced from sequence comparisons among retroid family members has been reviewed recently (McClure 1993). Therefore, we limit our discussion to the question of how the functional and spatial association between RT and RNase H activities may have evolved.

RNA:DNA hybrid-specific RNA hydrolysis was suggested to have originated as one of the functions of RT to facilitate removal of template by degradation rather than by displacement (Wintersberger 1990). Not all RT-like sequences, however, contain regions corresponding to an RNase H domain. For example, a typical RNase H domain is present only in one of the seven known retrons—the RT-encoding genetic elements in bacteria (Inouye and Inouye 1993). Telomerase, a specialized RT that carries its own template, has no associated RNase H activity, since its endogenous template RNA is not destroyed in the course of repeated syntheses of telomeric DNA (Blackburn 1993). This ribonucleoprotein could resemble an ancestral RT that may have acquired an RNase H domain by recombination, broadening the spectrum of usable templates. The capacity of RT-based replication systems to capture host sequences and incorporate them into viral genomes under appropriate selection pressure is well documented (Coffin 1993). The advantageous association of amino-terminal RNA-dependent DNA polymerase and carboxy-terminal RNase H, as seen in various members of retroid family, could thus have evolved independently more than once.

The RT from Mo-MLV may be regarded as an example of a relatively recent fusion of polymerase and RNase H domains connected by a tether region to form a single polypeptide chain. These domains still retain a relative independence and can exhibit their corresponding enzymatic activities when separated from each other. The heterodimeric RT from HIV-1, on the other hand, appears to be a result of extensive co-evolution after which the two activities are no longer separable. Since the selection operated on the functionality of the whole complex rather than its constituent parts, the RNase H domain represents a catalytic module that became totally dependent on extensive interactions with the polymerase domains for activity.

IV. EUKARYOTIC RNASES H: HUMAN RNASE H

RNases H isolated from eukaryotic cells have been grouped loosely into two classes on the basis of size, isoelectric point, and ability to utilize Mn^{++} (Vonwirth et al. 1991). Most cell types that have been studied contain both classes. Class I enzymes have molecular weights of 68,000 to 90,000, whereas class II RNases H are smaller (35,000–45,000). The class I enzymes have an acidic isoelectric point and can utilize Mn^{++} as a cofactor in place of Mg^{++}, although sometimes less efficiently. Class II RNases H, on the other hand, cannot substitute Mn^{++} for Mg^{++} and may even be inhibited by Mn^{++} in the presence of Mg^{++}. Eukaryotic RNases H have been purified from several organisms including yeast (Karwan and Wintersberger 1988; Cerritelli et al. 1993), the protozoon *Crithidia fasciculata* (Vonwirth et al. 1991), *Drosophila* (DiFrancesco and Lehman 1985), and calf thymus (Büsen 1980). However, no clue of a cellular function of RNase H has been presented in these studies. A presumed role in DNA replication is consistent with a predominantly nuclear localization of this enzyme shown in immunofluorescence studies with calf thymus RNase H (Büsen 1980).

Since the initial report of a partial purification of human RNase H from leukemic white blood cells (Sarngadharan et al. 1975), purification to homogeneity and subsequent characterization of the major RNase H from K562 erythroleukemia cells was reported by Eder and Walder (1991). This enzyme, designated RNase H1, has a native molecular weight of 89,000 and can use both Mg^{++} and Mn^{++} as cofactors, properties similar to those of class I RNases H (Vonwirth et al. 1991). Besides showing specificity for RNA of RNA:DNA duplexes, the enzyme cleaves preferentially at an RNA-DNA junction. The cleavage always occurs 5′ to the ribonucleotide (Eder et al. 1993). Unlike bacterial and retroviral RNases H, the human RNase H1 is able to cleave a substrate containing a single ribose residue in a double-stranded DNA. This suggests a role for the enzyme in excision repair of ribonucleotides misincorporated into DNA.

It is not clear at present whether human RNase H1 belongs to the same structural class as the well-characterized bacterial and retroviral enzymes. Additional studies of this enzyme would appear to be warranted on the basis of its potential importance in chemotherapy with antisense oligonucleotides and its role as the counterpart in humans to the antiviral target, RT-associated RNase H.

V. DETECTION OF RNASE H ACTIVITY

RNases H are defined operationally as enzymes that specifically cleave RNA only when hybridized to complementary DNA. To assign RNase H function to a particular protein species, unambiguous activity assays are necessary to control for RNA hydrolysis by nonenzymatic mechanisms or by contaminating enzymes unrelated to the primary protein of interest.

A. Activity of the Isolated RNase H Domain of RT

Identification of conserved motifs in amino acid sequence alignment of retroviral RTs with *E. coli* RNase HI led to the proposal that RNase H function resides at the carboxyl terminus of RT (Johnson et al. 1986). This was at variance with the earlier conclusions (Crouch and Dirksen 1982; Grandgenett et al. 1985) based on the amino-terminal origin of a 24-kD proteolytic fragment of avian myeloblastosis virus (AMV) RT, previously reported to have RNase H activity (Lai and Verma 1978). The carboxy-terminal localization of the RNase H domain in the Mo-MLV RT has been subsequently confirmed in mutational studies (Levin et al. 1988; Tanese and Goff 1988).

The assignment of enzymatic activity to the isolated carboxy-terminal domain of HIV-1 RT turned out to be more complex. Hansen et al. (1988) reported significant levels of a nonprocessive RNase H activity associated with a 15-kD protein isolated from HIV-1 lysate. On the basis of immunoprecipitation of this protein by a monoclonal antibody directed against a carboxy-terminal epitope of the p66 subunit of HIV-1 RT, this p15 appeared to be a product of proteolytic processing of p66.

These data can be compared and contrasted with no or only negligible RNase H activities found in recombinant preparations of the carboxy-terminal domain of HIV-1 RT (Becerra et al. 1990; Schatz et al. 1990a; Stammers et al. 1991; Hostomsky et al. 1991). In the latter report, however, the RNase H activity could be reconstituted upon addition of the isolated p51 domain of HIV-1 RT. On the other hand, Evans et al. (1991) reported an Mg^{++}-dependent enzymatic activity of the HIV-1 RNase H domain purified in one step via a histidine-rich extension attached to Tyr-427, although these results could not be confirmed by Smith and Roth (1993). Schulze et al. (1991) describe an active p15 RNase H released from the recombinant p66 form of HIV-1 RT by purified HIV-1 protease. Most surprisingly, Hafkemeyer et al. (1991a) report that the 15-kD protein carboxy-terminal domain of HIV-1 RT, resulting from processing of p66 in *E. coli*, possesses both RNase H and RT activities.

A detailed comparison of the three-dimensional structures of *E. coli* RNase HI with the RNase H domain of HIV-1 RT (Davies et al. 1991) clearly shows that the HIV-1 domain lacks the basic protrusion ("handle region"), implicated in substrate binding by the bacterial enzyme (cf. Figs. 1 and 2) (Kanaya et al. 1991). A recombinant form of *E. coli* RNase HI, in which the basic protrusion was replaced by the peptide junction as in the HIV-1 domain, was found to be devoid of RNase H activity (Z. Hostomska and Z. Hostomsky, unpubl.). The crystal structures of mature heterodimeric HIV-1 RT (Kohlstaedt et al. 1992; Jacobo-Molina et al. 1993) reveal that the polymerase domains, in addition to providing substrate binding determinants, are also important for directly stabilizing portions of the RNase H domain (Fig. 7). Comparative structural analysis thus offers a plausible explanation for the loss of activity of this domain upon separation from the heterodimer. This is also consistent with the observation that RNase H activity is confined exclusively to the dimeric form of HIV-1 RT (Restle et al. 1992).

Sequence alignment of retroviral RNases H indicate that Mo-MLV RNase H, in contrast to the corresponding enzyme from HIV and most other retroviruses, contains a region corresponding to the basic protrusion, which makes it similar in this respect to the bacterial enzyme. This structural similarity to *E. coli* RNase HI correlates with observations that various recombinant proteins comprising the carboxy-terminal domain of Mo-MLV RT exhibit RNase H activity even in the absence of the polymerase domain (Tanese and Goff 1988; Z. Hostomska and Z. Hostomsky, unpubl.).

Since it is almost always recombinant forms of retroviral enzymes that are used for both structural and functional studies, contamination by enzymes indigenous to the *E. coli* host could account for some of the discrepancies in the literature concerning catalytic activity. Close to 20 distinct RNases have been identified in *E. coli* so far and more are expected to be found (Deutscher 1993; see also Chapter 12 and Appendix C). Moreover, the specific activity of *E. coli* RNase HI is approximately two orders of magnitude higher than that for recombinant RT (Mizrahi et al. 1989; Z. Hostomska and Z. Hostomsky, unpubl.).

B. Altered Substrate Specificities

It has been observed that under nonstandard conditions, most typically when Mg^{++} is replaced by Mn^{++}, some retroviral RTs can hydrolyze a broader range of nucleic acid substrates than that implied by the strict definition of RNase H. For example, HIV-1 RT was shown to cleave double-stranded RNA in an in situ activity gel assay in the presence of Mn^{++}, but not Mg^{++} (Ben-Artzi et al. 1992b).[1]

This ribonuclease activity seems to reflect a relaxation of substrate specificity of RT-associated RNase H, which may be analogous to "star" activity of certain restriction endonucleases (Chapter 2). These enzymes can, under nonstandard conditions, cleave sequences similar but not identical to their defined recognition sequence. On the basis of the analogy with restriction endonucleases, it may be appropriate to name the activity of an RNase H with relaxed substrate specificity RNase H*, rather than RNase D, a term that has been already reserved for an *E. coli* ribonuclease (cf. Appendix C).

Mo-MLV RT was also found to be capable of degrading RNA in RNA:DNA as well as in RNA:RNA duplexes in the presence of Mn^{++}. However, the isolated RNase H domain of Mo-MLV RT, which can cleave an RNA:DNA substrate, was unable to cleave an RNA:RNA substrate (S. Blain and S. Goff, pers. comm.), which makes it similar in this

---

[1]This inherent Mn^{++}-dependent nuclease activity of HIV-1 RT should be distinguished from a double-stranded RNA-dependent "RNase D" originally reported as a new enzymatic activity associated with HIV-1 RT cleaving RNA at two positions within the double-stranded region of the tRNA primer-viral RNA template complex in the presence of Mg^{++} (Ben-Artzi et al. 1992a). Further study of this specific cleavage reaction with purified RNase III and with HIV-1 RT isolated from an RNase III⁻ strain of *E. coli* (Hostomsky et al. 1992b) showed that the observed "RNase D" activity was an artifact resulting from contamination of recombinant RT by *E. coli* RNase III.

respect to *E. coli* RNase HI. This observation suggests that the polymerase domain of Mo-MLV RT could contribute to the altered substrate specificity.

It has been observed that with certain polymerases, it is possible to alter the template or substrate specificity, when Mg^{++} is replaced by Mn^{++} (Lazcano et al. 1988). For example, with Mn^{++}, eubacterial DNA-dependent RNA polymerases can use RNA as a template in the absence of DNA and thus function as RNA replicases. Early studies with *E. coli* DNA polymerase I suggested that its limited inherent RT activity can be increased by using Mn^{++} as the divalent metal ion. The same applies also for RT activity of human DNA polymerases β and γ.

DNA polymerases from the thermophilic eubacteria *Thermus aquaticus* and *Thermus thermophilus* were recently shown to function as structure-specific RNases H that cleave at a single site opposite the 3' end of a complementary DNA in the presence of Mg^{++} (Lyamichev et al. 1993). Interestingly, DNA polymerase from *T. thermophilus* was unable to cleave the 5'-end-protruding RNA strand in the presence of Mn^{++}. The specificity of this enzyme for Mg^{++} may contribute to its ability to function as an efficient thermostable RT in the presence of Mn^{++} (Myers and Gelfand 1991).

In a different set of experiments, various recombinant forms of the HIV-1 RNase H domain were found to be active on an RNA:DNA hybrid substrate only in the presence of Mn^{++}, but not Mg^{++} (Smith and Roth 1993), in contrast with the report of Evans et al. (1991). The activity of these forms of the RNase H domain in the presence of Mn^{++} was dependent on an oligohistidine tag, which may facilitate substrate binding despite the loss of the polymerase domain, which apparently provides most of the important substrate-binding determinants in HIV-1 RT. Surprisingly, purified *E. coli* RNase III, known to be specific for double-stranded RNA, was also capable of cleaving RNA:DNA when Mn^{++} was present.

It remains to be established whether the relaxed or altered specificities observed in vitro in the presence of Mn^{++} have any relevance for the physiological roles of these proteins in vivo. But even if they turn out to be a mere biochemical curiosity, clarification of this issue may contribute to a general understanding of the roles that different metals play in substrate recognition and nuclease action.

VI. MECHANISM OF RNASE H ACTION

A. Endonucleolytic Character of RNase H Cleavage

Whether RNase H activity of RTs is exo- or endonucleolytic is not easily answered. The apparent inability of AMV RT to cleave certain circular

RNA:DNA hybrids, which were substrates of *E. coli* RNase HI, was interpreted to suggest that retroviral RNases H are exonucleases. However, using a relaxed, covalently closed circular plasmid with a 770-nucleotide RNA in one strand, Krug and Berger (1989) conclusively demonstrated that both HIV-1 and AMV RNases H are capable of authentic endonucleolytic cleavage. The same conclusion was reached by Oyama et al. (1989) using a circular RNA template hybridized to an oligodeoxynucleotide primer. Schatz et al. (1990a) have reported that HIV-1 RT displays both endonucleolytic and $3' \rightarrow 5'$ exonucleolytic cleavage. The presence of two activities in HIV-1 RT was subsequently supported in studies with RNA:DNA hybrids containing the polypurine tract (Wöhrl and Moelling 1990; Wöhrl et al. 1991). The notion of an exonucleolytic mechanism was based in part on studies of Starnes and Cheng (1989), who found that mono-, di-, and trinucleotides are the main products of cleavage of poly(dC)·poly(rG) by HIV-1 RT. However, Mizrahi (1989) found that the final cleavage products of HIV-1 RT hydrolysis of a more natural substrate, a hybrid derived from the *gag* region of HIV-1 genome, are fragments ranging from 4 to 15 nucleotides in length. To distinguish a situation in which the progression of RNase H cleavage is tightly coupled to the direction of DNA synthesis, as proposed by Oyama et al. (1989), the term "directional endonuclease" has been used to describe the RNase H activity of retroviral RTs (Hostomsky et al. 1991). Studies of Furfine and Reardon (1991) also suggested that the activities of HIV-1 RT are coordinated, with the distance between polymerase and RNase H catalytic sites corresponding to the length of 15–16 nucleotides of an RNA:DNA duplex. This distance was proposed to be 18–19 nucleotides in a different experiment using a heparin trap (Gopalakrishnan et al. 1992). In an analysis of the cleavage products of an internally labeled RNA:DNA substrate, which allows determination of directionality and processivity in a single enzyme-substrate encounter, the term "partially processive $3' \rightarrow 5'$ endonuclease" was used to describe the degradative activity of HIV-1 RT (DeStefano et al. 1991a).

B. Catalytic Role of Metals

Two alternative mechanisms have been proposed to explain hydrolysis of the RNA strand in RNA:DNA hybrids by RNase H: the carboxylate-hydroxyl relay mechanism (Nakamura et al. 1991; Oda et al. 1993) and the two-metal-ion mechanism (Yang et al. 1990; Davies et al. 1991).

A carboxylate-hydroxyl relay mechanism for RNase-H-mediated cleavage of RNA has been proposed based on certain similarities between the active sites of *E. coli* RNase HI and bovine DNase I (Suck and Oefner 1986; Weston et al. 1992). Kinetic and mutagenesis studies of *E.*

coli RNase HI have identified three conserved carboxylate side chains and a single histidine residue that are important in catalysis. Structural studies indicate that these residues are clustered together along with a fourth carboxylate side chain at the RNase H active site where they bind a single divalent metal ion (Katayanagi et al. 1990, 1992). One of the carboxylates (most likely Asp-70 in *E. coli* RNase HI) acts as a general base by abstracting a proton from an activated H$_2$O molecule. An in-line nucleophilic attack of the scissile phosphate by hydroxide then leads to a pentacoordinate intermediate. The resulting cleavage is accompanied by an inversion of configuration at the phosphorus. Kinetic analysis of mutants of His-124 in *E. coli* RNase HI suggested that this conserved residue enhances the catalytic efficiency of the RNase H cleavage by removing a proton from the catalytically essential Asp-70 (Oda et al. 1993), thus regenerating the general base. Positioning of the conserved histidine on a flexible loop in *E. coli*, *T. thermophilus*, and HIV-1 RNase H structures may support its proposed role as a proton shuttle. The close proximity of four acidic side chains in the RNase H active site suggests that at least one of the carboxylates may be protonated to reduce unfavorable electrostatic repulsions. Such a carboxyl group could then serve as a general acid for proton donation to the O3' leaving group (Nakamura et al. 1991). The role of a single metal ion in this scheme would then be for correct positioning of the phosphate group and for possibly stabilizing the pentacovalent intermediate.

A two-metal-ion mechanism has been proposed based in part on crystal structures of the Klenow fragment of *E. coli* DNA polymerase I complexed with a substrate and product in the catalytic site of the 3'→5' exonuclease domain (Beese and Steitz 1991). The two metals in this binuclear complex have been assigned specific roles. Metal ion A is proposed to facilitate the formation of an attacking hydroxide ion, metal ion B promotes the departure of the leaving 3' oxygen, whereas both ions activate the phosphodiester group. Mutagenesis of the four active site residues that contain carboxylate side chains and the pH dependence of the 3'→5' exonuclease reaction are consistent with a mechanism in which nucleophilic attack on the terminal phosphodiester bond is initiated by a hydroxide ion coordinated to one of the enzyme-bound metal ions (Derbyshire et al. 1991). A similar mechanism seems to operate in *E. coli* alkaline phosphatase (Kim and Wyckoff 1991), and a general two-metal-ion mechanism of phosphoryl transfer has been also proposed for the RNA-catalyzed reactions involved in RNA splicing and RNase P hydrolysis of precursor tRNA (Steitz and Steitz 1993).

The two Mn^{++} ions detected in the HIV-1 RNase H domain are approximately 4 Å apart, close to the distance observed between metal sites

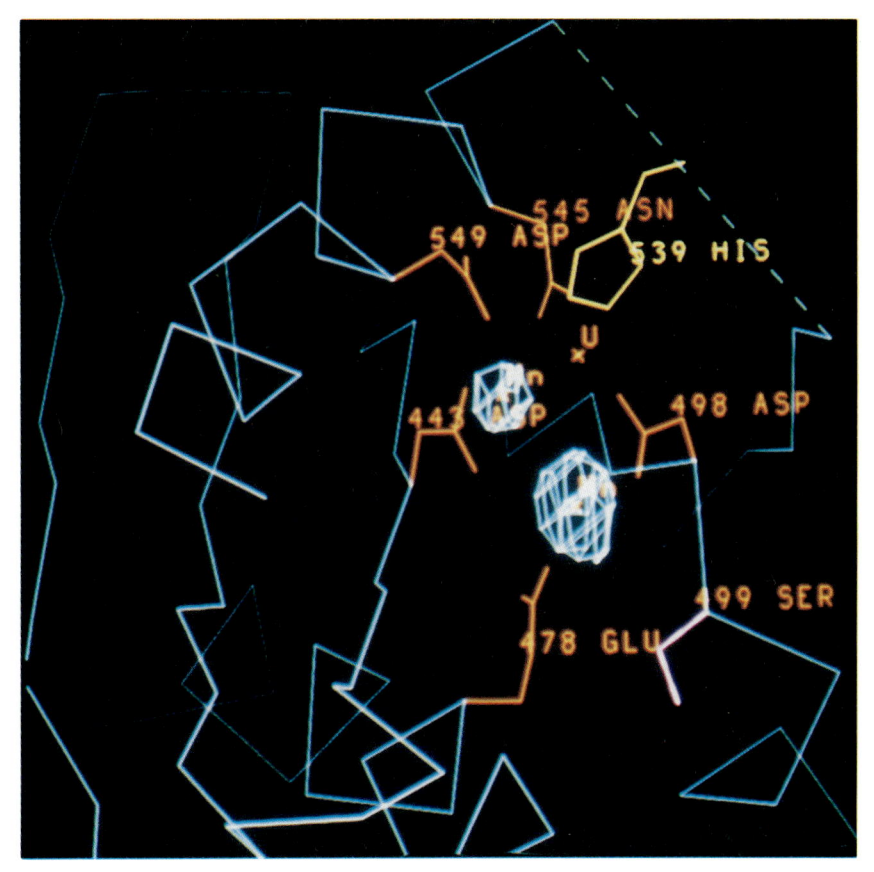

Figure 8 Difference electron density map showing positions of the two Mn^{++} cations with respect to the seven invariant residues that cluster near the RNase H catalytic site in HIV-1 RT. The loop containing His-539 is disordered in the structure of the isolated HIV-1 RNase H domain. The histidine (*yellow*) is positioned by analogy with its location in the *E. coli* RNase HI (Yang et al. 1990). Positions of the other side chains are from refined coordinates of the native structure. UO$_2$F$_5^{3-}$ position (U) is from the heavy atom refinement process. (Reprinted, with permission, from Davies et al. 1991 [copyright AAAS].)

A and B in the polymerase I exonuclease domain. Coordination of the metals is also similar. Two carboxylates interact exclusively with the metal at site A and one with the metal at site B, whereas the fourth carboxylate coordinates both metal ions (Fig. 8). The UO$_2$F$_5^{3-}$ anion, used for isomorphous replacement phasing, is 3 Å from both Mn^{++}-binding sites in the RNase H domain, similar to the distance of the phosphate of dTMP to the metal sites in the 3′→5′ exonuclease complex. Due to its geometrical similarity to the presumed pentacovalent phosphate intermediate, the $UO_2F_5^{3-}$ may be bound at the scissile phosphate-binding site of RNase H.

Despite the fact that 3'→5' exonuclease and RNase H are otherwise structurally unrelated, the striking similarity in the geometrical arrangement of the corresponding metal-carboxylate clusters—apparently a result of convergent evolution—is suggestive of a common mechanism of phosphodiester bond cleavage. The clear homology between bacterial and retroviral RNases H, on the other hand, and the almost identical arrangement of their catalytic sites make it difficult to envisage that these two RNases H would use different mechanisms for the catalysis of basically the same reaction. More research is necessary to resolve this issue.

C. Recognition of RNA in the Heteroduplex

A fascinating unanswered question concerns how RNase H distinguishes RNA:DNA hybrids as a specific substrate from RNA:RNA or DNA:DNA duplexes. RNase H displays only minimal sequence specificity, and it is presumed that it recognizes some general structural features of the heteroduplex.

Double-stranded DNA and RNA usually adopt different conformations. In DNA, the C2'-endo pucker of deoxyribose rings leads to formation of mostly B-type helices. In RNA, the O2' oxygen causes steric hindrance that alters the ribose conformation, resulting in a C3'-endo pucker. O2' oxygen is also involved in several hydrogen-bonding interactions, some of them water-mediated, all of which result in an exclusively A-type conformation for double-stranded RNA. In contrast to the well-characterized more limited conformations of homoduplexes (for a detailed review, see Saenger 1984), the literature on RNA:DNA hybrids suggests more structural heterogeneity for heteroduplexes. Most investigations conclude that both strands of RNA:DNA duplexes are in an A-form RNA-like conformation, although there are reports of B-form heteroduplexes. Zimmerman and Pheiffer (1981) first suggested that RNA:DNA hybrids can have different conformations depending on their degree of solvation. Arnott et al. (1986) concluded from X-ray fiber diffraction studies that synthetic hybrid duplexes have a heteromerous secondary structure: the RNA strands assuming A-like conformation and the DNA strand B-like conformation. Most recently, Salazar et al. (1993) determined, on the basis of J-coupling analysis and two-dimensional nuclear Overhauser effect (NOE) studies, that in solution, the RNA strand of the heteroduplex is in an A-form conformation, whereas the DNA strand assumes an intermediate conformation, which is neither A-form nor B-form. All of these results suggest that in general, the RNA strand has an A-form structure that perturbs the conformation of the DNA strand.

Progress in nucleotide chemistry, reflected in the development of compatible strategies for RNA and DNA syntheses, makes it possible to study structures of chimeric oligonucleotides—strands containing covalently linked RNA and DNA segments. These forms have relevance as models for natural substrates of RNases H, which occur in DNA replication. A stretch of four ribonucleotides in a chimeric DNA-RNA-DNA oligonucleotide hybridized to a complementary DNA is the shortest segment of RNA required for efficient cleavage by *E. coli* RNase HI (Hogrefe et al. 1990). In the recently determined crystal structure of a model Okazaki fragment, composed of a chimeric RNA-DNA oligonucleotide combined with the complementary DNA, a stretch of three ribonucleotides seems to have locked the complete double-stranded decamer into a mostly A-type conformation, without an A- to B-form transition at the RNA-DNA junction (Egli et al. 1992). It appears that a relatively short stretch of RNA can change the conformation of a duplex so that it can be recognized as substrate by RNase H.

Structural features of RNA:DNA duplexes have been investigated by immunological approaches using anti-hybrid antibodies (Sanford et al. 1988). These immunochemical studies suggest that the hybrid sections of the 36-base-pair duplex $rA_{12}d(GC)_6d(T)_{12}$, although having "some features of the A-helix," adopt a unique conformation that is distinct from both B-DNA and A-RNA. The ability of an antibody to recognize RNA:DNA duplexes specifically may be analogous to the mode of substrate recognition by single-domain RNases H. Site-directed mutagenesis of *E. coli* RNase HI and substrate titration studies using heteronuclear two-dimensional NMR have led to a model describing enzyme-substrate interaction in considerable detail (Nakamura et al. 1991). In this model, an RNA:DNA hybrid binds to a groove-like depression, forming hydrogen bonds between polar groups of the enzyme and 2′-hydroxyl groups of the RNA strand. Clearly, a crystal structure of an RNase H complexed with the RNA:DNA substrate would be highly informative in attempting to define further both the substrate conformation and the specific interaction important for binding and catalysis.

VII. INHIBITORS OF RETROVIRAL RNASE H

Since a functional RNase H is essential for retrovirus replication, inhibitors of this activity could, in principle, be developed into specific antiretroviral agents. Most published data concerning inhibition of

RNase H relate to inhibition of the HIV-1 enzyme. Several methods of inhibiting HIV RT-associated RNase H can be envisaged: (1) interfering with binding of the RNA:DNA substrate, (2) blocking the catalytic site, and (3) interfering with essential protein-protein interactions within the RT heterodimer.

As mentioned earlier, the DNA polymerase and RNase H domains of HIV-1 RT share a common binding site for template-primer substrate. Agents that interfere with binding of the nucleic acid substrate would be expected to inhibit both activities. This has been confirmed with a high-affinity synthetic RNA-pseudoknot, selected from a combinatorial library of ribonucleotide sequences, which was shown to be an inhibitor of the polymerase activity of HIV-1 RT at nanomolar concentrations (Tuerk et al. 1992). This tight binding ligand is also an equally potent inhibitor of RNase H at comparable concentrations (Z. Hostomsky and Z. Hostomska, unpubl.). A similar mode of action may underlie the reported inhibitory activity of polymeric molecules such as dextran sulfate and polysulfated anions (Moelling et al. 1989) which probably compete with the binding of nucleic acid substrates. Inhibition of the specific RNase H cleavage in the polypurine tract (PPT) region of the HIV-1 genome by an antisense oligonucleotide potentially involved in triple-helix formation (Volkmann et al. 1993) probably occurs by a similar mechanism. Ribonucleoside-vanadyl complexes also inhibit the RT-associated polymerase and RNase H activities competitively with respect to template-primer, but with K_i values ranging from 1 to 3 mM (Krug and Berger 1989).

A few small-molecule inhibitors have been reported to inhibit RNase H activity with little or no effect on the associated polymerase activity of RT. Captan, a known inhibitor of RNA and DNA polymerases, was reported to inhibit RNase H activity of AMV RT preferentially (Freeman-Wittig et al. 1986). Interestingly, whereas AMP did not inhibit RNase H activity at concentrations up to 10 mM, dAMP did selectively inhibit RNase H (but not polymerase) activity of RT, with an IC_{50} of 3 mM. AZTMP and other nucleoside 5′monophosphates inhibit the RNase H activity of a recombinant HIV RT. The concentration of inhibitor required to achieve 50% inhibition of virus replication in cultured cells was 0.03 mM in the presence of Mn^{++} (Tan et al. 1991). Although this concentration of AZTMP is much higher than the level of AZTTP required to inhibit HIV-1 RT, these levels might be reached following therapeutic doses of AZT. Although several potent RNase H inhibitors, such as vanadyl sulfate or polyguanylic acid, are known, cellular toxicity is a significant issue with many of these agents (Mitsuya et al. 1990). HP 0.35, a cephalosporin degradation product, was reported to be a specific inhibitor of lentiviral RNases H (Hafkemeyer et al. 1991b).

The inhibitory effect of the sulfhydryl-specific reagent N-ethylmaleimide (NEM) is perhaps surprising since HIV-1 RNase H contains no cysteine groups. Data from Hizi et al. (1991) and Buiser et al. (1991) show that although RNase H activity of HIV-1 RT was inactivated by NEM (50% inhibition at 0.6 mM), DNA polymerase function was unaffected by this chemical modification. Illimaquinone, a natural marine product, was reported to inhibit the RNase H activity of HIV-1 RT selectively (Loya et al. 1990). However, the inhibitory effect of illimaquinone on RNase H activity, similar to the effect of NEM, seems to be mediated by interaction with Cys-280 which is located in the polymerase domain of HIV-1 RT (Loya and Hizi 1993).

The availability of three-dimensional structural information on a target protein-ligand complex provides an opportunity for structure-based inhibitor design, in which "lead compounds" are improved by iterative cycles of design, chemical synthesis, and co-crystallization with the new generation of inhibitors (Appelt et al. 1991). With a single exception, no such complexes have been reported for any RNase H. In their X-ray structural study of the RNase H domain of HIV-1 RT, Davies et al. (1991) found that the complex anion $UO_2F_5^{3-}$ binds at a single site near the cluster of active site carboxylates about 3 Å from each of the two metal ions (Fig. 8). Uranyl pentafluoride inhibits cleavage of RNA:DNA heteroduplex in vitro. Unfortunately, these compounds or close structural analogs are unlikely to be useful therapeutically because they are inherently nonspecific since their mechanism of action is dependent on structural features present in many other enzymes.

Although the idea of disrupting dimerization of HIV-1 RT as an approach to interfere with polymerase and RNase H activities has been proposed (Restle et al. 1990), and dependence of RNase H activity on specific interactions with the polymerase domain in the RT heterodimer has been demonstrated (Hostomsky et al. 1991), no specific inhibitors acting by this mechanism have been reported.

VIII. ANTISENSE EFFECT

Antisense oligonucleotides can be used to block the translation or processing of specific mRNAs. The antisense inhibition of gene expression may occur via translational arrest, in which the formation of the mRNA:DNA duplex presents a direct steric block to translational machinery. The inhibitory effect of antisense oligodeoxynucleotides can be markedly enhanced when their reversible hybridization to the target mRNA is accompanied by subsequent RNase H cleavage, an essentially irreversible process (Walder and Walder 1988).

In addition to relying on the endogenous RNase H, antisense inhibition can also take advantage of the virus-encoded activity. It has been demonstrated that cDNA synthesis by RTs from AMV and Mo-MLV can be prevented by oligonucleotides hybridized to the RNA template downstream from the 3' end of the primer. RT-associated RNase H activity was shown to be responsible for this block of reverse transcription, with cleavage of the template at the antisense-binding site (Boiziau et al. 1992).

Involvement of RNase H in cleavage of mRNA in the presence of antisense oligonucleotides has been amply demonstrated in vitro. The only in vivo evidence is based on microinjection of oligonucleotides into *Xenopus* oocytes (for review, see Neckers et al. 1992). However, there is no clear evidence implicating RNase H in a mechanism of hybrid-arrested translation in mammalian cells. In one recent experiment addressing this issue, a leukemia cell line was constructed that expressed episomally encoded *E. coli* RNase HI in cytoplasm at a level tenfold higher than that of endogenous RNase H. Unexpectedly, neither increased sensitivity to the antisense oligonucleotide nor a specific cleavage of the Myc mRNA was detected (Rosolen et al. 1993). Thus, these results so far cannot support a role for RNase H in mediating the observed antisense-induced reduction of the Myc protein in that particular cell line. More experiments are needed to address the in vivo accessibility of the newly formed RNA:DNA duplexes to various intracellular RNases H.

IX. CONCLUSIONS AND PERSPECTIVES

A marriage of the tools of molecular biology and three-dimensional structural analysis has contributed to a wealth of new information and insights concerning RNase H in cellular and retroviral systems on a scale that could have been scarcely imagined a decade ago at the time of the first comprehensive review of the subject (Crouch and Dirksen 1982). High-resolution X-ray and NMR structures for bacterial and retroviral RNases H now allow us to replace vague conjecture with refined and testable hypotheses about the relationship between structure and function in these fascinating enzymes. As new discoveries continue to extend our knowledge, a number of fundamental questions remain unanswered. We need to better characterize and define the role of RNases H as part of the cellular machinery in eukaryotes. An explanation of the problem of RNA:DNA heteroduplex recognition by RNase H at the atomic level is eagerly awaited but will require high-resolution analysis of suitable co-crystals by X-ray diffraction or detailed multidimensional NMR studies of a relevant complex in solution. Additional insights are needed concerning the role of metals and the active site carboxylate cluster to resolve conflicting hypotheses pertaining to the enzyme's putative mechanism of action. On the more practical side, a major challenge in the years ahead is to begin using our rapidly increasing knowledge of these enzymes to discover or design small-molecule RNase H inhibitors that can discriminate between human and retroviral enzymes, thus possibly contributing to the urgent need for safe antiretroviral drugs that can control diseases such as AIDS.

Note Added in Proof

A detailed solution of a hybrid RNA:DNA duplex has been published recently (Fedoroff et al., Structure of a DNA:RNA hybrid duplex: Why RNase H does not cleave pure RNA. *J. Mol. Biol. 233:* 509 [1993]). The DNA strand sugars have a novel O4'-*endo* conformation leading to a model in which *E. coli* RNase H has important interactions with both strands of the heteroduplex.

ACKNOWLEDGMENTS

This work was supported in part by National Institutes of Health grants GM-39599 (to D.A.M.) and AI-33380 (to Z. Hy).

REFERENCES

Appelt, K., R.J. Bacquet, C.A. Bartlett, C.L.J. Booth, S.T. Freer, M.A.M. Fuhry, M.R. Gehring, S.M. Herrmann, E.F. Howland, C.A. Janson, T.R. Jones, C.-C. Kan, V. Kathardekar, K.K. Lewis, G.P. Marzoni, D.A. Matthews, C. Mohr, E.W. Moomaw, C.A. Morse, S.J. Oatley, R.C. Ogden, M.R. Reddy, S.H. Reich, W.A. Schoettlin, W.W. Smith, M.D. Varney, J.E. Villafranca, R.W. Ward, S. Webber, S.E. Webber, K.M. Welsh, and J. White. 1991. Design of enzyme inhibitors using iterative protein crystallographic analysis. *J. Med. Chem.* 34: 1925-1934.

Arnold, E., A. Jacobo-Molina, R.G. Nanni, R.L.Williams, X. Lu, J. Ding, A.D. Clark, Jr., A. Zhang, A.L. Ferris, P. Clark, A. Hizi, and S.H. Hughes. 1992. Structure of HIV-1 reverse transcriptase/DNA complex at 7Å resolution showing active site locations. *Nature* 357: 85-89.

Arnott, S., R.Chandrasekaran, R.P. Millane, and H.S. Park. 1986. DNA-RNA hybrid secondary structures. *J. Mol. Biol.* 188: 631-640.

Becerra, S.P., G.M. Clore, A.M. Gronenborn, A.R. Karlstrom, S.J. Stahl, S.H. Wilson, and P.T. Wingfield. 1990. Purification and characterization of the RNase H domain of HIV-1 reverse transcriptase expressed in recombinant *Escherichia coli. FEBS Lett.* 270: 76-80.

Beese, L.S. and T.A. Steitz. 1991. Structural basis for the 3'-5' exonuclease activity of *Escherichia coli* DNA polymerase I: A two metal ion mechanism. *EMBO J.* 10: 25-33.

Ben-Artzi, H., E. Zeelon, M. Gorecki, and A. Panet. 1992a. Double-stranded RNA-dependent RNase activity associated with human immunodeficiency virus type 1 reverse transcriptase. *Proc. Natl. Acad. Sci.* **89:** 927-931.

Ben-Artzi, H., E. Zeelon, S.F.J. Le-Grice, M. Gorecki, and A. Panet. 1992b. Characterization of the double stranded RNA dependent RNase activity associated with recombinant reverse transcriptases. *Nucleic Acids Res.* **20:** 5115-5118.

Blackburn, E.H. 1993. Telomerases. In *Reverse transcriptase* (ed. A.M. Skalka and S.P. Goff), pp. 411-424. Cold Spring Harbor Laboratory Press, Cold Spring Harbor, New York.

Boiziau, C., N.T. Thuong, and J.J. Toulme. 1992. Mechanism of the inhibition of reverse transcription by antisense oligonucleotides. *Proc. Natl. Acad. Sci.* **89:** 768-772.

Buiser, R.G., J.J. De Stefano, L.M. Mallaber, P.J. Fay, and R.A. Bambara. 1991. Requirements for the catalysis of strand transfer synthesis by retroviral DNA polymerases. *J. Biol. Chem.* **266:** 13103-13109.

Büsen, W. 1980. Purification, subunit structure and serological analysis of calf thymus ribonuclease H I. *J. Biol. Chem.* **255:** 9434-9443.

Cerritelli, S.M., D.Y. Shin, H.C. Chen, M. Gonzales, and R.J. Crouch. 1993. Proteolysis of *Saccharomyces cerevisiae* RNase H1 in *E. coli*. *Biochimie* **75:** 107-111.

Champoux, J.J. 1993. Roles of RNase H in reverse transcription. In *Reverse transcriptase* (ed. A.M. Skalka and S.P. Goff), pp. 103-117. Cold Spring Harbor Laboratory Press, Cold Spring Harbor, New York.

Chattopadhyay, D., D.B. Evans, M.R. Deibel, Jr., A.F. Vosters, F.M. Eckenrode, H.M. Einspahr, J.O. Hui, A.G. Tomaselli, H.A. Zurcher-Neely, R.L. Heinrikson, and S.K. Sharma. 1992. Purification and characterization of heterodimeric human immunodeficiency virus type 1 (HIV-1) reverse transcriptase produced by *in vitro* processing of p66 with recombinant HIV-1 protease. *J. Biol. Chem.* **267:** 14227-14232.

Coffin, J. 1993. Reverse transcriptase and evolution. In *Reverse transcriptase* (ed. A.M. Skalka and S.P. Goff), pp. 445-479. Cold Spring Harbor Laboratory Press, Cold Spring Harbor, New York.

Crouch, R.J. 1990. Ribonuclease H: From discovery to 3D structure. *New Biol.* **2:** 771-777.

Crouch, R.J. and M.L. Dirksen. 1982. Ribonucleases H. In *Nucleases* (ed. S.M. Linn and R.J. Roberts), pp. 211-241. Cold Spring Harbor Laboratory, Cold Spring Harbor, New York.

Davies, J.F., Z. Hostomska, Z. Hostomsky, S. Jordan, and D.A. Mathews. 1991. Crystal structure of the RNase H domain of HIV-1 reverse transcriptase. *Science* **252:** 88-95.

Derbyshire, V., N.D.F. Grindley, and C.M. Joyce. 1991. The 3'-5' exonuclease of DNA polymerase I of *Escherichia coli*: Contribution of each amino acid at the active site to the reaction. *EMBO J.* **10:** 17-24.

DeStefano, J.J., R.G. Buiser, L.M. Mallaber, R.A. Bambara, and P.J. Fay. 1991a. Human immunodeficiency virus reverse transcriptase displays a partially processive 3' to 5' endonuclease activity. *J. Biol. Chem.* **266:** 24295-24301.

DeStefano, J.J., R.G. Buiser, L.M. Mallaber, T.W. Myers, R.A. Bambara, and P.J. Fay. 1991b. Polymerization and RNase H activities of the reverse transcriptases from avian myeloblastosis, human immunodeficiency, and Moloney murine leukemia viruses are functionally uncoupled. *J. Biol. Chem.* **266:** 7423-7431.

Deutscher, M.P. 1993. Ribonuclease multiplicity, diversity, and complexity. *J. Biol. Chem.* **268:** 13011-13014.

DiFrancesco, R.A. and I.R. Lehman. 1985. Interaction of ribonuclease H from *Drosophila melanogaster* embryos with DNA polymerase-primase. *J. Biol. Chem.* **260:** 14764-14770.

Doolittle, R.F., D.-F. Feng, M.S. Johnson, and M.A. McClure. 1989. Origins and evolutionary relationships of retroviruses. *Q. Rev. Biol.* **64:** 1–30.

Eder, P.S. and J.A. Walder. 1991. Ribonuclease H from K562 human erythroleukemia cells. *J. Biol. Chem.* **266:** 6472–6479.

Eder, P.S., R.Y. Walder, and J.A. Walder. 1993. Substrate specificity of human RNase1 and its role in excision repair of ribose residues misincorporated in DNA. *Biochimie* **75:** 123–126.

Egli, M., N. Usman, S. Zhang, and A. Rich. 1992. Crystal structure of an Okazaki fragment at 2-Å resolution. *Proc. Natl. Acad. Sci.* **89:** 534–538.

Evans, D.B., K. Brawn, M.R. Deibel, Jr., W.G. Tarpley, and S.K. Sharma. 1991. A recombinant ribonuclease H domain of HIV-1 reverse transcriptase that is enzymatically active. *J. Biol. Chem.* **266:** 20583–20585.

Ferris, A.L., A. Hizi, S.D. Showalter, S. Pichuantes, L. Babe, C.S. Craik, and S.H. Hughes. 1990. Immunologic and proteolytic analysis of HIV-1 reverse transcriptase structure. *Virology* **175:** 456–464.

Freeman-Wittig, M.J., M. Vinocour, and R.A. Lewis. 1986. Differential effects of captan on DNA polymerase and ribonuclease activities of avian myeloblastosis virus reverse transcriptase. *Biochemistry* **25:** 3050–3055.

Furfine, E.S. and J.E. Reardon. 1991. Reverse transcriptase RNase H from the human immunodeficiency virus. Relationship of the DNA polymerase and RNA hydrolysis activities. *J. Biol. Chem.* **266:** 406–412.

Gopalakrishnan, V., J.A. Peliska, and S.J. Benkovic. 1992. Human immunodeficiency virus type 1 reverse transcriptase: Spatial and temporal relationship between the polymerase and RNase H activities. *Proc. Natl. Acad. Sci.* **89:** 10763–10767.

Grandgenett, D., T. Quinn, P.J. Hippenmeyer, and S. Oroszlan. 1985. Structural characterization of the avian retrovirus reverse transcriptase and endonuclease domains. *J. Biol. Chem.* **260:** 8243–8249.

Hafkemeyer, P., E. Ferrari, J. Brecher, and U. Hübscher. 1991a. The p15 carboxyl-terminal proteolysis product of the human immunodeficiency virus type 1 reverse transcriptase p66 has DNA polymerase activity. *Proc. Natl. Acad. Sci.* **88:** 5262–5266.

Hafkemeyer, P., K. Neftel, R. Hobi, A. Pfaltz, H. Lutz, K. Luthi, F. Focher, S. Spadari, and U. Hübscher. 1991b. HP 0.35, a cephalosporin degradation product is a specific inhibitor of lentiviral RNases H. *Nucleic Acids Res.* **19:** 4059–4065.

Hansen, J., T. Schulze, W. Mellert, and K. Moelling. 1988. Identification and characterization of HIV-1 specific RNase H by monoclonal antibody. *EMBO. J.* **7:** 239–243.

Hausen, P. and H. Stein. 1970. Ribonuclease H: An enzyme degrading the RNA moiety of DNA-RNA hybrids. *Eur. J. Biochem.* **14:** 279–283.

Hizi, A., R. Tal, M. Shaharabany, and S. Loya. 1991. Catalytic properties of the reverse transcriptase of human immunodeficiency virus type 1 and type 2. *J. Biol. Chem.* **266:** 6230–6239.

Hogrefe, H.H., R.I. Hogrefe, R.Y. Walder, and J.A. Walder. 1990. Kinetic analysis of *Escherichia coli* RNase H using DNA-RNA-DNA substrates. *J. Biol. Chem.* **265:** 5561–5566.

Hostomska, Z., D.A. Mathews, J.F. Davies, B.R. Nodes, and Z. Hostomsky. 1991. Proteolytic release and crystallization of RNase H domain of HIV-1 reverse transcriptase. *J. Biol. Chem.* **266:** 14697–14702.

Hostomsky, Z., Z. Hostomska, T.-B. Fu, and J. Taylor. 1992a. Reverse transcriptase of human immunodeficiency virus type 1: Functionality of subunits of the heterodimer in DNA synthesis. *J. Virol.* **66:** 3179–3182.

Hostomsky, Z., G.O. Hudson, S. Rahmati, and Z. Hostomska. 1992b. RNase D, a reported new activity associated with HIV-1 reverse transcriptase, displays the same cleavage specificity as *Escherichia coli* RNase III. *Nucleic Acids Res.* **20:** 5819-5824.

Hostomsky, Z., Z. Hostomska, G.O. Hudson, E.W. Moomaw, and B.R. Nodes. 1991. Reconstitution *in vitro* of RNase H activity by using purified N-terminal and C-terminal domains of HIV-1 reverse transcriptase. *Proc. Natl. Acad. Sci.* **88:** 1148-1152.

Inouye, S. and M. Inouye. 1993. Bacterial reverse transcriptase. In *Reverse transcriptase* (ed. A.M. Skalka and S.P. Goff), pp. 391-410. Cold Spring Harbor Laboratory Press, Cold Spring Harbor, New York.

Ishikawa, K., S. Kimura, S. Kanaya, K. Morikawa, and H. Nakamura. 1993a. Structural study of mutants of *Escherichia coli* ribonuclease HI with enhanced thermostability. *Protein Eng.* **6:** 85-91.

Ishikawa, K., M. Okumura, K. Katayanagi, S. Kimura, S. Kanaya, H. Nakamura, and K. Morikawa. 1993b. Crystal structure of ribonuclease H from *Thermus thermophilus* HB8 refined at 2.8 Å resolution. *J. Mol. Biol.* **230:** 529-542.

Itaya, M. 1990. Isolation and characterization of a second RNase H (RNase HII) of *Escherichia coli* K-12 encoded by the *rnhB* gene. *Proc. Natl. Acad. Sci.* **87:** 8587-8591.

Itaya, M. and R.J. Crouch. 1991. A combination of RNase H (*rnh*) and *rec*BCD or *sbc*B mutations in *Escherichia coli* K12 adversely affects growth. *Mol. Gen. Genet.* **227:** 424-432.

Itaya, M., D. McKelvin, S.K. Chatterjie, and R.J. Crouch. 1991. Selective cloning of genes encoding RNase H from *Salmonella typhimurium, Saccharomyces cerevisiae* and *Escherichia coli rnh* mutant. *Mol. Gen. Genet.* **227:** 438-445.

Itoh, T. and J. Tomizawa. 1980. Formation of an RNA primer for initiation of replication of *Col*E1 DNA by ribonuclease H. *Proc. Natl. Acad. Sci.* **77:** 2450-2454.

Johnson, M.S., M.A. McClure, D.F. Feng, J. Gray, and R.F. Doolittle. 1986. Computer analysis of retroviral *pol* genes: Assignment of enzymatic functions to specific sequences and homologies with nonviral enzymes. *Proc. Natl. Acad. Sci.* **83:** 7648-7652.

Jacobo-Molina, A., J. Ding, R.G. Nanni, A.D. Clark, Jr., X. Lu, C. Tantillo, R.L. Williams, G. Kamer, A.L. Ferris, P. Clark, A. Hizi, S.H. Hughes, and E. Arnold. 1993. Crystal structure of human immunodeficiency virus type 1 reverse transcriptase complexed with double-stranded DNA at 3.0 Å resolution shows bent DNA. *Proc. Natl. Acad. Sci.* **90:** 6320-6324.

Kanaya, S. and R.J. Crouch. 1983. DNA sequence of the gene coding for *Escherichia coli* ribonuclease H. *J. Biol. Chem.* **258:** 1276-1281.

Kanaya, S., C. Katsuda-Nakai, and M. Ikehara. 1991. Importance of the positive charge cluster in *Escherichia coli* ribonuclease HI for the effective binding of the substrate. *J. Biol. Chem.* **266:** 11621-11627.

Kanaya, S., A. Kohara, Y. Miura, A. Sekiguchi, S. Iwai, H. Inoue, E. Otsuka, and M. Ikehara. 1990. Identification of amino acid residues involved in an active site of *Escherichia coli* ribonuclease H by site-directed mutagenesis. *J. Biol. Chem.* **265:** 4615-4621.

Karwan, R. and W. Wintersberger. 1988. In addition to RNase H(70) two other proteins of *Saccharomyces cerevisiae* exhibit ribonuclease H activity. *J. Biol. Chem.* **263:** 14970-14977.

Katayanagi, K., M. Miyagawa, M. Matsushima, M. Ishikawa, S. Kanaya, M. Ikehara, T. Matsuzaki, and K. Morikawa. 1990. Three-dimensional structure of ribonuclease H from *E. coli*. *Nature* **347:** 306-309.

Katayanagi, K., M. Miyagawa, M. Matsushima, M. Ishikawa, S. Kanaya, H. Nakamura, M. Ikehara, T. Matsuzaki, and K. Morikawa. 1992. Structural details of ribonuclease H from *Escherichia coli* as refined to an atomic resolution. *J. Mol. Biol.* **223:** 1029-1052.

Kim, E.E. and H.W. Wyckoff. 1991. Reaction mechanism of alkaline phosphatase based on crystal structures. *J. Mol. Biol.* **218:** 449–464.

Kogoma, T. 1986. RNase H-defective mutants of *Escherichia coli. J. Bacteriol.* **166:** 361–363.

Kogoma, T., X. Hong, G.W. Cadwell, K.G. Barnard, and T. Asai. 1993. Requirement of homologous recombination functions for viability of the *Escherichia coli* cell that lacks RNase HI and exonuclease V activities. *Biochimie* **75:** 89–99.

Kohlstaedt, L.A., J. Wang, J.M. Freidman, P.A. Rice, and T.A. Steitz. 1992. Crystal structure at 3.5 Å resolution of HIV-1 reverse transcriptase complexed with an inhibitor. *Science* **256:** 1783–1790.

Krug, M.S. and S.L. Berger. 1989. Ribonuclease H activities associated with viral reverse transcriptases are endonucleases. *Proc. Natl. Acad. Sci.* **86:** 3539–3543.

Lai, M.-H.T. and I.M. Verma. 1978. Reverse transcriptase of RNA tumor viruses. V. *In vitro* proteolysis of reverse transcriptase from avian myeloblastosis virus and isolation of a polypeptide manifesting only RNase H activity. *J. Virol.* **25:** 652–653.

Lazcano, A., J. Fastag, P. Gariglio, C. Ramirez, and J. Oro. 1988. On the early evolution of RNA polymerase. *J. Mol. Evol.* **27:** 365–376.

Le Grice, S.F.J., J. Mills, R. Ette, and J. Mous. 1989. Comparison of the human immunodeficiency virus type 1 and 2 proteases by hybrid gene construction and *trans*-complementation. *J. Biol. Chem.* **264:** 14902–14908.

Le Grice, S.F.J., T. Naas, B. Wohlgensinger, and O. Schatz. 1991. Subunit-selective mutagenesis indicates minimal polymerase activity in heterodimer-associated p51 HIV-1 reverse transcriptase. *EMBO J.* **10:** 3905–3911.

Levin, J.G., R.J. Crouch, K. Post, S.C. Hu, D. McKelvin, M. Zweig, D.L. Court, and B.I. Gerwin. 1988. Functional organization of the murine leukemia virus reverse transcriptase: Characterization of a bacterially expressed AKR DNA polymerase deficient in RNase H activity. *J. Virol.* **62:** 4376–4380.

Lowe, D.M., A. Aitken, C. Bradley, G.K. Darby, B.A. Larder, K.L. Powell, D.J.M. Purifoy, M. Tisdale, and D.K. Stammers. 1988. HIV-1 reverse transcriptase: Crystallization and analysis of domain structure by limited proteolysis. *Biochemistry* **27:** 8884–8889.

Loya, S. and A. Hizi. 1993. The interaction of illimaquinone, a selective inhibitor of the RNase H activity, with the reverse transcriptases of human immunodeficiency and murine leukemia retroviruses. *J. Biol. Chem.* **268:** 9323–9328.

Loya, S., R. Tal, Y. Kashman, and A. Hizi. 1990. Illimaquinone, a selective inhibitor of the RNase H activity of human immunodeficiency virus type 1 reverse transcriptase. *Antimicrob. Agents Chemother.* **34:** 2009–2012.

Lyamichev, V., M.A.D. Brow, and J.E. Dahlberg. 1993. Structure-specific endonucleolytic cleavage of nucleic acids by eubacterial DNA polymerases. *Science* **260:** 778–783.

McClure, M.A. 1993. Evolutionary history of reverse transcriptase. In *Reverse transcriptase* (ed. A.M. Skalka and S.P. Goff), pp. 425–444. Cold Spring Harbor Laboratory Press, Cold Spring Harbor, New York.

Mitsuya, H., R. Yarchoan, and S. Broder. 1990. Molecular targets for AIDS therapy. *Science* **249:** 1533–1542.

Mizrahi, V. 1989. Analysis of the ribonuclease H activity of HIV-1 reverse transcriptase using RNA.DNA hybrid substrates derived from the *gag* region of HIV-1. *Biochemistry* **28:** 9088–9094.

Mizrahi, V., M.T. Usdin, A. Harington, and L.R. Dudding. 1990. Site-directed mutagenesis of the conserved Asp-443 and Asp-498 carboxy-terminal residues of HIV-

1 reverse transcriptase. *Nucleic Acids Res.* **18:** 5359-5363.
Mizrahi, V., G.M. Lazarus, L.M. Miles, C.A. Meyers, and C. Debouck. 1989. Recombinant HIV-1 reverse transcriptase: Purification, primary structure, and polymerase/ribonuclease H activities. *Arch. Biochem. Biophys.* **273:** 347-358.
Moelling, K., T. Schulze, and H. Diringer. 1989. Inhibition of human immunodeficiency virus type 1 RNase H by sulphated polyanions. *J. Virol.* **63:** 5489-5491.
Myers, T.W. and D.H. Gelfand. 1991. Reverse transcription and DNA amplification by a *Thermus thermophilus* DNA polymerase. *Biochemistry* **30:** 7661-7666.
Nakamura, H., Y. Oda, S. Iwai, H. Inoue, E. Ohtsuka, S. Kanaya, S. Kimura, C. Katsuda, K. Katayanagi, K. Morikawa, H. Miyashiro, and M. Ikehara. 1991. How does RNase H recognize a DNA-RNA hybrid? *Proc. Natl. Acad. Sci.* **88:** 11535-11539.
Neckers, L.M., L. Whitesell, A. Rosolen, and D.A. Geselowitz. 1992. Antisense inhibition of oncogene expression. *Crit. Rev. Oncog.* **3:** 175-231.
Oda, Y., M. Yoshida, and S. Kanaya. 1993. Role of histidine 124 in the catalytic function of ribonuclease H1 from *Escherichia coli. J. Biol. Chem.* **268:** 88-92.
Ogawa, T. and T. Okazaki. 1984. Function of RNase H in DNA replication revealed by RNase H defective mutants of *Escherichia coli. Mol. Gen. Genet.* **193:** 231-237.
Oyama, F., R. Kikuchi, R.J. Crouch, and T. Uchida. 1989. Intrinsic properties of reverse transcriptase in reverse transcription. Associated RNase H is essentially regarded as an endonuclease. *J. Biol. Chem.* **264:** 18808-18817.
Powers, R., G.M. Clore, S.J. Stahl, P.T. Wingfield, and A. Gronenborn. 1992. Analysis of the backbone dynamics of ribonuclease H domain of the human immunodeficiency virus reverse transcriptase using ^{15}N relaxation measurements. *Biochemistry* **31:** 9150-9157.
Radziwill, G., W. Tucker, and H. Schaller. 1990. Mutational analysis of the hepatitis B virus P gene product: Domain structure and RNase H activity. *J. Virol.* **64:** 613-620.
Repaske, R., J.W. Hartley, M.F. Kavlick, R.R. O'Neill, and J.B. Austin. 1989. Inhibitions of RNase H activity and viral replication by single mutations in the 3' region of Moloney murine leukemia virus reverse transcriptase. *J. Virol.* **63:** 1460-1464.
Restle, T., B. Müller, and R.S. Goody. 1990. Dimerization of human immunodeficiency virus type 1 reverse transcriptase. *J. Biol. Chem.* **265:** 8986-8988.
―――. 1992. RNase H activity of HIV reverse transcriptase is confined exclusively to the dimeric forms. *FEBS Lett.* **300:** 97-100.
Rosolen, A., E. Kyle, C. Chavany, R. Bergan, E.T. Kalman, R. Crouch, and L. Neckers. 1993. Effect of over-expression of bacterial ribonuclease H on the utility of antisense MYC oligonucleotides in the monocytic leukemia cell line U937. *Biochemie* **75:** 79-87.
Saenger, W. 1984. *Principles of nucleic acid structure.* Springer Verlag, New York.
Salazar, M., O.Y. Fedoroff, J.M. Miller, N.S. Ribeiro, and B.R. Reid. 1993. The DNA strand in DNA-RNA hybrid duplexes is neither B-form nor A-form in solution. *Biochemistry* **32:** 4207-4215.
Sanford, D. G., K.J. Kotkow, and B.D. Stollar. 1988. Immunochemical detection of multiple conformations within a 36 base pair oligonucleotide. *Nucleic Acids Res.* **16:** 10643-10655.
Sarngadharan, M.G., J.P. Leis, and R.C. Gallo. 1975. Isolation and characterization of a ribonuclease from human leukemic blood cells specific for ribonucleic acid of ribonucleic acid-deoxyribonucleic acid hybrid molecules. *J. Biol. Chem.* **250:** 365-373.
Schatz, O., J. Mous, and S.F.J. Le Grice. 1990a. HIV-1 RT-associated ribonuclease H displays both endonuclease and 3'-5' exonuclease activity. *EMBO J.* **9:** 1171-1176.
Schatz, O., F. Cromme, T. Naas, D. Lindemann, J. Mous, and S.F.J. Le Grice. 1989.

Point mutations within the C-terminal domain of HIV1 reverse transcriptase specifically repress RNaseH function. *FEBS Lett.* **257:** 311-314.

———. 1990b. Inactivation of the RNase H domain of HIV-1 reverse transcriptase block viral infectivity. In *Gene regulation and AIDS* (ed. T. Papas), pp. 293-303. Portfolio Publishing, Houston, Texas.

Schulze, T., M. Nawrath, and K. Moelling. 1991. Cleavage of the HIV-1 p66 reverse transcriptase/RNase H by the p9 protease *in vitro* generates active p15 RNase H. *Arch. Virol.* **118:** 179-188.

Skalka, A.M. and S.P. Goff, eds. 1993. *Reverse transcriptase.* Cold Spring Harbor Laboratory Press, Cold Spring Harbor, New York.

Smith, J.S. and M.J. Roth. 1993. Purification and characterization of an active human immunodeficiency virus type 1 RNase H domain. *J. Virol.* **67:** 4037-4049.

Stammers, D.K., M. Tisdale, S. Court, V. Parmar, C. Bradley, and C.K. Ross. 1991. Rapid purification and characterisation of HIV-1 reverse transcriptase and RNase H engineered to incorporate a C-terminal tripeptide α-tubulin epitope. *FEBS Lett.* **283:** 298-302.

Starnes, M.C. and Y.C. Cheng. 1989. Human immunodeficiency virus reverse transcriptase-associated RNase H activity. *J. Biol. Chem.* **264:** 7073-7077.

Stein, H. and P. Hausen. 1969. Enzyme from calf thymus degrading the RNA moiety of DNA-RNA hybrids: Effect on DNA-dependent RNA polymerase. *Science* **166:** 393-395.

Steitz, T.A. and J.A. Steitz. 1993. A general two-metal-ion mechanism for catalytic RNA. *Proc. Natl. Acad. Sci.* **90:** 6498-6502.

Suck, D. and C. Oefner. 1986. Structure of DNase I at 2.0 Å resolution suggests a mechanism for binding and cutting DNA. *Nature* **321:** 620-625.

Tan, C.K., R. Civil, A.M. Mian, A.G. So, and K.M. Downey. 1991. Inhibition of the RNase H activity of HIV reverse transcriptase by azidothymidylate. *Biochemistry* **30:** 4831-4835.

Tanese, N. and S.P. Goff. 1988. Domain structure of the Moloney murine leukemia virus reverse transcriptase: Mutational analysis and separate expression of the DNA polymerase and RNase H activities. *Proc. Natl. Acad. Sci.* **85:** 1777-1781.

Telesnitsky, A. and S.P. Goff. 1993. Strong-stop strand transfer during reverse transcription. In *Reverse transcriptase* (ed. A.M. Skalka and S.P. Goff), pp. 49-83. Cold Spring Harbor Laboratory Press, Cold Spring Harbor, New York.

Temin, H.M. 1993. Retrovirus variation and reverse transcription: Abnormal strand transfers result in retrovirus genetic variation (review). *Proc. Natl. Acad. Sci.* **90:** 6900-6903.

Tisdale, M., T. Schulze, B.A. Larder, and K. Moelling. 1991. Mutations within the RNase H domain of HIV-1 reverse transcriptase abolish virus infectivity. *J. Gen. Virol.* **72:** 59-66.

Tuerk, C., S. MacDougal, and L. Gold. 1992. RNA pseudoknots that inhibit human immunodeficiency virus type 1 reverse transcriptase. *Proc. Natl. Acad. Sci.* **89:** 6988-6992.

Volkmann, S., J. Dannull, and K. Moelling. 1993. The polypurine tract, PPT, of HIV as target for antisense and triple-helix-forming oligonucleotides. *Biochimie* **75:** 71-78.

von Meyenburg, K., E. Boye, K. Skarstad, L. Koppes, and T. Kogoma. 1987. Mode of initiation of constitutive stable DNA replication in RNase H-defective mutants of *Escherichia coli* K-12. *J. Bacteriol.* **169:** 2650-2658.

Vonwirth, H., J. Kock, and W. Büsen. 1991. Class I and class II ribonuclease H activities in *Crithidia fasciculata* (Protozoa). *Experientia* **47:** 92-95.

Walder, R.Y. and J.A. Walder. 1988. Role of RNase H in hybrid arrested translation by antisense oligonucleotides. *Proc. Natl. Acad. Sci.* **85:** 5011-5015.

Weston, S.A., A. Lahm, and D. Suck. 1992. X-ray structure of the DNase I-d(GGTATACC)$_2$ complex at 2.3 Å resolution. *J. Mol. Biol.* **226:** 1237-1256.

Whitcomb, J.M. and S.H. Hughes. 1992. Retroviral reverse transcription and integration: Progress and problems. *Annu. Rev. Cell Biol.* **8:** 275-306.

Wintersberger, U. 1990. Ribonucleases H of retroviral and cellular origin. *Pharmacol. Ther.* **48:** 259-280.

Wintersberger, U. and E. Wintersberger. 1987. RNA makes DNA: A speculative view of the evolution of DNA replication mechanisms. *Trends Genet.* **3:** 198-202.

Wöhrl, B.M. and K. Moelling. 1990. Interaction of HIV-1 ribonuclease H with polypurine tract containing RNA-DNA hybrids. *Biochemistry* **29:** 10141-10147.

Wöhrl, B.M., S. Volkmann, and K. Moelling. 1991. Mutations of a conserved residue within HIV-1 ribonuclease H affect its exo- and endonuclease activities. *J. Mol. Biol.* **220:** 801-818.

Yamazaki, T., M. Yoshida, S. Kanaya, H. Nakamura, and K. Nagayama. 1991. Assignments of backbone ^{1}H, ^{13}C, and ^{15}N resonances and secondary structure of ribonuclease H from *Escherichia coli* by heteronuclear three-dimensional NMR spectroscopy. *Biochemistry* **30:** 6036-6047.

Yang, W., W.A. Hendrickson, R.J. Crouch, and Y. Satow. 1990. Structure of ribonuclease H phased at 2 Å resolution by MAD analysis of the selenomethionyl protein. *Science* **249:** 1398-1405.

Zimmerman, S.B. and B.H. Pheiffer. 1981. A RNA.DNA hybrid that can adopt two conformations: An x-ray diffraction study of poly(rA).poly(dT) in concentrated solution or in fibers. *Proc. Natl. Acad. Sci.* **78:** 78-82.

12
RNA Maturation Nucleases

Murray P. Deutscher
Department of Biochemistry
University of Connecticut Health Center
Farmington, Connecticut 06030-3305

I. Introduction
II. The Study of RNA Maturation
III. Pathways of RNA Maturation
   A. Maturation of Ribosomal RNA
   B. Maturation of Transfer RNA
   C. Maturation of Messenger RNA
IV. RNA Maturation Nucleases
   A. RNase III
   B. RNase P
   C. RNase E
   D. RNase M5
   E. RNase II
   F. RNase D
   G. RNase BN
   H. RNase T
   I. RNase PH
V. Conclusions

I. INTRODUCTION

Most, if not all, RNAs are synthesized as precursor molecules that must undergo a series of maturation reactions to generate the functional form of the RNA. These include, first of all, reactions in which cotranscribed RNAs are separated from one another (e.g., the various species of ribosomal RNA, multimeric transfer RNAs, and messenger RNA-tRNA dual function transcripts). Second, the mature 5′ and 3′ termini of the RNA molecules are formed by cleavage (endonucleolytic) or trimming (exonucleolytic) reactions that remove precursor-specific sequences. In some cases, for RNA transcripts that contain introns, internal RNA regions also are removed by complex splicing reactions. Finally, some RNAs are not completely matured until base or sugar modifications or nucleotide additions (e.g., poly[A], –CCA) are incorporated into the molecule. All of these types of reactions, carried out on an ever-increasing number of diverse RNA molecules, have made RNA maturation an exceedingly complex, albeit exciting, area of research. Most of the reactions encompassing RNA maturation require the action of a ribonuclease. As the complexity and diversity of RNA maturation have unfolded, there has been a corresponding increase in our awareness of the many RNases necessary to carry out these reactions. This chapter focuses on RNA maturation pathways and on those RNases for which clear evidence, genetic or enzymological, exists for their involvement in RNA maturation. The many other RNases for which a definite role in maturation has not yet been established are not considered here or are mentioned only briefly.

II. THE STUDY OF RNA MATURATION

To elucidate the details of an RNA maturation pathway, one would want to determine all of the processing intermediates that are formed between the initial transcript and the functional RNA and to identify the enzymes catalyzing each of the processing reactions. These two aspects of the problem are, in fact, complementary. Knowledge of the processing intermediates helps to define the substrates and products of the relevant enzymes, and identification of a potential maturation enzyme and the reaction it catalyzes can help to define the processing intermediates.

Both of these approaches have been employed over the years to elucidate maturation pathways. In eukaryotic cells, studies focused initially on identification of intermediates, e.g., those present during rRNA maturation (Perry 1976), that accumulated in sufficient amounts to be characterized. In contrast, much of the initial information from prokaryotic cells came from examination of mutants defective in particular RNases (Gegenheimer and Apirion 1981). When such mutations led to disruption of an RNA maturation pathway, they not only established the involvement of the enzyme in the pathway, but the accumulation of the enzyme's substrate identified a processing intermediate. It is now clear that both biochemical and genetic approaches are needed to elucidate RNA maturation pathways. The availability of simple genetics in prokaryotes has greatly facilitated studies of RNA maturation in these systems. Consequently, much more is known about RNases and their involvement in RNA maturation in prokaryotic systems, particularly *Escherichia coli* (Deutscher 1985; King et al. 1986). On the other hand, certain RNA maturation pathways and RNases exist that are unique to eukaryotic cells, and studies of those systems are required. In such instances, the genetic confirmation of an RNase's involvement in an RNA maturation pathway will likely have to come from examination of yeast mutants, but these are not yet available.

The study of RNA maturation, particularly at the enzymatic level, has been hampered by the difficulty of obtaining appropriate substrates for identifying and assaying the relevant RNases. This problem has been alleviated to some degree by the use of in vitro transcripts from cloned genes, but in these cases, only radioactive levels of substrates are generally available. In some situations, artificial substrates have proven useful for the isolation and characterization of RNases.

Isolation and purification of maturation nucleases have also been difficult because of the cumbersome nature of the assays. Generally, processing reactions must be followed by direct visualization of substrate and product on acrylamide gels. This type of assay does not lend itself to extensive purification procedures. Specifically labeled artificial substrates that permit acid-soluble-type assays have been especially helpful in this regard (Deutscher 1984).

An additional difficulty that has become apparent as maturation RNases have been purified and studied is that cells contain many RNases with overlapping specificities in vitro (Deutscher 1988). In addition, an enzyme's action in vitro may not accurately reflect its in vivo function. Thus, one may be left with the difficult task of establishing which of several possible nucleases actually carries out a particular maturation step in the cell. It is in these instances that genetic approaches have been particularly useful. Unfortunately, in some situations, even these multiple approaches have proven unsatisfactory because mutations eliminating an RNase have not led to an observable phenotype (Deutscher 1990). Nevertheless, quite a few maturation nucleases have now been identified and purified; but it is also clear that many more remain to be found, inasmuch as there are numerous RNA maturation steps for which no nuclease has yet been identified.

The study of maturation nucleases has also been complicated by the "one laboratory—one substrate" phenomenon. Generally, different laboratories investigating maturation pathways have employed different substrates. The identification of maturation nucleases under these circumstances has often led to multiple names being assigned to the same enzyme and to insufficient information to ascertain their common identity. On the other side of the coin, there have also been instances in which RNases have been named when they were insufficiently purified, such that their presumed catalytic properties were actually due to combinations of enzymes. In both of the aforementioned situations, some clarification of the problem could be afforded by the availability of mutants. As the catalog of known RNases expands, and as more RNase-deficient mutants become available, many of these problems should be resolved. Furthermore, the identification and isolation of additional maturation RNases should be made simpler.

III. PATHWAYS OF RNA MATURATION

Although the general outlines of RNA maturation are now known, we are still a long way from understanding a maturation pathway in detail. Thus, in no case is it certain that all of the processing intermediates have been identified and all of the maturation enzymes isolated. Some information is available regarding the ordering of reactions in various pathways, but again, it is not clear how obligatory such ordering is and whether there are any general rules that apply to processing of all transcripts of a particular class. Moreover, almost nothing is known about whether RNA maturation pathways may be regulated and whether the expression of particular RNA species can be controlled at the level of RNA maturation. Nevertheless, a significant amount of information has been obtained about RNA maturation pathways and some of the RNases that are involved in these processes.

A. Maturation of Ribosomal RNA

In essentially all organisms examined, the different rRNA species are completely or almost completely cotranscribed into long precursor molecules (Srivastava and Schlessinger 1991). Thus, in *E. coli*, for example, each of the seven known rRNA cistrons encodes the 16S, 23S, and 5S RNAs, in that order, that are cotranscribed as a 30S precursor (King et al. 1986). In addition, a tRNA gene is found in the spacer between the 16S and 23S RNAs, and an additional tRNA gene is often found downstream from the 5S gene. A similar arrangement occurs in *Bacillus subtilis*, although tRNA genes are present less frequently in spacer regions.

Cotranscription of rRNAs is also observed in eukaryotic cells. In the yeast, *Saccharomyces cerevisiae*, the 18S, 5.8S, and 25S rRNAs are cotranscribed into a 35S precursor (Crouch 1984; Reeder 1990). The 5S RNA genes are located between these tandem transcription units and are transcribed separately. Similarly, in higher eukaryotes, 18S, 5.8S, and 28S rRNAs are cotranscribed into a 45S precursor molecule, with 5S RNA genes generally clustered in a separate region of the genome. The cotranscription of the various rRNA species indicates that endonucleolytic cleavages must be necessary to convert the precursor transcript to the separate RNA molecules present in the ribosome.

Early studies of rRNA maturation in mammalian cells revealed that the initial 45S nucleolar pre-rRNA transcript is converted to mature rRNA through a complex series of reactions (Perry 1976). Pulse-chase experiments and viral infection of HeLa cells, which causes nucleolar accumulation of rRNA processing intermediates, indicated that the 45S transcript is first cleaved to a 41S molecule that is then converted to 32S and 20S. The 32S ultimately goes on to the mature 5.8S and 28S rRNAs and the 20S goes on to the 18S mature form. During the course of this maturation, more than 40% of the initial RNA is lost. More detailed studies in recent years using a variety of eukaryotic systems, such as mouse, *Drosophila*, and yeast, have confirmed the general outlines of rRNA processing in all of these systems and have also identified cleavage sites (Crouch 1984; Reeder 1990). However, it is still not clear how many cleavages actually are involved in rRNA maturation and whether processing occurs in a ribonucleoprotein (RNP) structure.

Attempts have been made to demonstrate rRNA maturation reactions in vitro using transcripts of cloned rRNA genes or fragments thereof. These types of studies demonstrate clearly that the 3' terminus of the rRNA precursor is generated by a processing event (Yip and Holland 1989) and that various cleavages can be identified in the 5' region of the initial transcript or of pre-18S rRNA (Craig et al. 1987; Hannon et al. 1989). Of particular interest is the observation that a U3-containing small nuclear RNP (snRNP) participates in a cleavage in the external transcribed spacer of the mouse rRNA precursor (Kass et al. 1990). Despite this information, relatively little is known about the molecular details of eukaryotic rRNA processing.

The most promising approach for getting at this information appears to be the yeast system in which some interesting mutants affecting rRNA maturation have been isolated. The general outlines of yeast rRNA processing follow those of other eukaryotes. The 35S transcript is cleaved to generate 27S and 20S species; the former ultimately leads to the 5.8S and 25S rRNAs, and the 20S species is the precursor to the mature 18S (Crouch 1984). One yeast mutant is known in which the 27S to 25S rRNA conversion is blocked and in which the 27S species is ultimately degraded (Andrew et al. 1976). A second mutation, *rrp2*, interferes with the conversion of 35S to 27S and 20S rRNAs, leading to the accumulation of several aberrant RNAs (Shuai and Warner 1991). Studies of other yeast mutants have also implicated at least two snRNAs in the maturation pathway (Li et al. 1990). One can expect that this approach will ultimately identify genes encoding nucleases involved in eukaryotic rRNA maturation.

Considerably more information is available about rRNA maturation in prokaryotic cells (Srivastava and Schlessinger 1990). Sequencing of rRNA cistrons in *E. coli* revealed regions flanking the 16S and 23S sequences that were complementary and had the potential to form double-stranded stems (Young and Steitz 1978; Bram et al. 1980). A similar situation was also observed in other prokaryotic systems (King et al.

1986). These stems are the sites of the initial cleavages that separate the individual RNA species present on the transcript. Studies of a mutant lacking the endoribonuclease, RNase III, revealed that this enzyme is primarily responsible for the cleavages that generate precursors to the individual rRNA species (Dunn and Studier 1973). In the absence of this enzyme, a 30S precursor containing all rRNAs accumulates, and this precursor can be cleaved by RNase III in vitro.

Cleavage by RNase III leaves a precursor of 16S RNA with 115 extra 5'nucleotides and 33 3'nucleotides that must be removed by other nucleases. Likewise, for 23S RNA, the RNase III cleavage leads to a precursor with three to seven extra 5'nucleotides and seven to nine extra nucleotides at the 3'end (King et al. 1986). Also released by RNase III is a 9S RNA that is a precursor to the mature 5S RNA. Although RNase III clearly participates in rRNA maturation in wild-type cells, it is not essential for the process, inasmuch as cells lacking the enzyme are viable and can still produce functional rRNAs at a reduced rate (Gegenheimer et al. 1977). It is believed that other endonucleases that participate in the final processing of 16S, 23S, 5S, and also tRNA can carry out separation of the various cotranscribed species even in the absence of RNase III.

Very little is known about the enzymes that remove the remaining extra nucleotides on the precursor p16S and p23S RNase III products or even about how many intermediate steps are involved in their maturation. A mutant has been isolated that is defective in processing the 5'end of 16S RNA (Dahlberg et al. 1978). In this strain, a 16.3S species with 66 extra 5'nucleotides accumulates. which can be converted to 16S RNA by wild-type extracts. Pulse-chase experiments suggest that the 16.3S RNA may be a true intermediate, rather than an aberrant product generated by a nonspecific cleavage (Dahlberg et al. 1978; King and Schlessinger 1983). On the basis of in vitro conversion of 16.3S to 16S RNA, it is likely that for this reaction to occur the RNA must already be present in an RNP particle (Dahlberg et al. 1978). If so, isolation of the maturation nuclease(s) responsible for the conversion will be more difficult since an RNP substrate will have to be used. An enzyme that may be involved in the 3'maturation of 16S RNA has been reported (Hayes and Vasseur 1976), but it has not been sufficiently studied to know for certain. Enzymes responsible for the final maturation of 23S rRNA have not been reported.

The 9S precursor of 5S RNA in *E. coli* is cleaved by an enzyme, termed RNase E, to generate a 5S precursor with three extra nucleotides at each end (Misra and Apirion 1979), and mutant strains lacking this enzyme accumulate the 9S species (Roy et al. 1983). It is not known how the three extra nucleotides are removed from each end of the p5S molecule, although based on intermediates observed in cells, it is likely that exoribonucleases are involved. In *B. subtilis*, the precursor to 5S RNA is converted to the mature form by a single enzyme, RNase M5, that carries out a concerted reaction which generates both mature ends of the molecule (Meyhack et al. 1977). The substrate for this reaction is an RNP that contains the pre-5S RNA and a single ribosomal protein, BL16, associated with 5S RNA in the ribosome (Stahl et al. 1984).

B. Maturation of Transfer RNA

tRNA genes are found in a variety of different contexts and, consequently, are present in various types of transcripts. In *E. coli*, these include transcripts of individual tRNAs, multimeric transcripts containing several tRNAs, and complex transcripts in which tRNAs are associated with rRNAs or mRNAs (Komire et al. 1990). In *B. subtilis*, a similar diverse arrangement is found, but there is more clustering of tRNA genes which would lead to large multimeric precursors (Vold 1985). The eight tRNA genes of bacteriophage T4 are also clustered and are thought to be synthesized on a single transcript (Goldfarb et al. 1978); however, this multimeric precursor has not yet been isolated from phage-infected cells (Guthrie et al. 1975). Nuclear eukaryotic tRNA transcripts tend to encompass only single tRNA molecules, although a few larger ones have been detected as well; in contrast, tRNAs in mitochondria and chloroplasts are often part of larger transcription units (Deutscher 1984). The general strategy for maturation of tRNAs from these diverse transcripts is to extract individual tRNA precursors by endonucleolytic cleavages and subsequently to generate the mature 5' and 3' ends of the functional tRNAs (Deutscher 1990).

Much of what we know about tRNA maturation comes from studies of *E. coli* and bacteriophage-T4-infected *E. coli* (Deutscher 1993). This is due largely to the availability of mutants in this system that interfere with the tRNA maturation process and the consequent identification of nucleases that are affected by the mutations. On the basis of these genetic and enzymological studies, it has been possible to propose a general model for tRNA maturation in *E. coli* (Deutscher 1993), although there are still major gaps that need to be filled in.

As noted above, tRNAs present in complex transcripts are first separated from their cotranscribed partners to generate monomeric precursors containing extra 3' and 5' residues. If the 3' trailer of a monomeric precursor is long, it may be shortened further. The nucleases responsible for these steps in many transcripts have not been well characterized. It is clear that the enzymes separating cotranscribed RNAs must be endoribonucleases, and it is known that RNase III serves this role in the case of tRNAs cotranscribed with the rRNAs (Gegenheimer and Apirion 1981; King et al. 1986). RNase III also separates a tRNA encoded by *metY* from the mRNA encoding NusA and IF2 (Regnier and Grunberg-Manago 1989). RNases, termed P2 and 0, have been implicated in cleavages in intercistronic regions of multimeric precursors, but these activities are still poorly defined (Deutscher 1984, 1985). Mutations in the *rne* gene, presumably encoding RNase E, also affect tRNA maturation (Misra and Apirion 1979). Shortening of the 3' trailer of monomeric tRNA precursors has been attributed to an enzyme termed RNase PIV, thought to be an endoribonuclease (Bikoff et al. 1975). However, rapid shortening of 3' trailer sequences can also be carried out by the processive exoribonucleases, RNase II, and polynucleotide phosphorylase (PNPase) (Z. Li and M.P. Deutscher, unpubl.).

Generation of the mature 5' end of all tRNAs is carried out by the ribonucleoprotein enzyme, RNase P, an endoribonuclease (Altman et al. 1982). Mutant strains devoid of this enzyme accumulate tRNA precursors retaining their 5' leader sequences and also in many cases, some 3' precursor sequences. Thus, 5' processing may be necessary for 3' processing to be completed. However, since some 3' processing occurs even in the absence of RNase P action, it is likely that the removal of the 3' precursor nucleotides is a multistep process, only part of which is dependent on prior 5' maturation. Inasmuch as 3' processing of *E. coli* tRNA precursors is thought to be an exonucleolytic trimming event, several RNases of this type could be involved.

It has been known for some time that *E. coli* contains at least seven distinct exoribonucleases, of which six can act on the 3' terminus of tRNA precursors in vitro (Deutscher 1990). Mutations removing any one of these RNases, termed RNase II, D, BN, T, PH, and PNPase, have essentially no effect on *E. coli* growth or on tRNA maturation (Kelly et al. 1992). Even cells lacking four of the RNases, II, D, BN, and T, show only small growth defects and only about a 50% decrease in synthesis of a suppressor tRNA; however, elimination of a fifth exoribonuclease, RNase PH, leads to inviability (Kelly et al. 1992; Reuven and Deutscher 1993). The presence of any one of these five RNases is sufficient to support the viability of *E. coli* with the following order of effectiveness: RNase T>RNase PH>RNase D>RNase II>RNase BN (Kelly and Deutscher 1992a). Thus, there is a high degree of functional overlap among these enzymes in vivo, with any one of them capable of taking over the functions of all the others to some degree. Use of a quantitative in vivo suppression assay with a variety of RNase-deficient strains revealed that all five of the RNases, and perhaps PNPase as well, can participate in tRNA processing and that at least for tRNA$^{Tyr}su_3{}^+$, the largest contribution to the overall level of suppressor synthesized is made by RNases T and PH (Reuven and Deutscher 1993). Whether the same order of importance of RNases holds for other tRNA precursors as well is not yet known.

The actual role of each of the exoribonucleases in 3' maturation of *E. coli* tRNA was examined with an in vitro tRNA processing system in which a precursor to tRNA$^{Tyr}su_3{}^+$ was used as substrate (Z. Li and M.P. Deutscher, unpubl.). These studies showed that removal of the long 3' trailer sequence is carried out most effectively by RNase II or PNPase, leaving a precursor with two to four 3' residues remaining; RNase PH is most effective in trimming further to the +1 residue; RNase T works best in removing the final residue to expose the –CCA sequence. Thus, the different RNases display somewhat different specificities. Yet, there is sufficient overlap in their catalytic properties that any one of the five exoribonucleases (presumably together with PNPase, which remains in the mutant cells) can carry out complete 3' processing to generate functional tRNAs.

Many *E. coli* bacteriophages also encode a population of tRNAs that are expressed upon infection. For example, eight new tRNAs are synthesized upon infection by bacteriophage T4, and this system has served as a simple model for tRNA maturation in the uninfected cell (Guthrie et al. 1975). Although the eight T4 tRNAs are probably made as a single transcript, only three dimeric molecules and two monomers can be isolated from cells, suggesting that some endonucleolytic cleavages take place rapidly. A detailed analysis of the maturation of the dimeric pre-tRNA$^{Pro+Ser}$ using mutant strains and some purified enzymes has shown that this precursor can be completely processed by RNase P, RNase BN, and tRNA nucleotidyltransferase (McClain 1977). RNase BN first removes some extra residues at the 3' terminus of the dimeric precursor, tRNA nucleotidyltransferase adds the –CCA sequence, and cleavage of the dimer ensues, releasing mature tRNA^{Ser}. A similar series of events leads to maturation of the now monomeric tRNA^{Pro} precursor. For some of the other T4 tRNAs, as well as for host tRNAs, the –CCA sequence is already encoded in the tRNA genes, and tRNA nucleotidyltransferase action is not needed (Deutscher 1990). The maturation of pre-tRNA$^{Pro+Ser}$ is probably the most completely understood RNA processing pathway.

tRNA maturation in eukaryotic cells is less well understood (Deutscher 1993). The absence of tRNA processing mutants has made identification of processing intermediates and of maturation enzymes much more difficult in these systems. In addition, the presence of introns in some eukaryotic tRNA genes, the universal absence of an encoded –CCA sequence, and the multiple compartments in which tRNA is made all contribute to the greater complexity of the maturation process in eukaryotic cells. What has emerged from studies in a number of systems is that the order of maturation events (5' and 3' cleavage, intron removal, –CCA addition) can vary in different systems and that different precursor tRNAs may be processed differently even in the same extract (Deutscher 1984, 1993). A major caveat of these studies, which do not have the benefit of genetic confirmation, is that some of the differences observed may simply reflect the efficiency of extraction and stability of various maturation enzymes. Changes in ratios of enzymes present in an extract could influence the particular maturation pathway followed.

At the enzymatic level, eukaryotic cells do contain RNase-P-like enzymes to generate the 5' termini of tRNAs. However, the structure of the enzyme, and whether it functions as an RNP, is in dispute. Maturation at the 3' terminus is also not clear. In some systems, an endonucleolytic cleavage occurs at a position that would allow –CCA addition, but exonucleolytic trimming has also been reported (Deutscher 1984, 1990, 1993). Since the eukaryotic maturation enzymes have generally not been well characterized, and mutants blocked at specific steps of the maturation pathway are not available, it is not yet possible to come to firm conclusions about eukaryotic tRNA processing.

Removal of introns from eukaryotic tRNA precursors requires specific endonucleolytic cleavages and ligation reactions carried out by a specific splicing endonuclease (Gandini-Attardi et al. 1990; Green and Abelson 1990) and ligase (Xu et al. 1990). In *S. cerevisiae*, all of the tRNA introns are located in the same position, one base 3' to the anticodon. The splicing endonuclease cleaves pre-tRNA at both ends of the intron, generating a 2',3'-cyclic phosphoryl terminus on the 5' exon and a 5'-hydroxyl terminus on the 5' exon (Peebles et al. 1983). Studies in vitro (Miao and Abelson 1993) and in vivo (O'Connor and Peebles 1991) suggest that the cleavage reactions at the two sites occur independently of each other but that the 3' splice site cleavage may proceed somewhat faster. The yeast endonuclease has been purified to homogeneity (Rauhut et al. 1990). It is a heterotrimeric enzyme containing subunits of 31, 42, and 51 kD. The enzyme is thought to be membrane-bound and to be a very minor component, present at about 100 molecules per cell. A cold-sensitive yeast mutant, *sen2-3*, affects the endonuclease activity, most likely through the 42-kD subunit (Ho et al. 1990). Exactly how the endonuclease recognizes and interacts with intron-containing precursors is not completely understood, but both the mature tRNA domain and the intron itself play a role (Abelson 1992; Baldi et al. 1992).

C. Maturation of Messenger RNA

Maturation of mRNA, as distinct from processes related to its degradation, has been difficult to discern in prokaryotic systems because of the very short half-life of most messages. The clearest examples, where one knows a maturation event must occur, are in those transcripts in which a stable RNA, such as tRNA, is associated with a message. Among these dual-function transcripts is one that encodes tRNA$^{Tyr}_1$ and a small, basic peptide (Rossi et al. 1981), one in which several tRNAs are cotranscribed with the message for EFTu (Hudson et al. 1981), and another that contains a tRNA^{Met} and the messages for NusA and IF2 (Ishii et al. 1984). In the latter case, it is known that RNase III is responsible for separating the tRNA portion of the transcript from the mRNAs (Regnier and Grunberg-Managò 1989). Several examples are known in which polycistronic mRNAs are cleaved in intercistronic regions, presumably as part of a maturation event, but a relationship to mRNA decay cannot be excluded completely (Regnier and Grunberg-Managò 1990). Both RNase III and RNase E have been implicated in these events (Faubladier et al. 1990).

mRNA maturation events, dependent on RNase III cleavages, have also been observed upon phage infection of *E. coli*. Thus, RNase III can cleave the phage T7 early mRNA precursor at five sites to generate functional mRNAs (Rosenberg et al. 1974). In at least one case, that of gene *0.3*, RNase III maturation has a major effect on expression of the gene. RNase III also participates in removing leader RNA from the bacteriophage λ N gene transcript (Steege et al. 1987). Infection of *E. coli* by the filamentous phage, f1, leads to production of a number of major phage mRNAs that are the products of maturation events (Kokoska et al. 1990). The maturation activity is encoded by the host genome. It is distinct from the endoribonucleases, RNase III, and RNase P, and although RNase E was also thought not to be involved, the *ams* locus, presumably encoding RNase E, does have an effect. RNase E has also been suggested to promote stabilization of phage T4 gene *32* mRNA by a cleavage event, but inasmuch as the enzyme has a global effect on decay of T4 mRNA, it is not clear whether this is really a maturation event (Mudd et al. 1988).

Maturation of mRNA in eukaryotes usually involves the splicing out of introns and poly(A) addition. The latter process requires an endonucleolytic cleavage at a site 10–30 nucleotides downstream from a polyadenylation signal, AAUAAA, in mammals (Proudfoot 1991). Four factors are now known to be involved in the cleavage reaction (Takagaki et al. 1989). The factors have been purified and their structure and mode of action are under study (Takagaki and Manley 1992). It is already clear from what is known thus far that even the cleavage reaction is a complex multicomponent process.

IV. RNA MATURATION NUCLEASES

Although quite a number of processing systems have been developed for the maturation of particular RNA precursors, in only a handful of these have relevant enzymes been isolated, purified, and characterized. There are even fewer examples in which a putative maturation nuclease has been conclusively shown to participate in a specific processing event in vivo. The properties of some of the best characterized maturation nucleases are described briefly in the material to follow.

A. RNase III

E. coli RNase III was initially identified as an endoribonuclease that specifically degrades double-stranded RNA (Robertson et al. 1968; Dunn 1982; Robertson 1990) and, as discussed earlier, was subsequently shown to be involved in the maturation of T7 mRNA and rRNA (Dunn and Studier 1973). With the availability of clones overexpressing RNase III to extremely high levels, the enzyme has been purified to homogeneity with high yields (Chen et al. 1990; March and Gonzalez 1990). Thus, it has been possible to isolate 3 mg of RNase III with 99% purity from approximately 40 mg of *E. coli* (Chen et al. 1990). RNase III is a homodimeric enzyme with a native molecular mass of approximately 50 kD (Robertson et al. 1968; March and Gonzalez 1990). Interestingly, it was found that RNase III can exist in two distinct forms, one of which is dependent on the presence of the Era protein, with which it is translationally coupled (March and Gonzalez 1990). It was suggested that Era may be involved in phosphorylation of RNase III because of the increased activity of the second form of the enzyme (March and Gonzalez 1990) and the previous demonstration that RNase III is phosphorylated and activated by T7 protein kinase upon phage infection (Mayer and Schweiger 1983).

RNase III requires both a divalent and a monovalent cation for activity (Robertson et al. 1968). On synthetic double-stranded RNAs, the enzyme can carry out extensive double-strand cleavages to generate oligonucleotides with 5'-phosphate and 3'-hydroxyl termini (Dunn 1982). With natural RNA precursors, RNase III is much more selective, cleaving at specific sites due to ill-defined processing signals. Cleavage specificity is markedly dependent on ionic strength as well as the identity of the ions present. It is clear that double-strandedness alone is not sufficient to confer cleavage specificity, and although some conserved sequence elements at cleavage sites have been suggested, there is also evidence against a role for such sequences (Chelladurai et al. 1991). The use of small defined RNAs carrying defined RNase III processing sites

(March and Gonzalez 1990) should help to define the identity elements important for RNase III recognition and cleavage.

RNase III is encoded by the *rnc* gene located at 55 minutes on the *E. coli* chromosome. The *rnc* gene has been cloned and sequenced (March et al. 1985), and it encodes an open reading frame of 227 amino acids, corresponding to a monomer molecular mass of 25,218 daltons, in excellent agreement with the protein studies. Upstream of the coding region on the *rnc* transcript is a stem-loop structure that has the features of an RNase III cleavage site, and RNase III is known to autoregulate its own expression as well as that of Era protein (Bardwell et al. 1989). RNase III translation also regulates translation of *era* (Chen et al. 1990). Both proteins are expressed at relatively low levels, amounting to about 0.01% of total cell protein.

B. RNase P

RNase P is an endoribonuclease that cleaves the 5′ leader sequence from tRNA precursors to generate the 5′-phosphoryl terminus of mature tRNAs (Robertson et al. 1972). RNase-P-like enzymes have now been identified in every cell and every cellular compartment examined that synthesize RNA (Altman 1989; Darr et al. 1992). The interesting and unusual aspect of RNase P is that in many organisms, it is clear that the enzyme is a ribonucleoprotein, and it is the RNA component that is responsible for catalysis in vitro (Guerrier-Takada et al. 1983). Whether this is true in all organisms and organelles has not yet been settled (Darr et al. 1992). In vivo, on the other hand, both the RNA and protein subunits are necessary inasmuch as temperature-sensitive mutants of *E. coli* are known that are affected in either subunit (Altman et al. 1987). The RNA components from the bacterial enzymes vary in size but are in the 330–440-nucleotide range; the RNAs from other organisms and organelles also vary considerably, and in these cases, they have not yet been shown to have catalytic activity independent of the protein (Darr et al. 1992). The protein subunit has been studied in much less detail; in *E. coli*, it has a molecular mass of approximately 14 kD (Hansen et al. 1985; Brown and Pace 1992).

Considerable effort has gone into attempts to generate a structural model for the RNase P RNA using phylogenetic comparisons, mutagenesis studies, and fragments of the RNA, and some general structures have emerged (Brown and Pace 1992; Darr et al. 1992; Altman et al. 1993). In addition, various other studies have attempted to identify the site on the RNA that interacts with tRNA precursors and the site that interacts with the protein subunit (Altman et al. 1993). At present, no definitive conclusions are yet available, although regions of importance are being identified. The complexity of the structure of the RNase P RNA, the RNase P protein, and the substrate itself makes this a formidable task.

The actual function of the protein subunit of RNase P is still a matter for speculation. The *E. coli* and *B. subtilis* proteins and RNAs interact very strongly with affinities in the 1 nM range (Altman et al. 1993). One suggestion has been that the protein may function to shield charges on the interacting RNAs. Studies have shown that the RNase P protein can influence the V_{max} or the K_m values of the reaction depending on the RNA substrate. These kinds of findings suggest that the protein may function to stabilize the conformation of RNase P RNA, from among the many conformations possible, which is most effective for binding to substrate and catalyzing cleavage. These conformations clearly must be different for different RNA precursors, implying that the RNA subunit must have flexibility. However, once the appropriate conformation for a particular substrate is found, it needs to be stabilized for the reaction to proceed efficiently. Clearly, since the RNA subunit is active by itself, it can attain the correct conformation on its own, but stabilization by the protein would be expected to enhance catalytic efficiency.

How RNase P recognizes substrates and identifies sites of cleavage has also received considerable attention. It is known that the three-dimensional structure of the precursor plays a role, but specific nucleotides, as well as the –CCA residues, may also be important, especially for the reaction catalyzed by the RNA alone (Altman et al. 1993). On the other hand, model substrates containing only an acceptor stem and a T stem of tRNA can be cleaved (Forster and Altman 1990). Location of the cleavage site has not been explained. It is not simply nucleotide sequence or distance from a fixed point. Yet, many small changes in the RNA substrate can alter the location or efficiency of cleavage (Altman et al. 1993). Much more work will be necessary to sort out this difficult problem.

C. RNase E

RNase E was originally identified as an endoribonuclease that could cleave the 9S RNA released by RNase III action on pre-rRNA and convert it to p5S RNA (Misra and Apirion 1979). The enzyme cleaves both the 5′ and 3′ termini of the 9S intermediate, generating a 5S precursor with three extra residues at each end and containing a 5′-phosphoryl and a 3′-hydroxyl group. The original partial purification of RNase E suggested a molecular weight of 70,000 (Misra and Apirion 1979), but this has been called into question (see below). A mutation in a gene, termed *rne*, rendered RNase E thermolabile and prevented maturation of 5S rRNA (Misra and Apirion 1980). The *rne* mutation also had more pleiotropic effects, suggesting involvement in processes other than 5S RNA maturation as well (Gitelman and Apirion 1980; Goldblum and Apirion 1981).

The role of RNase E in RNA metabolism was greatly expanded when it was found that the *rne* mutation affected the stability of T4 and *E. coli* mRNAs and that *rne* was identical to *ams,* a gene known to affect mRNA stability (Ehretsmann et al. 1992a). The situation has been further complicated by disagreement on the size of *rne/ams*, since open reading frames of 91 and 62 kD have been reported, neither of which agrees with the original protein size of 70 kD. Moreover, overexpression of *rne/ams* yielded a 110-kD gene product (Ehretsmann et al. 1992a). One additional complication is that another enzyme, RNase K, which also acts on mRNA, is affected by the *rne* mutation (Lundberg et al. 1990) and that RNase K has some properties distinct from those of RNase E. Recently, it has been shown that RNase E, but not RNase K, interacts functionally with the heat-shock protein, GroEL (Sohlberg et al. 1993).

Possible resolution of these confusing data has recently become available. Sequencing of the *E. coli* gene, *hmpl*, whose product cross-reacts with monoclonal antibody to yeast myosin heavy chain, has shown it to be the *rne/ams* gene (Casaregola et al. 1992). The gene in fact encodes a protein of 114 kD, which runs anomalously on SDS-PAGE as a protein with an apparent mass of 180 kD. This study showed that the earlier sequencing efforts included errors that led to apparent early termination. Additional information has suggested that RNase K may represent an active proteolytic fragment of the larger protein (A.J. Carpoussis, pers. comm.).

The cleavage specificity of RNase E may be influenced by both secondary structure and sequence. Cleavages often occur upstream of stable secondary structures (Tomcsanyi and Apirion 1985), yet they can be blocked by stem-loop structures that eliminate unpaired nucleotides at the 5′ end of an RNA (Bouvet and Belasco 1992). On the other hand, analysis of RNase E cleavage sites suggests that a consensus sequence (G/A) AUU (A/U) may be relevant for cleavage, where cleavage can follow either of the first two purines, when a stable secondary structure is nearby downstream (Ehretsmann et al. 1992b).

D. RNase M5

Maturation of 5S RNA in *Bacillus* differs from that in *E. coli* in that there is an endoribonuclease, termed RNase M5, that generates both the mature 5' and 3' termini of 5S RNA directly from the 5S RNA precursor (Sogin et al. 1977). The enzyme requires a divalent cation and forms 5'-phosphoryl and 3'-hydroxyl groups as a result of the two cleavages made in the precursor molecule. Maturation activity is dependent on two components, termed α and β (Sogin et al. 1977). The catalytic subunit is α, which has a monomer molecular weight of 24,000 (Pace et al. 1984). Its oligomeric state in the native enzyme is not known, but considering that both cleavages are made simultaneously, it might be expected to function as a dimer. The α subunit can cleave p5S RNA in the absence of the β subunit when dimethylsulfoxide is present in the reaction mixture (Pace et al. 1984). Under normal circumstances, the RNA substrate is placed in the proper conformation for cleavage by binding to the β subunit, which is a basic protein with a molecular weight of 15,000 and which appears to be identical to ribosomal protein BL16 (Stahl et al. 1984). Inasmuch as BL16 also binds to mature 5S RNA, it is clear that the mature sequence and structure are important for substrate recognition by RNase M5 (Meyhack and Pace 1978).

Cleavage at both termini of the 5S RNA precursor can occur during a single binding event because the folded structure of the substrate places the two cleavage sites opposite each other in a double-stranded stem (Meyhack and Pace 1978). Intermediates matured at only one end of the precursor are not detected during the reaction, although they can act as substrates to some degree when prepared artificially (Meyhack et al. 1977). Examination of the recognition of RNase M5 for its substrate indicated that the helical region of the cleavage site, rather than its specific sequence, is important for action by the enzyme (Stahl et al. 1980). Because of the high degree of specificity of RNase M5 for its natural substrate, other features of the precursor structure also must be important, but these remain to be determined.

E. RNase II

RNase II is a nonspecific, hydrolytic exoribonuclease that was one of the first ribonucleases to be identified in *E. coli* (Shen and Schlessinger 1982). The enzyme has been purified to homogeneity and studied extensively for its action on various RNA molecules. In particular, genetic and biochemical evidence has pointed to its involvement in mRNA degradation (Donovan and Kushner 1986; Guarneros and Portier 1990; McLaren et al. 1991). The enzyme generally acts processively, but because it is extremely sensitive to secondary structure, it will pause at hairpins in RNA (Shen and Schlessinger 1982; Guarneros and Portier 1990; McLaren et al. 1991). RNase II is a very active enzyme, particularly on homopolymer substrates, accounting for essentially all of the poly(A) hydrolytic activity in an *E. coli* extract (Zaniewski et al. 1984; Deutscher and Reuven 1991). In contrast, an enzyme with this specificity is not detectable in extracts of *B. subtilis*, where degradation of RNA is largely phosphorolytic (Deutscher and Reuven 1991).

RNase II is encoded by the *rnb* gene located at 28 minutes on the *E. coli* map (Shen and Schlessinger 1982). The gene has been isolated and recently sequenced (C.M. Arraiano, pers. comm.), and the protein has been overexpressed (Donovan and Kushner 1983). The molecular mass of RNase II determined by both SDS-PAGE and gel filtration is approximately 70 kD, indicating that the protein is a monomer (Cudny and Deutscher 1980; Shen and Schlessinger 1982).

It was suggested that RNase II removed extra residues following the –CCA sequence in *E. coli* tRNA precursors (Schedl and Primakoff 1973; Schedl et al. 1976). Elimination of RNase II was also shown to affect the processing of T4 tRNAs (Birenbaum et al. 1980). However, studies with natural (Bikoff et al. 1975) and artificial (Cudny and Deutscher 1980) tRNA precursors indicated that RNase II was ineffective in generating the mature 3' terminus, although it could remove some of the 3' trailer sequence. In addition, mutants lacking RNase II displayed no defect in tRNA maturation (Zaniewski et al. 1984; Kelly et al. 1992). On the other hand, recent studies with multiple RNase-deficient strains, in which only RNase II of five exoribonucleases is retained in the cells, indicate that this enzyme is sufficient for cell viability (Kelly and Deutscher 1992a). Thus, RNase II by itself, or perhaps in conjunction with PNPase or other unknown activities remaining in these cells, must be able to process all essential RNA precursors completely, although inefficiently.

F. RNase D

The original conclusion that RNase II was not a tRNA 3' processing enzyme led to a search for another exoribonuclease that might carry out such a reaction (Ghosh and Deutscher 1978a). Using an artificial precursor that contained an average of two to three radioactive C residues following the –CCA sequence, Ghosh and Deutscher (1978a) developed a simple acid-soluble assay that led to the identification of RNase D. This enzyme had already been identified a year earlier as a nuclease that could act on denatured tRNA, hence its name (Ghosh and Deutscher 1978b). RNase D had the properties expected for a 3' maturation nuclease in that it could remove extra 3' residues from the artificial precursor and regenerate amino acid acceptor activity (Cudny and Deutscher 1980), and its poor activity on mature tRNA (Cudny et al. 1981) was in keeping with earlier observations that repair of the –CCA sequence was not part of the normal maturation pathway for tRNA precursors (Deutscher et al. 1977).

RNase D is highly specific for molecules with tRNA-like structures, either containing extra 3′ residues or lacking all or part of the –CCA sequence, and it has no activity against homopolymers (Cudny et al. 1981). It requires a free 3′-hydroxyl group on the RNA substrate for activity. Its low activity against mature tRNA is not due to the presence of the –CCA sequence because a second such sequence can be readily removed (Cudny et al. 1981). Rather, it appears that the secondary structure of the tRNA, also shown by its action on denatured tRNA, is important (Ghosh and Deutscher 1978b). All of its properties suggest that RNase D has some role in tRNA metabolism in vivo. In fact, overexpression of RNase D has shown that it can act on tRNA molecules in vivo (Zhang and Deutscher 1988c). However, cells devoid of RNase D have not yet been found to display any altered properties in a variety of tests (Blouin et al. 1983). Recent studies with cells containing only one of five RNases have shown that those with only RNase D are viable; however, such cells grow relatively poorly, with a doubling time of 64 minutes in rich media (Kelly and Deutscher 1992a).

RNase D is encoded by the *rnd* gene located at about 40 minutes on the *E. coli* genetic map (Zaniewski and Deutscher 1982). The gene has been cloned, sequenced, and overexpressed (Zhang and Deutscher 1988a,b). The *rnd*-coding region encodes a protein of 42.7 kD, in close agreement with the native and denatured molecular mass of RNase D. The coding region is somewhat unusual in that it initiates with the infrequently used UUG codon, and conversion of this to AUG leads to tenfold elevation of RNase D activity in cells (Zhang and Deutscher 1988b, 1989). Overexpression of RNase D can be deleterious to cells, particularly ones that lack tRNA nucleotidyltransferase (Zhang and Deutscher 1988c). However, even in normal cells, elevation of activity of more than 20-fold cannot be tolerated (Zhang and Deutscher 1988a). Such cells grow slowly and there is strong selection for faster-growing cells that contain less of the enzyme.

In addition to the UUG initiation codon that downregulates RNase D expression, the *rnd* gene contains another feature that also greatly influences its expression at the translational level. Between the promoter and the Shine-Delgarno sequence is a GC-rich stem-loop followed by eight U residues in the mRNA (Zhang and Deutscher 1989). Deletion of this region has no effect on the levels of *rnd* mRNA but almost completely abolishes RNase D expression. Site-directed mutagenesis of the U-rich region that alters two to five of the uridine residues has the same effect (Zhang and Deutscher 1992). Moreover, these mutant transcripts bind to 30S ribosomal subunits with about tenfold lower affinity than their wild-type counterparts. Thus, this upstream sequence serves as a translational enhancer, presumably by influencing the binding of ribosomes to the initiation region of this message. Other studies showed that the Shine-Delgarno sequence also is necessary for translation of the *rnd* mRNA (Zhang and Deutscher 1992), indicating that at least three distinct sequences (initiation codon, S-D sequence, U-rich region) can influence translation of this message. Whether these observations indicate that RNase D levels are subject to regulation in vivo remains to be determined.

G. RNase BN

A series of studies on the processing of precursors to tRNAs specified by bacteriophage T4 indicated that several of them were unable to complete 3′ processing in a mutant *E. coli* strain, termed BN (Guthrie et al. 1975; Seidman et al. 1975). Identification of the enzyme inactivated by the BN mutation was simplified by the removal of RNases II and D using mutations in *rnb* and *rnd* (Asha et al. 1983). In this genetic background, "RNase BN" was identified as an activity present in extracts of *E. coli* B, but absent from extracts of strain BN, that could remove radioactive UMP from tRNA-C-U. Another mutant strain, termed CAN, a K12 derivative that was isolated in the same manner as strain BN, also lacked this ribonuclease (Asha and Deutscher 1983). RNase BN also works well on tRNA-C-A, but it is much less active on tRNA-C-C, which has the correct 3′ residues (Asha et al. 1983). In contrast to RNases II and D, RNase BN has a relatively low level of activity against the artificial tRNA precursor, tRNA-CCA-C_{2-3}; it is also inactive against poly(A).

RNase BN has been partially purified (D. Neri-Cortes and M.P. Deutscher, unpubl.). It has the unusual property that it is most active at comparatively low pH values (pH 6) and in the presence of the divalent cation, Co^{++}. On the basis of gel filtration, the molecular mass of purified RNase BN is about 60 kD. Further detailed examination of the catalytic and structural properties of RNase BN will require its complete purification. Although mutations affecting RNase BN are known (Maisurian and Buyanovskaya 1973), the gene encoding this enzyme has not been mapped.

Until recently, the role of RNase BN in cellular tRNA processing has been unclear since strains lacking this enzyme display no obvious phenotype (Zaniewski et al. 1984). However, using a tRNA$^{Tyr}su_3^+$ suppression assay, Reuven and Deutscher (1993) have shown that RNase

BN also contributes to the overall processing of the precursor to this tRNA. Reductions in mature suppressor tRNA upon removal of RNase BN are only observed if other RNases are also absent. The ability of RNase BN to process tRNA precursors was also shown by its ability to restore viability to a strain lacking five exoribonucleases, although growth in this case was extremely slow (doubling time >100 min in rich medium) (Kelly and Deutscher 1992a). The data suggest that RNase BN can contribute to tRNA processing but that it is generally ineffective in substituting completely for other exoribonucleases.

H. RNase T

RNase T was originally identified in extracts of a strain lacking RNases II, D, and BN as an activity that could remove the 3'-terminal AMP residue from intact tRNA-CCA (Deutscher et al. 1984). The enzyme was subsequently shown to be required for the end-turnover of tRNAs (Deutscher et al. 1985), a process that occurs in all cells and involves the removal of 3'-terminal residues and their resynthesis by tRNA nucleotidyltransferase. This process is almost completely abolished in cells lacking RNase T (Deutscher et al. 1985; Padmanabha and Deutscher 1991), suggesting that in *E. coli,* at least, essentially only this enzyme is able to act on intact tRNA molecules. Although the physiological significance of end-turnover has not been ascertained, it is known to require ongoing protein synthesis, uncharged tRNAs, and to involve tRNAs for only certain amino acids (Deutscher et al. 1977). Inactivation of the *rnt* gene encoding RNase T has shown that the enzyme is also essential for the normal growth of *E. coli* and for its recovery from starvation conditions (Padmanabha and Deutscher 1991). RNase T is the only exoribonuclease whose removal, by itself, leads to a reduction in *E. coli* growth rate, changing it from 24 minutes to 30-35 minutes (Padmanabha and Deutscher 1991; Kelly et al. 1992). Inasmuch as tRNA end-turnover does not appear to be an essential process in cells, this observation suggests that RNase T has another important function.

RNase T has been purified to homogeneity (Deutscher and Marlor 1985). The purified protein is an α_2 dimer of 50 kD. The sequence of the *rnt* gene indicates that the monomer molecular weight is 23,521 (Huang and Deutscher 1992), in close agreement with the protein work. Purified RNase T acts on several tRNA-type substrates, but the preferred one is intact tRNA-CCA (Deutscher et al. 1984). Other molecules such as tRNA-CCA-C_{2-3} (15%), tRNA-CC and tRNA-CU (<5%), and tRNA-CA (~20%) are much less active as substrates, and poly(A) and tRNA-CCp are inactive (Deutscher et al. 1984, 1985). In keeping with the action of

RNase T in vivo, aminoacyl-tRNA is not a substrate for the enzyme. RNase T is extremely sensitive to sulfhydryl oxidation (Deutscher et al. 1985), and interestingly, it contains twice as many cysteines in its sequence compared to the average *E. coli* protein of its size (Huang and Deutscher 1992).

The *rnt* gene encoding RNase T is located at approximately 36 minutes on the *E. coli* genetic map (Case et al. 1989). The gene has been cloned, sequenced, and overexpressed (Huang and Deutscher 1992). It shows no sequence similarity to other proteins in the database or to the exoribonucleases, RNase D, RNase PH, or PNPase. The *rnt* promoter, identified by primer extension, is located close to the coding region and, based on its homology score, is a reasonable σ^{70} promoter. Downstream from the *rnt*-coding region is a second open reading frame with sequence similarities to RNA helicases. This open reading frame is cotranscribed with *rnt*, indicating that *rnt* is the first gene of an operon (Huang and Deutscher 1992).

By several criteria, RNase T has turned out to be an important enzyme. First of all, as noted above, its removal by a single mutation affects growth. This implies a major reduction in the rate of some essential process. Second, cells containing only RNase T from among five exoribonucleases grow faster (35 min) than any of the other multiple RNase-deficient mutant strains containing only one RNase (Kelly and Deutscher 1992a). Even a cell with only 30% of the normal level of RNase T, in this RNase-deficient background, grows with a doubling time of 57 minutes, which is quite well considering the RNase deficiencies in this strain. These observations suggest that RNase T has the broadest specificity of the exoribonucleases in that it can very effectively take over the function of all the missing enzymes. However, none of the other enzymes can completely take over its functions.

RNase T plays a significant role in tRNA processing. On the basis of a tRNA^{Tyr}su$_3^+$ suppression assay, removal of RNase T can decrease the level of this tRNA dramatically in cells already lacking other RNases and particularly in cells lacking RNase PH (Reuven and Deutscher 1993). Studies of tRNA^{Tyr}su$_3^+$ maturation in vitro suggest that the main role of RNase T in the process may be in the removal of the last residue closest to the –CCA terminus (Z. Li and M.P. Deutscher, unpubl.). Further work is necessary to substantiate this point and to determine whether it holds for other precursors as well.

I. RNase PH

Extracts from a cell deficient in RNases II, D, BN, and T retained the ability to process accurately a precursor to tRNA^{Tyr}su$_3^+$ to mature tRNA

(Cudny and Deutscher 1988). The final 3'trimming in this system was exonucleolytic, indicating that yet another exoribonuclease must be present in the extract. Dialysis and re-addition experiments established that the final 3'trimming reaction was dependent on inorganic phosphate present in the extract and that nucleoside diphosphates were generated as products. The conclusion from these experiments was that 3'processing in these RNase-deficient extracts was dependent on a phosphorolytic nuclease, most likely PNPase, inasmuch as it was the only nuclease of this type known. Surprisingly, studies with purified PNPase and with extracts from a strain devoid of PNPase revealed that another activity was involved. This enzyme, termed RNase PH, was isolated from the mutant extracts and shown to differ from PNPase in size and substrate specificity and to be present at the same level in wild-type and PNPase⁻ extracts (Deutscher et al. 1988).

Purification of RNase PH and amino-terminal sequence analysis of the purified protein revealed that it was encoded by a previously identified open reading frame (*orfE*) upstream of *pyrE* located at 81.7 minutes on the *E. coli* genetic map (Poulsen et al. 1984; Ost and Deutscher 1991). This locus has been renamed *rph*. An *rph* gene highly homologous to that in *E. coli* has also been identified in *B. subtilis* at 251°, and this cloned gene suppresses a number of cold-sensitive *E. coli* mutants by an unknown mechanism (Craven et al. 1992). Studies of the *orfE* (*rph*) gene had earlier shown that it was cotranscribed with *pyrE* and that its translation was essential for pyrimidine regulation of the *pyr* attenuator (Bonekamp et al. 1985). Moreover, *orfE* (*rph*) could be inactivated with no deleterious effects on the cell (Poulsen et al. 1989).

The *orfE* (*rph*) gene product has been overexpressed and purified to homogeneity and shown to contain RNase PH activity (Jensen et al. 1992; Kelly and Deutscher 1992b). The purified protein displays unusual structural properties. DNA sequencing showed that *orfE* (*rph*) encodes a 238-amino-acid residue protein with a molecular mass of 25.5 kD (Poulsen et al. 1983, 1984). On SDS-PAGE, the protein runs as a 30–33-kD protein (Ost and Deutscher 1991; Jensen et al. 1992) and in some cases migrates as multiple distinct bands (Jensen et al. 1992). The native molecular mass, based on gel filtration, varies from 45–50 kD to greater than 200 kD, depending on the loading concentration (Kelly and Deutscher 1992b). Cross-linking studies showed that the purified protein can self-aggregate extensively. However, the smallest, active form appears to be the dimer, and it remains to be determined whether the self-aggregation serves any physiological role.

Purified RNase PH requires phosphate for activity, indicating that it is exclusively a phosphorolytic enzyme (Kelly and Deutscher 1992b). Consequently, RNase PH can also act as a synthetic enzyme, using nucleoside diphosphates and adding them to the 3' termini of RNA molecules (Ost and Deutscher 1990). The equilibrium constant for the enzyme is near unity, suggesting that at in vivo phosphate concentrations, it would participate primarily in RNA degradation (Ost and Deutscher 1990; Kelly and Deutscher 1992b). Among tRNA-type substrates, the enzyme is most active against those containing extra residues following the –CCA sequence, and it can act on these to generate mature tRNA with amino acid acceptor activity. However, it does so only with difficulty, compared to, for example, RNase D (Kelly and Deutscher 1992b).

Studies with cells containing only RNase PH from among five exoribonucleases indicate that it is second only to RNase T in supporting the viability of cells (Kelly and Deutscher 1992a). Such cells grow in rich media with a doubling time of 40 minutes, indicating that RNase PH can very effectively take over the function of all the missing ribonucleases. Studies of *rph* mutants indicate that RNase PH is required for tRNA processing when some other RNases are missing, particularly RNase T or PNPase (Kelly et al. 1992). Cells lacking RNase T and RNase PH grow with a doubling time of 62 minutes in rich media and, based on a suppression assay with tRNA$^{Tyr}su_3^+$, synthesize tRNA only 30% as well as wild-type cells. Likewise, removal of RNase PH can have major effects on tRNA processing in various other RNase-deficient cells, and in addition, such cells appear to accumulate tRNA precursors (Reuven and Deutscher 1993). In an in vitro tRNA processing system that uses a precursor to tRNA$^{Tyr}su_3^+$ as substrate, RNase PH is most effective in shortening the precursor from about four to six additional residues to about one or two (Kelly et al. 1992). At higher concentrations, it can complete the processing, but RNase T apparently removes the last few residues more effectively (Z. Li and M.P. Deutscher, unpubl.). When both enzymes are missing, 3' processing is quite poor.

V. CONCLUSIONS

There is no question that the area of RNA maturation has become extremely complex. Given the number of RNA molecules already known to be present in cells, and the number of different maturation events that these RNA molecules undergo, it is clear that the maturation RNases described here are only a few of what will turn out to be a large family of enzymes. Already numerous other activities have been identified that probably play a role in RNA maturation but have not yet been characterized in detail as enzymes. There are still other RNases that have been purified and characterized, but their involvement in RNA maturation has not been demonstrated. With the genetic and biochemical tools now being brought to this problem, and the number of RNA maturation events remaining to be described in detail, the next several years should greatly expand our knowledge of maturation nucleases.

ACKNOWLEDGMENTS

Work from the author's laboratory was supported by grant GM-16317 from the National Institutes of Health. I thank Allen W. Nicholson for providing information prior to publication.

REFERENCES

Abelson, J. 1992. Recognition of tRNA precursors: A role for the intron. *Science* **255:** 1390.

Altman, S. 1989. Ribonuclease P: An enzyme with a catalytic RNA subunit. *Adv. Enzymol.* **62:** 1–36.

Altman, S., L. Kirsebom, and S. Talbot. 1993. Recent studies of ribonuclease P. *FASEB J.* **7:** 7–14.

Altman, S., C. Guerrier-Takada, H.M. Frankfort, and H.D. Robertson. 1982. RNA processing nucleases. In *Nucleases* (ed S.M. Linn and R.J. Roberts), pp. 243–274. Cold Spring Harbor Laboratory Press, Cold Spring Harbor, New York.

Altman, S., M. Baer, H. Gold, C. Guerrier-Takada, L. Kirsebom, N. Lawrence, N. Lumelsky, and A. Vioque. 1987. Cleavage of RNA by RNase P from *Escherichia coli*. In *Molecular biology of RNA* (ed. M. Inouye and B. Dudock), pp. 3–15. Academic Press, San Diego.

Andrew, C., A.K. Hopper, and B.D. Hall. 1976. A yeast mutant defective in the processing of 27S rRNA precursor. *Mol. Gen. Gen.* **144:** 29–35.

Asha, P.K. and M.P. Deutscher. 1983. *Escherichia coli* strain CAN lacks a tRNA processing nuclease. *J. Bacteriol.* **156:** 419–420.

Asha, P.K., R.T. Blouin, R. Zaniewski, and M.P. Deutscher. 1983. Ribonuclease BN: Identification and partial characterization a new tRNA processing enzyme. *Proc. Natl. Acad. Sci.* **80:** 3301–3304.

Baldi, M.I., E. Mattoccia, E. Bufardeci, S. Fabbri, and G.P. Tocchini-Valentini. 1992. Participation of the intron in the reaction catalyzed by the *Xenopus* tRNA splicing endonuclease. *Science* **255:** 1404–1408.

Bardwell, J.C.A., P. Regnier, S.-M. Chen, M. Grunberg-Manago, Y. Nakamura, and D.L. Court. 1989. Autoregulation of RNase III operon by mRNA processing. *EMBO J.* **8:** 3401–3407.

Bikoff, E.K., B.F. LaRue, and M.L. Gefter. 1975. *In vitro* synthesis of transfer RNA II. Identification of required enzymatic activities. *J. Biol. Chem.* **250:** 6248–6255.

Birenbaum, M., D. Schlessinger, and Y. Ohnishi. 1980. Altered bacteriophage T4 ribonucleic acid metabolism in a ribonuclease II-deficient mutant of *Escherichia coli*. *J. Bacteriol.* **142:** 327–330.

Blouin, R.T., R. Zaniewski, and M.P. Deutscher. 1983. Ribonuclease D is not essential for the normal growth of *Escherichia coli* or bacteriophage T4 or for the biosynthesis of a T4 suppressor tRNA. *J. Biol. Chem.* **258:** 1423–1426.

Bonekamp, F., H.D. Andersen, T. Christensen, and K.F. Jensen. 1985. Codon defined ribosomal pausing in *Escherichia coli* detected by using the *pyrE* attenuator to probe the uncoupling between transcription and translation. *Nucleic Acids Res.* **13:** 4113–4123.

Bouvet, P. and J.G. Belasco. 1992. Control of RNase E-mediated RNA degradation by 5′ terminal base pairing in *E. coli. Nature* **360:** 488–491.

Bram, R.J., R.A. Young, and J.A. Steitz. 1980. The ribonuclease III site flanking 23S sequences in the 30S ribosomal precursor RNA of *E. coli. Cell* **19:** 393–401.

Brown, J.W. and N.R. Pace. 1992. Ribonuclease P RNA and protein subunits from bacteria. *Nucleic Acids Res.* **20:** 1451–1456.

Casaregola, S., A. Jacq, D. Laoudj, G. McGurk, S. Margarson, M. Tempete, V. Norris, and I.B. Holland. 1992. Cloning and analysis of the entire *Escherichia coli ams* gene. *J. Mol. Biol.* **228:** 30–40.

Case, L.M., X. Chen, and M.P. Deutscher. 1989. Localization of the *Escherichia coli rnt* gene encoding RNase T using a combination of physical and genetic mapping. *J. Bacteriol.* **171:** 5736–5737.

Chelladurai, B.S., H.L. Li, and A.W. Nicholson. 1991. A conserved sequence element in ribonuclease III processing signals is not required for accurate *in vitro* enzymatic cleavage. *Nucleic Acids Res.* **19:** 1759–1766.

Chen, S.-M., H.E. Takiff, A.M. Barber, G.C. Dubois, J.C.A. Bardwell, and D.L. Court. 1990. Expression and characterization of RNase III and era proteins. *J. Biol. Chem.* **265:** 2888–2895.

Craig, N., S. Kass, and B. Sollner-Webb. 1987. Nucleotide sequence determining the first cleavage site in the processing of mouse precursor rRNA. *Proc. Natl. Acad. Sci.* **84:** 629–633.

Craven, M.G., D.J. Henner, D. Alessi, A.T. Schauer, K.A. Ost, M.P. Deutscher, and D.I. Friedman. 1992. Identification of the *rph* (RNase PH) gene of *B. subtilis*: Evidence for suppression of *Cs* mutations in *E. coli. J. Bacteriol.* **174:** 4727–4735.

Crouch, R.J. 1984. Ribosomal RNA processing in eukaryotes. In *Processing of RNA* (ed. D. Apirion), pp. 214–226. CRC Press, Boca Raton, Florida.

Cudny, H. and M.P. Deutscher. 1980. Apparent involvement of RNase D in the 3′ processing of tRNA precursors. *Proc. Natl. Acad. Sci.* **77:** 837–841.

———. 1988. 3′ Processing of tRNA precursors in ribonuclease-deficient *Escherichia coli. J. Biol. Chem.* **263:** 1518–1523.

Cudny, H., R. Zaniewski, and M.P. Deutscher. 1981. *E. coli* RNase D. Catalytic properties and substrate specificity. *J. Biol. Chem.* **256:** 5633–5637.

Dahlberg, A.E., J.E. Dahlberg, E. Lund, H. Tokimatsu, A.B. Rabson, P.C. Calvert, F. Reynolds, and M. Zahalak. 1978. Processing of the 5′ end of *Escherichia coli* 16S ribosomal RNA. *Proc. Natl. Acad. Sci.* **75:** 3598–3602.

Darr, S.C., J.W. Brown, and N.R. Pace. 1992. The varieties of ribonuclease P. *Trends Biochem. Sci.* **17:** 178–182.

Deutscher, M.P. 1984. Processing of tRNA in prokaryotes and eukaryotes. *Crit. Rev. Biochem.* **17:** 45–71.

———. 1985. *E. coli* RNases: Making sense of alphabet soup. *Cell* **40:** 731–732.

———. 1988. The metabolic role of RNases. *Trends Biochem. Sci.* **13:** 136–139.

———. 1990. Ribonucleases, tRNA nucleotidyltransferase and the 3′ processing of tRNA. *Prog. Nucleic Acid Res. Mol. Biol.* **39:** 209–240.

———. 1993. tRNA processing nucleases. In *Transfer RNA* (ed. D. Söll and U.L. Rajbhandary). American Society of Microbiology, Washington, D.C. (In press.)

Deutscher, M.P. and C.W. Marlor. 1985. Purification and characterization of *E. coli*

RNase T. *J. Biol. Chem.* **260:** 7067-7071.
Deutscher, M.P. and N.B. Reuven. 1991. Enzymatic basis for hydrolytic versus phosphorolytic mRNA degradation in *Escherichia coli* and *Bacillus subtilis*. *Proc. Natl. Acad. Sci.* **88:** 3277-3280.
Deutscher, M.P., J.C. Lin, and J.A. Evans. 1977. Transfer RNA metabolism in *Escherichia coli* cells deficient in tRNA nucleotidyltransferase. *J. Mol. Biol.* **117:** 1081-1094.
Deutscher, M.P., C.W. Marlor, and R. Zaniewski. 1984. Ribonuclease T: A new exoribonuclease possibly involved in end-turnover of tRNA. *Proc. Natl. Acad. Sci.* **81:** 4290-4293.
―――. 1985. RNase T is responsible for the end-turnover of tRNA in *Escherichia coli*. *Proc. Natl. Acad. Sci.* **82:** 6427-6430.
Deutscher, M.P., G.T. Marshall, and H. Cudny. 1988. RNase PH: A new phosphate-dependent nuclease distinct from polynucleotide phosphorylase. *Proc. Natl. Acad. Sci.* **85:** 4710-4714.
Donovan, W.P. and S.R. Kushner. 1983. Amplication of ribonuclease II (*rnb*) activity in *Escherichia coli* K-12. *Nucleic Acids Res.* **11:** 265-275.
―――. 1986. Polynucleotide phosphorylase and ribonuclease II are required for cell viability and mRNA turnover in *Escherichia coli* K12. *Proc. Natl. Acad. Sci.* **83:** 120-124.
Dunn, J.J. 1982. Ribonuclease III. In *The enzymes* (ed. P.D. Boyer), vol. 15, part B, pp. 485-499. Academic Press, New York.
Dunn, J.J. and F.W. Studier. 1973. T7 early RNAs and *Escherichia coli* ribosomal RNAs are cut from large precursors *in vivo* by ribonuclease III. *Proc. Natl. Acad. Sci.* **70:** 3296-3300.
Ehretsmann, C.P., A.J. Carpousis, and H.M. Krisch. 1992a. mRNA degradation in prokaryotes. *FASEB J.* **6:** 3186-3192.
―――. 1992b. Specificity of *E. coli* endoribonuclease E: *In vivo* and *in vitro* analysis of mutants in a bacteriophage T4 mRNA processing site. *Genes Dev.* **6:** 149-159.
Faubladier, M., K. Cam, and J.P. Bouche. 1990. *E. coli* cell division inhibitor *DicF* RNA of the *dicB* operon. Evidence for its generation *in vivo* by transcription termination and by RNase III and RNase E dependent processing. *J. Mol. Biol.* **212:** 461-471.
Forster, A.C. and S. Altman. 1990. External guide sequence for an RNA enzyme. *Science* **249:** 783-786.
Gandini-Attardi, D., I.M. Boldi, E. Mattoccia, and G.P. Tocchini-Valentini. 1990. Transfer RNA splicing endonuclease from *Xenopus laevis*. *Methods Enzymol.* **181(B):** 510-517.
Gegenheimer, P. and D. Apirion. 1981. Processing of prokaryotic ribonucleic acid. *Microbiol. Rev.* **45:** 502-541.
Gegenheimer, P., N. Watson, and D. Apirion. 1977. Multiple pathways for primary processing of ribosomal RNA in *Escherichia coli*. *J. Biol. Chem.* **252:** 3064-3068.
Ghosh, R.K. and M.P. Deutscher. 1978a. Purification of potential 3' processing nucleases using synthetic tRNA precursors. *Nucleic Acids Res.* **5:** 3831-3842.
―――. 1978b. Identification of an *E. coli* nuclease acting on structurally altered transfer RNA molecules. *J. Biol. Chem.* **253:** 997-1000.
Gitelman, D.R. and D. Apirion. 1980. The synthesis of some proteins is affected in RNA processing mutants of *Escherichia coli*. *Biochem. Biophys. Res. Commun.* **96:** 1063-1070.
Goldblum, K. and D. Apirion. 1981. Inactivation of the ribonucleic acid processing enzyme ribonuclease E blocks cell division. *J. Bacteriol.* **146:** 128-132.

Goldfarb, A., E. Seaman, and V. Daniel. 1978. *In vitro* transcription and isolation of a polycistronic RNA product of the T4 tRNA operon. *Nature* 273: 562–564.

Green, P.R. and J.N. Abelson. 1990. Highly purified transfer RNA splicing endonuclease from *Saccharomyces cerevisiae*. *Methods Enzymol.* 181(B): 471–480.

Guarneros, G. and C. Portier. 1990. Different specificities of ribonuclease II and polynucleotide phosphorylase in 3′ mRNA decay. *Biochimie* 72: 771–777.

Guerrier-Takada, C., K. Gardiner, T. Marsh, N.R. Pace, and S. Altman. 1983. The RNA moiety of RNase P is the catalytic subunit of the enzyme. *Cell* 35: 849–857.

Guthrie, C., J.G. Seidman, M.M. Comer, R.M. Bock, F.J. Schmidt, B.G. Barrell, and W.H. McClain. 1975. The biology of bacteriophage T4 transfer RNAs. *Brookhaven Symp. Biol.* 26: 106–123.

Hannon, G.J., P.A. Maroney, A. Branch, B.J. Benenfield, H.D. Robertson, and T.W. Nilsen. 1989. Accurate processing of human pre-rRNA *in vitro*. *Mol. Cell. Biol.* 9: 4422–4431.

Hansen, F.G., E.G. Hansen, and T. Atlung. 1985. Physical mapping and nucleotide sequence of the *rnpA* gene that encodes the protein component of RNase P in *Escherichia coli*. *Gene* 38: 85–93.

Hayes, F. and M. Vasseur. 1976. Processing of the 17S *Escherichia coli* precursor RNA in the 27-S preribosomal particle. *Eur. J. Biochem.* 61: 433–442.

Ho, C.K., R. Rauhut, U. Vijayraghaven, and J. Abelson. 1990. Accumulation of pre-tRNA splicing 2′/3′ intermediates in a *Saccharomyces cerevisiae* mutant. *EMBO J.* 9: 1245–1252.

Huang, S. and M.P. Deutscher. 1992. Sequence and transcriptional analysis of the *Escherichia coli rnt* gene encoding RNase T. *J. Biol. Chem.* 267: 25609–25613.

Hudson, L., J. Rossi, and A. Landy. 1981. Dual function transcripts specifying tRNA and mRNA. *Nature* 294: 422–427.

Ishii, S., K. Kuroki, and F. Imamoto. 1984. tRNA^{Met} and IF2 genes in the leader region of the *nusA* operon in *Escherichia coli*. *Proc. Natl. Acad. Sci.* 81: 409–413.

Jensen, K.F., J.T. Andersen, and P. Poulsen. 1992. Overexpression and rapid purification of the *orfE/rph* gene product, RNase pH of *Escherichia coli*. *J. Biol. Chem.* 267: 17147–17152.

Kass, S., K. Tye, J.A. Steitz, and B. Sollner-Webb. 1990. The U3 small nucleolar ribonucleoprotein functions in the first step of preribosomal RNA processing. *Cell* 60: 897–908.

Kelly, K.O. and M.P. Deutscher. 1992a. The presence of only one of five exoribonucleases is sufficient to support the growth of *Escherichia coli*. *J. Bacteriol.* 174: 6682–6684.

———. 1992b. Characterization of *Escherichia coli* RNase PH. *J Biol. Chem.* 267: 17153–17158.

Kelly, K.O., N.B. Reuven, Z. Li, and M.P. Deutscher. 1992. RNase PH is essential for tRNA processing and viability in RNase-deficient *Escherichia coli* cells. *J. Biol. Chem.* 267: 16015–16018.

King, T.C. and D. Schlessinger. 1983. S1 nuclease mapping analysis of ribosomal RNA processing in wild type and processing deficient *Escherichia coli*. *J. Biol. Chem.* 258: 12034–12042.

King, T.C., R. Sirdeskmukh, and D. Schlessinger. 1986. Nucleolytic processing of ribonucleic acid transcripts in procaryotes. *Microbiol. Rev.* 50: 428–451.

Kokoska, R.J., K.J. Blumer, and D.A. Steege. 1990. Phage f1 mRNA processing in *Escherichia coli*: Search for the upstream products of endonuclease cleavage, requirement for the product of the altered mRNA stability (*ams*) locus. *Biochimie* 72: 803–811.

Komire, Y., A. Toshiharu, H. Inokuchi, and H. Ozeki. 1990. Genomic organization and physical mapping of the transfer RNA genes in *Escherichia coli*. *J. Mol. Biol.* **212:** 579-598.

Li, H.L., J. Zagorski, and M.J. Fournier. 1990. Depletion of U14 small nuclear RNA (SNR128) disrupts production of 18S rRNA in *Saccharomyces cerevisiae*. *Mol. Cell. Biol.* **10:** 1145-1152.

Lundberg, E., A. von Gabain, and O. Melefors. 1990. Cleavages in the 5' region of the ompA and bla mRNA control stability: Studies with an *E. coli* mutant altering mRNA stability and a novel endoribonuclease. *EMBO J.* **9:** 2731-2741.

Maisurian, A.N. and E.A. Buyanovskaya. 1973. Isolation of an *Escherichia coli* strain restricting bacteriophage suppressor. *Mol. Gen. Genet.* **120:** 227-229.

March, P.E. and M.A. Gonzalez. 1990. Characterization of the biochemical properties of recombinant ribonuclease III. *Nucleic Acids Res.* **18:** 3293-3299.

March, P.E., A. Joohong, and M. Inouye. 1985. The DNA sequence of the gene (*rnc*) encoding ribonuclease III of *Escherichia coli*. *Nucleic Acids Res.* **13:** 4677-4685.

Mayer, J.E. and M. Schweiger. 1983. RNase III is positively regulated by T7 protein kinase. *J. Biol. Chem.* **285:** 5340-5343.

McClain, W.H. 1977. Seven terminal steps in a biosynthetic pathway leading from DNA to transfer RNA. *Accts. Chem. Res.* **10:** 418-425.

McLaren, R.S., S.F. Newbury, G.S.C. Dance, H.C. Causton, and C.F. Higgins. 1991. mRNA degradation by processive 3'-5' exoribonucleases *in vitro* and the implications for prokaryotic mRNA decay *in vivo*. *J. Mol. Biol.* **221:** 81-95.

Meyhack, B. and N.R. Pace. 1978. Involvement of the mature domain in the *in vitro* maturation of *Bacillus subtilis* precursor 5S ribosomal RNA. *Biochemistry* **17:** 5804-5810.

Meyhack, B., B. Pace, and N.R. Pace. 1977. Involvement of precursor-specific segments in the *in vitro* maturation of *Bacillus subtilis* precursor 5S ribosomal RNA. *Biochemistry* **16:** 5009-5015.

Miao, F. and J. Abelson. 1993. Yeast tRNA-splicing endonuclease cleaves tRNA in a random pathway. *J. Biol. Chem.* **268:** 672-677.

Misra, T.K. and D. Apirion. 1979. RNase E, an RNA processing enzyme from *Escherichia coli*. *J. Biol. Chem.* **254:** 11154-11159.

———. 1980. Gene *rne* affects the structure of the ribonucleic acid processing enzyme ribonuclease E of *Escherichia coli*. *J. Bacteriol.* **142:** 359-361.

Mudd, E.A., P. Prentki, D. Balin, and H.M. Krisch. 1988. Processing of unstable bacteriophage T4 gene 32 mRNAs into a stable species requires *E. coli* ribonuclease E. *EMBO J.* **7:** 3601-3607.

O'Connor, P.J. and C.L. Peebles. 1991. *In vivo* pre-tRNA processing in *Saccharomyces cerevisiae*. *Mol. Cell. Biol.* **11:** 425-439.

Ost, K.A. and M.P. Deutscher. 1990. RNase PH catalyzes a synthetic reaction, the addition of nucleotides to the 3' end of tRNA. *Biochimie* **72:** 813-818.

———. 1991. *Escherichia coli orfE* (upstream of *pyrE*) encodes RNase PH. *J. Bacteriol.* **173:** 5589-5591.

Pace, B., D.A. Stahl, and N.R. Pace. 1984. The catalytic element of a ribosomal RNA processing complex. *J. Biol. Chem.* **259:** 11454-11458.

Padmanabha, K.P. and M.P. Deutscher. 1991. RNase T affects growth and recovery of *Escherichia coli* from metabolic stress. *J. Bacteriol.* **173:** 1376-1381.

Peebles, C.L., P. Gegenheimer, and J. Abelson. 1983. Precise excision of intervening sequences from precursor tRNAs by a membrane-associated yeast endonuclease. *Cell* **32:** 525-536.

Perry, R.P. 1976. Processing of RNA. *Annu. Rev. Biochem.* **45:** 605-629.

Poulsen, P., J.T. Andersen, and K.F. Jensen. 1989. Molecular and mutational analysis of three genes preceding *pyrE* on the *Escherichia coli* chromosome. *Mol. Microbiol.* **3:** 393-404.

Poulsen, P., F. Bonekamp, and K.F. Jensen. 1984. Structure of the *Escherichia coli* operon and control of *pyrE* expression by a UTP modulated intercistronic attenuation. *EMBO J.* **3:** 1783-1790.

Poulsen, P., K.F. Jensen, P. Valentin-Hansen, P. Carlsson, and L.G. Lundberg. 1983. Nucleotide sequence of the *Escherichia coli pyrE* gene and of the DNA in front of the protein-coding region. *Eur. J. Biochem.* **135:** 223-229.

Proudfoot, N. 1991. Poly(A) signals. *Cell* **64:** 671-674.

Rauhut, R., P.R. Green, and J. Abelson. 1990. Yeast tRNA-splicing endonuclease is a heterotrimeric enzyme. *J. Biol. Chem.* **265:** 18180-18184.

Reeder, R.H. 1990. rRNA synthesis in the nucleolus. *Trends Genet.* **6:** 390-395.

Regnier, P. and M. Grunberg-Manago. 1989. Cleavage by RNase III in the transcripts of the *metY-nusA-infB* operon of *Escherichia coli* releases the tRNA and initiates the decay of the downstream mRNA. *J. Mol. Biol.* **210:** 293-302.

―――. 1990. RNase III cleavages in non-coding leaders of *Escherichia coli* transcripts control mRNA stability and genetic expression. *Biochimie* **72:** 825-834.

Reuven, N.B. and M.P. Deutscher. 1993. Multiple exoribonucleases are required for the 3′ processing of *Escherichia coli* tRNA precursors *in vivo*. *FASEB J.* **7:** 143-148.

Robertson, H.D. 1990. *Escherichia coli* ribonuclease III. *Methods Enzymol.* **181:** 189-202.

Robertson, H.D., S. Altman, and J.D. Smith. 1972. Purification and properties of a specific *Escherichia coli* ribonuclease which cleaves a tyrosine transfer RNA. *J. Biol. Chem.* **247:** 5243-5251.

Robertson, H.D., R.E. Webster, and N.D. Zinder. 1968. Purification and properties of ribonuclease III from *Escherichia coli*. *J. Biol. Chem.* **243:** 82-91.

Rosenberg, M., R.A. Kramer, and J.A. Steitz. 1974. T7 early messenger RNAs are the direct products of ribonuclease III cleavage. *J. Mol. Biol.* **89:** 777-782.

Rossi, J., J. Egan, L. Hudson, and A. Landy. 1981. The *tyrT* locus: Termination and processing of a complex transcript. *Cell* **26:** 305-314.

Roy, M.K., B. Singh, B.K. Ray, and D. Apirion. 1983. Maturation of 5S rRNA: Ribonuclease E cleavages and their dependence on precursor sequences. *Eur. J. Biochem.* **131:** 119-127.

Schedl, P. and P. Primakoff. 1973. Mutants of *Escherichia coli* thermosensitive for the synthesis of transfer RNA. *Proc. Natl. Acad. Sci.* **70:** 2091-2095.

Schedl, P., J. Roberts, and P. Primakoff. 1976. *In vitro* processing of *E. coli* tRNA precursors. *Cell* **8:** 581-594.

Seidman, J.G., F.J. Schmidt, K. Foss, and W.H. McClain. 1975. A mutant of *Escherichia coli* defective in removing 3′ terminal nucleotides from some transfer RNA precursor molecules. *Cell* **5:** 389-400.

Shen, V. and D. Schlessinger. 1982. RNases I, II, and IV of *Escherichia coli*. In *The enzymes* (ed. P.D. Boyer), vol. 15, part B, pp. 501-515. Academic Press, New York.

Shuai, K. and J.R. Warner. 1991. A temperature sensitive mutant of *Saccharomyces cerevesiae* defective in pre-mRNA processing. *Nucleic Acids Res.* **19:** 5059-5064.

Sogin, M., B. Pace, and N.R. Pace. 1977. Partial purification and properties of a ribosomal RNA maturation endonuclease from *Bacillus subtilis*. *J. Biol. Chem.* **252:** 1350-1357.

Sohlberg, B., U. Lundberg, F.-U. Hartl, and A. von Gabain. 1993. Functional interaction of heat shock protein GroEL with an RNase E-like activity in *Escherichia coli. Proc. Natl. Acad. Sci.* **90:** 277-281.

Srivastava, A.K. and D. Schlessinger. 1990. Mechanism and regulation of bacterial ribosomal RNA processing. *Annu. Rev. Microbiol.* **44:** 105-129.

———. 1991. Structure and organization of ribosomal DNA. *Biochimie* **73:** 631-638.

Stahl, D.A., B. Meyhack, and N.R. Pace. 1980. Recognition of local nucleotide conformation in contrast to sequence by a rRNA processing endonuclease. *Proc. Natl. Acad. Sci.* **77:** 5644-5648.

Stahl, D.A., B. Pace, T. Marsh, and N.R. Pace. 1984. The ribonucleoprotein substrate for a ribosomal RNA processing nuclease. *J. Biol. Chem.* **259:** 11448-11453.

Steege, D.A., K.C. Cone, C. Queen, and M. Rosenberg. 1987. Bacteriophage λ N gene leader RNA. *J. Biol. Chem.* **262:** 17651-17658.

Takagaki, Y. and J.L. Manley. 1992. A human polyadenylation factor is a G protein β subunit homologue. *J. Biol. Chem.* **267:** 23471-23474.

Takagaki, Y., L.C. Ryner, and J.L. Manley. 1989. Four factors are required for 3′-end cleavage of pre-mRNAs. *Genes Dev.* **3:** 1711-1724.

Tomcsanyi, T. and D. Apirion. 1985. Processing enzyme ribonuclease E cleaves RNA 1 an inhibitor of primer formation in plasmid DNA synthesis. *J. Mol. Biol.* **185:** 713-720.

Vold, B. 1985. Structure and organization of genes for transfer RNA in *Bacillus subtilis. Microbiol. Rev.* **49:** 71-80.

Xu, Q., E.M. Phizicky, C.L. Greer, and J.N. Abelson. 1990. Purification of yeast transfer RNA ligase. *Methods Enzymol.* **181(B):** 463-471.

Yip, M.T. and M.J. Holland. 1989. In vitro RNA processing generates mature 3′ termini of yeast 35S and 25S ribosomal RNAs. *J. Biol. Chem.* **264:** 4045-4051.

Young, R.A. and J.A. Steitz. 1978. Complementary sequences 1700 nucleotides apart form a ribonuclease III cleavage site in *Escherichia coli* ribosomal precursor RNA. *Proc. Natl. Acad. Sci.* **75:** 3593-3597.

Zaniewski, R. and M.P. Deutscher. 1982. Genetic mapping of a mutation in *Escherichia coli* leading to a temperature-sensitive RNase D. *Mol. Gen. Genet.* **185:** 142-147.

Zaniewski, R., E. Petkaitis, and M.P. Deutscher. 1984. A multiple mutant of *Escherichia coli* lacking the exoribonucleases RNase II, RNase D and RNase BN. *J. Biol. Chem.* **259:** 11651-11653.

Zhang, J. and M.P. Deutscher. 1988a. Cloning, characterization and physiological consequences of overexpression of the *Escherichia coli rnd* gene encoding RNase D. *J. Bacteriol.* **170:** 522-527.

———. 1988b. *Escherichia coli* RNase D: Sequencing of the *rnd* structural gene and purification of the overexpressed protein. *Nucleic Acids Res.* **16:** 6265-6278.

———. 1988c. Transfer RNA is a substrate for RNase D *in vivo. J Biol. Chem.* **263:** 17909-17912.

———. 1989. Analysis of the upstream region of the *Escherichia coli rnd* gene encoding RNase D: Evidence for translational regulation of a putative tRNA processing enzyme. *J. Biol. Chem.* **264:** 18228-18233.

———. 1992. A uridine-rich sequence required for translation of prokaryotic mRNA. *Proc. Natl. Acad. Sci.* **89:** 2605-2609.

13
Nucleases That Are RNA

Michael D. Been
Department of Biochemistry
Duke University Medical Center
Durham, North Carolina 27710

I. Introduction
   A. Discovery of RNA-mediated RNA Cleavage and Ligation: Self-splicing Ribozymes
   B. Ribozymes That Are Enzymes
   C. Examples of RNAs That Contain Self-processing Sequences
   D. Exploiting the *Trans* Form of Ribozyme Reactions
   E. Overview of Splicing and Self-cleavage Mechanisms
II. Group I Introns
   A. General Features of the Self-splicing Mechanism
   B. Secondary Structure of Group I Introns
   C. The Guanosine-binding Site
   D. Structures Involved in Splice Site Selection
   E. The Intermolecular (*Trans*) Reaction and Site-specific Endonucleolytic Cleavage by Group I Ribozymes
   F. Kinetics of the *Trans*-reaction
   G. Enhancement of Activities of Group I Ribozymes by In Vitro Evolution
III. Ribozymes Derived from the Naturally Occurring Self-cleaving RNAs
   A. A Diverse Group of Ribozymes Self-cleave by a Similar Mechanism
   B. Hammerhead Ribozymes
   C. Hairpin/Paperclip Ribozyme
   D. Ribozymes from Hepatitis Delta Virus
IV. Summary and Perspectives

I. INTRODUCTION

A. Discovery of RNA-mediated RNA Cleavage and Ligation: Self-splicing Ribozymes

In some species of *Tetrahymena*, the gene for the large subunit ribosomal RNA (rRNA) is interrupted by an intron or intervening sequence (IVS) of about 400 nucleotides. The intron sequence is transcribed along with the exon sequences, and it is removed in a posttranscriptional processing event called RNA splicing. As with any precursor RNA that is processed by splicing, the RNA must be cleaved at the 5' and 3' ends of the intron (the 5' and 3' splice sites), the intron removed, and the exons (generally the structural or coding sequences) ligated to re-form the covalent backbone of the RNA. Cech et al. (1981) found that the *Tetrahymena* rRNA precursor (pre-rRNA) could be prepared in vitro for use as a substrate in the splicing reaction. Use of this pre-rRNA as a substrate to identify splicing factors led to the discovery of self-splicing RNA: RNA that spliced in vitro in the absence of proteins (Cech et al. 1981; Kruger et al. 1982). Thus, within 1 year of the first "Nuclease" meeting at Cold Spring Harbor Laboratory in 1981, the definition of "enzyme" was under revision. At that meeting, data demonstrating breakage and rejoining of a deproteinized RNA derived from the intron of the *Tetrahymena* pre-rRNA (Grabowski et al. 1981) were presented by P. Grabowski. Not only did the excised intron RNA catalyze its own breakage and rejoining, but the precursor RNA, from which the intron was excised, appeared to catalyze the intron excision (Cech et al. 1981). Although a number of laboratories were attempting to fractionate and purify splicing factors, at that point, none had suggested that the process might be catalyzed by anything but protein catalysts, and it is safe to say that there was some skepticism of the results reported from Cech's laboratory. In 1982, the definitive experiment demonstrating self-catalyzed intron excision from an in-vitro-synthesized RNA was reported (Kruger et al. 1982). The RNA sequence possessing the enzyme-like activity was the intron, and it was noted at the time that it catalyzed a single splicing event and that it was therefore not a true enzyme; hence, the term "ribozyme" was introduced. In retrospect, the intron sequence was sold short; circularization and reopening reactions of the intron after excision (Grabowski et al. 1981; Kruger et al. 1982; Zaug et al. 1984, 1985) were in fact additional rounds of cleavage and ligation catalyzed by the intron, and the intron catalyzed only a single round of splicing because the substrate, not the ribozyme, was depleted.

B. Ribozymes That Are Enzymes

Kruger et al. (1982) had suggested that some ribonuclear protein enzymes might be a form of a *trans*-acting ribozyme. RNase P is a ubiquitous enzyme, containing both protein and RNA components, that catalyzes the endonucleolytic processing of the 5' ends of precursor tRNAs. In 1983, the deproteinized RNA component of RNase P, from both *Escherichia coli* and *Bacillus subtilis*, was shown to contain the catalytic activity (Guerrier-Takada et al. 1983), and turnover was shown to be a property of the RNA component. Absolute proof that the RNase P RNA is an enzyme was demonstrated with the RNA transcribed in vitro (Guerrier-Takada and Altman 1984).

Multiple turnover by the excised *Tetrahymena* intron was demonstrated a few years later (Zaug and Cech 1986), when the ribozyme sequence was separated into an enzyme portion and a substrate portion. The term ribozyme is now usually applied to both the *cis*-acting self-processing RNAs and the *trans*-acting catalytic RNAs and is no longer used exclusively to denote the self-processing RNAs. In those cases where the *trans*-acting ribozymes turn over, they are enzymes by all criteria except composition (Guerrier-Takada et al. 1983; Zaug and Cech 1986; Uhlenbeck 1987).

C. Examples of RNAs That Contain Self-processing Sequences

The *Tetrahymena* intron is a member of a large class of introns, called group I introns. All group I introns appear to splice by the same chemical mechanism; however, not all group I introns can be demonstrated to self-splice. Group I introns have been found in the nuclei of lower eukaryotes, mitochondria, chloroplasts, eubacteria, and bacteriophages. The sequences form a common secondary structure (Fig. 1), and conserved long-range interactions are consistent with a common tertiary structure (Davies et al. 1982; Michel et al. 1982; Michel and Westhof 1990).

Group II introns, found in the organelles of plants and fungi and most recently in eubacteria (Ferat and Michel 1993), have a conserved secondary structure (Davies et al. 1982; Michel et al. 1982) that distinguishes them from the group I introns. Some group II introns are also capable of self-splicing in vitro (Peebles et al. 1986; Van der Veen et al. 1986) but by a mechanism distinctly different from that of the group I splicing introns. Because the group II introns and introns of pre-mRNAs in higher eukaryotes share a common splicing mechanism that produces a characteristic branched circle or lariat structure, it has been suggested the two are distantly related (Sharp 1985, 1991).

In addition to the self-splicing RNAs, a number of naturally occurring RNAs exist that self-cleave. RNA sequences capable of simple self-cleavage reactions were found in some plant virus satellite RNAs, virosoids, and a viroid (Buzayan et al. 1986a,b; Hutchins et al. 1986; Prody et al. 1986; Symons 1989, 1992). These RNAs, which are less than 400 nucleotides in length and encode no protein sequences, are likely dependent on the self-cleavage reaction to process multimeric sequences generated during the replication of these small RNAs to monomers (Symons 1992). In animals, there are two examples of self-cleaving sequences. A self-cleaving sequence was found in the transcript of a newt satellite DNA (Epstein and Gall 1987); the function of the transcript as well as the self-cleavage reaction is unknown. In the other example, each strand of the hepatitis delta virus (HDV) (Kuo et al. 1988; Sharmeen et al. 1988; Wu et al. 1989) contains self-cleavage sites that probably func-

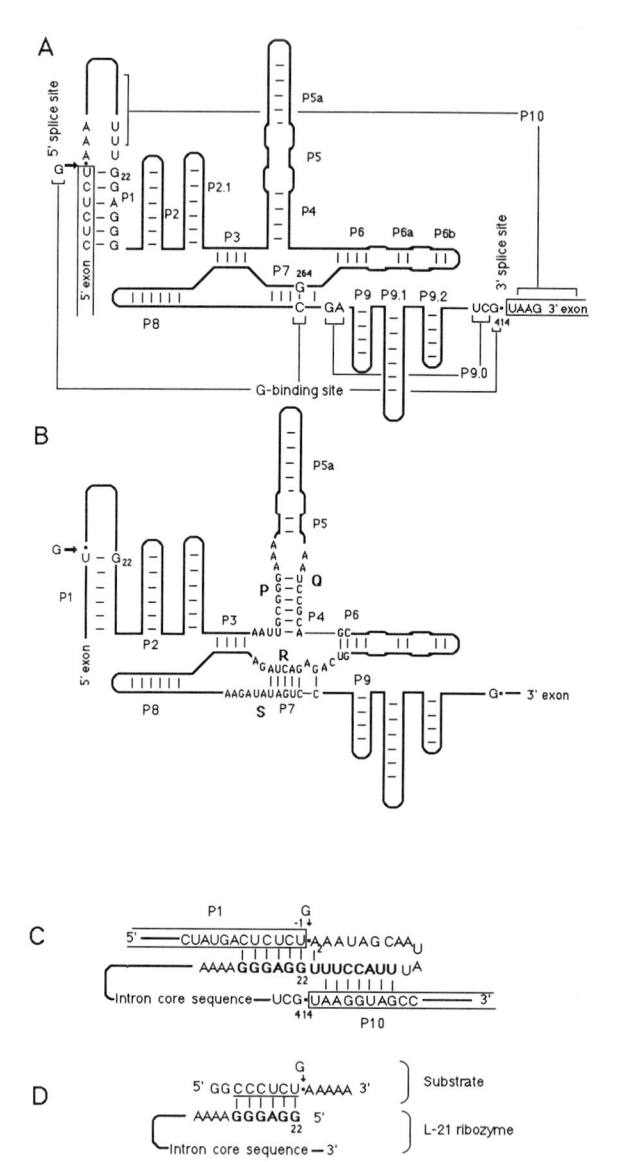

Figure 1 Group I intron ribozyme. (*A*) Line drawing of the secondary structure of the *Tetrahymena* intron. Interactions most directly involved in substrate selection, P1, P10, P9.0, and the G-binding site are labeled. (*B*) Location and sequence of the conserved sequence elements P, Q, R, and S in the *Tetrahymena* intron. (*C*) The precise alignment structure with the IGS in bold. (*D*) Base pairing between a 5' splice site analog and the L-21 ribozyme in a *trans*-cleavage reaction.

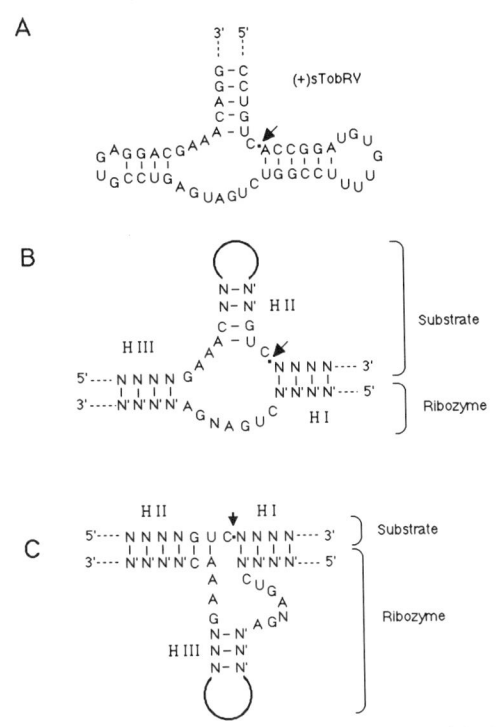

Figure 2 The hammerhead ribozyme. (*A*) Sequence from the cleavage site of the plus strand of the satellite of tobacco ringspot virus folded into the hammerhead motif (Forster and Symons 1987a). (*B*) One representation of a ribozyme-substrate complex in which helices I and III are formed from two RNA fragments in *trans* (Uhlenbeck 1987). (*C*) An alternative, but equivalent, representation of the hammerhead, but in this drawing the enzyme-substrate complex is formed with helices I and II in *trans* (Haseloff and Gerlach 1988). In *B* and *C*, the bases for which identity contributes to optimal activities are indicated as such, while N represents any base and N' its complement. For further discussion of minimal hammerhead structures, see Symons (1992).

tion in the replication of the RNA genome. Several of the self-cleaving sequences adopt a conserved three-stem structure called a hammerhead (see Fig. 2) (Forster and Symons 1987a). Exceptions to the hammerhead motif exist in the minus strand of the satellite of tobacco ringspot virus (sTobRV), which adopts a structure called the hairpin (Hampel et al. 1990) or paperclip (see Fig. 3) (Feldstein and Bruening 1993), and in the ribozymes from HDV and *Neurospora* mitochondria, which cannot be folded into either the hammerhead or hairpin/paperclip motif (Kuo et al. 1988; Sharmeen et al. 1988; Wu et al. 1989; Saville and Collins 1990).

Are there examples of ribozymes that catalyze reactions other than

```
              ↓
A  3' ------ U U U G U C  C U G·A  C A G U ---- 5'
             | | | | | |         | | | |
   5' ------ A A A C A G  A G A A  G U C A
                                        A - U  A C G ---- 3'
                                        C - G
        (-)sTobRV                       C - G
                                        A - U
                                        G - C
                                      A       C
                                      G       A
                                      A       U
                                      A       U
                                      A       A
                                      C       U
                                      A       A
                                      C       U
                                      A       G
                                      C       G
                                      G       U
                                        G U U
```

```
            H1              ↓       H2
B  3' ---- N N N N N B  Y H G·N  Y R N N ----- 5'
           | | | | | |          | | | |
   5' N' N' N' N' N' V          G Y N' N' N' - N N 3'
                      N G A A           N' - N
                                        N' - N  H3
                                        N' - N
                                        N' - N
                                         N   A
                                        G    C
                                        A    U
                                        A    A
                                        A   N  H
                                        C   N
                                        N
                                        N' - N
                                        N' - N  H4
                                        N' - N
                                         N   N
                                           N
```

Substrate

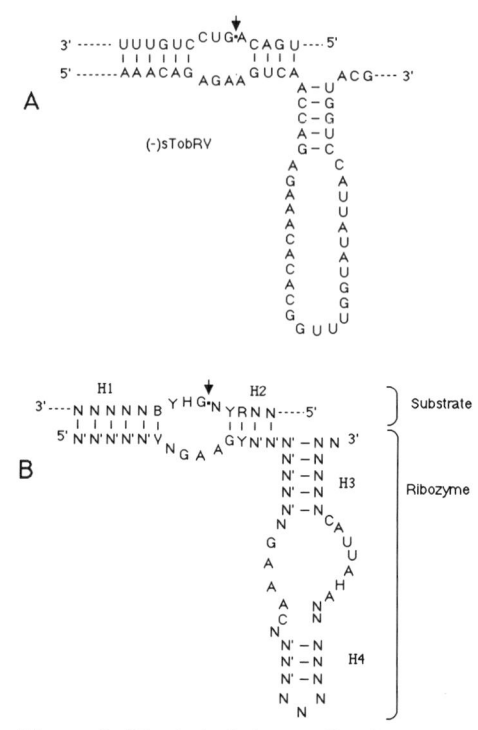

Ribozyme

Figure 3 The hairpin/paperclip ribozyme. (*A*) Sequence from the cleavage site of the minus strand of the satellite of tobacco ringspot virus in the hairpin/paperclip motif. (*B*) Structural and sequence requirements for hairpin/paperclip ribozyme activity as determined by in vitro mutagenesis and selection by Berzal-Herranz et al. (1993). N is A, G, U, or C; N' is a base complementary to N; B is G, U, or C; H is A, U, or C; Y is pyrimidine; R is purine. (Redrawn from Berzal-Herranz et al. 1993.)

RNA cleavage and ligation? Work by Noller and co-workers (Noller and Woese 1981; Noller et al. 1990; Noller 1991) has provided substantial evidence that the large subunit of rRNA contains the active site for peptidyl transferase. In a recent and dramatic example (Noller et al. 1992), the ribosomal large subunit, after proteinase digestion and phenol extraction, still catalyzed the formation of a model peptide bond by transferring fMet from a charged fragment of tRNA to the amino group of puromycin. Treatment with ribonuclease T1 abolishes the activity, suggesting that the RNA but perhaps not the protein is essential to form the active site of the peptidyl transferase activity. The authors do not claim to prove that the reaction is RNA-catalyzed, and they await demonstration that the same reaction can be catalyzed with an in vitro transcript, totally free of any ribosomal proteins or peptide fragments.

D. Exploiting the *Trans* Form of Ribozyme Reactions

Precursor RNAs, containing either the self-splicing or self-cleaving sequences, are most easily prepared by in vitro transcription with purified phage polymerases. The transcription conditions are usually favorable for the self-processing reaction, which can make it difficult to prepare large quantities of uncleaved precursor for subsequent analysis. As with the *Tetrahymena* intron, the hammerhead was separated into enzyme and substrate portions (Uhlenbeck 1987; Haseloff and Gerlach 1988) that facilitated further biochemical characterization and generated interest in the potential for ribozymes as small sequence-specific endonucleases (Haseloff and Gerlach 1988). Thus, a general approach to study ribozymes became established; ribozymes were divided into two fragments that reassociate through Watson-Crick pairing, thus forming the equivalent of an enzyme-substrate complex. By altering the sequences responsible for substrate binding, the potential existed to engineer the *trans*-acting ribozyme derivatives into sequence-specific endonucleases (Been and Cech 1986; Zaug et al. 1986; Uhlenbeck 1987; Haseloff and Gerlach 1988). Suggested uses have ranged from the immediately feasible (tools for in vitro analysis of RNA sequence and structure) to the exotic (drugs to cleave viral RNA sequences).

Ribozyme *trans*-reactions have been most useful, however, in the characterization of the mechanism of the reactions for the group I intron (Cech et al. 1992) and the hammerhead and hairpin/paperclip ribozymes (Long and Uhlenbeck 1993) and have also been fundamental for establishing crucial interactions in group II intron splice site selection (Michel et al. 1989). As a potential source of engineered sequence-specific endonucleases, the small self-cleaving sequences have received more attention than the group I or group II ribozymes, as they appear to be better suited for this purpose. This chapter focuses on the group I intron-derived ribozyme and the hammerhead and hairpin/paperclip ribozymes. These examples serve to demonstrate how the *trans*-reactions developed from the self-processing reactions have greatly facilitated the understanding of ribozymes. The broader field of ribozyme structure, mechanism, and potential application as well as the biology of group I and group II introns has been covered in detail in a number of excellent recent reviews (including, but not limited to Cech 1988, 1990, 1993; Michel et al. 1989; Altman 1990; Pace and Smith 1990; Belfort 1991; Castanotto et al. 1992; Cech et al. 1992; Long and Uhlenbeck 1993; Pyle 1993; Saldanha et al. 1993; von Ahsen and Schroeder 1993).

E. Overview of Splicing and Self-cleavage Mechanisms

Ribozymes cleave RNA by disrupting the phosphodiester backbone. As with protein nucleases, the nature of the end groups that are generated are characteristic of the specific ribozyme and its associated mechanism of cleavage (Fig. 4). All of the self-cleaving sequences generate a 2',3'-cyclic phosphate and a 5'-hydroxyl group on the ends of the RNA, suggesting that the most probable mechanism involves attack on the phosphorus by the adjacent 2'oxygen, with a transesterification, to generate the cyclic phosphodiester and 5'-hydroxyl leaving group. Group I introns also catalyze a transesterification reaction, but in the first step, the nucleophile is the 3'-hydroxyl group of the guanosine substrate, and cleavage results in a new 3'-hydroxyl group, whereas a new phosphodiester bond is formed between the guanosine and the 5' end of the intron. Water can substitute for guanosine as a nucleophile in the reaction with the group I ribozymes (Herschlag and Cech 1990b), but direct hydrolysis of the phosphodiester bond in a ribozyme derived from a self-cleaving RNA has not been observed (however, see Pan and Uhlenbeck 1992).

The group II introns and RNase P, like the group I introns, generate a 3'-hydroxyl group at the cleavage site. For RNase P, the nucleophile is H_2O (or OH^-) (Marsh and Pace 1985). For the group II introns, the nucleophile is a 2'-hydroxyl group on a specific adenosine within the intron (Peebles et al. 1986; Van der Veen et al. 1986). With the group II introns, a branched circular molecule (lariat) is generated (Peebles et al. 1986; Van der Veen et al. 1986). The branched circular form of the excised intron is also a characteristic of the splicing mechanism of the nuclear pre-mRNAs. This similarity is often noted as some of the evidence for the possibility that pre-mRNA spliceosome complexes may be ribozymes at heart (Sharp 1991).

The substrate for ribozymes is not limited to phosphorous centers in RNA. The group I intron ribozyme will cleave DNA (Robertson and Joyce 1990; Herschlag and Cech 1990c), and RNase P RNA will cleave substrates containing a deoxynucleotide at the cleavage site (Smith and Pace 1993). The group I intron will also catalyze hydrolysis of an aminoacyl ester (Piccirilli et al. 1992), providing further evidence that rRNA could catalyze reactions at carbon centers (Noller et al. 1992).

II. GROUP I INTRONS

A. General Features of the Self-splicing Mechanism

For those group I introns that self-splice in vitro, the required conditions are neutral pH, Mg^{++}, and guanosine (or guanosine nucleotide). Monovalent cations can stimulate the reaction under some conditions, and certain other divalent cations can substitute in all or part for Mg^{++} (Grosshans and Cech 1989). The basic splicing mechanism was recog-

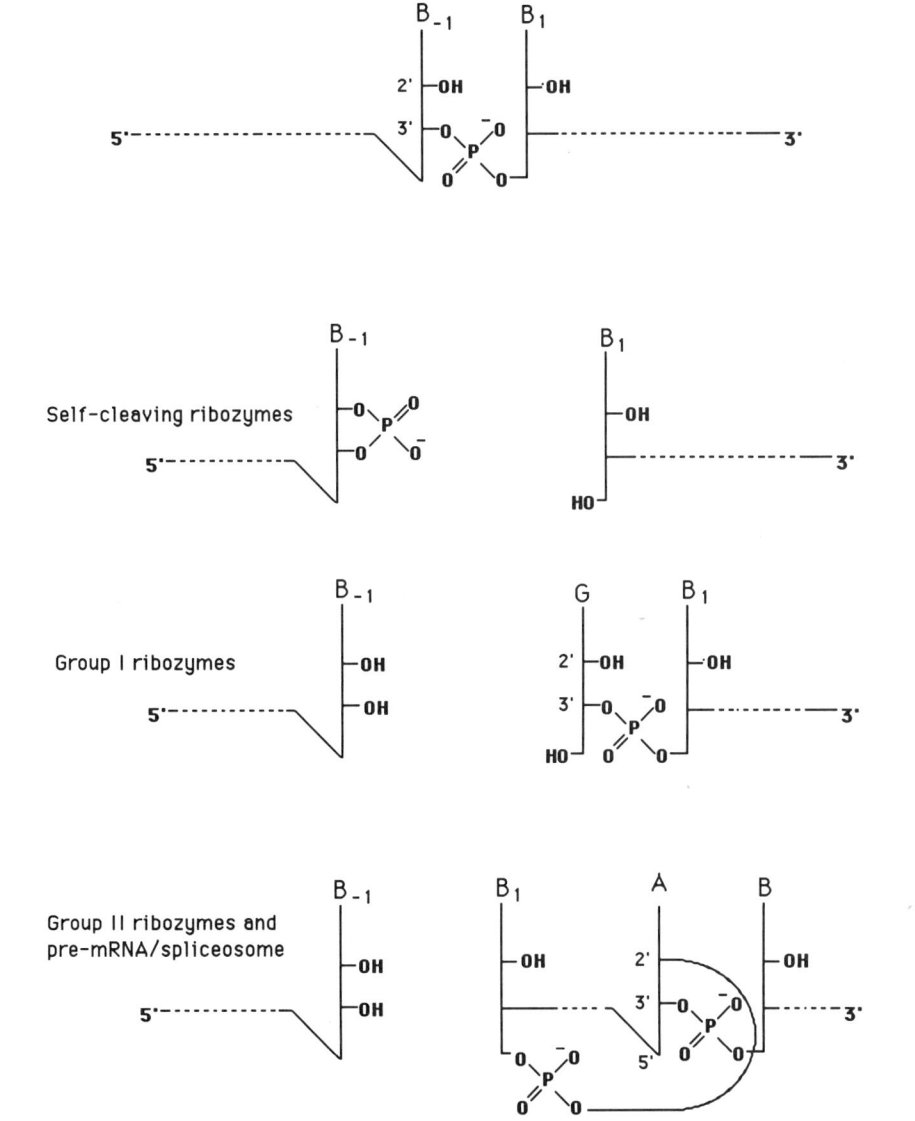

Figure 4 Products of self-cleavage, and reactions at the 5′ splice site with group I and group II ribozymes.

nized early on as comprising two phosphoester transfer or transesterification reactions (Cech et al. 1981; Cech 1987) in which there were two concerted cleavage and ligation steps. The first step is a nucleophilic attack by the 3′-hydroxyl group of guanosine on the phosphorus following the U in the 5′ exon (U_{-1} at the 5′ splice site). The guanosine becomes attached to the 5′ end of the intron and a 3′-hydroxyl group is generated on U_{-1}. Thus, one phosphodiester bond is broken but one is also made, so that the total number of phosphodiester bonds remains constant. This reaction has been shown to proceed through an inversion of configuration of the phosphate, consistent with the proposed mechanism and indicating that cleavage most likely occurs with a trigonal bipyramidal intermediate via attack by the 3'oxygen of G opposite and in-line with the 5'oxygen leaving group (McSwiggen and Cech 1989; Rajagopal et al. 1989).

In the exon ligation step, there is a second nucleophilic attack by the newly generated 3'-hydroxyl group of the 5'exon on the phosphorus following G_{414}, the 3'-terminal nucleotide of the intron at the 3'splice site. As this step is also a transesterification reaction, there is no requirement for ATP or any other energy-donating cofactor. This step also proceeds with inversion of configuration of the phosphate (Suh and Waring 1992). The first and second steps of the splicing reaction can be viewed as the simple forward and reverse of the same reaction (Kay and Inoue 1987): G_{OH} + UpN <==> U_{OH} + GpN, with the first step in splicing going to the right, and exon ligation going to the left.

In addition to the above-mentioned stereochemistry, the model for the mechanism has been further refined with data from studies using phosphorothioates replacing the splice site phosphates. Substitution of a nonbridging oxygen with sulfur generates two stereoisomers. Replacing the pro-Rp oxygen at the 5'splice site with sulfur does not inhibit the forward reaction (McSwiggen and Cech 1989; Rajagopal et al. 1989), but in the Sp configuration, the phosphorothioate is inhibitory (Herschlag 1991). Suh and Waring (1992) showed that the Rp phosphorothioate at the 3'splice site blocks the forward reaction (exon ligation) but not the reverse (G_{414} attack at the ligation junction). Thus, the stereochemistry is consistent with a single active site catalyzing both steps in the splicing reaction (Suh and Waring 1992). Substitution of the bridging 3'oxygen with sulfur in a 5'splice site analog results in a preference for Mn^{++} over Mg^{++}, consistent with a metal ion coordinated to the 3'oxygen (Piccirilli et al. 1993), which may stabilize a developing negative charge on this leaving group. Details of the proposed mechanism and a model for the transition state have been reviewed recently (Cech et al. 1992; Pyle 1993).

B. Secondary Structure of Group I Introns

Group I introns were first defined as a class because comparative sequence analysis revealed that the sequences could be folded into a distinctive secondary structure (Davies et al. 1982; Michel et al. 1982). Thus, the essential features of the secondary structure of the group I ribozymes had been defined by phylogenetic comparisons concurrently with the discovery of self-splicing (Davies et al. 1982; Michel et al. 1982). By convention (Burke et al. 1987), ten major pairing regions (P1 through P10) common to most group I introns are numbered 5′ to 3′, with the numerical value indicating a defined position relative to the consensus structure (see Fig. 1A). Because the sizes of group I introns can vary from slightly larger than 200 bases to well over a kilobase, allowances in the numbering system are made to accommodate the variations that are found. As a result, intron sequences missing a particular pairing (e.g., P2) or having additional pairings (e.g., between P2 and P3, or between P9 and P10) will still have a guanosine-binding site formed by P7 and the 3′ splice site specified in part by P10 (see below).

Despite extensive conservation of secondary structure, conservation of primary structure is mostly limited to four regions that form the core or catalytic center of the molecule (see Fig. 1B). These conserved sequences, or elements, are named as proposed by Waring and Davies (1984) and Burke et al. (1987) 5′ to 3′ as P, Q, R, and S. P4 is formed by pairing of portions of P and Q, P6 by portions of Q and R, and P7 by portions of R and S. For a compilation of these sequences, see Cech (1988) and Michel and Westhof (1990). Sequence conservation at the splice sites is very limited; a U (U_{-1}) specifies the end of the 5′ exon at the 5′ splice site and a G (G_{414}) specifies the end of the intron at the 3′ splice site.

As with protein enzymes, ribozymes are folded into a specific three-dimensional structure that is required for activity. Activity is lost when the RNA is denatured with heat or chemical denaturants. Although most ribozymes are purified by denaturing gel electrophoresis, the active structure often spontaneously re-forms or can be regenerated by a renaturation step (Walstrum and Uhlenbeck 1990). The most detailed three-dimensional model of the core or active center of a group I ribozyme is that produced by Michel and Westhof (1990). Additional tertiary interactions that imposed long-range constraints on the structure were obtained from an extensive phylogenetic analysis of covariant bases in 87 intron sequences (Michel and Westhof 1990). Using these constraints and stereochemical modeling, Michel et al. (1989) generated a three-dimensional model that brings together the guanosine-binding site, the helices containing the splice sites, and the conserved bases within the intron. The Michel and Westhof (1990) model provides a valuable framework for designing experiments and interpreting results with group I introns (Pyle et al. 1992; Wang et al. 1993).

C. The Guanosine-binding Site

Guanosine (or GMP, GDP, or GTP) is required for splicing precursor RNA and for substrate cleavage in *trans*. Kinetic analysis indicated a specific binding site within the precursor RNA (Bass and Cech 1984). Michel et al. (1989) identified a specific conserved base pair (G264: C311) within P7 that determined the specificity for the guanosine nucleoside for intermolecular reactions analogous to both the first and second steps in splicing. The effects of mutations at positions 264 and 311 on base specificity are consistent with that model; a G264A mutation generates a binding site for 2-aminopurine ribonucleoside (Michel et al. 1989), and a G264C:C311G mutation generates a binding site for adenosine or ATP (Been and Perrotta 1991). In one model for nucleoside binding, the guanosine substrate forms a triplet with G_{264}:C_{311} by forming two hydrogen bonds with G_{264} in the major groove of P7 (Michel et al. 1989). However, other models are also consistent with the mutagenesis data. "Axial" models, in which the guanosine binds sideways in the major groove, have been proposed, along with a hybrid model in which the guanosine substrate forms a twisted base triplet with two hydrogen bonds to G_{264} and a third hydrogen bond to the phosphate on A_{263} (Yarus et al. 1991).

D. Structures Involved in Splice Site Selection

Sequences capable of forming paired regions that could function to specify and select the splice sites were identified by phylogenetic comparisons (Davies et al. 1982; Michel et al. 1982; Michel et al. 1989; Burke et al. 1990) and have been confirmed in a variety of experiments. At the 5′ splice site, a short duplex between the last few bases of the 5′ exon and part of the internal guide sequence (IGS, see below) form a pairing now named P1 (Fig. 1A,B) (Davies et al. 1982; Michel et al. 1982). The portion of the IGS that pairs with the 5′ exon has also been called the 5′ exon-binding site (Cech 1990). Only the U at position −1 and G_{22}, with which it is hypothesized to pair, are conserved in this pairing among the group I introns. Experimental evidence for P1 was obtained by introducing mutations in both strands of P1 to generate non-Watson-Crick pairs and evaluating the effect on splicing relative to combinations that restored pairing (Been and Cech 1986; Waring et al. 1986). With the exception of U_{-1} and G_{22}, the identity of other bases in this structure can vary as long as the stability of the helix is maintained. Although for some introns there is the potential for additional pairing 3′ to the 5′ splice site, the interaction is not as highly conserved. In binding of the P1 helix to the core of the ribozyme, there is evidence for additional tertiary interactions with 2′-hydroxyl groups in P1 (Bevilacqua and Turner 1991; Pyle and Cech 1991; Pyle et al. 1992).

The first interaction identified to facilitate 3′ splice site selection was the P10 pairing (Davies et al. 1982). The 3′ side of this pairing is contributed by exon sequences near the 3′ splice site. The 5′ side of this pairing is contiguous with the sequence forming the 3′ side of the P1 pairing such that these two helices together would have the potential to form a coaxial structure referred to as the "precise alignment structure" (Fig. 1B) (Davies et al. 1982). The sequence in the intron that forms a continuous strand with both P1 and P10 is the IGS and is a characteristic feature of most group I introns. P10 contributes to 3′ splice site selection (Suh and Waring 1990), but it is not the only determinant; two other interactions have been identified as important for this process. They are the guanosine-binding site that binds G_{414} at the 3′ splice site (Michel et al. 1989; Been and Perrotta 1991) and P9.0, a short pairing between two intron bases adjacent to the terminal G at the splice site and two bases between P7 and P9 (Michel et al. 1989; Burke et al. 1990). As with P1, the contributions of P10, P9.0, and the guanosine-binding site in exon ligation have been tested by mutagenesis and confirmed with suppressor mutants (Michel et al. 1989; Burke et al. 1990; Suh and Waring 1990; Been and Perrotta 1991).

E. The Intermolecular (*Trans*) Reaction and Site-specific Endonucleolytic Cleavage by Group I Ribozymes

Much of the above data was obtained using one or more intermolecular forms of the reaction of group I introns. Using a shortened (L-19) form of the *Tetrahymena* intron missing 19 nucleotides from the 5′ end and 5 nucleotides from the 3′ end, but containing the core conserved region and the 5′ exon portion of the IGS (the L-21 in Fig. 1D is a later version of this ribozyme), Zaug et al. (1986) demonstrated cleavage of a 5′ splice site analog. The cleavage reaction required Mg^{++} and guanosine, which was covalently added to the 5′ end of the 3′ cleavage product, so that it appeared to mimic faithfully the first step in the splicing reaction. This form of the reaction made it possible to use small RNA substrates for mechanistic studies.

Using mutant ribozymes with alterations in the sequence of the exon-binding site (positions 23 and 24), Zaug et al. (1986) demonstrated that a certain level of specificity was achieved in which only the matched substrates (those with compensatory changes corresponding to positions −2 and −3 in the exon) were cleaved. Discrimination against mismatches, however, required the addition of 2.5 M urea to the reaction; in its absence, both matched and single-base mismatched oligonucleotides were cleaved. It was hypothesized that the urea destabilized the interaction between mismatched substrate and ribozyme, resulting in increased discrimination against mismatches (Zaug et al. 1986, 1988). Kinetic analysis of the cleavage reaction with matched and mismatched sequences indicated that this interpretation was in essence correct, but more importantly, it defined those kinetic parameters that determined specificity (Herschlag and Cech 1990a,b). Several other sequence variants in the 5' exon-binding site were generated and tested for specificity (Murphy and Cech 1989). For changes at positions 22, 23, and 24, 10 of the 64 possible sequence variants have been tested and are active in *trans* cleavage, with each particular sequence having optimal Mg^{++} and salt conditions for cleavage and specificity which corresponded to the predicted stabilities of the P1 helix (Murphy and Cech 1989).

Another form of a *trans*-reaction has been used in a number of studies, including template-directed assembly of nucleotide sequences by group I introns (Szostak 1986; Doudna and Szostak 1989). Rather than providing only the 5' side of the P1 duplex in *trans*, the entire duplex is provided as a substrate (Szostak 1986), and guanosine participates in the cleavage reaction as described above. This form of a *trans*-reaction is, in a sense, fundamentally different because the substrate is not bound to the ribozyme by base pairing. In the reverse of this reaction, two oligonucleotides in which there are juxtaposed ends, one with a 3'-hydroxyl group and the other with an extra G at the 5' end, can be ligated and the guanosine released (Doudna and Szostak 1989). In this form, the 3' side of P1 is, in a sense, acting in a manner analogous to that of an exogenous template in a template-directed polymerase-catalyzed reaction (Doudna and Szostak 1989).

F. Kinetics of the *Trans* Reaction

A detailed kinetic scheme has been developed for the reaction, GGCCC UCUA$_5$ + G <==> GGCCCUCU + GA$_5$, catalyzed by a *Tetrahymena* group I ribozyme (Herschlag and Cech 1990a,b). With saturating guanosine, the k_{cat}/K_m value for the cleavage reaction at 50°C under presteady-state conditions approaches 10^8 $M^{-1}min^{-1}$. This value is close to that expected for the formation of a duplex from two oligonucleotides and is thought to represent the rate-limiting step of substrate binding. Under steady-state conditions, turnover (k_{cat}) equals 0.1 min^{-1}; this is slower than the rate constant, estimated at approximately 350 min^{-1}, for cleavage in the ternary complex (ribozyme/oligo/guanosine). The lower k_{cat} value represents the off-rate for the product, GGCCCUCU, the free 5′ exon analog that has to dissociate before the ribozyme can bind a second substrate.

The fact that the free exon is released slowly is important for the splicing mechanism because it will reduce the probability that the exon could be released prior to the ligation step and thus minimizes abortive reactions following 5′ splice site cleavage. Of significance to the design of site-specific ribozymes, the kinetic studies with a substrate oligonucleotide containing a single mismatch revealed a turnover number 100-fold higher than that seen with the matched substrate. This increase is because, although the rates for substrate binding and cleavage are fast and little affected by the mismatch, the rate-limiting step for turnover, product release, is faster with the mismatched substrate.

G. Enhancement of Activities of Group I Ribozymes by In Vitro Evolution

The fact that RNA is genetic material with catalytic potential makes it possible to produce some truly novel enzymes by in vitro mutagenesis and selection. Joyce and co-workers (Robertson and Joyce 1990; Beaudry and Joyce 1992; Lehman and Joyce 1993) have designed and refined an in vitro cycle of mutagenesis, selection, and amplification that has allowed them to evolve group I ribozymes with enhanced ability to cleave DNA, and others with altered metal ion requirements. An excised form of the intron can catalyze the covalent addition of a new sequence to its own 3′ end (Robertson and Joyce 1990). In a reaction that is essentially the reverse of the exon ligation step, a substrate oligonucleotide with the 5′ exon splice site sequence, followed by the primer-binding site sequence (a ligated exon sequence analog), is bound in the 5′ exon-binding site and attacked by G_{414}, transferring the primer-binding site to the 3′ end of the intron. Joyce and co-workers made use of this reaction by selecting randomly mutated ribozymes able to add a primer-binding sequence to their 3′ end; only ligation-competent ribozymes could be reverse-transcribed in a subsequent step. The new sequence was then subjected to selective isothermal RNA amplification (Lehman and Joyce 1993) or a combination of reverse transcription, polymerase chain reaction (PCR) amplification, and transcription (Beaudry and Joyce 1992). With repeated cycles of mutagenesis, selection, and amplification, it was possible to select sequences that were increasingly active with DNA as a substrate or with calcium as the divalent cation.

III. RIBOZYMES DERIVED FROM THE NATURALLY OCCURRING SELF-CLEAVING RNAS

A. A Diverse Group of Ribozymes Self-cleave by a Similar Mechanism

Self-cleavage or autolytic cleavage of sequences from the RNA pathogens of plants (viroids and RNA satellites of plant viruses) is thought to be required for the processing of multimeric copies of the RNA generated during replication (Buzayan et al. 1986b; Hutchins et al. 1986; Prody et al. 1986). Transcripts of dimeric copies of the plus- and minus-strand sequences of avocado sunblotch viroid (ASBV) self-cleaved at two sites during transcription in vitro to generate a monomeric size fragment (Hutchins et al. 1986). Purified transcripts incubated at neutral pH in Mg^{++} at 37°C also self-cleaved, and thus the conditions and factors required for cleavage were minimal. Cleavage was specific to a single site in the 247-nucleotide sequence and generated a $2',3'$-cyclic phosphate and a $5'$-hydroxyl group (Hutchins et al. 1986). Although sequences surrounding the cleavage sites in the plus and minus strands were not identical, some similarity was noted, and a secondary structure containing three pairings that placed the cleavage site in a similar location relative to the shared sequences was described (Fig. 2) (Hutchins et al. 1986). Forster and Symons (1987a,b) presented convincing evidence for the association of this structure, which they called the hammerhead, with self-cleaving sequences from a number of sources. They also provided experimental evidence for a minimal sequence of 52 nucleotides that was sufficient for cleavage (for review, see Symons, 1989, 1992). Self-cleaving hammerhead ribozymes were also discovered in transcripts of newt satellite DNA (Epstein and Gall 1987), but self-cleaving RNA sequences have not been reported for other vertebrate genome sequences.

In another example of a self-cleaving motif, Prody et al. (1986) demonstrated autolytic cleavage of a dimer of satellite plus-strand RNA of sTobRV. As with ASBV RNA, cleavage was specific; only one site in the 359-nucleotide sequence of this RNA was cleaved, and it resulted in a $2',3'$-cyclic phosphate and a $5'$-hydroxyl group. RNAs transcribed in vitro from plasmids containing cDNA clones of the autolytic cleaving sequence also self-cleaved, providing additional evidence for the autocatalytic nature of this reaction (Buzayan et al. 1986a). The minus strand of sTobRV also contains a self-cleaving sequence, and in this case, the reverse reaction, auto-ligation, was also seen to proceed efficiently (Buzayan et al. 1986b). Although the plus strand of sTobRV is another example of the hammerhead ribozyme, the minus strand has distinct sequence requirements and structure which has been described as a hairpin (Hampel et al. 1990) or paperclip (see Fig. 3) (Feldstein et al. 1990; Feldstein and Bruening 1993).

Two other self-cleaving RNAs are less well characterized than the hammerhead and hairpin/paperclip. The only self-cleaving RNA sequences identified so far that have any association with humans are those found in the genomic (viral) and the complementary, antigenomic, sequences of HDV (Kuo et al. 1988; Sharmeen et al. 1988; Wu et al. 1989). As with the plant pathogenic RNAs, self-cleavage is thought to function in processing multimers of the two polarities of RNAs generated during rolling circle replication (Kuo et al. 1988; Sharmeen et al. 1988; Wu et al. 1989). The sequences flanking the cleavage sites are not able to assume either the hammerhead or the hairpin/paperclip motif (Kuo et al. 1988; Sharmeen et al. 1988; Wu et al. 1989), and the most active form of the self-cleaving sequence forms a pseudoknot structure containing four paired regions (Fig. 5) (Perrotta and Been 1991; Been et al. 1992). The transcript of the *Neurospora* mitochondrial plasmid contains a self-cleavage site (Saville and Collins 1990), but the sequence is not compatible with previously described motifs (Saville and Collins 1990; Collins and Olive 1993). Definition of the boundaries of the sequence required for cleavage indicates that a single nucleotide 5' to the cleavage site is sufficient for cleavage, a property it shares with the HDV ribozyme (Collins and Olive 1993; Guo et al. 1993); however, it does not appear to fold into the HDV ribozyme secondary structure (R. Collins, pers. comm.).

B. Hammerhead Ribozymes

The secondary structure of the hammerhead ribozyme is usually represented in one of two equivalent fashions (Fig. 2B,C). Three stems or helices come together with 11 nonpaired nucleotides to form the core. A uniform numbering system has been adopted recently for the hammerhead motif (Hertel et al. 1992). In an elaboration of this motif, two identical sequences can associate to form a double hammerhead, which is more active than the monomer structure, especially in a hammerhead ribozyme with a short helix II (Forster et al. 1988). In those cases, the double hammerheads (Forster et al. 1988; Epstein and Pabón-Peña 1991) have interesting biological implications for the regulation of cleavage.

Assignment of nucleotides to paired regions (helices I, II, and III) is supported by covariation in natural sequences (Forster and Symons 1987a), the effect of base substitutions in *trans* constructs (Fedor and Uhlenbeck 1990; Ruffner et al. 1990), and imino proton nuclear magnetic resonance (NMR) (Heus et al. 1990; Pease and Wemmer 1990). Of the three conserved base pairs in the helices, only one (A15-U16) is ab-

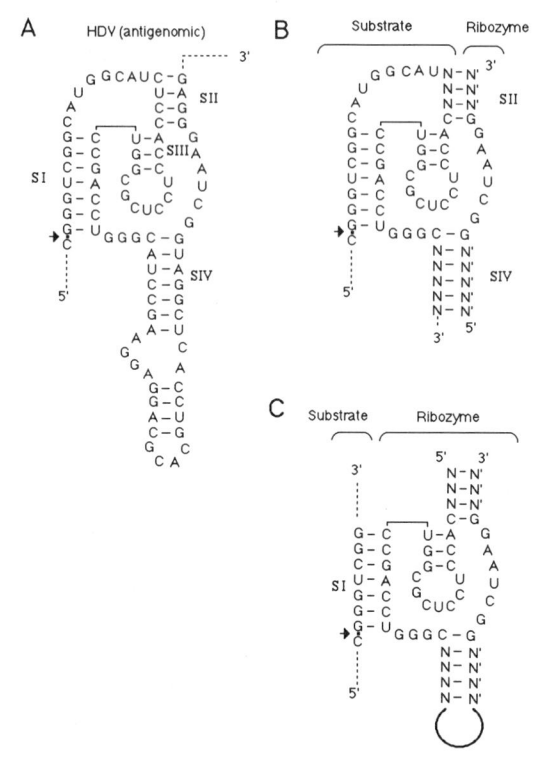

Figure 5 The HDV ribozyme. (*A*) Secondary structure of the antigenomic HVD self-cleaving sequence as proposed by Perrotta and Been (1991). (*B,C*) Two versions of intermolecular cleavage complexes with the HDV ribozyme (Perrotta and Been 1992, 1993).

solutely necessary for self-cleavage (Ruffner et al. 1990). The CG base pair in helix II may also contribute to higher activity in some versions, but it is not essential, and tolerance for sequence variations in other helical regions is one of the reasons that the hammerhead is an attractive candidate for a "designer" ribozyme. Nevertheless, large effects on the kinetics of cleavage are seen with some sequences, and these effects are mainly attributed to the secondary structure in the substrate RNAs (Ruffner et al. 1989; Fedor and Uhlenbeck 1990).

Of the 11 nonhelical nucleotides making up the core, 9 are conserved (Forster and Symons 1987a), and base substitutions at these nine positions result in a substantial reduction in activity (greater than tenfold) (Ruffner et al. 1990; Long and Uhlenbeck 1993). Substitutions of the nonhelical base at the cleavage site (C17) suggest that C is preferred over U and A, whereas G at this position results in very low activity (Ruffner et al. 1990; Perriman et al. 1992). In an effort to identify possible compensatory changes, 243 double mutants were tested and found to have less activity than either single mutant (Ruffner et al. 1990). The above studies suggest that a *trans*-acting hammerhead ribozyme, of the Haseloff and Gerlach design (Haseloff and Gerlach 1988) (Fig. 2C), prefers to cut after the sequence GUC but could be targeted to cut after the sequence GUH or even UH (H being C, A, or U) with flanking sequences specified by helices I and II.

Chemical synthesis of these sequences has allowed the evaluation of the effects of specific functional groups of bases and sugars in the nonhelical region. Removing a 2-amino group, by substitution of inosine for G at positions 5 (Fu and McLaughlin 1992) and 12 (Slim and Gait 1992), resulted in substantial loss of activity, whereas substitution of inosine for G8 or purine for A6 and A9 had little effect. The effects of modification of the 2'-hydroxyl (substitution of 2'-deoxy, 2'-amino, 2'-fluoro, 2'-O-allyl, or 2'-O-methyl) have also been examined (Perreault et al. 1990; Pieken et al. 1991; Fu and McLaughlin 1992; Paolella et al. 1992; Yang et al. 1992). A 2'-hydroxyl at the cleavage site is essential (Perreault et al. 1990), as would be predicted by the proposed mechanism. The introduction of multiple substitutions with 2' derivatives can result in an active ribozyme, providing key positions remain ribose (Perreault et al. 1990; Pieken et al. 1991; Paolella et al. 1992; Yang et al. 1992). The ribozymes with extensive 2'-amino, 2'-fluoro, and 2'-allyl modifications may be of practical use because they are resistant to nucleases but are still active (Pieken et al. 1991; Paolella et al. 1992). Substitutions with 2'-deoxy (Taylor et al. 1992) or 2'-O-methyl (Goodchild 1992) in the helical regions of the ribozyme enhance resistance to nucleolytic degradation, do not decrease cleavage activity, and can enhance turnover in the *trans*-reaction, presumably by increasing product off rates.

Modification of phosphates, by substitution of a thiophosphate with a nonbridging oxygen, has been used to examine details of the chemistry of cleavage (Dahm and Uhlenbeck 1991; Slim and Gait 1991) and to map phosphates that play an essential role in the cleavage reaction (Ruffner and Uhlenbeck 1990). Phosphorothioate substitutions identified four phosphates, including the cleavage site phosphate, that are important for activity (Ruffner and Uhlenbeck 1990). At the cleavage site, the introduction of a phosphorothioate affects activity because the sulfur reduces the affinity for Mg^{++} (Dahm and Uhlenbeck 1991; Slim and Gait 1991). It is not known why the other phosphorothioate substitutions have an effect, but two possibilities are that metal ion coordination or hydrogen-bonding potential is altered (Ruffner and Uhlenbeck 1990). Slim and Gait (1991) examined the effect of phosphorothioates at the cleavage site in both Rp and Sp configurations using chemically synthesized RNA. With Mg^{++} as the cation, the effect of the Sp isomer on cleavage was minimal, whereas the rate at which the Rp isomer was cleaved dropped dramatically; substitution of Mn^{++} restored a higher rate of cleavage. Dahm and Uhlenbeck (1991) described similar results with the Rp isomer made by transcription. Because Mn^{++} coordinates with both oxygen and sulfur, the effects of phosphorothioate at the cleavage site suggest that coordination of a metal ion to the cleavage site phosphate is required for cleavage (Dahm and Uhlenbeck 1991; Slim and Gait 1991). The reaction at the cleavage site containing a phosphorothioate in the Rp configuration proceeds with inversion to the Sp configuration (van Tol et al. 1990; Slim and Gait 1991). Thus, a model for the transesterification reaction would be direct attack by the 2'oxygen on the phosphorus and displacement of the 5'oxygen, generating a 2',3'-cyclic phosphate on the other cleavage product. A metal ion coordinated to the pro-R oxygen would stabilize the pentacovalent phosphorus in the trigonal bipyramidal intermediate or transition state (Long and Uhlenbeck 1993). Key to a more complete description of the mechanism will be the solution of a three-dimensional structure, because we have yet to understand how the folded RNA functions to generate the 10^6-fold enhancement over the estimated noncatalyzed rate (Long and Uhlenbeck 1993).

The kinetics of the intermolecular reaction have been treated like an enzymatic reaction. The association of two RNA fragments were taken to be analogous to substrate binding by the enzyme and, following cleavage, dissociation of three fragments analogous to product release (Uhlenbeck 1987; Fedor and Uhlenbeck 1992). It was anticipated that many of the same physical parameters that describe helix-coil transition of complementary sequences would be applicable to this interaction as well (Fedor and Uhlenbeck 1992). Rate constants were determined under steady-state and pre-steady-state conditions with substrates of various lengths and sequences, and a minimal kinetic scheme was developed (Fedor and Uhlenbeck 1992). It was found that rate constants for substrate association and product dissociation were in close agreement with the predicted behavior of small RNA helices and varied with the sequence and number of base pairs that could form between ribozyme and substrate. Thus, an effect of changing the substrate specificity is seen in the steady-state kinetics. Under pre-steady-state conditions, it was found that the rate constants for cleavage were similar for different ribozyme/substrate sequences (~1 min^{-1}), which may indicate that they represent the rate constant for the chemical step. A somewhat surprising result is that the substrate complex dissociates at a higher rate than expected (Fedor and Uhlenbeck 1992); the implications of this finding for catalysis and specificity are discussed by Fedor and Uhlenbeck (1992; see also Herschlag 1991).

C. Hairpin/Paperclip Ribozyme

The minus strand of the satellite of (−)sTovRV contains a self-cleavage site that is distinct in a number of properties from the hammerhead ribozyme. However, cleavage generates a 5′-hydroxyl and a 2′,3′-cyclic phosphate (Buzayan et al. 1986b), and the reaction results in inversion of configuration of the phosphate (van Tol et al. 1990). Important differences are the novel structure (Haseloff and Gerlach 1989) and the ability of this ribozyme to catalyze ligation efficiently by the reverse reaction (Buzayan et al. 1986b). The reverse reaction has been exploited extensively by Burke and co-workers in selection schemes to evaluate the contribution to catalysis of specific sequences in this ribozyme (Berzal-Herranz et al. 1992, 1993; Joseph et al. 1993; Burke and Berzal-Herranz 1993). Another interesting difference is in the metal ion requirements (Chowrira et al. 1993); in the absence of other divalent cations, rapid cleavage requires Mg^{++}, Sr^{++}, or Ca^{++}. However, slow cleavage occurs in the presence of spermidine alone (Prody et al. 1986), and cleavage in spermidine can be enhanced by the addition of Mn^{++} or Co^{++} (Chowrira et al. 1993). These results can be interpreted as evidence for at least two distinct classes of cation-binding sites in the ribozyme. The cation requirements of the ligation reaction are the same as those for cleavage, which is consistent with the model in which ligation is exactly the reverse of cleavage (Chowrira et al. 1993).

Secondary structures proposed for the hairpin/paperclip ribozyme all consisted of a four-helix structure organized into two domains, with the cleavage site located in an internal loop region of one of the domains (Fig. 3) (Haseloff and Gerlach 1989; Hampel and Tritz 1989; Feldstein et al. 1990). Two names have been proposed for the motif—hairpin (Hampel et al. 1990) and paperclip (Feldstein and Bruening 1993). The second name emphasizes the idea that the tertiary interactions between the two domains are required for ribozyme activity, but, as with the other ribozymes, a three-dimensional model is needed. The limited examples for phylogenetic comparisons (Haseloff and Gerlach 1989) have slowed defining additional details of the secondary structure. However, development of efficient intermolecular reactions (Feldstein et al. 1990; Hampel et al. 1990), use of in vitro mutagenesis (Feldstein et al. 1990; Hampel et al. 1990), and, most recently, the in vitro selection procedure (Berzal-Herranz et al. 1992, 1993; Joseph et al. 1993; Burke and Berzal-Herranz 1993) have resulted in a refined structure (Fig. 3) (Berzal-Herranz et al.

1993). Each of the two domains contain two paired sequences (helices H1-H4) that flank internal loops (Fig. 3A). The cleavage site is located between a pyrimidine (position –1) in H2 and an essential G in the loop between H1 and H2 (Joseph et al. 1993). Of the 15 other positions in the ribozyme where substitutions result in at least a tenfold drop in activity, only one position, G11, is part of a helix (H2) (Berzal-Herranz et al. 1993). Eleven of these essential nucleotides are in the internal loop between H3 and H4, emphasizing the fact that this region must fold back and form an intimate association with the other domain that contains the cleavage site (Fig. 3B).

As with the hammerhead, the development of an intermolecular reaction (Feldstein et al. 1990; Hampel et al. 1990) has facilitated the biochemical studies of the mechanism (Feldstein et al. 1990; van Tol et al. 1990; Chowrira et al. 1993; Feldstein and Bruening 1993) and generated interest in development of the hairpin/paperclip motif as a sequence-specific nuclease. In the most detailed analysis of substrate sequence requirements, Joseph et al. (1993) determined the target sequence for efficient cleavage to be $5'-N_2(G/A)(U/C)N*G(A/U/C)(U/C)(G/U/C)N_5-3'$, where the asterisk indicates the cleavage site. This information will facilitate the identification of potential targets in mRNAs.

D. Ribozymes from Hepatitis Delta Virus

HDV is a satellite of hepatitis B virus and contains a circular single-stranded 1700-nucleotide RNA genome that is thought to replicate via a double rolling circle mechanism (Taylor 1990). The identification of Mg^{++}-dependent self-cleavage sites in the genomic and antigenomic strand RNAs of the virus revealed that the ribozyme structures associated with this virus would represent novel structures (Kuo et al. 1988; Sharmeen et al. 1988; Wu et al. 1989).

Less is known about the HDV ribozymes than either the hammerhead or the hairpin/paperclip; however, they have been shown to have some unique features. The self-cleaving forms of these sequences have robust structures such that cleavage proceeds efficiently in moderately high concentrations of denaturants (>5 M urea or >10 M formamide) (Rosenstein and Been 1990; Belinsky and Dinter-Gottlieb 1991; Perrotta and Been 1991; Smith and Dinter-Gottlieb 1991). Because there is a tendency for this ribozyme to fold into inactive structures, a property that is exaggerated by flanking sequences (Perrotta and Been 1991), the addition of denaturants can enhance the rate of self-cleavage of some transcripts (Rosenstein and Been 1990).

A number of models for the secondary structures of both the genomic and antigenomic ribozymes have been proposed. Several of these include sequences 5' to the cleavage site that are not required for either the intra- or intermolecular cleavage reactions and therefore are unlikely to be accurate. An 85-nucleotide sequence of either the genomic or antigenomic sequence, which includes 1 nucleotide 5' to the cleavage site and 84 nucleotides 3' to the cleavage site, is the minimal sequence required for efficient cleavage under a variety of conditions (Perrotta and Been 1990, 1991). At least 20 nucleotides can be deleted from the antigenomic sequence with no decrease in the rate of cleavage (Been et al. 1992), so the size of this internally deleted ribozyme is similar to the size of the hair pin/paperclip ribozyme. Site-specific mutagenesis (Perrotta and Been 1991; Been et al. 1992), insertion and deletion studies (Been et al. 1992; Wu and Huang 1992; Wu et al. 1992), nuclease digestion (Rosenstein and Been 1991; Wu et al. 1992), and chemical probing (Wu et al. 1992; Belinsky et al. 1993) are consistent with a 4-stem structure (Fig. 5A) (Perrotta and Been 1991; Rosenstein and Been 1991; Been et al. 1992). The cleavage site is at the base of stem I, and stem II is formed from the 3'-end pairing with a sequence within loop I; together, these two pairings form a pseudoknot. Disruption of stem II, either by deletion or by mutagenesis (Perrotta and Been 1990, 1991; Been et al. 1992), reduces cleavage activity without destroying it, suggesting that this interaction may help to stabilize the structure forming the catalytic center. Stem IV, a large hairpin stem and loop, may also stabilize the overall conformation without providing essential catalytic residues (Been et al. 1992; Wu and Huang 1992; Wu et al. 1992; Thill et al. 1993). Mutations in stem III (Been et al. 1992; Kumar et al. 1992; Thill et al. 1993) and stemloop III (Kumar et al. 1992; Thill et al. 1993) have more dramatic effects on cleavage activity, suggesting that they may be intimately associated with the catalytic center.

Two forms of an intermolecular reaction have been developed. In one form, the ribozyme is separated in stem IV (Branch and Robertson 1991; Wu et al. 1992; Perrotta and Been 1993) and the cleavable complex is formed by intermolecular formation of stem IV and stem II (Fig. 5B) (Wu et al. 1992; Perrotta and Been 1993). In the other form, the ribozyme has been separated between stems I and II, generating a substrate that is bound by virtue of the pairing in stem I (Fig. 5C) (Been et al. 1992; Perrotta and Been 1992). The latter form more closely resembles an enzyme substrate combination in that most of the essential sequences and structures required for cleavage activity reside in the ribozyme portion. Available data, which are much less than that available for the hammerhead and hairpin/paperclip ribozymes, indicate that the base pairs in stems II and IV can be varied (this is indicated by N, N' pairing in Fig. 5B,C). The sequence in stem I can also be varied, although the substitution of the proposed G:U interaction at the base of stem I with other base pairs or mismatches substantially reduces activity. Both forms of the intermolecular reaction will be useful in the further characterization of the HDV ribozyme (Perrotta and Been 1992), although at this point, it appears that neither form of the HDV *trans*-reaction would serve as well as the hammerhead or hairpin/paperclip for in vivo applications.

IV. SUMMARY AND PERSPECTIVES

Ribozymes emerged in the past decade as a new class of enzymes. The most common reaction catalyzed by ribozymes involves the making and breaking of phosphodiester bonds in RNA, and so many may be considered nucleases (in fact, those discussed in this chapter more closely resemble topoisomerases in that they catalyze phosphoester transfer reactions rather than hydrolysis). The only naturally occurring enzyme that is clearly a ribozyme and a nuclease is RNase P, but the self-processing RNAs have proven to be a good source of molecules that can be manipulated as enzymes to cleave RNA in *trans*. Because substrate binding and specificity are determined largely through normal base-pairing interaction, it has proven to be much easier to manipulate target specificity with the artificial *trans*-acting ribozymes than it has been, for example, with a restriction endonuclease. Whether this application will eventually reach the stage of having therapeutic applications remains to be seen. Nevertheless, the major utility of the intermolecular forms of the ribozyme reactions has been that it simplifies the elucidation of the structure and mechanism of this class of biological catalysts. The major impact of ribozymes in biology has been to revise the way we view the processes of gene expression, as well as the origin and evolution of self-replicating systems. Understanding how RNA functions as an enzyme in these simple systems will provide us with insight into how more complex systems containing essential RNA components, such as the ribosome and spliceosome, carry out catalysis.

ACKNOWLEDGMENTS

I thank S. Rosenstein, A. Perrotta, M. Puttaraju, and M. Canady for comments on the manuscript, L. LeMosy for valuable assistance on the manuscript, and numerous colleagues for providing reprints and preprints of recent work. Support was received from National Institutes of Health (GM-40689 and GM-47233).

REFERENCES

Altman, S. 1990. Ribonuclease P. *J. Biol. Chem.* **265:** 20053–20056.
Bass, B.L. and T.R. Cech. 1984. Specific interaction between the self-splicing RNA of *Tetrahymena* and its guanosine substrate: Implications for biological catalysis by RNA. *Nature* **308:** 820–826.
Beaudry, A.A. and G.F. Joyce. 1992. Directed evolution of an RNA enzyme. *Science* **257:** 635–641.
Been, M.D. and T.R. Cech. 1986. One binding site determines sequence specificity of *Tetrahymena* pre-rRNA self-splicing, *trans*-splicing, and RNA enzyme activity. *Cell* **47:** 207–216.
Been, M.D. and A.T. Perrotta. 1991. Group I intron self-splicing with adenosine: Evidence for a single nucleoside-binding site. *Science* **252:** 434–437.
Been, M.D., A.T. Perrotta, and S.P. Rosenstein. 1992. Secondary structure of the self-cleaving RNA of hepatitis delta virus: Applications to catalytic RNA design. *Biochemistry* **31:** 11843–11852.
Belfort, M. 1991. Self-splicing introns in prokaryotes: Migrant fossils? *Cell* **64:** 9–11.
Belinsky, M.G. and G. Dinter-Gottlieb. 1991. Non-ribozyme sequences enhance self-cleavage of ribozymes derived from hepatitis delta virus. *Nucleic Acids Res.* **19:** 559–564.
Belinsky, M.G., E. Britton, and G. Dinter-Gottlieb. 1993. Modification interference analysis of a self-cleaving RNA from hepatitis delta virus. *FASEB J.* **7:** 130–136.
Berzal-Herranz, A., S. Joseph, and J.M. Burke. 1992. In vitro selection of active hairpin ribozymes by sequential RNA-catalyzed cleavage and ligation reactions. *Genes Dev.* **6:** 129–134.
Berzal-Herranz, A., S. Joseph, B.M. Chowrira, S.E. Butcher, and J.M. Burke. 1993. Essential nucleotide sequences and secondary structure elements of the hairpin ribozyme. *EMBO J.* **12:** 2567–2574.
Bevilacqua, P.C. and D.H. Turner. 1991. Comparison of binding of mixed ribose-deoxyribose analogues of CUCU to a ribozyme and to GGAGAA by equilibrium dialysis: Evidence for ribozyme specific interactions with 2'OH groups. *Biochemistry* **30:** 10632–10640.
Branch, A.D. and H.D. Robertson. 1991. Efficient *trans* cleavage and a common structural motif for the ribozymes of the human hepatitis δ agent. *Proc. Natl. Acad. Sci.* **88:** 10163–10167.
Burke, J.M. and A. Berzal-Herranz. 1993. In vitro selection and evolution of RNA: Applications for catalytic RNA, molecular recognition, and drug discovery. *FASEB J.* **7:** 106–112.
Burke, J.M., J.S. Esherick, W.R. Burfeind, and J.L. King. 1990. A 3' splice site-binding sequence in the catalytic core of a group I intron. *Nature* **344:** 80–82.
Burke, J.M., M. Belfort, T.R. Cech, R.W. Davies, R.J. Schweyen, D.A. Shub, J.W. Szostak, and H.F. Tabak. 1987. Structural conventions for group I introns. *Nucleic Acids Res.* **15:** 7217–7221.
Buzayan, J.M., W.L. Gerlach, and G. Bruening. 1986a. Satellite tobacco ringspot virus RNA: A subset of the RNA sequence is sufficient for autolytic processing. *Proc. Natl. Acad. Sci.* **83:** 8859–8862.
———. 1986b. Non-enzymatic cleavage and ligation of RNAs complementary to a plant virus satellite RNA. *Nature* **323:** 349–353.
Castanotto, D., J.J. Rossi, and J.O. Deshler. 1992. Biological and functional aspects of catalytic RNAs. *Crit. Rev. Eukaryotic Gene Expression* **2:** 331–357.

Cech, T.R. 1987. The chemistry of self-splicing RNA and RNA enzymes. *Science* **236**: 1532–1539.

———. 1988. Conserved sequences and structures of group I introns—Building an active site for RNA catalysis—A review. *Gene* **73**: 259–271.

———. 1990. Self-splicing of group I introns. *Annu. Rev. Biochem.* **59**: 543–568.

———. 1993. Structure and mechanism of the large catalytic RNAs: Group I and group II introns and ribonuclease P. In *The RNA world* (ed. R.F. Gesteland and J.F. Atkins), pp. 239–269. Cold Spring Harbor Laboratory Press, Cold Spring Harbor, New York.

Cech, T.R., A.J. Zaug, and P.J. Grabowski. 1981. *In vitro* splicing of the ribosomal RNA precursor of *Tetrahymena*: Involvement of a guanosine nucleotide in the excision of the intervening sequence. *Cell* **27**: 487–496.

Cech, T.R., D. Herschlag, J.A. Piccirilli, and A.M. Pyle. 1992. RNA catalysis by a group I ribozyme. Developing a model for transition state stabilization. *J. Biol. Chem.* **267**: 17479–17482.

Chowrira, B.M., A. Berzal-Herranz, and J.M. Burke. 1993. Ionic requirements for RNA binding, cleavage, and ligation by the hairpin ribozyme. *Biochemistry* **32**: 1088–1095.

Collins, R.A. and J.E. Olive. 1993. Reaction conditions and kinetics of self-cleavage of a ribozyme derived from *Neurospora* VS RNA. *Biochemistry* **32**: 2795–2799.

Dahm, S.C. and O.C. Uhlenbeck. 1991. Role of divalent metal ions in the hammerhead RNA cleavage reaction. *Biochemistry* **30**: 9464–9469.

Davies, R.W., R.B. Waring, J.A. Ray, T.A. Brown, and C. Scazzocchio. 1982. Making ends meet: A model for RNA splicing in fungal mitochondria. *Nature* **300**: 719–724.

Doudna, J.A. and J.W. Szostak. 1989. RNA-catalysed synthesis of complementary-strand RNA. *Nature* **339**: 519–522.

Epstein, L.M. and J.G. Gall. 1987. Self-cleaving transcripts of satellite DNA from the newt. *Cell* **48**: 535–543.

Epstein, L.M. and L.M. Pabón-Peña. 1991. Alternative modes of self-cleavage by newt satellite 2 transcripts. *Nucleic Acids Res.* **19**: 1699–1705.

Fedor, M.J. and O.C. Uhlenbeck. 1990. Substrate sequence effects on "hammerhead" RNA catalytic efficiency. *Proc. Natl. Acad. Sci.* **87**: 1668–1672.

———. 1992. Kinetics of intermolecular cleavage by hammerhead ribozymes. *Biochemistry* **31**: 12042–12054.

Feldstein, P.A. and G. Bruening. 1993. Catalytically active geometry in the reversible circularization of "mini-monomer" RNAs derived from the complementary strand of tobacco ringspot virus satellite RNA. *Nucleic Acids Res.* **21**: 1991–1998.

Feldstein, P.A., J.M. Buzayan, H. van Tol, J. DeBear, G.R. Gough, P.T. Gilham, and G. Bruening. 1990. Specific association between an endoribonucleolytic sequence from a satellite RNA and a substrate analogue containing a 2′-5′ phosphodiester. *Proc. Natl. Acad. Sci.* **87**: 2623–2627.

Ferat, J.-L. and F. Michel. 1993. Group II self-splicing introns in bacteria. *Nature* **364**: 358–361.

Forster, A.C. and R.H. Symons. 1987a. Self-cleavage of plus and minus RNAs of a virusoid and a structural model for the active sites. *Cell* **49**: 211–220.

———. 1987b. Self-cleavage of virusoid RNA is performed by the proposed 55-nucleotide active site. *Cell* **50**: 9–16.

Forster, A.C., C. Davies, C.C. Sheldon, A.C. Jeffries, and R.H. Symons. 1988. Self-cleaving viroid and newt RNAs may only be active as dimers. *Nature* **334**: 265–267.

Fu, D.-J. and L.W. McLaughlin. 1992. Importance of specific purine amino and hydroxyl groups for efficient cleavage by a hammerhead ribozyme. *Proc. Natl. Acad. Sci.* **89**: 3985–3989.

Goodchild, J. 1992. Enhancement of ribozyme catalytic activity by a contiguous oligodeoxynucleotide (facilitator) and by 2′-O-methylation. *Nucleic Acids Res.* **20:** 4607-4612.

Grabowski, P.J., A.J. Zaug, and T.R. Cech. 1981. The intervening sequence of the ribosomal RNA precursor is converted to a circular RNA in isolated nuclei of *Tetrahymena*. *Cell* **23:** 467-476.

Grosshans, C.A. and T.R. Cech. 1989. Metal ion requirements for sequence-specific endoribonuclease activity of the *Tetrahymena* ribozyme. *Biochemistry* **28:** 6888-6894.

Guerrier-Takada, C. and S. Altman. 1984. Catalytic activity of an RNA molecule prepared by transcription in vitro. *Science* **233:** 285-286.

Guerrier-Takada, C., K. Gardiner, T. Marsh, N. Pace, and S. Altman. 1983. The RNA moiety of ribonuclease P is the catalytic subunit of the enzyme. *Cell* **35:** 849-857.

Guo, H.C.T., D.M. De Abreu, E.R.M. Tillier, B.J. Saville, J.E. Olive, and R.A. Collins. 1993. Nucleotide sequence requirements for self-cleavage of *Neurospora* VS RNA. *J. Mol. Biol.* **231:** 351-361.

Hampel, A. and R. Tritz. 1989. RNA catalytic properties of the minimum (-)sTRSV sequence. *Biochemistry* **28:** 4929-4933.

Hampel, A., R. Tritz, M. Hicks, and P. Cruz. 1990. "Hairpin" catalytic RNA model: Evidence for helices and sequence requirement for substrate RNA. *Nucleic Acids Res.* **18:** 299-304.

Haseloff, J. and W.L. Gerlach. 1988. Simple RNA enzymes with new and highly specific endoribonuclease activities. *Nature* **334:** 585-591.

———. 1989. Sequences required for self-catalysed cleavage of the satellite RNA of tobacco ringspot virus. *Gene* **82:** 43-52.

Herschlag, D. 1991. Implications of ribozyme kinetics for targeting the cleavage of specific RNA molecules *in vivo*: More isn't always better. *Proc. Natl. Acad. Sci.* **88:** 6921-6925.

Herschlag, D. and T.R. Cech. 1990a. Catalysis of RNA cleavage by the *Tetrahymena thermophila* ribozyme. 1. Kinetic description of the reaction of an RNA substrate complementary to the active site. *Biochemistry* **29:** 10159-10171.

———. 1990b. Catalysis of RNA cleavage by the *Tetrahymena thermophila* ribozyme. 2. Kinetic description of the reaction of an RNA substrate that forms a mismatch at the active site. *Biochemistry* **29:** 10172-10180.

———. 1990c. DNA cleavage catalysed by the ribozyme from *Tetrahymena*. *Nature* **344:** 405-409.

Hertel, K.J., A. Pardi, O.C. Uhlenbeck, M. Koizumi, E. Ohtsuka, S. Uesugi, R. Cedergren, F. Eckstein, W.L. Gerlach, R. Hodgson, and R.H. Symons. 1992. Numbering system for the hammerhead. *Nucleic Acids Res.* **20:** 3252-3253.

Heus, H.A., O.C. Uhlenbeck, and A. Pardi. 1990. Sequence-dependent structural variations of hammerhead RNA enzymes. *Nucleic Acids Res.* **18:** 1103-1108.

Hutchins, C.J., P.D. Rathjen, A.C. Forster, and R.H. Symons. 1986. Self-cleavage of plus and minus RNA transcripts of avocado sunblotch viroid. *Nucleic Acids Res.* **14:** 3627-3640.

Joseph, S., A. Berzal-Herranz, B.M. Chowrira, S.E. Butcher, and J.M. Burke. 1993. Substrate selection rules for the hairpin ribozyme determined by in vitro selection, mutation, and analysis of mismatched substrates. *Genes Dev.* **7:** 130-138.

Kay, P.S. and T. Inoue. 1987. Catalysis of splicing-related reactions between dinucleotides by a ribozyme. *Nature* **327:** 343-346.

Kruger, K., P.J. Grabowski, A.J. Zaug, J. Sands, D.E. Gottschling, and T.R. Cech. 1982. Self-splicing RNA: Autoexcision and autocyclization of the ribosomal RNA interven-

ing sequence of *Tetrahymena. Cell* **31:** 147–157.
Kumar, P.K.R., Y.-A. Suh, H. Miyashiro, F. Nishikawa, J. Kawakami, K. Taira, and S. Nishikawa. 1992. Random mutations to evaluate the role of bases at two important single-stranded regions of genomic HDV ribozyme. *Nucleic Acids Res.* **20:** 3919–3924.
Kuo, M.Y.-P., L. Sharmeen, G. Dinter-Gottlieb, and J. Taylor. 1988. Characterization of self-cleaving RNA sequences on the genome and antigenome of human hepatitis delta virus. *J. Virol.* **62:** 4439–4444.
Lehman, N. and G.F. Joyce. 1993. Evolution *in vitro* of an RNA enzyme with altered metal dependence. *Nature* **361:** 182–185.
Long, D.M. and O.C. Uhlenbeck. 1993. Self-cleaving catalytic RNA. *FASEB J.* **7:** 25–30.
Marsh, T.L. and N.R. Pace. 1985. Ribonuclease P catalysis differs from ribosomal RNA self-splicing. *Science* **229:** 79–81.
McSwiggen, J.A. and T.R. Cech. 1989. Stereochemistry of RNA cleavage by the *Tetrahymena* ribozyme and evidence that the chemical step is not rate-limiting. *Science* **244:** 679–683.
Michel, F. and E. Westhof. 1990. Modelling of the three-dimensional architecture of group I catalytic introns based on comparative sequence analysis. *J. Mol. Biol.* **216:** 585–610.
Michel, F., A. Jacquier, and B. Dujon. 1982. Comparison of fungal mitochondrial introns reveals extensive homologies in RNA secondary structure. *Biochimie* **64:** 867–881.
Michel, F., M. Hanna, R. Green, D.P. Bartel, and J.W. Szostak. 1989. The guanosine binding site of the *Tetrahymena* ribozyme. *Nature* **342:** 391–395.
Murphy, F.L. and T.R. Cech. 1989. Alteration of substrate specificity for the endoribonucleolytic cleavage of RNA by the *Tetrahymena* ribozyme. *Proc. Natl. Acad. Sci.* **86:** 9218–9222.
Noller, H.F. 1991. Ribosomal RNA and translation. *Annu. Rev. Biochem.* **60:** 191–227.
Noller, H.F. and C.R. Woese. 1981. Secondary structure of 16S ribosomal RNA. *Science* **212:** 403–411.
Noller, H.F., V. Hoffarth, and L. Zimniak. 1992. Unusual resistance of peptidyl transferase to protein extraction procedures. *Science* **256:** 1416–1424.
Noller, H.F., D. Moazed, S. Stern, T. Powers, P.N. Allen, J.M. Robertson, B. Weiser, and K. Triman. 1990. Structure of rRNA and its functional interactions in translation. In *The ribosome: Structure, function, and evolution* (ed. W.E. Hill et al.), pp. 73–92. American Society for Microbiology, Washington, D.C.
Pace, N.R. and D. Smith. 1990. Ribonuclease P: Function and variation. *J. Biol. Chem.* **265:** 3587–3590.
Pan, T. and O.C. Uhlenbeck. 1992. A small metalloribozyme with a two-step mechanism. *Nature* **358:** 560–563.
Paolella, G., B.S. Sproat, and A.I. Lamond. 1992. Nuclease resistant ribozymes with high catalytic activity. *EMBO J.* **11:** 1913–1919.
Pease, A.C. and D.E. Wemmer. 1990. Characterization of the secondary structure and melting of a self-cleaved RNA hammerhead domain by ^{1}H NMR spectroscopy. *Biochemistry* **29:** 9039–9046.
Peebles, C.L., P.S. Perlman, K.L. Mecklenburg, M.L. Petrillo, J.H. Tabor, K.A. Jarrell, and H.-L. Cheng. 1986. A self-splicing RNA excises an intron lariat. *Cell* **44:** 213–223.
Perreault, J.-P., T. Wu, B. Cousineau, K.K. Ogilvie, and R. Cedergren. 1990. Mixed deoxyribo- and ribo- oligonucleotides with catalytic activity. *Nature* **344:** 565–567.
Perriman, R., A. Delves, and W.L. Gerlach. 1992. Extended target-site specificity for a hammerhead ribozyme. *Gene* **113:** 157–163.
Perrotta, A.T. and M.D. Been. 1990. The self-cleaving domain from the genomic RNA of

hepatitis delta virus: Sequence requirements and the effects of denaturant. *Nucleic Acids Res.* **18:** 6821-6827.

———. 1991. A pseudoknot-like structure required for efficient self-cleavage of hepatitis delta virus RNA. *Nature* **350:** 434-436.

———. 1992. Cleavage of oligoribonucleotides by a ribozyme derived from the hepatitis δ virus RNA sequence. *Biochemistry* **31:** 16-21.

———. 1993. Assessment of disparate structural features in three models of the hepatitis delta virus ribozyme. *Nucleic Acids Res.* **21:** 3959-3965.

Piccirilli, J.A., J.S. Vyle, M.H. Caruthers, and T.R. Cech. 1993. Metal ion catalysis in the *Tetrahymena* ribozyme reaction. *Nature* **361:** 85-88.

Piccirilli, J., T.S. McConnell, A.J. Zaug, H.F. Noller, and T.R. Cech. 1992. Aminoacyl esterase activity of the *Tetrahymena* ribozyme. *Science* **256:** 1420-1424.

Pieken, W.A., D.B. Olsen, F. Benseler, H. Aurup, and F. Eckstein. 1991. Kinetic characterization of ribonuclease-resistant 2'-modified hammerhead ribozymes. *Science* **253:** 314-317.

Prody, G.A., J.T. Bakos, J.M. Buzayan, I.R. Schneider, and G. Bruening. 1986. Autolytic processing of dimeric plant virus satellite RNA. *Science* **231:** 1577-1580.

Pyle, A.M. 1993. Ribozymes: A distinct class of metalloenzymes. *Science* **261:** 709-714.

Pyle, A.M. and T.R. Cech. 1991. Ribozyme recognition of RNA by tertiary interactions with specific ribose 2'-OH groups. *Nature* **350:** 628-631.

Pyle, A.M., F.L. Murphy, and T.R. Cech. 1992. RNA substrate binding site in the catalytic core of the *Tetrahymena* ribozyme. *Nature* **358:** 123-128.

Rajagopal, J., J.A. Doudna, and J.W. Szostak. 1989. Stereochemical course of catalysis by the *Tetrahymena* ribozyme. *Science* **244:** 692-694.

Robertson, D.L. and G.F. Joyce. 1990. Selection *in vitro* of an RNA enzyme that specifically cleaves single-stranded DNA. *Nature* **344:** 467-468.

Rosenstein, S.P. and M.D. Been. 1990. Self-cleavage of hepatitis delta virus genomic strand RNA is enhanced under partially denaturing conditions. *Biochemistry* **29:** 8011-8016.

———. 1991. Evidence that genomic and antigenomic RNA self-cleaving elements from hepatitis delta virus have similar secondary structures. *Nucleic Acids Res.* **19:** 5409-5416.

Ruffner, D.E. and O.C. Uhlenbeck. 1990. Thiophosphate interference experiments locate phosphates important for the hammerhead RNA self-cleavage reaction. *Nucleic Acids Res.* **18:** 6025-6029.

Ruffner, D.E., S.C. Dahm, and O.C. Uhlenbeck. 1989. Studies on the hammerhead RNA self-cleaving domain. *Gene* **82:** 31-41.

Ruffner, D.E., G.D. Stormo, and O.C. Uhlenbeck. 1990. Sequence requirements of the hammerhead RNA self-cleavage reaction. *Biochemistry* **29:** 10695-10702.

Saldanha, R., G. Mohr, M. Belfort, and A.M. Lambowitz. 1993. Group I and group II introns. *FASEB J.* **7:** 15-24.

Saville, B.J. and R.A. Collins. 1990. A site-specific self-cleavage reaction performed by a novel RNA in *Neurospora* mitochondria. *Cell* **61:** 685-696.

Sharmeen, L., M.Y.-P. Kuo, G. Dinter-Gottlieb, and J. Taylor. 1988. Antigenomic RNA of human hepatitis delta virus can undergo self-cleavage. *J. Virol.* **62:** 2674-2679.

Sharp, P.A. 1985. On the origin of RNA splicing and introns. *Cell* **42:** 397-400.

———. 1991. Five easy pieces. *Science* **254:** 663.

Slim, G. and M.J. Gait. 1991. Configurationally defined phosphorothioate-containing oligoribonucleotides in the study of the mechanism of cleavage of hammerhead ribozymes. *Nucleic Acids Res.* **19:** 1183-1188.

———. 1992. The role of the exocyclic amino groups of conserved purines in hammerhead ribozyme cleavage. *Biochem. Biophys. Res. Commun.* **183:** 605–609.
Smith, D. and N.R. Pace. 1993. Multiple magnesium ions in the ribonuclease P reaction mechanism. *Biochemistry* **32:** 5273–5281.
Smith, J.B. and G. Dinter-Gottlieb. 1991. Antigenomic hepatitis delta virus ribozymes self-cleave in 18 M formamide. *Nucleic Acids Res.* **19:** 1285–1289.
Suh, E.R. and R.B. Waring. 1990. Base pairing between the 3′ exon and an internal guide sequence increases 3′ splice site specificity in the *Tetrahymena* self-splicing rRNA intron. *Mol. Cell. Biol.* **10:** 2960–2965.
———. 1992. A phosphorothioate at the 3′ splice-site inhibits the second splicing step in a group I intron. *Nucleic Acids Res.* **20:** 6303–6309.
Symons, R.H. 1989. Self-cleavage of RNA in the replication of small pathogens of plants and animals. *Trends Biochem. Sci.* **14:** 445–450.
———. 1992. Small catalytic RNAs. *Annu. Rev. Biochem.* **61:** 641–671.
Szostak, J.W. 1986. Enzymatic activity of the conserved core of a group I self-splicing intron. *Nature* **322:** 83–86.
Taylor, J.M. 1990. Hepatitis delta virus: *cis* and *trans* functions required for replication. *Cell* **61:** 371–373.
Taylor, N.R., B.E. Kaplan, P. Swiderski, H. Li, and J.J. Rossi. 1992. Chimeric DNA-RNA hammerhead ribozymes have enhanced *in vitro* catalytic efficiency and increased stability *in vivo*. *Nucleic Acids Res.* **20:** 4559–4565.
Thill, G., M. Vasseur, and N.K. Tanner. 1993. Structural and sequence elements required for the self-cleaving activity of the hepatitis delta virus ribozyme. *Biochemistry* **32:** 4254–4262.
Uhlenbeck, O.C. 1987. A small catalytic oligoribonucleotide. *Nature* **328:** 596–600.
Van der Veen, R., A.C. Arnberg, G. Van der Horst, L. Bonen, H.F. Tabak, and L.A. Grivell. 1986. Excised group II introns in yeast mitochondria are lariats and can be formed by self-splicing in vitro. *Cell* **44:** 225–234.
van Tol, H., J.M. Buzayan, P.A. Feldstein, F. Eckstein, and G. Bruening. 1990. Two autolytic processing reactions of a satellite RNA proceed with inversion of configuration. *Nucleic Acids Res.* **18:** 1971–1975.
von Ahsen, U. and R. Schroeder. 1993. RNA as a catalyst: Natural and designed ribozymes. *BioEssays* **15:** 299–307.
Walstrum, S.A. and O.C. Uhlenbeck. 1990. The self-splicing RNA of *Tetrahymena* is trapped in a less active conformation by gel purification. *Biochemistry* **29:** 10573–10576.
Wang, J.-F., W.D. Downs, and T.R. Cech. 1993. Movement of the guide sequence during RNA catalylsis by a group I ribozyme. *Science* **260:** 504–508.
Waring, R.B. and R.W. Davies. 1984. Assessment of a model for intron RNA secondary structure relevant to RNA self-splicing—A review. *Gene* **28:** 277–291.
Waring, R.B., P. Towner, S.J. Minter, and R.W. Davies. 1986. Splice-site selection by a self-splicing RNA of *Tetrahymena*. *Nature* **321:** 133–139.
Wu, H.-N. and Z.-S. Huang. 1992. Mutagenesis analysis of the self-cleavage domain of hepatitis delta virus antigenomic RNA. *Nucleic Acids Res.* **20:** 5937–5941.
Wu, H.-N., Y.-J. Wang, C.-F. Hung, H.-J. Lee, and M.M.C. Lai. 1992. Sequence and structure of the catalytic RNA of hepatitis delta virus genomic RNA. *J. Mol. Biol.* **223:** 233–245.
Wu, H.-N., Y.-J. Lin, F.-P. Lin, S. Makino, M.-F. Chang, and M.M.C. Lai. 1989. Human hepatitis δ virus RNA subfragments contain an autocleavage activity. *Proc. Natl. Acad. Sci.* **86:** 1831–1835.

Yang, J.-H., N. Usman, P. Chartrand, and R. Cedergren. 1992. Minimum ribonucleotide requirement for catalysis by the RNA hammerhead domain. *Biochemistry* **31:** 5005–5009.

Yarus, M., M. Illangesekare, and E. Christian. 1991. An axial binding site in the *Tetrahymena* precursor RNA. *J. Mol. Biol.* **222:** 995–1012.

Zaug, A.J. and T.R. Cech. 1986. The intervening sequence RNA of *Tetrahymena* is an enzyme. *Science* **231:** 470–475.

Zaug, A.J., M.D. Been, and T.R. Cech. 1986. The Tetrahymena ribozyme acts like an RNA restriction endonuclease. *Nature* **324:** 429–433.

Zaug, A.J., C.A. Grosshans, and T.R. Cech. 1988. Sequence-specific endoribonuclease activity of the *Tetrahymena* ribozyme: Enhanced cleavage of certain oligonucleotide substrates that form mismatched ribozyme-substrate complexes. *Biochemistry* **27:** 8924-8931.

Zaug, A.J., J.R. Kent, and T.R. Cech. 1984. A labile phosphodiester bond at the ligation junction in a circular intervening sequence RNA. *Science* **224:** 574-578.

———. 1985. Reactions of the intervening sequence of the *Tetrahymena* rRNA precursor: pH dependence of cyclization and site-specific hydrolysis. *Biochemistry* **24:** 6211-6218.

APPENDIX A
The Restriction Enzymes

Richard J. Roberts and Dana Macelis
New England Biolabs
Beverly, Massachusetts 01915

The restriction enzyme database, REBASE, is a collection of information about restriction enzymes and DNA methylases. More than 2400 restriction enzymes are now known, including 17 different type I specificities, 187 different type II specificities, and 4 different type III specificities. The table contains a listing of all prototype restriction enzymes (types I, II, and III), and some statistics about the numbers of isoschizomers and neoschizomers (enzymes that cleave at a position different from that of the prototype) for each prototype. The most common patterns recognized by type II restriction enzymes are the tetranucleotide and hexanucleotide palindromes. The enzymes recognizing 14 of the 16 possible tetranucleotide palindromes are now known; the two missing sequences are ATAT and TATA. The enzymes recognizing 55 of the 64 possible hexanucleotide palindromes are now known; the missing nine sequences are AAATTT, ATATAT, CGCGCG, CTATAG, TAATTA, TAGCTA, TATATA, TTATAA, and TTGCAA.

REBASE is updated daily. Each month, a set of REBASE data files are released publically and distributed to the scientific community, at no charge, via e-mail. They can also be retrieved by anonymous ftp from vent.neb.com (192.138.220.2). These data files are flat ASCII text files, many of which are designed specifically for use with a variety of software packages such as GCG, IGSuite, GENEPRO, Staden, DNA Strider, Pro-Cite, and PC/Gene. Other data files include a complete set of references, including abstracts, to papers on restriction enzymes and methylases, a list of all commercial suppliers of restriction enzymes and methylases, complete with contact information for each, and a list of enzymes currently being sold. New data files are constantly being added and each release of REBASE includes a monthly release note, indicating that the files at the ftp site have been updated, and listing new enzymes, newly available formats, enzyme name changes, etc. To join the mailing list or for more information, send a request to R.J. Roberts via e-mail to roberts@neb.com, telephone (508)927-3382 or fax (508)921-1527.

Restriction Enzymes

		Numbers of isoschizomers cleaving at			
	Prototypes	same	different	unknown	
enzyme	recognition sequence	site	site	site	total

TYPE I

CfrAI	GCANNNNNNNNGTGG				1
EcoAI	GAGNNNNNNNGTCA				1
EcoBI	TGANNNNNNNNTGCT				1
EcoDI	TTANNNNNNNGTCY				1
EcoDR2I	TCANNNNNNGTCG				1
EcoDR3I	TCANNNNNNNATCG				1
EcoDXXI	TCANNNNNNNRTTC				1
EcoEI	GAGNNNNNNATGC				1
EcoKI	AACNNNNNNGTGC				1
EcoR124I	GAANNNNNNRTCG				1
EcoR124/3I	GAANNNNNNNRTCG				1
EcoRD2I	GAANNNNNRTTC				1
EcoRD3I	GAANNNNNNRTTC				1
StySBI	GAGNNNNNNRTAYG				1
StySJI	GAGNNNNNNGTRC				1
StySPI	AACNNNNNNGTRC				1
StySQI	AACNNNNNNRTAYG				1

TYPE II

AatII	GACGT↑C	1		1	3
AccI	GT↑MKAC	1		1	3
AciI	CCGC(-3/-1)				1
AclI	AA↑CGTT	1			2
AcyI	GR↑CGYC	12		6	19
AflII	C↑TTAAG	4		13	18
AflIII	A↑CRYGT				1
AgeI	A↑CCGGT	1			2
AhaIII	TTT↑AAA	1			2
AluI	AG↑CT	1		6	8
AlwNI	CAGNNN↑CTG				1
ApaBI	GCANNNNN↑TGC				1
ApaI	GGGCC↑C		1	8	10
ApaLI	G↑TGCAC	3		8	12
ApoI	R↑AATTY	2			3
AscI	GG↑CGCGCC				1
AsuI	G↑GNCC	15		44	60
AsuII	TT↑CGAA	18		12	31
AvaI	C↑YCGRG	8	1	16	26
AvaII	G↑GWCC	22		56	79
AvaIII	ATGCAT		8	5	14
AvrII	C↑CTAGG	2			3

Restriction Enzymes

enzyme	Prototypes recognition sequence	same site	different site	unknown site	total
*Bae*I	ACNNNNGTAYC				1
*Bal*I	TGG↑CCA	3			4
*Bam*HI	G↑GATCC	10		66	77
*Bbv*I	GCAGC(8/12)	3		4	8
*Bbv*II	GAAGAC(2/6)	4		3	8
*Bcc*I	CCATC				1
*Bce*83I	CTTGAG(16/14)				1
*Bce*fI	ACGGC(12/13)				1
*Bcg*I	GCANNNNNNTCG(12/10)				1
*Bcl*I	T↑GATCA	4		15	20
*Bet*I	W↑CCGGW	2		1	4
*Bgl*I	GCCNNNN↑NGGC			1	2
*Bgl*II	A↑GATCT	4			5
*Bin*I	GGATC(4/5)	1		3	5
*Bpu*10I	CCTNAGC(-5/-2)				1
*Bsa*AI	YAC↑GTR	1		3	5
*Bsa*BI	GATNN↑NNATC	3		2	6
*Bsc*GI	CCCGT				1
*Bse*PI	GCGCGC		1	10	12
*Bse*RI	GAGGAG(10/8)				1
*Bsg*I	GTGCAG(16/14)				1
*Bsi*I	CTCGTG(-5/-1)				1
*Bsi*YI	CCNNNNN↑NNGG	1		4	6
*Bsm*AI	GTCTC(1/5)	1			2
*Bsm*I	GAATGC(1/-1)	9		2	12
*Bsp*1407I	T↑GTACA	2		2	5
*Bsp*GI	CTGGAC				1
*Bsp*HI	T↑CATGA	2		1	4
*Bsp*LU11I	A↑CATGT				1
*Bsp*MI	ACCTGC(4/8)				1
*Bsp*MII	T↑CCGGA	12		11	24
*Bsr*BI	GAGCGG(-3/-3)				1
*Bsr*DI	GCAATG(2/0)				1
*Bsr*I	ACTGG(1/-1)	3		3	7
*Bst*EII	G↑GTNACC	10		9	20
*Bst*XI	CCANNNNN↑NTGG			3	4
*Cac*8I	GCN↑NGC				1
*Cau*II	CC↑SGG	5	2	29	37
*Cfr*10I	R↑CCGGY	3		2	6
*Cfr*I	Y↑GGCCR	1		16	18
*Cla*I	AT↑CGAT	22		55	78
*Cvi*JI	RG↑CY			5	6
*Cvi*RI	TG↑CA				1

Restriction Enzymes

enzyme	recognition sequence	same site	different site	unknown site	total
DdeI	C↑TNAG				1
DpnI	GA↑TC	1		6	8
DraII	RG↑GNCCY	1	1	2	5
DraIII	CACNNN↑GTG				1
DrdI	GACNNNN↑NNGTC				1
DrdII	GAACCA				1
DsaI	C↑CRYGG				1
Eam1105I	GACNNN↑NNGTC	2		3	6
EciI	TCCGCC				1
Eco31I	GGTCTC(1/5)	2		36	39
Eco47III	AGC↑GCT	2			3
Eco57I	CTGAAG(16/14)			4	5
EcoNI	CCTNN↑NNNAGG			8	9
EcoRI	G↑AATTC	4		7	12
EcoRII	↑CCWGG	2	29	93	125
EcoRV	GAT↑ATC	4		10	15
Esp3I	CGTCTC(1/5)			3	4
EspI	GC↑TNAGC	3		4	8
FauI	CCCGC(4/6)				1
FinI	GGGAC		1		2
Fnu4HI	GC↑NGC	4		3	8
FnuDII	CG↑CG	10	1	20	32
FokI	GGATG(9/13)		1	1	3
FseI	GGCCGG↑CC				1
GdiII	YGGCCG(-5/-1)				1
GsuI	CTGGAG(16/14)	1		6	8
HaeI	WGG↑CCW				1
HaeII	RGCGC↑Y	1	2	6	10
HaeIII	GG↑CC	22		96	119
HgaI	GACGC(5/10)				1
HgiAI	GWGCW↑C	5		3	9
HgiCI	G↑GYRCC	6		22	29
HgiEII	ACCNNNNNNGGT				1
HgiJII	GRGCY↑C	9		36	46
HhaI	GCG↑C	2	3	7	13
HindII	GTY↑RAC	2		4	7
HindIII	A↑AGCTT	4		24	29
HinfI	G↑ANTC	3		8	12
HpaI	GTT↑AAC	1		3	5
HpaII	C↑CGG	8		22	31
HphI	GGTGA(8/7)			1	2
KpnI	GGTAC↑C		4	24	29
Ksp632I	CTCTTC(1/4)	5		5	11
MaeI	C↑TAG	3		2	6

Restriction Enzymes

		Numbers of isoschizomers cleaving at			
Prototypes		same	different	unknown	
enzyme	recognition sequence	site	site	site	total
MaeII	A↑CGT				1
MaeIII	↑GTNAC				1
MboI	↑GATC	28		92	121
MboII	GAAGA(8/7)			2	3
McrI	CGRY↑CG	3		1	5
MfeI	C↑AATTG	1			2
MluI	A↑CGCGT			2	3
MlyI	GACTC(5/5)				1
MmeI	TCCRAC(20/18)				1
MnlI	CCTC(7/6)				1
MseI	T↑TAA	2			3
MslI	CAYNN↑NNRTG				1
MstI	TGC↑GCA	5		9	15
MwoI	GCNNNNN↑NNGC	1		1	3
NaeI	GCC↑GGC	6	3	26	36
NarI	GG↑CGCC	5	4	7	17
NcoI	C↑CATGG	1		2	4
NdeI	CA↑TATG				1
NheI	G↑CTAGC				1
NlaIII	CATG↑	1	1	1	4
NlaIV	GGN↑NCC	2		7	10
NotI	GC↑GGCCGC				1
NruI	TCG↑CGA	4		6	11
NspBII	CMG↑CKG	1			2
NspI	RCATG↑Y	1		1	3
PacI	TTAAT↑TAA				1
Pfl1108I	TCGTAG				1
PflMI	CCANNNN↑NTGG	5			6
PleI	GAGTC(4/5)			1	2
PmaCI	CAC↑GTG	4		4	9
PmeI	GTTT↑AAAC				1
PpuMI	RG↑GWCCY	2		1	4
PshAI	GACNN↑NNGTC				1
PstI	CTGCA↑G	15		103	119
PvuI	CGAT↑CG	8	1	24	34
PvuII	CAG↑CTG	9		12	22
RleAI	CCCACA(12/9)				1
RsaI	GT↑AC	1	3	4	9
RsrII	CG↑GWCCG	2			3
SacI	GAGCT↑C	1	3	4	9
SacII	CCGC↑GG	17	1	77	96
SalI	G↑TCGAC	6		16	23
SapI	GCTCTTC(1/4)	1			2
SauI	CC↑TNAGG	11		12	24

Restriction Enzymes

	Prototypes	Numbers of isoschizomers cleaving at			
enzyme	recognition sequence	same site	different site	unknown site	total
ScaI	AGT↑ACT	3		8	12
ScrFI	CC↑NGG	2	3	13	19
SduI	GDGCH↑C	5		3	9
SecI	C↑CNNGG	1		2	4
SexAI	A↑CCWGGT				1
SfaNI	GCATC(5/9)		1	1	3
SfeI	C↑TRYAG	3		1	5
SfiI	GGCCNNNN↑NGGCC	1			2
SgrAI	CR↑CCGGYG				1
SmaI	CCC↑GGG	2	7	2	12
SnaBI	TAC↑GTA	1		6	8
SnaI	GTATAC		4		5
SpeI	A↑CTAGT				1
SphI	GCATG↑C	2		6	9
SplI	C↑GTACG	5		4	10
SrfI	GCCC↑GGGC				1
Sse8387I	CCTGCA↑GG				1
SspI	AAT↑ATT				1
StuI	AGG↑CCT	7		14	22
StyI	C↑CWWGG	4		8	13
SwaI	ATTT↑AAAT				1
TaqI	T↑CGA	1		1	3
TaqII	CACCCA(11/9)				1
	GACCGA(11/9)				
TfiI	G↑AWTC				1
Tsp45I	GTSAC				1
TspEI	AATT				1
Tth111I	GACN↑NNGTC	1		8	10
Tth111II	CAARCA(11/9)				1
VspI	AT↑TAAT	2			3
XbaI	T↑CTAGA			3	4
XcmI	CCANNNNN↑NNNNTGG				1
XhoI	C↑TCGAG	14	1	63	79
XhoII	R↑GATCY	4		10	15
XmaIII	C↑GGCCG	5		1	7
XmnI	GAANN↑NNTTC	2		1	4

TYPE III

EcoPI	AGACC				1
EcoP15I	CAGCAG				1
HinfIII	CGAA			1	2
StyLTI	CAGAG				1

APPENDIX B
Some Well-characterized DNA-repair Nucleases/Glycosylases

R. Stephen Lloyd
Sealy Center for Molecular Science
University of Texas Medical Branch
Galveston, Texas 77555-0852

The following table lists some of the well-characterized DNA-repair nucleases, DNA glycosylases, and combined DNA glycosylase/AP lyases. The DNA glycosylases and DNA glycosylase/AP lyases are categorized and presented relative to the type of modified bases on which they function.

This table does not attempt to cover comprehensively all of the DNA-repair enzymes that have appeared in the literature, but rather to serve as a general guide to the complexity that is associated with the maintenance of an organism's genetic integrity. References were chosen to document the most salient features of the individual enzymes.

SOME WELL-CHARACTERIZED DNA-REPAIR NUCLEASES/GLYCOSYLASES

Enzyme	Gene	Molecular mass (kD)	Properties	References
I. GLYCOSYLASES, GLYCOSYLASE/AP LYASES				
A. Oxidized Pyrimidines				
E. coli endonuclease III	*nth*	23.5	DNA glycosylase/AP lyase specific for thymine glycols and other pyrimidine radiolysis products; crystal structure solved; a 4Fe-4S enzyme	1–4
Murine UV Endonucleases I,II	— —	43 (I) 28 (II)	DNA glycosylase/AP lyases specific for thymine glycols and other pyrimidine radiolysis products	5
B. Oxidized Purines				
E. coli Fpg DNA glycosylase (or 8-oxoguanine DNA glycosylase, or *mutM* gene product)	*fpg* *mutM*	30	DNA glycosylase/AP lyase specific for Fapy-A, Fapy-G, their methylated derivatives and 8-oxodG; catalyzes δ-elimination; zinc finger motif	6–11
C. Deaminated Bases				
E. coli uracil DNA glycosylase	*udg*	26	DNA glycosylase specific for the release of uracil from ssDNA and dsDNA; mammalian cells have both nuclear and mitochondrial forms	12–14
Mammalian G/T thymine DNA glycosylase	—	?	thymine-specific DNA glycosylase with associated incision activities that leave a one-base gap; thymine/guanine mismatches occur from the spontaneous deamination of 5-methylcytosine	15,16

5-Hydroxymethyl-uracil (5-HmU) DNA glycosylase	—	38	DNA glycosylase and possible AP lyase activity specific for 5-hydroxymethyl-uracil; 5-HmU arises in DNA by sequential oxidation of 5-methylcytosine to hydroxymethyl cytosine and subsequent deamination	17
Hypoxanthine DNA glycosylase	—	31	DNA glycosylase without an associated AP lyase specific for hypoxanthine; spontaneous deamination of adenine yields hypoxanthine	18,19

D. Alkylated Bases

E. coli 3-methyladenine (3-meA) DNA glycosylase I	tag	21	glycosylase specific for 3-meA; accounts for ~90% of 3-meA glycosylase activity in E. coli; constitutively expressed	20,21
E. coli methyladenine (3-meA) DNA glycosylase II	alkA	31	glycosylase with broad substrate specificity that catalyzes the removal of a variety of alkylated purines and pyrimidines; induced in the adaptive response	21,22
Mammalian 3-methyl-adenine glycosylases	—	42 (calf thymus) 27 (calf thymus)	glycosylases with broad substrate specificity like E. coli alkA gene product; not inducible	23–27

E. Cyclobutane Pyrimidine Dimers

T4 endonuclease V	denV	16	DNA glycosylase/AP lyase specific for cyclobutane pyrimidine dimers; locates its substrate by a processive scanning mechanism; X-ray crystal structure known; active site residue identified	28–33

(Continued on following page.)

Table continued

Enzyme	Gene	Molecular mass (kD)	Properties	References
Micrococcus luteus UV endonuclease	—	18	DNA glycosylase/AP lyase specific for cyclobutane pyrimidine dimers; locates its substrate by processive scanning mechanism	34,35
Yeast pyrimidine dimer endonuclease	—	16–20	DNA glycosylase/AP lyase specific for cyclobutane pyrimidine dimers	36
F. Mismatched Bases				
E. coli adenine DNA glycosylase	*mutY*	39	an adenine-specific DNA glycosylase at A/G, A/8-oxodG, and A/C mismatches; strong sequence homology with *E. coli* endonuclease III; 4Fe/4S cluster enzyme	37–39
II. ENDONUCLEASES				
E. coli exonuclease III (endonuclease VI)	*xth*	31	85% of the cell's AP nicking activity; $3' \rightarrow 5'$ exonucleolytic activity on dsDNA; RNase H activity; crucial for survival following oxidative challenge	40–42
E. coli endonuclease IV	*nfo*	32	EDTA-resistant AP endonuclease; hydrolyzes 3'-phosphoglycoaldehydes, 3'-phosphates, and 3-phosphate esters	43–46

E. coli Vsr endonuclease (G/T mismatch endonuclease)	vsr	18	endonucleolytic activity with no associated glycosylase; incision 5' to G/T; specific for 5'-CTWGG-3' 3'-GGWCC-5' arising by spontaneous deamination in *dcm* methyltransferase recognition sequences	47
Yeast apurinic endonuclease APN1	APN1	37	catalytic activities very similar to E. coli endonuclease IV; 55% homology with E. coli *nfo* gene product	48–51
Drosophila apurinic endonuclease (I)	—	63 (I)	class II AP endonuclease	52–54
Drosophila apurinic endonuclease (II)	—	66 (II)	class III AP endonuclease	52–54
Drosophila recombinational repair protein	rrp1	74	class II AP endonuclease; 3' exonuclease activity associated with recombinational repair; shares sequence homology with E. coli exonuclease III	55
Drosophila apurinic endonuclease AP3	AP3	35	class II AP endonuclease with a high homology with human AP endonuclease P0, which also serves as a ribosomal protein	56, 57
Human apurinic endonuclease	ape, hap2	41	class II AP endonuclease	58–62

(*Continued on following page.*)

Table continued

Enzyme	Gene	Molecular mass (kD)	Properties	References
III. NUCLEOTIDE EXCISION REPAIR NUCLEASES				
E. coli UvrABC	*uvrA* *uvrB* *uvrC*	104 76 66	multisubunit complex makes two breaks in damaged DNA strand, 12 bases apart; UvrA$_2$B scans DNA for a variety of structural damages; UvrB makes the 3′ cut, and UvrC makes the 5′ incision	63–66
S. cerevisiae RAD1/RAD10	*RAD1* *RAD10*	126 24	two-subunit complex that binds and degrades circular ssDNA; no preferential binding to UV-irradiated DNA; endonucleolytic activity on dsDNA increases with negative superhelix density	67–69
IV. METHYL-DIRECTED MISMATCH REPAIR				
E. coli MutH	*mutH*	25	recognizes a hemimethylated d(GATC) sequence and catalyzes a single-strand break in the unmethylated strand	70,71

References:
1. Gates, F.T., III and S. Linn. 1977. Endonuclease from *Escherichia coli* that acts specifically upon duplex DNA damaged by ultraviolet light, osmium tetroxide, acid, or X-rays. *J. Biol. Chem.* **252:** 2802–2807.
2. Katcher, H.L. and S.S. Wallace. 1983. Characterization of the *Escherichia coli* X-ray endonuclease, endonuclease III. *Biochemistry* **22:** 4071–4081.

3. Asahara, H., P.M. Wistor, J.F. Bank, R.H. Bakerian, and R.P. Cunningham. 1989. Purification and characterization of *Escherichia coli* endonuclease III from the cloned *nth* gene. *Biochemistry* **28:** 4444–4449.
4. Kuo, C.-F., D.E. McRee, C.L. Fisher, S.F. O'Handley, R.P. Cunningham, and J.A. Tainer. 1992. Atomic structure of the DNA repair [4Fe-4S] enzyme endonuclease III. *Science* **258:** 434–440.
5. Kim, J. and S. Linn. 1989. Purification and characterization of UV endonucleases I and II from murine plasmacytoma cells. *J. Biol. Chem.* **264:** 2739–2745.
6. Chetsanga, C.J. and T. Lindahl. 1979. Release of 7-methylguanine residues whose imidazole rings have been opened from damaged DNA by a DNA glycosylase from *Escherichia coli*. *Nucleic Acids Res.* **6:** 3673–3684.
7. Boiteux, S., E. Gajewski, J. Laval, and M. Dizdaroglu. 1992. Substrate specificity of the *Escherichia coli* FPG protein (formamidopyrimidine-DNA glycosylase): Excision of purine lesions in DNA produced by ionizing radiation or photosensitization. *Biochemistry* **31:** 106–110.
8. Boiteux, S., T.R. O'Connor, and J. Laval. 1987. Formamidopyrimidine-DNA glycosylase of *Escherichia coli*: Cloning and sequencing of the *fpg* structural gene and overproduction of the protein. *EMBO J.* **6:** 3177–3183.
9. O'Connor, T.R., R.J. Graves, G. de Murcia, B. Castaing, and J. Laval. 1993. Fpg protein of *Escherichia coli* is a zinc finger protein whose cysteine residues have a structural and/or functional role. *J. Biol. Chem.* **268:** 9063–9070.
10. Michaels, M.L. and J.H. Miller. 1992. The GO system protects organisms from the mutagenic effect of the spontaneous lesion 8-hydroxyguanine (7,8-dihydro-8-oxoguanine). *J. Bacteriol.* **174:** 6321–6325.
11. Tchou, J., H. Kasai, S. Shibutani, M.-H. Chung, J. Laval, A.P. Grollman, and S. Nishimura. 1991. 8-Oxoguanine (8-hydroxyguanine) DNA glycosylase and its substrate specificity. *Proc. Natl. Acad. Sci.* **88:** 4690–4694.
12. Lindahl, T. 1974. An *N*-glycosidase from *Escherichia coli* that releases free uracil from DNA containing deaminated cytosine residues. *Proc. Natl. Acad. Sci.* **71:** 3649–3653.
13. Varshney, U., T. Hutcheon, and J.H. van de Sande. 1988. Sequence analysis, expression, and conservation of *Escherichia coli* uracil DNA glycosylase and its gene (*ung*). *J. Biol. Chem.* **263:** 7776–7784.
14. Tomilin, N.V. and O.N. Aprelikova. 1989. Uracil-DNA glycosylases and DNA uracil repair. *Int. Rev. Cytol.* **114:** 125–179.
15. Brown, T.C. and J. Jiricny. 1987. A specific mismatch repair event protects mammalian cells from loss of 5-methylcytosine. *Cell* **50:** 945–950.
16. Wiebauer, K. and J. Jiricny. 1990. Mismatch-specific thymine DNA glycosylase and DNA polymerase β mediate the correction of G/T mispairs in nuclear extracts from human cells. *Proc. Natl. Acad. Sci.* **87:** 5842–5845.
17. Cannon-Carlson, S.V., H. Gokhale, and G.W. Teebor. 1989. Purification and characterization of 5-hydroxymethyluracil-DNA glycosylase from calf thymus. *J. Biol. Chem.* **264:** 13306–13312.
18. Karran, P. and T. Lindahl. 1978. Enzymatic excision of free hypoxanthine from polydeoxynucleotides and DNA containing deoxyinosine monophosphate residues. *J. Biol. Chem.* **253:** 5877–5879.
19. Dianov, G. and T. Lindahl. 1991. Preferential recognition of I:T base-pairs in the initiation of excision-repair by hypoxanthine-DNA glycosylase. *Nucleic Acids Res.* **19:** 3829–3833.
20. Riazuddin, S. and T. Lindahl. 1978. Properties of 3-methyladenine-DNA glycosylase from *Escherichia coli*. *Biochemistry* **17:** 2110–2118.
21. Clarke, N.D., M. Kvaal, and E. Seeberg. 1984. Cloning of *Escherichia coli* genes encoding 3-methyladenine DNA glycosylases I and II. *Mol. Gen. Genet.* **197:** 368–372.
22. Samson, L.D. and J. Cairns. 1977. A new pathway for DNA repair in *Escherichia coli*. *Nature* **267:** 281–282.

23. Male, R., D.E. Helland, and K. Kleppe. 1985. Purifications and characterization of 3-methyladenine-DNA glycosylase from calf thymus. *J. Biol. Chem.* **260:** 1623–1629.
24. Samson, L., B. Derfler, M. Boosalis, and K. Call. 1991. Cloning and characterization of a 3-methyladenine DNA glycosylase cDNA from human cells whose gene maps to chromosome 16. *Proc. Natl. Acad. Sci.* **88:** 9127–9131.
25. O'Connor, T.R. and F. Laval. 1990. Isolation and structure of a cDNA expressing a mammalian 3-methyl-adenine-DNA glycosylase. *EMBO J.* **9:** 3337–3342.
26. Chakravarti, D., G.C. Ibeanu, K. Tano, and S. Mitra. 1991. Cloning and expression in *Escherichia coli* of a human cDNA encoding of the DNA repair protein *N*-methylpurine DNA glycosylase. *J. Biol. Chem.* **266:** 15710–15715.
27. Vickers, M.A., P. Vyas, P.C. Harris, D.L. Simmons, and D.R. Higgs. 1993. Structure of the human 3-methyladenine DNA glycosylase gene and localization close to the 16p telomere. *Proc. Natl. Acad. Sci.* **90:** 3437–3441.
28. Lloyd, R. S., P.C. Hanawalt, and M.L. Dodson. 1980. Processive action of T4 endonuclease V on ultraviolet-irradiated DNA. *Nucleic Acids Res.* **8:** 5113–5127.
29. Gordon, L.K. and W.A. Haseltine. 1980. Comparison of the cleavage of pyrimidine dimers by the bacteriophage T4 and *Micrococcus luteus* UV-specific endonucleases. *J. Biol. Chem.* **24:** 12047–12050.
30. Valerie, K., E.E. Henderson, and J.K. deRiel. 1984. Identification, physical map location and sequence of the *denV* gene from bacteriophage T4. *Nucleic Acids Res.* **12:** 8085–8097.
31. Gruskin, E.A. and R.S. Lloyd. 1986. The DNA scanning mechanism of T4 endonuclease V. *J. Biol. Chem.* **261:** 9607–9613.
32. Morikawa, K., O. Matsumoto, M. Tsujimoto, K. Katayanagi, M. Ariyoshi, T. Doi, M. Ikehara, T. Inaoka, and R. Ohtsuka. 1992. X-ray structure of T4 endonuclease V: An excision repair enzyme specific for a pyrimidine dimer. *Science* **256:** 523–526.
33. Schrock, R.D., III and R.S. Lloyd. 1991. Reductive methylation of the N-terminus of endonuclease V eradicates catalytic activities—Evidence for an essential role of the N-terminus in the chemical mechanisms of catalysis. *J. Biol. Chem.* **266:** 17631–17639.
34. Haseltine, W.A., L.K. Gordon, C.P. Lindan, R.H. Grafstrom, N.L. Shaper, and L. Grossman. 1980. Cleavage of pyrimidine dimers in specific DNA sequences by a pyrimidine dimer DNA-glycosylase of *M. luteus*. *Nature* **285:** 634–640.
35. Hamilton, R.W. and R.S. Lloyd. 1989. Modulation of the DNA scanning activity of the *Micrococcus luteus* UV endonuclease. *J. Biol. Chem.* **264:** 17422–17427.
36. Hamilton, K.K., P.M.J. Kim, and P.W. Doetsch. 1992. A eukaryotic DNA glycosylase/lyase recognizing ultraviolet light-induced pyrimidine dimers. *Nature* **356:** 725–728.
37. Au, K.G., S. Clark, J.H. Miller, and P. Modrich. 1989. *Escherichia coli mutY* gene encodes an adenine glycosylase active on G-A mispairs. *Proc. Natl. Acad. Sci.* **86:** 8877–8881.
38. Lu, A.-L. and D.-Y. Chang. 1988. A novel nucleotide excision repair for the conversion of an A/G mismatch to C/G base pair in *E. coli*. *Cell* **54:** 805–812.
39. Michaels, M.L., L. Pham, Y. Ngheim, C. Cruz, and J.H. Miller. 1990. MutY, an adenine glycosylase active on G-A mispairs, has homology to endonuclease III. *Nucleic Acids Res.* **18:** 3841–3845.
40. Warner, H.R., B.F. Demple, W.A. Deutsch, C.M. Kane, and S. Linn. 1980. Apurinic/apyrimidinic endonucleases in repair of pyrimidine dimers and other lesions in DNA. *Proc. Natl. Acad. Sci.* **77:** 4602–4606.
41. Demple, B., A. Johnson, and D. Fund. 1986. Exonuclease III and endonuclease IV remove 3' blocks from DNA synthesis primers in H_2O_2-damaged *Escherichia coli*. *Proc. Natl. Acad. Sci.* **83:** 7731–7735.
42. Saporito, S.M., B.J. Smith-White, and R.P. Cunningham. 1988. Nucleotide sequence of the *xth* gene of *Escherichia coli* K-12. *J. Bacteriol.* **170:** 4542–4547.
43. Ljungquist, S. 1977. A new endonuclease from *Escherichia coli* acting at apurinic sites in DNA. *J. Biol. Chem.* **252:** 2808–2814.

44. Levin, J.D., A.W. Johnson, and B. Demple. 1988. Homogenous *Escherichia coli* endonuclease IV. *J. Biol. Chem.* **263:** 8066–8071.
45. Levin, J.D., R. Shapiro, and B. Demple. 1991. Metalloenzymes in DNA repair. *J. Biol. Chem.* **266:** 22893–22898.
46. Saporito, S.M. and R.P. Cunningham. 1988. Nucleotide sequence of the *nfo* gene of *Escherichia coli* K-12. *J. Bacteriol.* **170:** 5141–5145.
47. Hennecke, F., H. Kolmar, K. Bründl, and H.-J. Fritz. 1991. The *vsr* gene product of *E. coli* K-12 is a strand- and sequence-specific DNA mismatch endonuclease. *Nature* **353:** 776–778.
48. Armel, P.R. and S.S. Wallace. 1984. DNA repair in *Saccharomyces cerevisiae*: Purification and characterization of apurinic endonucleases. *J. Bacteriol.* **160:** 895–902.
49. Johnson, A.W. and B. Demple. 1988a. Yeast DNA diesterase for 3′-fragments of deoxyribose: Purification and physical properties of a repair enzyme for oxidative DNA damage. *J. Biol. Chem.* **263:** 18009–18016.
50. Johnson, A.W. and B. Demple. 1988b. Yeast 3′-repair diesterase is the major cellular/apyrimidinic endonuclease: Substrate specificity and kinetics. *J. Biol. Chem.* **263:** 18017–18022.
51. Popoff, S.C., A.I. Spira, A.W. Johnson, and B. Demple. 1990. Yeast structural gene (*APNI*) for the major apurinic endonuclease: Homology to *Escherichia coli* endonuclease IV. *Proc. Natl. Acad. Sci.* **87:** 4193–4197.
52. Spiering, A.L. and W.A. Deutsch. 1981. Apurinic DNA endonucleases from *Drosophila melanogaster* embryos. *Mol. Gen. Genet.* **183:** 171–174.
53. Spiering, A.L. and W.A. Deutsch. 1986. *Drosophila* apurinic/apyrimidinic DNA endonucleases. *J. Biol. Chem.* **261:** 3222–3228.
54. Venugopal, S., S.N. Guzder, and W.A. Deutsch. 1990. Apurinic endonuclease activity from wild-type and repair-deficient *mei-9 Drosophila* ovaries. *Mol. Gen. Genet.* **221:** 421–426.
55. Sander, M., K. Lowenhaupt, and A. Rich. 1991. *Drosophila* Rrp1 protein: An apurinic endonuclease with homologous recombination activities. *Proc. Natl. Acad. Sci.* **88:** 6780–6784.
56. Kelley, M.R., S. Venugopal, J. Harless, and W.A. Deutsch. 1989. Antibody to a human DNA repair protein allows for cloning of a *Drosophila* cDNA that encodes an apurinic endonuclease. *Mol. Cell. Biol.* **9:** 965–973.
57. Grabowski, D.T., W.A. Deutsch, D. Derda, and M.R. Kelley. 1991. *Drosophila* AP3, a presumptive DNA repair protein, is homologous to human ribosomal associated protein PO. *Nucleic Acids Res.* **19:** 4297.
58. Kane, C.M. and S. Linn. 1981. Purification and characterization of an apurinic/apyrimidinic endonuclease from HeLa cells. *J. Biol. Chem.* **256:** 2405–3414.
59. Chen, D.S., T. Herman, and B. Demple. 1991. Two distinct human DNA diesterases that hydrolyze 3′-blocking deoxyribose fragments from oxidized DNA. *Nucleic Acids Res.* **19:** 5907–5914.
60. Demple, B., T. Herman, and D.S. Chen. 1991. Cloning and expression of APE, the cDNA encoding the major human apurinic endonuclease: Definition of a family of DNA repair enzymes. *Proc. Natl. Acad. Sci.* **88:** 11450–11454.
61. Robson, C.N. and I.D. Hickson. 1991. Isolation of cDNA clones encoding a human apurinic/apyrimidinic endonuclease that corrects DNA repair and mutagenesis defects in *E. coli xth* (exonuclease III) mutants. *Nucleic Acids Res.* **19:** 5519–5523.
62. Robson, C.N., D. Hochhauser, K. Craig, K. Rack, V.J. Buckle, and I.D. Hickson. 1992. Structure of the human DNA repair gene *HAP1* and its localization to chromosome 14q 11.2-12. *Nucleic Acids Res.* **20:** 4417–4421.
63. Myles, G.M. and A. Sancar. 1989. DNA repair. *Chem. Res. Toxicol.* **2:** 197–226.
64. Lin, J.-J. and A. Sancar. 1992. Active site of (A)BC excinuclease. *J. Biol. Chem.* **267:** 17688–17692.

65. Lin, J.-J., A.M. Phillips, J.E. Hearst, and A. Sancar. 1992. Active site of (A)BC excinuclease. I. Binding, bending and catalysis mutants of UvrB reveal a direct role in 3′ and an indirect role in 5′ incision. *J. Biol. Chem.* **267:** 17693–17700.
66. Van Houten, B. and A. Snowden. 1993. Mechanism of action of the *Escherichia coli* UvrABC nuclease: Clues to the damage recognition problem. *BioEssays* **15:** 51–59.
67. Sung, P., L. Prakash, and S. Prakash. 1992. Renaturation of DNA catalyzed by yeast DNA repair and recombination protein RAD10. *Nature* **355:** 743–745.
68. Sung, P., P. Reynolds, L. Prakash, and S. Prakash. 1993. Purification and characterization of the *Saccharomyces cerevisiae* RAD1/RAD10 endonuclease. *J. Biol. Chem.* (in press).
69. Tomkinson, A.E., A.J. Bardwell, L. Bardwell, N.J. Rappe, and E.C. Friedberg. 1993. Yeast DNA repair and recombination proteins Rad1 and Rad10 are subunits of a single-stranded DNA endonuclease. *Nature* **361:** 860–862.
70. Welsh, K.M., A.L. Lu, S. Clark, and P. Modrich. 1987. Isolation and characterization of the *Escherichia coli mutH* gene product. *J. Biol. Chem.* **262:** 15624–15629.
71. Au, K.G., K. Welsh, and P. Modrich. 1992. Initiation of methyl-directed mismatch repair. *J. Biol. Chem.* **267:** 12142–12148.

APPENDIX C
The Nucleases of *Escherichia coli*

Stuart Linn
BMB Division, Barker Hall, University of California
Berkeley, California 94720

Murray P. Deutscher
Department of Biochemistry, University of Connecticut Health Center
Farmington, Connecticut 06030-3305

As an indication of the number of nucleases in a single organism, the tables list those nucleases that have been documented from *Escherichia coli*. These tables also serve as references for researchers looking for the paradigm of a number of enzyme homologs.

One or two recent references are provided for each enzyme for an entrée into the literature. Recent references can also be found in the other chapters of this book where the individual enzymes are discussed with regard to their function; older references can be found in the chapter on *E. coli* DNases or in the appendix on RNases in the first edition of *Nucleases* (S.M. Linn and R.J. Roberts; Cold Spring Harbor Laboratory [1982, 1985]). For additional references, see Bachmann's 8th edition of the *E. coli* linkage map (*Microbiol. Rev. 54:* 130 [1990]) or K. Rudd's EcoBase.

In the Deoxyribonucleases of *E. coli* table, molecular weights are based on the translated gene sequence where available, and map positions are according to Bachmann (*Microbiol. Rev. 54:* 130 [1990]) for the 100-min *E. coli* linkage map. All known *E. coli* DNases that act as simple phosphodiesterases on undamaged DNA yield 3'-hydroxyl and 5'-phosphate termini, with the exception of *Eco*KI/*Eco*BI which yield 3'-hydroxy and unknown 5' termini.

Included in the table for Ribonucleases of *E. coli* are RNases that have been reported to be present in *E. coli*, except for several that are now known to have been mixtures of enzymes or manifestations of other RNases. It is still possible that some of the RNases listed in the table are also not distinct proteins, particularly the poorly characterized endonucleases, and possibly also oligoribonuclease. In addition, enzymes that may be related have been grouped together. The physiological functions listed for some of the enzymes are those known to be affected by mutations affecting the RNase. However, other roles for many of these enzymes are also likely.

Nucleases, 2nd Edition
© 1993 Cold Spring Harbor Laboratory Press 0-87969-426-2/93 $5 + .00

DEOXYRIBONUCLEASES OF *ESCHERICHIA COLI*

Enzyme	Gene	Map position	Subunit MW (kD)	Substrate	Major characteristics and products	References
Exonuclease I	*sbcB* (*xonA*)	44	53.2	ssDNA; 3' single-stranded tails	3'→5'; mononucleotides and 5' dinucleotide; processive; removes 3'-sugar phosphate fragments; acts in methylation-directed mismatch repair; *sbcB* mutants suppress *recB* and *recC* mutations	1
DNA polymerase I (exonuclease II)	*polA*	87 min	103	3'→5'; mismatched double-stranded 3' termini; ssDNA; oligonucleotides. 5'→3'; dsDNA; RNA-DNA hybrid	3'→5': mononucleotides. 5'→3': oligonucleotides that can include damaged nucleotides	2
DNA polymerase II	*polB* (*dinA*)	2 min	88	ssDNA	3'→5'	2
DNA polymerase III, ε subunit	*dnaQ* (*mutD*)	5 min	27.5	ssDNA; mismatched 3'-primer termini	3'→5'; mononucleotides, 5'-terminal dinucleotide; non-processive; activity stimulated when complexed to α-subunit	2

Exonuclease III (Endo.II, Endo.VI)	xthA	38	30.9	dsDNA; AP sites; DNA 3'-sugar phosphate fragment; RNA-DNA hybrid	dsDNA: 3'→5'; mononucleotides and residual single strands; acts from ends and nicks. 3' DNA phosphate or 3'-sugar phosphate fragment: Pi or sugar phosphate fragment. RNA:DNA hybrid; dNMPs, rNMPs, and residual single strands. AP DNA; nicks with 3' nucleoside and 5'-deoxyribose phosphate	2,3
Exonucleases IVA and IVB	?	?	?	oligonucleotides	mononucleotides	4
RecBCD (exonuclease V)	recB recC recD	61 61 61	134 129 67	dsDNA; ssDNA; ATP	dsDNA; ATP required; duplex unwound; 3'→5' and 5'→3' exonuclease; oligonucleotides; processive; endonuclease at Chi sites: 5'-GCTGGTGG. ssDNA; ATP required for 3'→5' and 5'→3' exonuclease; oligonucleotides; ATP stimulates endonuclease; large fragments	5,6
Exonuclease VII	xseA xseB	54 10	51.8 ~10.5	ssDNA: single-stranded termini or displaced single strands	3'→5' and 5'→3'; oligonucleotides; processive; 5'→3' activity acts in methylation-directed mismatch repair	7

(Continued on following page.)

Table continued

Enzyme	Gene	Map position	Subunit MW (kD)	Substrate	Major characteristics and products	References
Exonuclease VIII[a]	recE recT	30	126 (Exo VIII) 92.2 (RecE) 29.5 (RecT)	dsDNA	$5' \rightarrow 3'$; mononucleotides; processive; regulated by the *recE* (*sbcA*) gene product. carboxyl terminus of RecE, RecT promotes renaturation of homologous ssDNAs; amino terminus catalyzes DNase	8
RecJ	recJ	62	63	ssDNA tails and displaced single strands	$5' \rightarrow 3'$ and possibly $3' \rightarrow 5'$ to a lesser extent; $5' \rightarrow 3'$ activity acts in methylation-directed mismatch repair	9,10
dRpase	?	?	~50–55	DNA with 2-deoxyribose-5-phosphate termini at nicks or ends	deoxyribose-5-phosphate	11,12
Endonuclease I	endA	64	~12	dsDNA	double-strand breaks; oligonucleotides; inhibited by dsRNA; nicks dsDNA lightly when complexed to tRNA; periplasmic	13,14

Endonuclease III	*nth*	36	23.5	DNA with AP sites or oxidized pyrimidines	DNA glycosylase removes oxidized pyrimidines; AP site cleaved by β lyase mechanism; inhibited by tRNA	15,16
Endonuclease IV	*nfo*	47	31.6	dsDNA; AP sites; DNA 3'-phosphate; DNA 3'-sugar phosphate fragment	DNA 3'-phosphate or 3'-sugar phosphate; Pi or 3'-sugar phosphate fragment. AP DNA; nicks with 3'-nucleoside and 5'-deoxyribose phosphate	17,18
Endonuclease V	?	?	~27 (native)	ssDNA; damaged or uracil-containing duplex DNA	recognizes some, but not all, bulky damages and oxidized damages in duplex DNA; forms one double-strand break per 8 single-strand nicks; processive	19
Endonuclease VII	?	?	~56 (native)	ssDNA with AP sites	active in EDTA; characterization preliminary	20
Endonuclease VIII	?	?	~25 (native)	dsDNA (relaxed) with oxidized pyrimidines or AP sites	DNA glycosylase removes oxidized pyrimidines; AP sites cleaved by β,δ-elimination mechanism; characterization preliminary	21

(Continued on following page.)

Table continued

Enzyme	Gene	Map position	Subunit MW (kD)	Substrate	Major characteristics and products	References
Fpg	*fpg* (*mutM*)	82	30.2	duplex DNA with oxidized or ring-opened purines (most notably formamidopyrimidines and 8-hydroxyguanine) or AP sites; 5′-terminal deoxyribose phosphate	DNA glycosylase removes oxidized purines; AP sites cleaved by concerted β,δ-elimination mechanism; terminal deoxyribose 5′-phosphate formed	22–24
UvrABC	*uvrA* *uvrB* *uvrC*	92 18 42	103.8 76.1 66.0	damaged duplex DNA, most notably bulky damages	excises DNA damage in a 12–13-mer with damage 7–8 nucleotides from 5′ end. requires ATP; A_2B complex translocates DNA to find damage; B binds damage; BC excises damage, the 3′ nick made by B, followed by the 5′ nick by C; has ATPase for a 5′→3′ translocation helicase and protein complex reactivation	25–27
MutH	*mutH* (*mutR*, *prv*)	61	25.4	duplex DNA hemimethylated at GATC Dam methylation sites and containing a nucleotide pair mismatch	nicks DNA either 3′ or 5′ to the mismatch on the unmodified strand	28,29

Vsr endonuclease	vsr	43	18.0	G/T mismatch in: CTAGG or CTTGG GGTCC GGACC	involved in Vsp repair; nicks 5′ to the mismatched T	30,31
RuvC	ruvC	41	19	Holliday junction in recombination intermediates	introduces symmetrically related nicks into two strands of the same polarity, each on the 3′ side of a thymidine; stimulated by homologous oligonucleotide; may act with RuvAB	32–34
EcoK, EcoB[b]	hsdR (hsr) hsdM (hsm) hsdS (hss)	99 99 99	Hsr=α 130 Hsm=β 60 Hss=γ 50	B: 5′ TGA (N8) TGCT K: 5′ AAC (N6) GTCG in unmodified duplex DNA	requires Ado-met and ATP; does not turn over; translocates to form loops between recognition and cleavage sites; ATPase activity; leaves blocked 5′ termini (structure unknown) and 3′ tails of ~100 nucleotides or gaps (intermediates), of ~100 nucleotides; forms oligonucleotides to create gaps and tails	35–37
McrBC	mcrB (rglB) mcrC	98 25	53.9 41.7	cis or trans fully or hemi-methylated sites separated by 40–80 nucleotides of the type R(m)C N$_{40-80}$ R(m)C where mC is 5-me-, 5-hydroxymethyl-, or N^4-methylcytosine	multiple cleavages within the N$_{40-80}$ spacer region; requires GTP to activate McrB to bind McrC; restricts nonglycosylated T-even phage in vivo	38,39

(Continued on following page.)

Table continued

Enzyme	Gene	Map position	Subunit MW (kD)	Substrate	Major characteristics and products	References
McrA	mcrA (rglA)	25	?	DNA containing methyl-cytosine modifications	restricts T-even phage; in vitro characterization not reported	40
Mrr	mrr	99	33.5	DNA containing N6-methyl-adenine or cytosine modifications	in vitro characterization not reported	41,42
Eco topoisomerase I (ω protein)	topA (supX)	28	97.4	duplex negatively super-coiled DNA	type I topoisomerase; single-strand nicks with 5′-phosphate termini bound to Y319 of the enzyme; cleaves DNA when reaction disrupted by chemicals or heat	43
Eco topoiso-merase II (DNA gyrase)	gyrA (nalA; parD) gyrB (acrB, cov, himB nalC, pcbA hopA, parA)	48 83	97 90	duplex DNA	type II topoisomerase; unique ability to introduce supercoils; double-strand breaks with 4-nucleotide, 5′ tails; 5′-phosphates bound to Y122 of GyrA; cleaves DNA when reaction disrupted by drugs or heat	44

Eco topoisomerase III	topB	39	73.2	duplex, super-coiled DNA	type I topoisomerase; when disrupted cleaves DNA *and RNA* at the same sequences	45,46
Eco topoisomerase IV	parC parE	65 65	81.2 66.7	duplex DNA	type II topoisomerase associated with nucleoid resolution; double-strand breaks	47,48

[a]Exonuclease VIII appears to be the product of a fusion between the adjacent *recE* (amino-terminal) and *recT* (carboxy-terminal) overlapping open reading frames.
[b]Many other strains have a member of the A or other family of type I restriction enzymes in lieu of the K family (see Chapter 3).

References:

1. Philips, G.J. and S.R. Kushner. 1987. Determination of the nucleotide sequence for the exonuclease I structural gene (*sbcB*) of *Escherichia coli* K12. *J. Biol. Chem.* **262**: 455–459.
2. Kornberg, A. and T. Baker. 1992. *DNA replication*, 2nd edition. W.H. Freeman, New York.
3. Saporito, S.M., B.J. Smith-White, and R.P. Cunningham. 1988. Nucleotide sequence of the *xth* gene of *Escherichia coli* K-12. *J. Bacteriol.* **170**: 4542–4547.
4. Jorgensen, S.E. and J.F. Koerner. 1966. Separation and characterization of deoxyribonucleases of *Escherichia coli* B. I. Chromatographic separation and properties of two deoxyribonucleotidases. *J. Biol. Chem.* **241**: 3090–3096.
5. Dixon, D.A. and S.C. Kowalczykowski. 1993. The recombination hotspot χ is a regulatory sequence that acts by attenuating the nuclease activity of *E. coli* RecBCD enzyme. *Cell* **73**: 87–96.
6. Masterson, C., P.E. Boehmer, F. McDonald, S. Chauduri, I.D. Hickson, and P.T. Emerson. 1992. Reconstitution of the activities of the RecBCD holoenzyme of *Escherichia coli* from purified subunits. *J. Biol. Chem.* **267**: 13564–13572.
7. Chase, J.W., B.A. Rubin, J.B. Murphy, K.L. Stone, and K.R. Williams. 1986. *Escherichia coli* exonuclease VII. Cloning and sequencing of the gene encoding the large subunit (*xseA*). *J. Biol. Chem.* **261**: 14929–14935.
8. Hall, S.D., M.F. Kane, and R.D. Kolodner. 1993. Identification and characterization of the *Escherichia coli* RecT protein, a protein encoded by the recE region that promotes renaturation of homologous single-stranded DNA. *J. Bacteriol.* **175**: 277–287.
9. Lovett, S.T. and R.D. Kolodner. 1989. Identification and purification of a single-stranded-DNA-specific exonuclease encoded by the *recJ* gene of *Escherichia coli*. *Proc. Natl. Acad. Sci.* **86**: 2627–2631.
10. Lovett, S.T. and R.D. Kolodner. 1991. Nucleotide sequence of the *Escherichia coli recJ* chromosomal region and construction of RecJ-overexpression plasmids. *J. Bacteriol.* **173**: 353–364.
11. Franklin, W.A. and T. Lindahl. 1988. DNA deoxyribophosphodiesterase. *EMBO J.* **7**: 3617–3622.
12. Bernelot-Moens, C. and B. Demple. 1989. Multiple DNA repair activities for 3'-deoxyribose fragments in *Escherichia coli*. *Nucleic Acids Res.* **17**: 587–600.
13. Lehman, I.R. 1971. Bacterial deoxyribonucleases. In *The enzymes* (ed. P.D. Boyer), vol. 4, p. 251–270. Academic Press, New York.
14. Durwald, H. and H. Hoffmann-Berling. 1969. Endonuclease I-deficient and ribonuclease I-deficient *Escherichia coli* mutants. *J. Mol. Biol.* **34**: 331–346.

(*Continued on following page.*)

15. Asahara, H., P.M. Wistort, J.F. Bank, R.H. Bakerian, and R.P. Cunningham. 1989. Purification and characterization of *Escherichia coli* endonuclease III from the cloned *nth* gene. *Biochemistry* **28**: 4444–4449.
16. Kuo, C.-F., D.E. McRee, C.C. Fisher, S.F. O'Handley, R.P. Cunningham, and J.A. Tainer. 1992. Atomic structure of the DNA repair [4 Fe-4S] enzyme endonuclease III. *Science* **258**: 434–440.
17. Levin, J.D., I. Shapiro, and B. Demple. 1991. Metalloenzymes in DNA repair. *Escherichia coli* endonuclease IV and *Saccharomyces cerevisiae* Apn I. *J. Biol. Chem.* **266**: 22893–22898.
18. Saporito, S.M. and R.P. Cunningham. 1988. Nucleotide sequence of the *nfo* gene of *Escherichia coli* K-12. *J. Bacteriol.* **170**: 5141–5145.
19. Demple, B. and S. Linn. 1982. On the recognition and cleavage mechanism of *Escherichia coli* endonuclease V, a possible DNA repair enzyme. *J. Biol. Chem.* **257**: 2848–2855.
20. Bonura, T., R. Schultz, and E.C. Friedberg. 1982. An enzyme activity from *Escherichia coli* that attacks single-stranded deoxyribopolymers and single-stranded deoxyribonucleic acid containing apyrimidinic sites. *Biochemistry* **21**: 2548–2556.
21. Wallace, S.S. 1988. AP endonucleases and DNA glycosylases that recognize oxidative DNA damage. *Environ. Mol. Mutagen.* **12**: 431–477.
22. Boiteux, S., R. O'Connor, and J. Laval. 1987. Formamido-pyrimidine-DNA glycosylase of *Escherichia coli*; cloning and sequencing of *fpg* structural gene and overproduction of the protein. *EMBO J.* **6**: 3177–3183.
23. Castaing, B., A. Geiger, H. Seliger, P. Nehls, J. Laval, C. Zelwer, and S. Boiteux. 1993. Cleavage and binding of a DNA fragment containing a single 8-oxyguanine by wild type and mutant *fpg* proteins. *Nucleic Acids Res.* **21**: 2899–2905.
24. Chung, M.H., H. Kasai, D.S. Jones, H. Inoue, H. Ishikawa, E. Ohtsuka, and S. Nishimura. 1991. An endonuclease activity from *Escherichia coli* that specifically removes 8-hydroxyguanine residues from DNA. *Mutat. Res.* **254**: 1–12.
25. Lin, J.J., A.M. Phillips, J.E. Hearst, and A. Sancar. 1992. Active site of (A)BC exinuclease. II. Binding, bending and catalysis mutants of UvrB reveal a direct role in 5' and an indirect role in 5' incision. *J. Biol. Chem.* **267**: 17693–17700.
26. Grossman, L.I. and A.T. Yeung. 1990. The uvrABC endonuclease system—A view from Baltimore. *Mutat. Res.* **236**: 213–221.
27. Van Houten, L. 1990. Nucleotide excision repair in *Escherichia coli*. *Microbiol. Rev.* **54**: 18–51.
28. Grafstrom, R.H. and R.H. Hoess. 1987. Nucleotide sequence of the *Escherichia coli mutH* gene. *Nucleic Acids Res.* **15**: 3073–3083.
29. Cooper, D.L., R.S. Lahve, and P. Modrich. 1993. Methyl-directed mismatch repair is bidirectional. *J. Biol. Chem.* **268**: 11823–11829.
30. Hennecke, F., H. Kolmar, K. Bründl, and H.-J. Fritz. 1991. The *vsr* gene product of *E. coli* K-12 is a strand- and sequence-specific DNA mismatch endonuclease. *Nature* **353**: 776–778.
31. Sohail, A., M. Lieb, M. Dar, and A.S. Bhagwat. 1990. A gene required for very short patch repair in *Escherichia coli* is adjacent to the DNA cytosine methylase gene. *J. Bacteriol.* **172**: 4214–4221.
32. Chapter 5, this volume.
33. Sharples, G.L. and R.G. Lloyd. 1991. Resolution of Holliday junctions in *E. coli*: Identification of the *ruvC* gene product as a 19kDa protein. *J. Bacteriol.* **173**: 7711–7715.
34. Iwasaki, H., M. Takahagi, T. Shiba, A. Nakata, and H. Shinagawa. 1991. *Escherichia coli* RuvC protein is an endonuclease that resolves Holliday structure. *EMBO J.* **10**: 4381–4389.
35. Chapter 3, this volume.
36. Loenen, W.A.M., A.S. Daniel, H.D. Braymer, and N.E. Murray. 1987. Organization and sequence of the *hsd* genes of *Escherichia coli* K-12. *J. Mol. Biol.* **198**: 159–170.

37. Endlich, B. and S. Linn. 1985. The DNA restriction endonuclease of *Escherichia coli* B. II. Further studies of the structure of DNA intermediates and products. *J. Biol. Chem.* **260:** 5729–5738.
38. Sutherland, E., L. Coe, and E.A. Raleigh. 1992. McrBC: A multisubunit GTP-dependent restriction endonuclease. *J. Mol. Biol.* **225:** 327–348.
39. Dila, D., E. Sutherland, L. Moran, B. Slatko, and E.A. Raleigh. 1990. Genetic and sequence organization of the *mcrBC* locus of *Escherichia coli* K-12. *J. Bacteriol.* **172:** 4888–4900.
40. Raleigh, E.A., R. Trimarchi, and H. Revel. 1989. Genetic and physical mapping of *mcrA* (*rglA*) and *mcrB* (*rglB*) loci of *Escherichia coli* K-12. *Genetics* **122:** 279–296.
41. Waite-Rees, P.A., C.J. Keating, L.S. Moran, B.E. Slatko, L.J. Hornstra, and J.S. Benner. 1991. Characterization and expression of the *Escherichia coli* Mrr restriction system. *J. Bacteriol.* **173:** 5207–5219.
42. Kretz, P.L., S.W. Kohler, and J.M. Short. 1991. Identification and characterization of a gene responsible for inhibiting propagation of methylated DNA sequences in *mcrA mcrB1 Escherichia coli* strains. *J. Bacteriol.* **73:** 4707–4716.
43. Lynn, R.M. and J.C. Wang. 1989. Peptide sequencing and site-directed mutagenesis identifying tyrosine 319 as the active site tyrosine of *Escherichia coli* DNA topoisomerase I. *Proteins* **6:** 231–239.
44. Horowitz, D.S. and J.C. Wang. 1987. Mapping the active site tyrosine of *Escherichia coli* DNA gyrase. *J. Biol. Chem.* **262:** 5339–5344.
45. Digate, R.J. and K.J. Marians. 1992. *Escherichia coli* topoisomerase III - catalyzed cleavage of RNA. *J. Biol. Chem.* **267:** 20532–20535.
46. Digate, R.J. and K.J. Marians. 1989. Molecular cloning and DNA sequence analysis of *Escherichia coli topB*, the gene encoding topoisomerase III. *J. Biol. Chem.* **264:** 17924–17930.
47. Kato, J., Y. Nishimura, R. Imamura, H. Niki, S. Hiraga, and H. Suzuki. 1990. New topoisomerase essential for chromosome segregation in *E. coli*. *Cell* **63:** 393–404.
48. Adams, D.E., E. M. Schektman, E.L. Zechiedrich, M.B. Schmid, and N.R. Cozzarelli. 1992. The role of topoisomerase IV in partitioning bacterial replicons and the structure of catenated intermediates in DNA replication. *Cell* **71:** 277–288.

RIBONUCLEASES OF *ESCHERICHIA COLI*

RNase	Gene	Map position (min)	Protein structure	Preferred substrates	Function	References
Endoribonucleases						
I, I*	*rna*	14	27 kD	most RNAs	—	1,2
M	—	—	27 kD	most RNAs	—	3
R	—	—	24 kD	most RNAs	—	4
III	*rnc*	55	α_2 dimer, 50 kD	dsRNAs	rRNA, mRNA maturation	5
P	*rnpA* *rnpB*	83 70	RNP, 14-kD protein, 377-nucleotide RNA	tRNA precursors	tRNA maturation	6
E,K	*rne, ams, hmpI*	24	114-kD(E), 55–60 kD(K)	5S RNA precursors, mRNA	5S RNA maturation, mRNA degradation	7,8
H	*rnh*	5.1	17.6 kD	RNA-DNA hybrids	DNA replication, repair	9
HII	*rnhB*	4.5	23.2 kD	RNA-DNA hybrids	—	10
IV,F	—	—	31 kD(F)	most RNAs	—	11,12
N	—	—	120 kD	most RNAs	—	13
P2, 0, PC, PIV	—	—	41 kD(O)	tRNA precursors	—	14–17

Exoribonucleases						
polynucleotide-phosphorylase	pnp	69	α₃(260 kD) or α₃β₂(360 kD)	unstructured RNAs	mRNA degradation	18
II	rnb	28	~70 kD	unstructured RNAs	mRNA degradation, tRNA maturation	19
D	rnd	40	42.7 kD	denatured and tRNA precursors	tRNA maturation	20
BN	—	—	~60 kD	tRNA precursors	tRNA maturation	21
T	rnt	36	α₂ dimer, 47 kD	tRNA, tRNA precursors	tRNA end-turnover and maturation	22
PH	rph	81.7	25.5-kD subunit (aggregates)	tRNA precursors	tRNA maturation	23
R	—	—	~80 kD	rRNA, mRNA	—	24
oligoribonuclease	—	—	~38 kD	oligoribonucleotides	—	25

References:
1. Meador, J., III, B. Cannon, V.J. Cannistraro, and D. Kennell. 1990. Purification and characterization of *Escherichia coli* RNase I. *Eur. J. Biochem.* **187:** 549–553.
2. Cannistraro, V.J. and D. Kennell. 1991. RNase I*, a form of RNase I, and mRNA degradation in *Escherichia coli*. *J. Bacteriol.* **173:** 4653–4659.
3. Cannistraro, V.J. and D. Kennell. 1989. Purification and characterization of ribonuclease M and mRNA degradation in *Escherichia coli*. *Eur. J. Biochem.* **181:** 363–370.
4. Srivastava, S.K., V.J. Cannistraro, and D. Kennell. 1991. Broad specificity endoribonucleases and mRNA degradation in *Escherichia coli*. *J. Bacteriol.* **174:** 56–62.

(Continued following page.)

5. Chelladurai, B.S., H.L. Li, and A.W. Nicholson. 1991. A conserved sequence element in ribonuclease III processing signals is not required for *in vitro* enzymatic cleavage. *Nucleic Acids Res.* **19:** 1759–1766.
6. Altman, S., L. Kirsebom, and S. Talbot. 1993. Recent studies of ribonuclease P. *FASEB J.* **7:** 7–14.
7. Casaregola, S., A. Jacq, D. Laoudj, G. McGurk, S. Margarson, M. Tempete, V. Norris, and I.B. Holland. 1992. Cloning and analysis of the entire *Escherichia coli ams* gene. *J. Mol. Biol.* **228:** 30–40.
8. Sohlberg, B., U. Lundberg, F-U. Hartl, and A. von Gabain. 1993. Functional interaction of heat shock protein GroEL with an RNase-like activity in *Escherichia coli. Proc. Natl. Acad. Sci.* **90:** 277–281.
9. Kanaya, S. and R.J. Crouch. 1983. DNA sequence of the gene coding for *Escherichia coli* ribonuclease H. *J. Biol. Chem.* **258:** 1276–1281.
10. Itaya, M. 1990. Isolation and characterization of a second RNase H (RNase HII) of *Escherichia coli* K-12 encoded by the *rnhB* gene. *Proc. Natl. Acad. Sci.* **87:** 8587–8591.
11. Spahr, P.F. and R.F. Gesteland. 1968. Specific cleavage of bacteriophage R17 RNA by an endonuclease isolated from *E. coli* MRE-600. *Proc. Natl. Acad. Sci.* **59:** 876–883.
12. Gurevitz, M., N. Watson, and D. Apirion. 1982. A cleavage site of ribonuclease F. *Eur. J. Biochem.* **124:** 553–559.
13. Misra, T.K. and D. Apirion. 1978. Characterization of an endoribonuclease, RNase N, from *Escherichia coli. J. Biol. Chem.* **253:** 5594–5599.
14. Schedl, P., J. Roberts, and P. Primakoff. 1976. *In vitro* processing of *E. coli* tRNA precursors. *Cell* **8:** 581–594.
15. Shimura, Y., H. Sakano, and F. Nagawa. 1978. Specific ribonucleases involved in processing of tRNA precursors of *Escherichia coli. Eur. J. Biochem.* **86:** 267–281.
16. Goldfarb, A. and V. Daniel. 1980. An *Escherichia coli* endonuclease responsible for primary cleavage of *in vitro* transcripts of bacteriophage T4 tRNA gene cluster. *Nucleic Acids Res.* **8:** 4501–4516.
17. Bikoff, E.K. and M.L. Gefter. 1975. *In vitro* synthesis of transfer RNA I. Purification of required components. *J. Biol. Chem.* **250:** 6240–6247.
18. Regnier, P., M. Grunberg-Manago, and C. Portier. 1987. Nucleotide sequence of the *pnp* gene of *Escherichia coli* encoding polynucleotide phosphorylase. *J. Biol. Chem.* **262:** 63–68.
19. Guarneros, G. and C. Portier. 1990. Different specificities of ribonuclease II and polynucleotide phosphorylase in 3′ mRNA decay. *Biochimie* **72:** 771–777.
20. Zhang, J. and M.P. Deutscher. 1988. *Escherichia coli* RNase D: Sequencing of the *rnd* structural gene and purification of the overexpressed protein. *Nucleic Acids Res.* **16:** 6265–6278.
21. Asha, P.K., R.T. Blouin, R. Zaniewski, and M.P. Deutscher. 1983. Ribonuclease BN: Identification and characterization of a new transfer RNA processing enzyme. *Proc. Natl. Acad. Sci.* **80:** 3301–3304.
22. Huang, S. and M.P. Deutscher. 1992. Sequence and transcriptional analysis of the *Escherichia coli rnt* gene encoding RNase T. *J. Biol. Chem.* **267:** 25609–25613.
23. Kelly, K.O. and M.P. Deutscher. 1992. Characterization of *Escherichia coli* RNase PH. *J. Biol. Chem.* **267:** 17153–17158.
24. Kasai, T., R.S. Gupta, and D. Schlessinger. 1977. Exoribonucleases in wild type *Escherichia coli* and RNase II-deficient mutants. *J. Biol. Chem.* **252:** 8950–8956.
25. Niyogi, S.K. and A.K. Datta. 1975. A novel oligoribonuclease of *Escherichia coli* I. Isolation and properties. *J. Biol. Chem.* **250:** 7307–7312.

APPENDIX D
Compilation of Commercially Available Nucleases

Ira Schildkraut
New England Biolabs, Inc.
Beverly, Massachusetts 01915

Commercially available nucleases (omitting the restriction endonucleases) have been compiled here in tabular form. The restriction endonucleases are dealt with separately in Appendix A. This appendix was created to assist the researcher in locating a specific biochemical tool for manipulating nucleic acids. The enzymes are divided into five groups.

The *Endonucleases* table lists various properties of nucleases that act on DNA and RNA (except DNase I, which does not act on RNA). Unlike type II restriction endonucleases, these endonucleases are nonsequence-specific and have varying requirements for strandedness of the DNA and for divalent cation.

The *Exonucleases* have been useful in manipulating and analyzing DNA and RNA. All of the exonucleases act on DNA molecules. The only two available exonucleases that also act on RNA molecules are phosphodiesterases I and II; they are sugar-nonspecific.

The *Homing endonucleases* are a relatively newly discovered group of highly sequence-specific deoxyribonucleases. The sequence recognized by these enzymes has proved elusive to define (see Chapter 4). Typically, they cleave DNA less frequently than 1 in 250,000 base pairs on the basis of the number of cleavage sites observed on digestion of bacterial and yeast chromosomal DNA. The sequences presented in the table are those sequences that are the smallest defined "natural" target site for each nuclease.

Most commercially available *DNA polymerases* contain one or more exonucleolytic activities. The DNA polymerases are included here for completeness.

The *Ribonucleases* table lists enzymes that exclusively cleave RNA molecules; they all cleave endonucleolytically.

ENDONUCLEASES

Enzyme	Properties substrate DNA	RNA	ss	ds	Application	Commercial sources	References
Bal31 nuclease (*Alteromonas espejiana*)	X	X	X	(X)	requires Mg^{++} and Ca^{++} cations; generates terminal 5′ monophosphates; can be inactivated by addition of EGTA; slow and fast forms	progressive shortening of duplex DNA; generate nested deletions; termini can be ligated	B, D, E, F, G, J	8,14
DNase I (bovine pancreas)	X		X	X	generates terminal 5′ monophosphates; only commercial DNA-specific non-sequence-specific endonuclease	DNA-specific; nick dsDNA; activation of dsDNA for nick translation; reduce nick viscosity; DNase footprinting	B, C, E, K, L	3,13
Micrococcal nuclease (staphylococcal nuclease) (*Staphylococcus aureus*)	X	X	X	X	generates terminal 3′ monophosphates; most active on ssDNA vs. dsDNA; requires Ca^{++}	preparation of mRNA-free protein synthesis system from rabbit reticulocyte lysates; digestion of chromatin for prep of nucleosomes	E, H, I, M	2,4
Mungbean nuclease (mung bean sprouts)	X	X	X		generates terminal 5′ monophosphates; Zn^{++} cofactor requirement	removal of single-stranded extensions (3′ and 5′) to leave blunt ligatable ends; transcription mapping; cleavage of hairpin loops	A, B, C, D, E, F, G, H, J, L	9,11,12,18

(*Continued on following page.*)

Table continued

Enzyme	substrate DNA	substrate RNA	Properties ss	Properties ds	Properties	Application	Commercial sources	References
Neurospora crassa endonuclease	x	x	x		generates terminal 5' monophosphates; very highly specific for single-stranded nucleic acids; active in the absence of divalent cation	removal of nonpaired single strands and loops	E	10,15,16
Nuclease P1 (*Penicillium citrinum*)	x	x	x		generates terminal 5' monophosphates; mixed endo and exo activity; initially produces oligonucleotides; requires Zn^{++}; heat-stable; readily digests damaged DNA	detection of 5' terminal nucleotide of RNA or DNA; RNA sequencing by mobility shift assay	B, E, H, L	5–7,17
Nuclease S1 (*Aspergillus oryzae*)	x	x	x		generates terminal 5' monophosphates; mixed endo and exo activity; initially produces oligonucleotides; requires Zn^{++}; heat stable	removal of single-stranded extensions (3' and 5') to leave blunt ends; "S1 mapping"; transcriptional mapping; cleavage of hairpin loops	A, B, D, E, H, I, J, K, L, M	1,17,19

ss and ds indicate single-stranded and double-stranded.

Commercial Sources: (A) Amersham; (B) Life Technologies Inc., GIBCO-BRL; (C) Stratagene; (D) Takara; (E) Boehringer-Mannheim; (F) New England BioLabs; (G) Toyobo; (H) Pharmacia P-L Biochemicals; (I) Molecular Biology Resources; (J) Promega Corporation; (K) Sigma; (L) United States Biochemical Corporation; (M) Worthington.

References:

1. Ando, T. 1966. A nuclease specific for heat-denatured DNA isolated from a product of *Aspergillus oryzae*. *Biochim. Biophys. Acta* **114:** 158-168.
2. Anfinsen, C., P. Cuatrecasas, and H. Taniuchi. 1971. Staphylococcal nuclease, chemical properties and catalysis. in *The enzymes* (ed. P.D. Boyer), vol. 4, pp. 177-204. Academic Press, New York.
3. Clark, P. and G.L. Eichorn. 1974. A predictable modification of enzyme specificity. Selective alteration of DNA bases by metal ions to promote cleavage specificity by deoxyribonuclease. *Biochemistry* **13:** 5098-5102.
4. Cotton, F.A. and E.E. Hazen. 1971. Staphylococcal nuclease X-ray structure. In *The enzymes* (ed. P.D. Boyer), vol. 4, pp. 153-175. Academic Press, New York.
5. Fujimoto, M., A. Kuninaka, and H. Yoshino. 1974. Substrate specificity of nuclease P1. *Agric. Biol. Chem.* **38:** 1555-1557.
6. Fujimoto, M., A. Kuninaka, and H. Yoshino. 1974. Purification of a nuclease from *Penicillium citrinum*. *Agric. Biol. Chem.* **38:** 777-784.
7. Fujimoto, M., A. Kuninaka, and H. Yoshino. 1974. Identity of phosphodiesterase and phosphomonoesterase activities with nuclease P1 (a nuclease from *Penicillium citrinum*). *Agric. Biol. Chem.* **38:** 785.
8. Gray, H.B., D.A. Ostrander, J.L. Hodnett, R.J. Legerski, and D.L. Robberson. 1975. Extracellular nucleases of *Pseudomonas Bal* 31. I. Characterization of the single-strand-specific deoxyriboendonuclease and double-strand deoxyriboexonuclease. *Nucleic Acids Res.* **2:** 1459-1492.
9. Hanson, D.M. and J.L. Fairley. 1969. Enzymes of nucleic acid metabolism from wheat seedlings. I. Purification and general properties of associated deoxyribonuclease, ribonuclease and 3′ nucleotidase activities. *J. Biol. Chem.* **244:** 2440-2449.
10. Kato, A.C., K. Bartok, M.J. Fraser, and D. Denhardt. 1973, Sensitivity of superhelical DNA to a single-strand specific endonuclease. *Biochim. Biophys. Acta* **308:** 68-70.
11. Kowalski, P., W.D. Kroeker, and M. Laskowski. 1976. Mung bean nuclease I. Physical chemical and catalytic properties. *Biochemistry* **15:** 4457-4463.
12. Kroeker, W.D., P. Kowalski, and M. Laskowski. 1976. Mung bean nuclease I. Terminally directed hydrolysis of native DNA. *Biochemistry* **15:** 4463-4467.
13. Laskowski, M. 1971. Deoxyribonuclease I. In *The enzymes* (ed. P.D. Boyer), vol. 4, pp. 289-311. Academic Press, New York.
14. Legerski, R.J., J.L. Hodnett, and H.B. Gray. 1978. Extracellular nucleases of *Pseudomonas Bal* 31 III. Use of the double-strand deoxyriboexonuclease activity as the basis of a convenient method for the mapping of fragments of DNA produced by cleavage with restriction enzymes. *Nucleic Acids Res.* **5:** 1455-1464.
15. Linn, S. 1967. An endonuclease from *Neurospora crassa* specific for polynucleotides lacking an ordered structure. *Methods Enzymol.* **12:** 247-255.
16. Linn, S. and I.R. Lehman. 1965. An endonuclease from *Neurospora crassa* specific for polynucleotides lacking an ordered structure. *J. Biol. Chem.* **240:** 1287-1293.
17. Shishido, K. and T. Ando. 1982. Single-strand-specific nucleases. In *Nucleases* (ed. S.M. Linn and R.J. Roberts), pp. 155-185. Cold Spring Harbor Laboratory, Cold Spring Harbor, New York.
18. Sung, S.C. and M. Laskowski. 1962. A nuclease from mung bean sprouts. *J. Biol. Chem.* **237:** 506-511.
19. Vogt, V.M. 1973. Purification and further properties of single-stranded specific nuclease from *Aspergillus oryzae*. *Eur. J. Biochem.* **33:** 192-200.

EXONUCLEASES

	Properties						Applications and comments	Commercial sources	References
	DNA	RNA	5'→3' ss	5'→3' ds	3'→5' ss	3'→5' ds			
Escherichia coli exonuclease I	x					x	removes 3' extensions; produces termini with 3'-hydroxyl; products are mononucleotides; leaves the 5' dinucleotide intact	J	7,9
Escherichia coli exonuclease III	x					x	unidirectional 3'→5' exonuclease; removes 3'-phosphates; generates nested deletions; site-directed mutagenesis; generates single-stranded from duplex; produces termini with 3'-hydroxyl-associated AP endonuclease activity	A, B, C, D, E, F, G, H, I, J	12,13
Escherichia coli exonuclease VII	x		x		x		solubilizes ssDNA; removes single-stranded termini	B	3,4
λ exonuclease	x			x			generates single-stranded DNA from duplex by a 5'→3' exonuclease activity; does not attack at nicks; produces termini with 5' monophosphates	B	8,10
Micrococcus luteus ATP-dependent DNase (exonuclease V)	x		x	x	x	x	removes oligonucleotides from 3' and 5' ends of duplex and ssDNAs	G, J	1,14

Phosphodiesterase I (snake venom)	x	x	oligonucleotide sequencing; produces termini with 3'-hydroxyl, 5'-phosphate; does not degrade duplexes, hairpins, etc.	E, H, J, K	11
Phosphodiesterase II (bovine spleen)	x	x	x	oligonucleotide sequencing; identification of a 3'-hydroxyl terminus; produces termini with 5'-hydroxyl, 3'-phosphate	E, J, K	2,5
T7 exonuclease	x	x	generates ssDNA from duplex; capable of initiating at nicks; produces termini with 5' monophosphates	K	6

Commercial Sources: (A) Amersham; (B) Life Technologies Inc., GIBCO-BRL; (C) Stratagene; (D) Takara; (E) Boehringer-Mannheim; (F) New England BioLabs; (G) Toyobo; (H) Pharmacia P-L Biochemicals; (I) Promega Corporation; (J) United States Biochemical Corporation; (K) Worthington.

References:

1. Anai, M., T. Hirahashi, and Y. Takagi. 1970. A deoxyribonuclease which requires nucleoside triphosphate from *Micrococcus lysodeikticus*. I. Purification and characterization of the deoxyribonuclease activity. *J. Biol. Chem.* **245:** 767-774.
2. Bernardi, A. and G. Bernardi. 1971. Spleen acid deoxyribonuclease. In *The enzymes* (ed. P.D. Boyer), vol. 4, pp. 271-287. Academic Press, New York.
3. Chase, J.W. and C.C. Richardson. 1974. Exonuclease VII of *Escherichia coli*. Purification and properties. *J. Biol. Chem.* **249:** 4545-4552.
4. Chase, J.W. and C.C. Richardson. 1974. Exonuclease VII of *Escherichia coli*. Mechanism of action. *J. Biol. Chem.* **249:** 4553-4561.
5. Hilmoe, R.J. 1961. Spleen phosphodiesterase. *Biochem. Prep.* **8:** 105-109.
6. Kerr, C. and P.D. Sadowski. 1972. Gene 6 exonuclease of bacteriophage T7. II. Mechanism of the reaction. *J. Biol. Chem.* **247:** 311-318.
7. Lehman, I.R. and A.L. Nussbaum. 1964. The deoxyribonucleases of *Escherichia coli*. *J. Biol. Chem.* **239:** 2628-2636.
8. Little, J.W. 1967. An exonuclease induced by bacteriophage lambda. II. Nature of the enzymatic reaction. *J. Biol. Chem.* **243:** 679-686.
9. Mackay, V. and S. Linn. 1974. Molecular structure of exonuclease I from *Escherichia coli*. *Biochim. Biophys. Acta* **349:** 131-134.
10. Radding, C.M. 1966. Regulation of λ exonuclease. I. Properties of λ exonuclease purified from lysogens of λ T11 and wild type. *J. Mol. Biol.* **18:** 235-250.
11. Razell, W.E. and H.G. Khorana. 1959. Studies on polynucleotides III enzymic degradation. Substrate specificity and properties of snake venom phosphodiesterase. *J. Biol. Chem.* **234:** 2105-2113.
12. Richardson, C.C., I.R. Lehman, and A. Kornberg. 1964. A deoxyribonucleic acid phosphatase-exonuclease from *Escherichia coli*. *J. Biol. Chem.* **239:** 251-258.
13. Weiss, B. 1981. Exodeoxyribonucleases of *Escherichia coli*. In *The enzymes* (ed. P.D. Boyer), vol. 14A, pp. 203-231. Academic Press, New York.
14. Yamagishi, H., T. Tsuda, S. Fujimoto, M. Toda, K. Kato, Y. Maekawa, M. Umeno, and M. Anai. 1983. Purification of small polydisperse circular DNA of eukaryotic cells by use of ATP-dependent deoxyribonuclease. *Gene* **26:** 317-321.

HOMING ENDONUCLEASES

Enzyme	Cognate target sequence	Application	Comment	Commercial sources	References
I-*Ceu*I (*Chlamydomonas eugametos*)	TAACTATAACGGTC_CTAA↑GGTAGCGA	rare site mapping	double-stranded cleavage yields four-base CTAA 3' extension; some sequence degeneracy is tolerated within this sequence; however, minimal degeneracy is found within a core 19-base sequence; seven cleavage sites are observed on *E. coli* chromosome	B	3,4
I-*Ppo*I (*Physarum polycephalum*)	ATGACTCTC_TTAA↑GGTAGCCAAA	rare site mapping	tolerates degeneracy within target sequence; digestion of bacterial and yeast chromosomal DNA indicates sequence specificity is 8–10 nucleotides; 12 cleavage sites are observed on *E. coli* chromosome; cleaves dsDNA to leave four-base TTAA 3' extension	B, C	5,6
I-*Sce*I (*Saccharomyces cerevisiae*)	TAGGG_ATAA↑CAGGGTAAT	rare site mapping	cleaves dsDNA to leave four-base ATAA 3' extension	A	8,9
I-*Tli*I (*Thermococcus litoralis*)	GGTTCTTTATGCGG_ACAC↑TGACGGCTTTATG	rare site mapping	tolerates degeneracy within target sequence; digestion of bacterial and yeast chromosomal	B	2,7

| VDE (*Saccharomyces cerevisiae*) | TCTATGTCGG_GTGC↑GGAGAAAGAGGTAATG | rare site mapping | DNA indicates sequence specificity is 8–10 nucleotides; ten cleavage sites are observed on *Rhodobacter sphaeroides* chromosome; cleaves dsDNA to leave four-base ACAC 3′ extension tolerates degeneracy within target sequence; digestion of bacterial and yeast chromosomal DNA indicates sequence specificity is 11 bases; cleaves dsDNA to leave four-base GTGC 3′ extension | B | 1 |

Commercial Sources: (A) Boehringer-Mannheim; (B) New England BioLabs; (C) Promega Corporation.

References:
1. Bremer, M.C., F.S. Gimble, J. Thorner, and C.L. Smith. 1992. VDE endonuclease cleaves *Saccharomyces cerevisiae* genomic DNA at a single site: Physical mapping of the *VMA1* gene. *Nucleic Acids Res.* **20:** 5484.
2. Davis, T. 1993. New England Biolabs Catalog, p. 51.
3. Marshall, P. and C. Lemieux. 1991. Cleavage pattern of the homing endonucleases encoded by the fifth intron in the chloroplast large subunit rRNA-encoding gene of *Chlamydomonas eugametos*. *Gene* **104:** 241-245.
4. Marshall, P. and C. Lemieux. 1992. The I-CeuI endonuclease recognizes a sequence of 19 base pairs and preferentially cleaves the coding strand of the *Chlamydomonas moewusii* chloroplast large subunit rRNA gene. *Nucleic Acids Res.* **20:** 6401-6407.
5. Muscarella, D.E. and V.M. Vogt. 1989. A mobile group I intron in the nuclear DNA of *Physarum polycephalum*. *Cell* **56:** 443-454.
6. Muscarella, D.E., E.L. Ellison, B.M. Ruoff, and V.M. Vogt. 1990. Characterization of I-Ppo, an intron-encoded endonuclease that mediates homing of a group I intron in the ribosomal DNA of *Physarum polycephalum*. *Mol. Cell. Biol.* **10:** 3386-3396.
7. Perler, F.B., D.G. Comb, W.E. Jack, L.S. Moran, B. Qiang, R.B. Kucera, J. Benner, B.E. Slatko, D.O. Nwankwo, S.K. Hempstead, C.K.S. Carlow, and H. Jannasch. 1992. Intervening sequences in an *Archaea* DNA polymerase gene. *Proc. Natl. Acad. Sci.* **89:** 5577-5581.
8. Thierry, A. and B. Dujon. 1992. Nested chromosomal fragmentation in yeast using the meganuclease I-SceI: A new method for physical mapping of eukaryotic genomes. *Nucleic Acids Res.* **20:** 5625-5631.
9. Thierry, A., A. Perrin, J. Boyer, C. Fairhead, B. Dujon, B. Frey, and G. Schmitz. 1991. Cleavage of yeast and bacteriophage T7 genomes at a single site using the rare cutter endonuclease I-Sce I. *Nucleic Acids Res.* **19:** 189-190.

DNA POLYMERASES WITH ASSOCIATED EXONUCLEASE ACTIVITIES

Enzyme	Specificity 5'→3' ss	Specificity 5'→3' ds	Specificity 3'→5' ss	Specificity 3'→5' ds	Application	Commercial sources	References
Deep vent DNA polymerase (*Pyrococcus species*)				x	labeling duplex DNA at 3' terminus; replication of DNA with proofreading exo; primer extension; polymerase chain reaction; fill in 5' extension	I	11
Dynazyme (*Thermus brockianus*)		x			DNA sequencing; primer extension; polymerase chain reaction	D	
Escherichia coli DNA polymerase I		x		x	labeling duplex DNA at 3' terminus; replication of DNA with proofreading exo; nick translation	A, G, I, K, M	9
Escherichia coli DNA polymerase I Klenow fragment				x	remove 3' extensions; DNA sequencing; fill in 5' extension	A, G, H, I, K, M	6
Pfu DNA polymerase (*Pyrococcus furiosus*)				x	labeling duplex DNA at 3' terminus; replication of DNA with proofreading exo; polymerase chain reaction	E	10
T4 DNA polymerase (bacteriophage T4)				x	labeling duplex DNA at 3' terminus; replication of DNA with proofreading exo; remove 3' extensions; second strand synthesis in site-directed mutagenesis; fill in 5' extension	A, G, I, K	13
T7 DNA polymerase (bacteriophage T7)				x	labeling duplex DNA at 3' terminus; replication of DNA with proofreading exo; "Sequenase" is a form of T7 DNA polymerase which lacks 3'→5' exo-nuclease activity	I	3,5,12

Taq DNA polymerase (*Thermus aquaticus*)	x	polymerase chain reaction; DNA sequencing; nick translation	B, C, E, F, G, H J, K, L	2,4,7
Thermatoga maritima DNA polymerase	x	labeling duplex DNA at 3' terminus; replication of DNA with proofreading exo; polymerase chain reaction	C	
Tth DNA polymerase (*Thermus thermophilus*)	x	primer extension; polymerase chain reaction; DNA sequencing	A, C	1
Vent DNA polymerase (*Thermococcus litoralis*)	x	labeling duplex DNA at 3' terminus; replication of DNA with proofreading exo; polymerase chain reaction; fill in 5' extension	I, K	8,14

Commercial Sources: (A) Amersham; (B) Life Technologies Inc., GIBCO-BRL; (C) Perkin Elmer; (D) Finnzymes; (E) Stratagene; (F) Fermentas MBI; (G) Takara; (H) Boehringer-Mannheim; (I) New England BioLabs; (J) Pharmacia P-L Biochemicals; (K) Promega Corporation; (L) United States Biochemical Corporation; (M) Worthington.

References:

1. Carballeira, N., M. Nazabal, J. Brito, and O. Garcia. 1990. Purification of a thermostable DNA polymerase from *Thermus thermophilus* HB8, useful in the polymerase chain reaction. *Biotechniques* **9**: 276-281.
2. Chien, A., D.B. Edgar, and J.M. Trela. 1976. Deoxyribonucleic acid polymerase from the extreme thermophile *Thermus aquaticus*. *J. Bacteriol.* **127**: 1550-1557.
3. Engler, M.J., R.L. Lechner, and C.C. Richardson. 1983. Two forms of the DNA polymerase of bacteriophage T7. *J. Biol. Chem.* **258**: 11165-11173.
4. Holland, P.M., R.D. Abramson, R. Watson, and D.H. Gelfand. 1991. Detection of specific polymerase chain reaction product by utilizing the 5' to 3' exonuclease activity of *Thermus aquaticus* DNA polymerase. *Proc. Natl. Acad. Sci.* **88**: 7276-7280.
5. Hori, K., D.F. Mark, and C. Richardson. 1979. Deoxyribonucleic acid polymerase of bacteriophage T7. *J. Biol. Chem.* **254**: 11598-11604.
6. Jacobsen, H., H. Klenow, and K. Overgaard-Hansen. 1974. The N-terminal amino-acid sequences of DNA polymerase I from *Escherichia coli* and of the large and the small fragments obtained by a limited proteolysis. *Eur. J. Biochem.* **45**: 623-627.
7. Kaledin, A.S., A.G. Sliusarenko, and S.I. Gorodetskii. 1980. Isolation and properties of DNA polymerase from extreme thermophilic bacteria *Thermus aquaticus* YT-1. *Biokhimiya* **45**: 644-651.
8. Kong, H., R.B. Kucera, and W.E. Jack. 1993. Characterization of a DNA polymerase from the hyperthermophile Archaea *Thermococcus litoralis*. *J. Biol. Chem.* **268**: 1965-1975.
9. Lehman, I.R. 1981. DNA polymerase I of *Escherichia coli*. In *The enzymes* (ed. P.D. Boyer), vol 14A, pp. 15-37. Academic Press, New York.
10. Mathur, E.J., M.W. Adams, W.N. Callen, and J.M. Cline. 1991. The DNA polymerase gene from the hyperthermophilic marine archaebacterium *Pyrococcus furiosus*, shows sequence homology with α-like DNA polymerase. *Nucleic Acids Res.* **19**: 6952.
11. New England Biolabs 1993 Catalog, p. 62.

12. Nordstrom, B., H. Randahl, and C.C. Richardson. 1981. Characterization of bacteriophage T7 DNA polymerase purified to homogeneity by Antithioredoxin immuno adsorbent chromatography. *J. Biol. Chem.* **256:** 3112-3117.
13. Panet, A., J.H. van de Sande, P.L. Loewen, and H.G. Khorana. 1973. Physical characterization and simultaneous purification of bacteriophage T4 induced polynucleotide kinase, polynucleotide ligase and deoxyribonucleic acid polymerase. *Biochemistry* **12:** 5045-5050.
14. Perler, F.B., D.G. Comb, W.E. Jack, L.S. Moran, B. Qiang, R.B. Kucera, J. Benner, B.E. Slatko, D.O. Nwankwo, S.K. Hempstead, C.K.S. Carlow, and H. Jannasch. 1992. Intervening sequences in an *Archaea* DNA polymerase gene. *Proc. Natl. Acad. Sci.* **89:** 5577-5581.

RIBONUCLEASES

Enzyme	Properties	Application	Commercial sources	References
Hammerhead ribozyme	sequence-specific ribonuclease recognizing ACGGUCUCACGAGC and cleaving after CUC	site-specific RNA cleavage	G	12
Ribozyme Tet1.0 (*Tetrahymena*)	sequence-specific ribonuclease recognizing CUCU and cleaving at the 3' end	site-specific RNA cleavage	G	17,18
RNase A (bovine pancreas)	cleaves single-stranded RNA Cp↓N, Up↓N	remove RNA from DNA preps; degrades RNA to oligonucleotides	E, F, G, H	7,8
RNase (*Bacillus cereus*)	cleaves at Up↓N, Cp↓N	RNA sequencing	E, G	11
RNase CL3 (chicken liver)	cleaves Cp↓N; does not cleave UpN	RNA sequencing	C, G	10
RNase H (*Escherichia coli*)	degrades RNA in RNA/DNA hybrids	remove RNA for cDNA synthesis	A, B, C, D, E	1,2
RNase PhyM (*Physarum polycephalum*)	cleaves at Ap↓N, Up↓N	RNA sequencing	E, G	3
RNase T1 (*Aspergillus oryzae*)	cleaves at Gp↓N	RNA sequencing	B, E, G, H	9,14
RNase T2 (*Aspergillus oryzae*)	cleaves nonspecifically; preference at Ap↓N	RNA sequencing	B	6,13,15

(*Continued on following page.*)

| RNase U2 (*Ustilago sphaerogena*) | cleaves at Ap↓N | RNA sequencing | E, G | 5,6 |
| RNase V1 (*cobra venom*) | cleaves doubled-stranded regions of RNA; leaves 3′-hydroxyl and 5′-phosphate | analyze RNA structure | E, G | 16 |

Commercial Sources: (A) Amersham; (B) Life Technologies Inc., GIBCO-BRL; (C) Boehringer-Mannheim; (D) Toyobo; (E) Pharmacia P-L Biochemicals; (F) Sigma; (G) United States Biochemical Corporation; (H) Worthington.

References:

1. Crouch, R.J. and M.-L. Dirksen. 1982. Ribonucleases H. In *Nucleases* (ed. S. Linn and R.J. Roberts), pp. 211-241. Cold Spring Harbor Laboratory, Cold Spring Harbor, New York.
2. Crouch, R.J. 1981. Analysis of nucleic acid structure by RNase. In *Gene amplification and analysis* (ed. J.G. Chirikjian and T.S. Papas), vol. 2, pp. 217. Elsevier/North-Holland, New York.
3. Donis-Keller, H. 1980. Phy M: An RNase activity specific for U and A residues useful in RNA sequence analysis. *Nucleic Acids Res.* **8:** 3133-3142.
4. Glitz, D.G. and C.A. Decker. 1964. Studies on a ribonuclease from *Ustilago sphaerogena*. I. Purification and properties of the enzyme. *Biochemistry* **3:** 1391-1398.
5. Glitz, D.G. and C.A. Decker. 1964. Studies on a ribonuclease from *Ustilago sphaerogena*. II. Specificity of the enzyme. *Biochemistry* **3:** 1399.
6. Heppel, L.A. 1966. Pig liver nuclei ribonuclease (1). In *Procedures in nucleic acid research* (ed. G.L. Cantoni and D.R. Davies), p. 31. Harper and Row, New York.
7. Hillenbrand, G. and W. Studenbauer. 1982. Discriminatory function of ribonuclease H in the selective initiation of plasmid DNA replication. *Nucleic Acids Res.* **10:** 833-853.
8. Kalnitsky, G., J.P. Hummel, and C. Dierks. 1959. Some factors which affect the enzymatic digestion of ribonucleic acid. *J. Biochem.* **234:** 1512-1516.
9. Kasai, K., T. Uchida, F. Egami, K. Yoshida, and N. Nomoto. 1969. Purification and crystallization of ribonuclease N1 from *Neurospora crassa*. *J. Biochem.* **66:** 389-396.
10. Levy, C.C. and T.P. Karpetsky. 1980. The purification and properties of chicken liver RNase: An enzyme which is useful in distinguishing between cytidylic and uridylic acid residues. *J. Biol. Chem.* **255:** 2153-2159.
11. Lockard, R.E., B. Alzner-Deweerd, J.E. Heckman, J. MacGee, M.W. Tabor, and U.L. RajBhandary. 1978. Sequence analysis of 5′ [^{32}P] labelled mRNA and tRNA using polyacrylamide gel electrophoresis. *Nucleic Acids Res.* **5:** 37-56.
12. Perrault, J.-P., T.F. Wu, B. Cousineau, K.K. Oligivie, and R. Cedergren. 1990. Mixed deoxyribo- and ribo-oligonucleotides with catalytic activity. *Nature* **344:** 565-567.

13. Reddi, K.K. and L.J. Mauser. 1965. Studies on the formation of tobacco mosaic virus ribonucleic acid. VI. Mode of degradation of host ribonucleic acid to ribonucleosides and their conversion to ribonucleoside 5' phosphates. *Proc. Natl. Acad. Sci.* **53:** 607-613.
14. Takahashi, K. 1961. The structure and function of ribonuclease T$_1$. *J. Biochem.* **49:** 1-8.
15. Uchida, T. and F. Egami. 1967. The specificity of ribonuclease T2. *J. Biochem.* **61:** 44-53.
16. Wyatt, J.R. and G.T. Walker. 1989. Deoxynucleotide containing oligoribonucleotide duplexes: Stability and susceptibility to RNase V1 and RNase H. *Nucleic Acids Res.* **17:** 7833-7842.
17. Zaug, A.G., C.A. Grosshans, and T.R. Cech. 1988. Sequence-specific endoribonuclease activity of the *Tetrahymena* ribozyme: Enhanced cleavage of certain oligonucleotide substrates that form mismatched ribozyme-substrate complexes. *Biochemistry* **27:** 8924-8931.
18. Zaug, A.J., M.D. Been, and T.R. Cech. 1986. The *Tetrahymena* ribozyme acts like an RNA restriction endonuclease. *Nature* **324:** 429-433.

Index

Adenine DNA glycosylase (*E. coli*)
  activity, 277
  gene, 277, 448
  iron-sulfur cluster, 278, 448
  molecular weight, 277, 448
  sequence homology, 448
  substrate specificity, 277, 448
Alkaline phosphatase, reaction mechanism, 8–9
Alkylating agents. *See also specific agents*, 319–320
Antisense oligonucleotide, 366–367
AP endonucleases. *See also specific AP endonucleases*
  classification, 278–280
    class I, 279
    class II, 279–280
    class III, 280
    class IV, 280
Apurinic endonuclease (calf thymus), substrates, 283
Apurinic endonuclease (human)
  activity, 283
  genes, 449
  inhibitor, 283
  molecular weight, 283, 499
Apurinic endonuclease, AP3 (*Drosophila*)
  gene, 449
  molecular weight, 282, 449
  properties, 449
  sequence homology, 282

Apurinic endonuclease, APN1 (yeast)
  activities, 281
  gene, 282, 449
  inactivation, 282
  metal ion requirement, 282
  molecular weight, 281, 449
  mutants, 282
  sequence homology, 449
Apurinic endonuclease I (*Drosophila*)
  molecular weight, 282, 449
  reaction mechanism, 282
Apurinic endonuclease II (*Drosophila*)
  molecular weight, 282, 449
  reaction mechanism, 282
Artificial nucleases
  cleavage chemistry, 318–321
  DNA-binding proteins and peptides as targeting ligands, 330–332
  oligonucleotides as targeting ligands, 323–330
    oligonucleotide-directed restriction endonuclease, 329–330
    oligonucleotide-nonhydrolytic nucleolytic agents, 325–327
    oligonucleotide-staphylococcal nuclease conjugates, 327–328
  substrate recognition, 323–330
  small molecules, 321–323
Azidophenacyl derivatives, 320

Bal31 nuclease (*A. espejiana*)
  application, 471

commercial sources, 471
properties, 471
protease sensitivity, 178
substrate, 471
BAP I protein
  activities, 290
  sequence homology, 290
Base excision repair of. *See also specific enzymes involved*
  alkylated bases, 273-275
  cyclobutane pyrimidine dimers, 275-277
  deaminated bases, 270-273
  mismatched bases, 277-278
  oxidized purines, 269-270
  oxidized pyrimidines, 267-269
*Bcg*I
  cleavage site, 39
  cofactor, 39
  recognition sequence, 39
β-Elimination
  endonuclease-catalyzed reactions, 24-30. *See also specific endonucleases*
  reaction mechanism, 24
  reaction rate, 30-31
Bleomycin
  cleavage site, 318
  structure, 319
N-Bromoacetyl
  alkylation site, 320
  structure, 320
N-Bromoacetyl distamycin, 321

Chi site, 150-152
p-(N-2-Chloroethyl-N-methylamino)benzyl, 320
Correxonuclease, 290

Dcm methyltransferase, 271-272
Deamination of DNA, 271
Deep vent DNA polymerase (*Pyrococcus*)
  application, 478
  commercial sources, 478
  specificity, 478
Distamycin analogs
  activity, 322
  recognition sequence, 322
  sequence specificity, 322

Distamycin-EDTA•Fe(II)
  DNA binding, 321
  DNA cleavage, 321
  structure, 322
D-loop, 328-329
DNA damages, 265-266
DNA-damaging agents, 265-266
DNA glycosylases. *See specific glycosylases*
DNA glycosylase/AP lyases. *See specific glycosylase/AP lyases*
DNA gyrase. *See* Eco topoisomerase II
DNA polymerase (T4)
  activities, 236
  application, 478
  commercial sources, 478
  correct nucleotide turnover, 245
  gene, 237
  lesion bypass, 250-251
  misincorporation rate, 246
  mutants
    antimutator, 238-241, 244-246, 248-249
    mutator, 238-241, 244-246, 249-251
  nuclease/polymerase (N/P) ratio, 239, 249
  primer melting, 246
  specificity, 478
DNA polymerase (T7)
  application, 478
  commercial sources, 478
  specificity, 478
DNA polymerase (*Thermatoga maritima*)
  application, 479
  commercial sources, 479
  specificity, 479
DNA polymerase δ, 236
DNA polymerase ε, 236
DNA polymerase γ, 236
DNA polymerase I (*E. coli*)
  activities, 236, 289
  application, 478
  characteristics, 456
  commercial sources, 478
  exonuclease site, 11-13
  gap filling, 272
  gene, 456
  map location, 456
  molecular weight, 456
  mutant, 290

products, 456
specificity, 478
substrate, 456
DNA polymerase II (*E. coli*)
characteristics, 456
gene, 456
map location, 456
molecular weight, 456
substrate, 456
DNA polymerase III (*E. coli*)
correct nucleotide turnover, 245
in methyl-directed mismatch repair, 298
mutation rate, 247
DNA polymerase III, σ subunit
characteristics, 456
gene, 239, 247, 456
map location, 456
mutants, 239, 247
products, 456
proofreading activity, 236, 246
substrate, 456
subunit molecular weight, 456
DNA polymerases. *See also specific DNA polymerases*
fidelity study, 235–236
misinsertion frequencies, 236
DNase A (*A. nidulans*), 176
DNase A (*A. oryzae*), 176
DNase A (*N. crassa*), 176
DNase B (*N. crassa*), 176
DNase I (bovine pancreas)
application, 471
commercial sources, 471
properties, 471
substrate, 471
DNase IV (mammalian), activities, 290
DNase V (mammalian), activities, 290
DNase-4 (*A. oryzae*), 176
dRPase (*E. coli*)
activity, 289
molecular weight, 458
products, 458
substrate, 458
Dynazyme (*T. brockianus*)
application, 478
commercial sources, 478
specificity, 478

*Eco*K, *Eco*B (*E. coli*)
characteristics, 461
genes, 461
map location, 461
products, 461
substrate, 461
subunit molecular weight, 461
*Eco*RI
active site, 67
crystal structure, 60–61, 65–66
gene location, 47
mutagenesis, 70–71
specificity, 50–53
star activity, 58
*Eco*RV
active site, 67–68
crystal structure, 61–67
mutagenesis, 70–71
specificity, 51–52
Eco topoisomerase I (ω *protein*)
active site, 216
activity, 209, 215
characteristics, 462
gene, 462
map location, 462
molecular weight, 462
mutants, 220
products, 462
substrate, 462
Eco topoisomerase II (DNA gyrase)
activity, 210, 220
characteristics, 462
genes, 462
map locations, 462
products, 462
substrate, 462
subunit molecular weight, 462
Eco topoisomerase III
characteristics, 463
gene, 463
homology, 210
map location, 463
molecular weight, 463
substrates, 215, 463
Eco topoisomerase IV
characteristics, 463
function, 211
genes, 463
map locations, 463
products, 463
substrate, 463
subunit, 210
subunit molecular weight, 463

Ellipticine derivatives, 320
Endo-exonucleases (*A. nidulans*), 179
Endo-exonucleases (*N. crassa*)
  cellular localization, 178–179
  mitochondrial form, 178–179
  molecular weight, 178–179
  nuclear form, 178
  protease activation, 177–178
  regulation, 179–180
  role in DNA repair and recombination, 181
Endo-exonucleases (*S. cerevisiae*)
  cellular localization, 179
  genes, 183–184
  predicted amino acid sequence, 183–184
  proteolysis, 179
  role in DNA repair and recombination, 180–181
Endonuclease (bovine heart mitochondria)
  activity, 195–196
  localization, 195
  molecular weight, 195
  specificity, 196
Endonuclease (*C. fasciculata*), 198
Endonuclease (*Drosophila*), 194
Endonuclease (*S. cerevisiae*)
  cellular level, 386
  function in tRNA maturation, 386
  mutant, 386
  subunit molecular weight, 386
Endonuclease G (calf thymus nuclei), 193, 196
Endonuclease I (*E. coli*)
  characteristics, 458
  gene, 458
  map location, 458
  molecular weight, 458
  products, 458
  substrate, 458
Endonuclease I (T7)
  activity, 159
  gene, 159
  molecular weight, 160
  mutants, 159
  substrate specificity, 159–160
Endonuclease III (*E. coli*)
  active site, 27
  activities, 446, 459
  crystal structure, 287–288
  gene, 268, 446, 459
  iron-sulfur cluster, 268, 287
  map location, 459
  molecular weight, 267, 446, 459
  nucleic acid interaction site, 288
  reaction mechanism, 25–26, 268,
  substrates, 267–268, 459
Endonuclease IV (*E. coli*)
  activities, 281, 289, 448, 459
  gene, 281, 448, 459
  map location, 459
  metal ion requirement, 281
  molecular weight, 281, 448, 459
  mutants, 281
  products, 459
  regulation, 297
  substrate, 459
Endonuclease IVA and IVB (*E. coli*)
  products, 457
  substrate, 457
Endonuclease V (*E. coli*)
  activity, 295
  molecular weight, 459
  products, 459
  substrate, 295, 459
Endonuclease V (T4)
  active site, 26–27
  active site residues involved in
    catalysis, 285–286
    nontarget DNA binding, 286–287
  activity, 275–276, 447
  crystal structure, 283–285
  DNA binding, 275
  gene, 447
  molecular weight, 447
  processive DNA scanning, 275
  reaction mechanism, 25–26, 285–286, 447
  site-directed mutagenesis, 285–287
  substrate specificity, 275, 447
Endonuclease VII (*E. coli*)
  molecular weight, 459
  substrate, 459
Endonuclease VII (T4)
  activity, 158–159
  gene, 158
  molecular weight, 158
  mutants, 158
  substrate specificity, 159–160
Endo.*Sce*I, sequence motif, 123–125
Enediynes, 318–319

1,N6-Ethenoadenine DNA glycosylase
  activity, 278
  molecular weight, 278
N-Ethylmaleimide, 367
Exonuclease (T7)
  application, 475
  commercial sources, 475
  properties, 475
  substrate, 475
Exonuclease I (*E. coli*)
  activities, 289, 456, 474
  application, 474
  commercial sources, 474
  gene, 456
  map location, 456
  in methyl-directed mismatch repair, 298
  molecular weight, 456
  products, 456, 474
  substrate, 456, 474
Exonuclease I (*S. cerevisiae*), 191
Exonuclease I (*S. pombe*)
  activity, 155
  gene, 155
  molecular weight, 155
Exonuclease II (*S. pombe*)
  activity, 149, 155
  gene, 155
  molecular weight, 155
Exonuclease II (yeast), 192
Exonuclease III (*E. coli*)
  activities, 280, 289, 448, 457, 474
  application, 474
  biological role, 281
  commercial sources, 474
  gene, 280, 448-457
  map location, 457
  molecular weight, 448, 457
  mutant, 280-281
  products, 457, 474
  regulation, 297
  substrate, 457, 474
Exonuclease III (*S. cerevisiae*), 191
Exonuclease IV (*S. cerevisiae*), 192
Exonuclease V (*M. luteus*)
  application, 474
  commercial sources, 474
  properties, 474
  substrate, 474
Exonuclease V (*S. cerevisiae*), 192
Exonuclease VII (*E. coli*)
  activity, 289, 457, 474
  application, 474
  commercial sources, 474
  genes, 290, 457
  map locations, 457
  in methyl-directed mismatch repair, 298
  mutants, 289-290
  products, 457
  specificity, 289
  substrate, 457, 474
  subunit molecular weight, 457
Exonuclease VIII (*E. coli*)
  activity, 148, 153, 458
  genes, 458
  map location, 153, 458
  molecular weight, 153
  products, 458
  regulation, 458
  substrate, 458
  subunit molecular weight, 153, 458
Exonuclease 3′-5′
  activity, 236-239, 242
  kinetic study
    effect of sequence context, 252-254
    effect of stability of a base pair, 251-252
    k_{exo}, 252-254
    rapid flow analysis, 251-255
  next nucleotide effect, 242
  primer-template configuration, 244
  proofreading model, 242-244
  proofreading specificities, 242, 247-248
Exonuclease 5′-3′, activity, 236
Exteins, characteristics, 117

Ferrous-EDTA
  cleavage chemistry, 319
  structure, 320
FPG DNA glycosylase (*E. coli*)
  activities, 446, 460
  AP lyase activity, 270
  gene, 269, 446, 460
  inhibitor, 270
  map location, 460
  molecular weight, 446, 460
  products, 460
  reaction mechanism, 25-26

FPG DNA glycosylase (*Continued.*)
site-directed mutagenesis, 270
substrate, 269, 460
zinc finger motif, 270

GlyGlyHis-cupric ion
cleavage chemistry, 319
structure, 320
Group I introns
axial model, 418
conserved sequence element, 410, 417
distribution, 409
enhancement of activities, 421
guanosine-binding site, 418
hybrid model, 418
kinetics, 420–421
mutagenesis, 418
reaction condition, 414
reaction mechanism, 414–416
secondary structure, 410, 416–417
sizes, 417
structure involved in splicing site selection, 410, 418–419
trans-reaction, 419–420
Group II introns, 409
G/T mismatch repair
E. coli, 272
mammalian cell, 272
G/T thymine DNA glycosylase (mammalian), activities, 446

Hairpin/paperclip ribozyme
cleavage site, 428
essential nucleotides, 428
metal ion requirement, 427
reverse reaction, 427
secondary structure, 412, 427–428
TovRV, 427
Hammerhead ribozyme
active phosphates, 425–426
application, 481
cleavage site, 481
commercial sources, 481
functional groups, 425
kinetics, 426–427
recognition sequence, 481
sequence requirement for self-cleavage, 424–425
specificity, 281

structural requirement for self-cleavage, 423–424
structure, 411, 423
HAP I protein
activities, 290
sequence homology, 290
Hepatitis delta virus ribozyme
cleavage site, 429
effect of denaturants, 428
genome size, 428
intermolecular reaction, 429
minimal sequence requirement, 429
secondary structure, 424, 429–430
HO endonuclease, sequence motif, 122–123, 125
Homing endonuclease
activity, 111, 114, 118–119
characteristics, 112–113, 118
cleavage site, 113, 127–128
codon usage, 113, 132
cutting frequency, 128
distribution, 112, 114–115, 124
DNA binding, 129–131
double-strand break-repair model, 115–116, 118
evolution
cross-species transfer, 135
down-regulation, 134–135
evidence for mobility, 131–133
invasion, 133–134
expression
group I intron ORF, 119–121
intein ORF, 120–123
gene location, 111
history, 115–118
ORF location, 113, 119–120
protein-splicing, 125
sequence motif
GIY-YIG motif, 126
LAGLI-DADG motif, 113, 122–125
unclassified, 126–128
zinc finger motif, 126
sequence specificity, 128–129
target site, 111, 114, 128–129
*Hpa*II
effect of methylation, 44–45
recognition sequence, 44
5-Hydroxymethyluracil DNA glycosylase
activity, 273, 447
molecular weight, 273, 447
Hypoxanthine DNA glycosylase, 273

Index   491

activity, 447
molecular weight, 447

Illimaquinone, 367
Inteins
　activity, 118
　characteristics, 117
　expression, 120-123
Intron maturase, 123, 125
5-Iodoacetyl-1, 10-phenanthroline, 321-322
I-CeuI
　application, 476
　characteristics, 112, 476
　cleavage site, 128
　cognate target sequence, 476
　commercial sources, 476
　sequence specificity, 128-129
I-DmoI
　characteristics, 112
　expression, 121-122, 134
　sequence motif, 122-123
I-IliI
　application, 476
　characteristics, 476
　cognate target sequence, 476
　commercial sources, 476
I-PpoI
　activity, 122
　application, 476
　characteristics, 112, 476
　cleavage site, 128
　cognate target sequence, 476
　commercial sources, 476
　cutting frequency, 128
　DNA binding, 130-131
　expression, 122
　sequence specificity, 128-129
　target site, 128, 130
I-SceI
　activity, 115, 119
　application, 476
　characteristics, 112, 476
　cognate target sequence, 476
　commercial sources, 476
　cutting frequency, 128
　expression, 119, 134
　gene, 112, 115
　intron, 112, 125
　sequence motif, 122

sequence specificities, 128-129
I-SceII
　characteristics, 112
　cutting frequency, 128
　DNA binding, 130-131
　sequence motif, 122-123, 126
　sequence specificity, 129
I-TevI
　characteristics, 112
　cutting frequency, 128
　DNA binding, 130-131
　sequence motif, 122-123, 126
　sequence specificity, 129
　target site, 128
I-TevII
　characteristics, 112
　DNA binding, 131

Klenow fragment, 237, 245, 478
　application, 478
　commercial sources, 478
　specificity, 478

λ Exonuclease
　activity, 148, 153-154, 474
　application, 474
　commercial sources, 474
　gene, 153
　molecular weight, 153
　properties, 474
　substrate, 474

McrA (E. coli)
　activity, 462
　gene, 462
　map location, 462
　substrate, 462
McrBC (E. coli)
　characteristics, 461
　genes, 461
　map locations, 461
　substrate, 461
　subunit molecular weight, 461
Metalloporphyrin
　cleavage chemistry, 319
　structure, 320
3-Methyladenine DNA glycosylase (bovine)
　molecular weight, 447

3-Methyladenine DNA glycosylase
(*Continued.*)
properties, 447
3-Methyladenine DNA glycosylase I
(*E.coli*)
activity, 274
expression, 274, 447
gene, 274, 447
molecular weight, 447
properties, 447
substrate, 274
3-Methyladenine DNA glycosylase II (*E. coli*)
activity, 274
expression, 274, 296, 447
gene, 274, 447
molecular weight, 447
properties, 447
substrate, 274
Methyl-directed mismatch repair
correction efficiency, 247
proteins involved, 298-299
reaction mechanism, 297
strand specificity, 297-298
substrate specificity, 298
N-Methylpurine DNA glycosylases (mammalian)
molecular weights, 274
substrate, 274
5-[3-{(3-methylthio)propionyl]amino}-*trans*-1-propenyl] 2'-deoxyuridine, 320
Micrococcal nuclease (*S. aureus*)
application, 471
commercial sources, 471
properties, 471
substrate, 471
Mitochondrial nuclease. *See also specific nucleases*
biological role, 196-198
damaged sites, specific, 198-199
history, 193
homologies among eukaryotes, 193-194
mutants, 197-198
Mitomycin, 319
Mrr (*E. coli*)
gene, 462
map location, 462
molecular weight, 462
substrate, 462

*Msp*I
effect of methylation, 44-45
recognition sequence, 44
Mungbean nuclease (mung bean sprouts)
application, 471
commercial source, 471
properties, 471
substrate, 471
MutH
activation, 299
activity, 298-299, 450, 460
ATP requirement, 299
gene, 450, 460
map location, 460
molecular weight, 298, 450, 460
substrate, 450, 460
substrate specificity, 299
MutL
molecular weight, 298
properties, 298
MutS
molecular weight, 298
properties, 298

Neurospora crassa endonuclease
application, 472
commercial sources, 472
properties, 472
substrate, 472
Nitrogen mustards, 319
Nuclease (fungal)
extracellular nuclease, 172-176. *See also specific nucleases*
intercellular nuclease. *See also specific nucleases*
inhibitors, 183
metal ion requirement, 182
mode of action, 182
purification, 181
role in DNA repair and recombination, 180-181
substrate, 181-182
Nuclease β (*U. maydis*), 191
Nuclease γ (*U. maydis*), 191
Nuclease O (*A. oryzae*), 180
Nuclease P1 (*P. citrinum*)
activities, 173, 472
amino acid sequence, 173
application, 472
biological role, 175

commercial sources, 472
homologies with other nucleases, 175–176
molecular weight, 173–174
pH optimum, 173
posttranslational modification, 174
proposed reaction mechanism, 175
substrate, 472
three-dimensional structure, 174–175
Nuclease S1 (*A. oryzae*)
activities, 173, 472
amino acid sequence, 173
application, 472
biological role, 175
commercial sources, 472
homologies with other nucleases, 175–176
molecular weight, 173
pH optimum, 173
posttranslational modification, 173
substrate, 472
three-dimensional structure, 174
Nucleotide excision repair
E. coli, 291–294
mammalian, 294
yeast, 294–295

Octahedral tris(phenanthroline) cobalt(III), 320, 323
Octahedral tris(phenanthroline) ruthenium(II), 320, 323
Oligonucleotide-Fe(II)·EDTA, 325
ω Protein. *See* Eco topoisomerase I (*E. coli*)
Organometallic complex, 322–323
Oxidative DNA-cleaving agents, 318–319. *See slso specific agents*

Pfu DNA polymerase (*P. furiosus*)
application, 478
commercial sources, 478
specificity, 478
1, 10-phenanthroline-cuprous ion
cleavage chemistry, 319
structure, 320
Phosphodiesterase I (snake venom)
application, 475
commercial sources, 475
properties, 475
substrate, 475

Phosphodiesterase II (bovine spleen)
application, 475
commercial sources, 475
properties, 475
substrate, 475
Phosphodiester bond cleavage
eliminative (C–O) cleavage. *See* β-Elimination
hydrolytic (P–O) cleavage. *See* $S_N1(P)$ and $S_N2(P)$
rate of hydrolysis, 1–2
enzyme catalyzed, 2
spontaneous, 2
Photoendonuclease, 326
Photosensitizers, 320. *See also specific agents*
PI-*Sce*I
characteristics, 112
cutting frequency, 128
expression, 135
molecular weight, 117
sequence motif, 122–123
target site, 117
PI-*Tli*I
active site, 125
characteristics, 112
mutant, 125
sequence motif, 122–123
Py-Py correndonuclease I
pI, 276
substrate specificity, 276
Py-Py correndonuclease II
pI, 276
substrate specificity, 276
Pyrimidine dimer endonuclease (yeast)
activity, 277, 448
molecular weight, 277, 448
pI, 277

RAD1, specificity, 294
RAD1/RAD10 (*S. cerevisiae*)
activity, 294–295, 450
genes, 450
molecular weight, 450
RAD10, specificity, 294
RecBCD (exonuclease V)
activity, 149–150, 289, 457
genes, 149, 457
map locations, 457
mutants, 150, 152, 290

RecBCD (exonuclease V) (*Continued.*)
  products, 457
  reaction mechanism, 150–152
  substrate, 457
  subunit molecular weight, 149, 457
  subunits, 149, 152
RecE protein. *See* Exonuclease VIII (*E. coli*)
RecJ protein
  activity, 148, 152–153, 290, 458
  expression, 296
  gene, 458
  map location, 458
  in methyl-directed mismatch repair, 298
  molecular weight, 152, 458
  mutant, 152
  substrate, 458
Recombinational repair protein (*Drosophila*)
  gene, 449
  molecular weight, 449
  properties, 449
Recombination models
  single-strand reannealing model, 148–149
  strand assimilation model, 146–148
RecQ helicase, 148, 153
Resolvases
  mammalian, 161
  S. cerevisiae, 160–161
Restriction-modification system, 36–37, 89–90
Reverse gyrase
  activity, 210, 217
  homology, 217
Reverse transcriptase, 347
Ribonucleases (*E. coli*), 466–467. *See also specific ribonucleases*
Ribonucleotide-vanadyl complexes, 366
Ribozyme. *See also specific ribozymes*
  activities, 412
  definition, 408–409
  reaction mechanism
    group I intron, 414–415
    group II intron, 414–415
    self-cleaving sequences, 414
  trans-reactions, 413
Ribozyme Tet1.0 (*Tetrahymena*)
  application, 481
  commercial sources, 481

properties, 481
RNA maturation
  mRNA maturation, 387
  rRNA maturation
    mammalian, 380–381
    prokaryotes, 381–383
    yeast, 381
  tRNA maturation
    E. coli, 383–385
    eukaryotes, 385–386
    T4, 385
RNA maturation nucleases. *See specific nucleases*
RNA-pseudoknot, 366
RNase (*B. cereus*)
  application, 481
  commercial sources, 481
  function, 467
  molecular weight, 467
  properties, 481
  substrate, 467
RNase A (bovine)
  active site, 16–18
  application, 481
  commercial source, 481
  properties, 481
  reaction mechanism, 7–8, 17
RNase BN
  cofactor, 395
  function in tRNA maturation, 384–385, 395–396
  molecular weight, 395
  mutant, 385
  pH optimum, 395
RNase CL3 (chicken liver)
  application, 481
  commercial sources, 481
  properties, 481
RNase D (*E. coli*)
  codon usage, 394
  expression, 394–395
  function in tRNA maturation, 384, 393–394
  gene, 394, 467
  map location, 394, 467
  molecular weight, 394, 467
  mutant, 384
  properties, 393–394
  substrate, 467
RNase E (*E. coli*)
  cleavage specificity, 391

function in
  rRNA maturation, 382, 390
  tRNA maturation, 384
gene, 384, 391, 466
map location, 466
molecular weight, 390-391, 466
mutants, 382, 384
products, 382
RNase H
  altered substrate specificity, 359-360
  detection of activity, 357-359
  effect of antisense oligonucleotides, 367-368
  properties, 341-342
  reaction mechanism, 360-364
    carboxylate-hydroxyl relay mechanism, 361-362
    two-metal-ion mechanism, 361-362
  substrate recognition, 364-365
RNase H (*E. coli*)
  application, 481
  commercial sources, 481
  function, 466
  gene, 466
  map location, 466
  molecular weight, 466
  properties, 481
  substrates, 466
RNase H (HIV-1 RT)
  active site, 351-353
  altered substrate specificity, 359-360
  asymmetric processing of homodimeric RT, 354
  difference electron density map, 363
  location in RT, 358
  maturation of heterodimeric RT, 354-355
  origin and evolution, 355-356
  processing, 354-355
  recombinant, 360
  structural role, 353-354, 358
  structure, 347, 349-353
  substrate-binding site, 350-351
  subunit composition, 349-350
RNase H (HIV-2 RT)
  homology, 353
  structure, 347, 353
RNase H (human)
  activities, 357
  cellular localization, 357
  classes, 356

cofactor, 356-357
molecular weight, 356-357
pI, 356
RNase H (Mo-MLV RT)
  altered substrate specificity, 359-360
  location in RT, 357
  mutants, 349
  recombinants, 359
  structure, 358-359
RNase H (retroviral). *See also specific RNase H*
  as antiretroviral target, 349
  inhibitors, 365-367
  origin and evolution, 355-356
  processing, 354-355
  reaction mechanism, 360-361
  roles during reverse transcription, 347, 378-379
  structure, 349-354
RNase H (*T. thermophilus*)
  sequence homology, 345
  structure, 345-347
RNase HI (*E. coli*)
  crystal structure, 343-345
    active site residues, 344-345
    handle region, 344
    RNA:DNA-binding site, 344-345
  functions, 342-343
  gene, 342
  molecular weight, 342
  mutants, 343
  recombinant, 358
  RNA:DNA hybrid interaction model, 346
RNase HII (*E. coli*)
  gene, 342
  map location, 466
  molecular weight, 342
  substrates, 466
RNase M5
  characteristics, 392
  component, 392
  functions in
    rRNA maturation, 392
    tRNA maturation, 382
  substrate recognition, 382
  subunit molecular weight, 392
RNase P (*E. coli*)
  function in tRNA maturation, 384-385, 389
  genes, 466

RNase P (*E. coli*) (*Continued.*)
  map locations, 466
  mutants, 384
  properties, 408
  protein subunit, 389-390, 446
  recognition and cleavage of substrate, 390
  RNA component, 389, 466
  substrate, 466
RNase PH
  cofactor, 398
  function in tRNA maturation, 384-385, 399
  gene, 398, 466
  map location, 398, 466
  molecular weight, 398, 466
  mutants, 384, 399
  substrate, 446
RNase PhyM (*P. polycephalum*)
  application, 481
  commercial sources, 481
  properties, 481
RNase PIV
  function in tRNA maturation, 384
  molecular weight, 466
  substrate, 466
RNase T
  function in tRNA maturation, 384, 396-397
  gene, 396, 467
  map location, 397, 467
  molecular weight, 396, 467
  mutants, 384, 396-397
  specificity, 385
  substrate, 396-397, 467
RNase T1 (*A. oryzae*)
  active site, 18-20
  application, 481
  commercial sources, 481
  properties, 481
RNase T2 (*A. oryzae*)
  application, 481
  commercial sources, 481
  properties, 481
RNase U2 (*U. sphaerogena*)
  application, 482
  commercial sources, 482
  properties, 482
RNase II
  characteristics, 392-393
  function in tRNA maturation, 384, 393
  gene, 393, 467
  map location, 393, 467
  molecular weight, 393, 467
  mutants, 384
  specificity, 385
  substrate, 467
RNase III
  activities, 388
  cation requirement, 388
  cleavage site, 389
  cleavage specificity, 388
  expression, 389
  function in
    mRNA maturation, 387
    rRNA maturation, 382
    tRNA maturation, 383-384
  gene, 389, 466
  map location, 389, 466
  molecular weight, 388-389, 466
  mutants, 382
  products, 382
  substrate, 466
RNA splicing, 407-408
RNPase
  function in tRNA maturation, 384
  mutants, 384
  specificity, 385
Rrp1 protein (*Drosophila*)
  activity, 149, 155-156, 282
  gene, 282
  molecular weight, 155
RuvAB protein (*E. coli*), 156
RuvC protein (*E. coli*)
  activity, 156, 461
  expression, 296
  gene, 156, 461
  map location, 156, 461
  molecular weight, 156, 461
  mutant, 156
  reaction mechanism, 156-158
  specificity, 157-158
  substrate, 461

*Sau*3AI, effect of mehylation, 44
*Seg*A, sequence motif, 123-124, 126
Self-cleaving sequences
  avocado sunblotch viroid (ASBV), 422
  minimal sequence requirement for splicing, 422
  minimal structural requirement for splicing, 422

properties, 409, 411
pseudoknot structure, 423
sTobRV, 422
structure, 411–412
Self-splicing RNA. See Ribozyme
Semisynthetic nucleases. See Artificial nucleases
Sep1 protein (*S. cerevisiae*)
  activity, 149, 154–155
  gene, 154
  molecular weight, 154
  mutants, 154–155
*Sgr*20I
  cleavage, 43
  recognition sequence, 39
Single-stranded DNA-binding protein, activity, 298
Single-stranded endonuclease (murine), 194–195
Snake venom phophodiesterase, reaction mechanism, 9
$S_N1(P)$ mechanism
  reaction intermediate, 3–4
  reaction mechanism, 3–4
$S_N2(P)$ mechanism
  concerted reaction, 4–5
  reaction intermediate, 3–10, 13
  reaction mechanism, 3–4
  reaction rate, 20–24
  stepwise reaction, 5
  stereochemical studies, 4–10
Spleen exonuclease, reaction mechanism, 9
Staphylococcal nuclease, active site, 13–16
Stress responses (*E. coli*)
  adaptive response, 296
  oxidative stress response, 296–297
  SOS response, 296

T7 exonuclease
  activity, 148, 154
  gene, 154
Taq polymerase (*T. aquaticus*)
  application, 479
  commercial sources, 479
  RNase H activity, 360
  specificity, 479
Telomerase, activity, 356
Topoisomerase. *See also specific topoisomerases*
  activity, 209–210
  assays, 212
  biological function, 210, 220–222
  drug sensitivity, 222–224
  gene, 210–211
  structural homology
    eubacterial topoisomerase I, 216–217
    eukaryotic topoisomerase I, 217–219
    type II topoisomerase, 218–220
Topoisomerase I (*Drosophila*), 222
Topoisomerase I (*E. coli*). *See* Eco topoisomerase I
Topoisomerase I (yeast)
  activity, 211
  mutants, 219–221, 223
Topoisomerase IIα (vertebrates)
  expression, 211
  homology, 211
  molecular weight, 211
Topoisomerase IIβ (vertebrates)
  expression, 211
  homology, 211
  molecular weight, 211
Topoisomerase II (yeast), mutants, 220–221, 223
Topoisomerase III (*E. coli*). *See* Eco topoisomerase III
Topoisomerase III (yeast)
  activity, 211
  biological function, 221
  gene, 211
  mutants, 221
  specificity, 215
Topoisomerase IV. *See* Eco topoisomerase IV
Triple helix, 323–325
tRNA nucleotidyltransferase, function in tRNA maturation, 385
Tth DNA polymerase (*T. thermophilus*)
  application, 479
  commercial sources, 479
  RNase H activity, 360
  specificity, 479
Type I restriction enzymes
  allosteric effectors, 98–99
  evolution of DNA sequence specificity
    by homologous recombination, 94
    by transposition, 96–97
    by unequal crossing over, 94–96
  family relationships

Type I restriction enzymes (*Continued.*)
  type IA, 92, 94, 97–100
  type IB, 92, 97, 100
  type IC, 92, 94–97, 100
  homologies between families, 93
  homologies within families, 93
  reaction mechanism, 98–99
  structural genes, 91, 93, 95–97
    hsdM, 91, 93, 97
    hsdR, 91, 93, 97
    hsdS, 91, 93, 95–97
Type I topoisomerase. *See also specific topoisomerases*
  active site, 217
  activity, 209, 215
  biological function, 210, 216, 220–222
  drug sensitivity, 217, 223
  mutants, 217–218
  strand cleavage and transfer, 216
  substrate specificity, 215
Type II DNA methyltransferase
  cloning, 45–47
  gene sequence, 48–50
  genetic location, 47–48
  substrate preference, 37
Type II DNA topoisomerase. *See also specific topoisomerases*
  active site, 219–220
  activity, 210
  ATP dependence, 214–215
  biological function, 220
  cleavage consensus sequence, 213
  covalent intermediate, 213
  DNA binding, 211–212
  drug sensitivity, 222–223
  protease sensitivity, 219
  strand cleavage and religation, 213–214
Type II restriction enzymes
  catalysis, 73–74
  cleavage sites, 38–39, 42–43
  cloning, 45–47
  crystallography, 59–67
    DNA-protein interactions, 59–67
    DNA structure, 63–65
    protein structure, 59–63
  DNA binding
    mechanism, 52–53
    specificity, 50–52
  DNA cleavage, 53–54, 67–69, 102–103

DNA recognition, 69–73
  altered enzymes, 70–71
  altered substrate, 71–73
effect of methylation, 44–45
enzyme structure, 97–98
evolution, 74–76
gene sequences, 48–50
genetic location, 47–48
genetics, 100–101
isoschizomers, 38–39, 44–45
neoschizomers, 38, 42
star activity, 58
substrate
  kinetic constant, 56–57
  oligonucleotides, 56–57
  plasmid, 54–56
  specificity, 57–59
unusual type II enzymes, 39, 43. *See also specific enzymes*
Type III restriction enzymes
  allosteric effectors, 101–102
  reaction mechanism, 101–102
  recognition sequence, 38–41, 43, 101–102
  single-stranded DNA cleavage, 45
  structural genes
    mod, 100–101
    res, 100–101

Uracil DNA glycosylase (*E. coli*)
  characteristics, 271
  gene, 271, 446
  inhibitor, 271
  molecular weight, 446
  properties, 446
Uranyl pentafluoride, 367
UV endonuclease (*M. luteus*)
  molecular weight, 448
  properties, 448
UV endonuclease I and II (mammalian)
  molecular weight, 269
  substrate, 268–269
UV endonuclease I and II (murine)
  genes, 446
  molecular weight, 446
  properties, 446
UV endonuclease V. *See* Endonuclease V (T4)
UvrABC
  active site residues, 293
  ATP-binding domain, 291–292

characteristics, 450, 460
gene, 450, 460
helix-turn-helix motif, 293
induction, 296
map location, 460
mutants, 293–294
reaction mechanism, 292
site-directed mutagenesis, 292
substrate specificity, 291, 460
subunit molecular weight, 450, 460
zinc finger motif, 292
UvrABC (mammalian), 294
UvrC
activity, 293
molecular weight, 293
mutants, 293
UvrD (DNA helicase II), in methyl-directed mismatch repair, 298

VDE (*S. cerevisiae*)
application, 477
characteristics, 477
cognate target site, 477
commercial sources, 477
Vent DNA polymerase (*T. litoralis*)
application, 479
commercial sources, 479
specificity, 479
Very short patch (VSP) repair, 272
Vsr endonuclease (G/T mismatch endonuclease)
activity, 272–273, 449, 461
gene, 272, 449, 461
map location, 461
molecular weight, 272, 449, 461
substrate, 272, 449, 462